QUANTUM MECHANICS

VOLUME I

FROM THE SERIES IN PHYSICS

General Editors:

J. DE BOER, Professor of Physics, University of Amsterdam
H. BRINKMAN, Professor of Physics, University of Groningen
H. B. G. CASIMIR, Director of the Philips Research Laboratories, Eindhoven

Monographs:

H. C. BRINKMAN, Application of Spinor Invariants in Atomic Physics
H. G. VAN BUEREN, Imperfections in Crystals
S. R. DE GROOT, Thermodynamics of Irreversible Processes
E. A. GUGGENHEIM, Thermodynamics
E. A. GUGGENHEIM and J. E. PRUE, Physicochemical Calculations
H. JONES, The Theory of Brillouin Zones and Electronic States in Crystals
H. A. KRAMERS, Quantum Mechanics
H. A. KRAMERS, The Foundations of Quantum Theory
J. G. LINHART, Plasma Physics
J. MCCONNELL, Quantum Particle Dynamics
P. ROMAN, Theory of Elementary Particles
J. L. SYNGE, Relativity: The Special Theory
J. L. SYNGE, Relativity: The General Theory
J. L. SYNGE, The Relativistic Gas
A. VAŠÍČEK, Optics of Thin Films

Edited Volumes:

D. A. BROMLEY and E. W. VOGT (editors), Proceedings of the International Conference on Nuclear Structure (Kingston, 1960)
P. M. ENDT and M. DEMEUR (editors), Nuclear Reactions, Volume I
C. J. GORTER (editor), Progress in Low Temperature Physics, Volumes I–III
H. KALLMAN BIJL (editor), Space Research (Proceedings of the First International Space Science Symposium, Nice, 1960)
H. J. LIPKIN (editor), Proceedings of the Rehovoth Conference on Nuclear Structure
N. R. NILSSON (editor), Proceedings of the Fourth International Conference on Ionization Phenomena in Gases (Uppsala, 1959)
K. SIEGBAHN (editor), Beta- and Gamma-Ray Spectroscopy
Turning Points in Physics. A series of lectures given at Oxford University in Trinity Term 1958
J. G. WILSON and S. A. WOUTHUYSEN (editors), Progress in Elementary Particle and Cosmic Ray Physics. Volumes I–V

BALTH. VAN DER POL, Selected Scientific Papers
P. EHRENFEST, Collected Scientific Papers

QUANTUM MECHANICS

VOLUME I

ALBERT MESSIAH

Saclay, France

TRANSLATED FROM THE FRENCH

BY

G. M. TEMMER

RUTGERS — THE STATE UNIVERSITY

NEW BRUNSWICK, N.J.

NORTH-HOLLAND PUBLISHING COMPANY–AMSTERDAM

JOHN WILEY & SONS

New York • Chichester • Brisbane • Toronto • Singapore

Reproduction or translation of any part of this work beyond that permitted by Sections 107 or 108 of the 1976 United States Copyright Act without the permission of the copyright owner is unlawful. Requests for permission or further information should be addressed to the Permissions Department, John Wiley & Sons, Inc.

ORIGINAL TITLE: MÉCANIQUE QUANTIQUE, VOL. I

PUBLISHERS:

NORTH-HOLLAND PUBLISHING COMPANY, AMSTERDAM

JOHN WILEY & SONS, INC. — New York • Chichester • Brisbane • Toronto

ISBN 0 471 59766 X

Printed in the United States of America

30 29 28 27 26 25 24 23 22

PREFACE

Nowadays, there hardly exists a branch of physics which one can seriously approach without a thorough knowledge of Quantum Mechanics. Its presentation, which is given in this work is, I hope, simple enough to be accessible to the student, and yet sufficiently complete to serve as a reference book for the working physicist.

This book resulted from a course given at the Center of Nuclear Studies at Saclay since 1953. Numerous discussions with students as well as with my colleagues, have helped me considerably in clarifying its presentation. Several people to whom I had transmitted certain parts of the manuscript, have kindly given me their criticism; among them I should like to mention Messrs. Edmond Bauer and Jean Ullmo, to whom I am indebted for interesting remarks concerning the presentation of principles. I am more particularly grateful to Mr. Roger Balian for having critically examined a large portion of the manuscript, and for having suggested to me a large number of improvements. Finally, I wish to thank those of my students who were kind enough to check over the text and the calculations of the various chapters, and to help me with the correction of the proofs.

The problems which occur at the end of each chapter were chosen not only for their educational value, but also to point out certain properties worthy of interest; this may explain the relative difficulty of certain ones among them.

The several works or articles cited as references have the purpose of aiding the reader to complete or round out certain passages. It was out of the question to give a complete bibliography of the various subjects treated here. An entire volume would not have sufficed for that.

October, 1958 ALBERT MESSIAH

OUTLINE

VOLUME I

THE FORMALISM AND ITS INTERPRETATION

SIMPLE SYSTEMS

VOLUME II

SYMMETRIES AND INVARIANCE

METHODS OF APPROXIMATION

ELEMENTS OF RELATIVISTIC QUANTUM MECHANICS

CONTENTS

OF VOLUME I

PART ONE

THE FORMALISM AND ITS INTERPRETATION

CHAPTER I

THE ORIGINS OF THE QUANTUM THEORY

CHAPTER II

MATTER WAVES AND THE SCHRÖDINGER EQUATION

CHAPTER III

ONE-DIMENSIONAL QUANTIZED SYSTEMS

CHAPTER IV

STATISTICAL INTERPRETATION OF THE WAVE-CORPUSCLE DUALITY AND THE UNCERTAINTY RELATIONS

CHAPTER V

DEVELOPMENT OF THE FORMALISM OF WAVE MECHANICS AND ITS INTERPRETATION

CHAPTER VI

CLASSICAL APPROXIMATION AND THE WKB METHOD

CHAPTER VII

GENERAL FORMALISM OF THE QUANTUM THEORY

(A) MATHEMATICAL FRAMEWORK

CHAPTER VIII

GENERAL FORMALISM
(B) DESCRIPTION OF PHYSICAL PHENOMENA

PART TWO

SIMPLE SYSTEMS

CHAPTER IX

SOLUTION OF THE SCHRÖDINGER EQUATION BY SEPARATION OF VARIABLES. CENTRAL POTENTIAL

CHAPTER X

SCATTERING PROBLEMS

CENTRAL POTENTIAL AND PHASE-SHIFT METHOD

CHAPTER XI

THE COULOMB INTERACTION

CHAPTER XII

THE HARMONIC OSCILLATOR

. . . "Il lui proposa de fai...
voyage de Copenhague, et lui en
facilita les moyens" (*Candide*)

PART ONE

THE FORMALISM AND ITS INTERPRETATION

CHAPTER I

THE ORIGINS OF QUANTUM THEORY

1. Introduction

According to the *classical doctrine* – generally adopted by physicists until the beginning of the 20th century – one associates with physical systems whose evolution one wishes to describe, a certain number of quantities or dynamical variables; each of these variables possesses at each instant a *well-defined value*, and the specification of this set of values defines the dynamical state of the system at that instant. One further postulates that the evolution in time of the physical system is entirely determined if one knows its state at a given initial instant. Mathematically this fundamental axiom is expressed more precisely by the fact that the dynamical variables satisfy a system of differential equations of the first order, as a function of time. The program of Classical Theoretical Physics thus consists in enumerating the dynamical variables of the system under study, and then in discovering the equations of motion which predict its evolution in accord with experimental observation.

From the formulation of Rational Mechanics by Newton until the end of the 19th century, this program was carried out with considerable success, each new experimental discovery being carried over to the theoretical plane either by introducing new variables and new equations, or by modifying the old equations, thereby allowing the newly observed phenomenon to be incorporated into the general scheme. During that entire period no experimental fact, no discovery led to any doubt concerning the soundness of the program itself. On the contrary, Classical Physics constantly progressed toward greater simplicity and greater unity. This happy evolution continued until about 1900; subsequently, as our knowledge of phenomena on the microscopic scale [1]) becomes more precise, Classical Theory runs

[1]) It is important to define the terms "microscopic" and "macroscopic" of which we shall make frequent use throughout this book. We define the "microscopic" scale as the one of atomic or subatomic phenomena, where the lengths which enter into consideration are at most of the order of several ångstroms (1 Å = 10^{-8} cm). The "macroscopic" scale is the one of phenomena observable with the naked eye or with the ordinary microscope, i.e. a resolution of the order of one micron (10^{-4} cm) at best.

into more and more difficulties and contradictions. It rapidly becomes evident that phenomena on the atomic and subatomic scale do not fit into the framework of classical doctrine itself, and that their explanation must be based upon entirely new principles. The discovery of these new principles will occur in stages, at the expense of numerous groping attempts; only around 1925, with the founding of Quantum Mechanics, will we have at our disposal a coherent theory of microscopic phenomena. The origins of this theory constitute the subject of the present chapter.

After sketching an overall picture of Classical Theoretical Physics, we shall discuss the main phenomena which justify the abandonment of the classical ideas. The phenomena are supposed familiar to the reader [1]); we shall therefore merely recall their essential features, emphasizing above all the points of contradiction with Classical Theory. The end of the chapter is devoted to a brief discussion of the first attempts at explaining these phenomena, known as the Old Quantum Theory.

I. THE END OF THE CLASSICAL PERIOD

2. Classical Theoretical Physics

At the end of the classical period, the various branches of physics are integrated in a general and coherent theoretical construct whose main features are as follows. In the universe, one distinguishes two categories of objects, *matter* and *radiation*. Matter is made up of perfectly localizable corpuscles subject to Newton's laws of Rational Mechanics; the state of each corpuscle is defined at any instant by its position and its velocity (or its momentum), that is six dynamical variables in all. Radiation obeys Maxwell's laws of electromagnetism; its dynamical variables – infinite in number – are the components of the electric and magnetic fields at each point of space. In contrast to matter, it is not possible to split radiation into corpuscles which can be localized in space and maintain this localized character during their evolution in the course of time; quite to the contrary, it exhibits a wave-like behavior which manifests itself particularly in the well-known phenomena of interference and diffraction.

---

[1]) One may find a detailed discussion of these phenomena in the works dealing with Atomic Physics, for instance: M. Born, *Atomic Physics*, 6th ed. (Blackie, Glasgow, 1957).

The *corpuscular theory of matter* continues to develop during the course of the 19th century. While limited at first to the mechanics of heavenly bodies and of solid bodies of macroscopic dimensions, it emerges more and more as the basic theory governing the evolution of matter on the microscopic scale to the extent that the atomic hypothesis, proposed by the chemists, is confirmed. Without being able to verify this hypothesis directly by isolating the molecules and studying their mutual interactions, one can justify it indirectly by showing that the macroscopic properties of material bodies derive from the laws of motion of the molecules of which they are composed. Mathematically, we are dealing with a very complex problem. Under this hypothesis, in fact, macroscopic quantities appear as the mean values of certain dynamical variables of a system having a very large number of degrees of freedom [1]; there is no hope of solving the equations of evolution of such a system exactly, and one must have recourse to statistical methods of investigation. Thus a new discipline originated and developed, Statistical Mechanics, whose results, particularly in the study of gases (Kinetic Theory of Gases) and in Thermodynamics (Statistical Thermodynamics) enable us to verify qualitatively, and within the limits set by the possibilities of calculation, quantitatively, the foundation of a corpuscular theory of matter [2].

At the same time, the *wave theory of radiation* becomes solidly established. In the field of optics, the old controversy on the wave nature or corpuscle nature of light is cut short in the first half of the 19th century, when decisive progress in the handling of problems of wave propagation (Fresnel) permits the exploration of all the consequences of the wave hypothesis. All the known light phenomena, including geometrical optics can now be based on this hypothesis. Meanwhile, the study of electric and magnetic phenomena develops rapidly. The decisive step forward is taken by Maxwell when he establishes, in 1855, the fundamental electromagnetic equations.

---

[1] We recall that the number N of molecules per mole (Avogadro's number) is $N = 6.02 \times 10^{23}$. The first precise determination of N, due to Loschmidt (1865), was based on the kinetic theory of gases.

[2] It is well to note that in all reasoning of Statistical Mechanics, there underlies a hypothesis of a statistical nature, the hypothesis of molecular chaos, from which one cannot escape without renouncing the statistical method itself. Although this hypothesis seems intuitively correct, its rigorous justification (ergodic theorem) turned out to be particularly delicate and is still the subject of controversy.

On the basis of these equations, he foresees the existence of electromagnetic waves – a prediction ultimately confirmed in spectacular fashion by the discovery of radio waves (Hertz) – and likens the light wave to a particular kind of electromagnetic wave, thus achieving the synthesis of optics and electricity.

Toward the end of the 19th century, the success of the classical program is impressive. All known physical phenomena, it seems, find their explanation in a general theory of matter and radiation; in all the cases where this explanation could not be found, one may reasonably attribute the failure to mathematical difficulties in the solution of the problem, without jeopardizing the form of the basic equations. What is most striking in this theory is its remarkable degree of unity. The desire to unify the various branches of their science has always been one of the most fruitful preoccupations of the physicists. In fact, the physicists of that era attribute to the classical theory more unity than it actually possesses. Indeed, wave propagation is not a phenomenon peculiar to electromagnetism. The study of vibrations was first made in connection with vibrations of matter (vibrating strings, surface waves of a liquid, etc.) and the wave nature of acoustical phenomena was detected before that of light phenomena. Moreover, the existence of waves within matter by no means contradicts the corpuscular theory; in fact, one deals here with a macroscopic phenomenon which can easily be deduced from the microscopic laws of motion with a suitable law of force. By analogy, the classical physicists endow electromagnetic waves with a supporting structure, a kind of material fluid which they call "ether" and whose structure and mechanical properties remain to be specified. Thus matter appears as the fundamental entity, subject to the principles of Newton's Rational Mechanics, and obeying force laws such that it can, under suitable conditions, experience various wave phenomena, of which electromagnetic vibration is one example.

This conception, which is to be completely abandoned later, suggests, at the period we are considering, a whole series of experiments which do not reveal many facts concerning the nature of the ether; however, one of these is to provoke a rather profound upset of classical physics. This is the famous Michelson-Morley experiment (1887), designed to reveal the motion of the earth relative to the ether by seeking to detect how the velocity of propagation of light with respect to the earth varies with the direction of propagation. The negative result of this experiment is well known. After several more or less artificial attempts at an explanation, this apparent paradox is definitely explained by Einstein in 1905 within the framework of the *Theory of Relativity*, following a critical analysis of the concepts of space and time which calls for the rejection of the notion of absolute time, and of some of the axioms of Newtonian Mechanics. In fact, the latter is but an approximation of Relativistic Mechanics, which is valid only in the limit where the velocities of the particles are negligible compared to the velocity of light c. We shall not elaborate on the principle of relativity here; we shall have occasion to return to it at the end of this book, when we study Relativistic Quantum Mechanics. The essential point to note is that this principle puts in doubt neither the doctrine nor the classical program such as they were defined earlier.

3. Progress in the Knowledge of Microscopic Phenomena and the Appearance of Quanta in Physics

At the turn of the century, the efforts of experimenters follow along two closely related lines; first, to make a precise analysis of the microscopic structure of matter; second, to determine the mutual interaction of material corpuscles, and their interactions with the electromagnetic field.

The first facts concerning the structure of matter are furnished by the study of the rays obtained from the discharge in rarefied gases, cathode rays and canal rays, which are correctly interpreted as beams of electrically charged particles travelling more or less swiftly. Thus the *electron* is discovered (J. J. Thomson, 1897), the particle of the cathode rays; its behavior in the presence of an electromagnetic field is determined experimentally, and a complete theory of the interaction between electrons and electromagnetic waves is established (electron theory of Lorentz) [1]).

Gradually, the very existence of atoms and molecules, for a long time considered to be a fruitful working hypothesis, becomes accepted reality. Its most convincing proof is furnished by the study of Brownian motion, the disordered motion of tiny particles suspended in a liquid or a gas; this motion is attributed to the frequent collisions which these particles undergo with surrounding molecules. It is to some

[1]) Cf. L. Rosenfeld, *Theory of Electrons* (North-Holland Publishing Co., Amsterdam, 1951).

extent a large-scale reproduction of molecular agitation and can be quantitatively related (Einstein, Smoluchowski, 1905) to the statistical laws of motion of the molecules themselves. The systematic measurements of Perrin (1908) confirm this hypothesis and furnish several precise and concordant determinations of Avogadro's number [1]. After this decisive progress, physicists no longer doubt the existence of atomic or subatomic particles, and we witness the perfection of more and more experimental techniques permitting observation of individual microscopic phenomena, or to count microscopic particles one by one (measurement of the elementary electric charge by Millikan in 1910; first observation of trajectories of charged particles with the Wilson cloud chamber in 1912; first Geiger counter in 1913). These techniques of "direct" observation have continued to develop, and constitute today nearly the entire arsenal at the disposal of experimenters in the exploration of microscopic phenomena.

However, a new chapter of physics is opened with the discovery of radioactivity (1896), the first manifestation of the properties of atomic nuclei. Important in itself, this discovery puts into the hands of the physicists a potent means for the investigation of atomic structure, namely alpha radiation, consisting of helium nuclei moving at high velocity. By exposing various targets to the alpha rays, Rutherford (1911) studies systematically the scattering of alpha particles by atoms and thus succeeds in extracting the first modern picture of the atom.

The *Rutherford atom* is formed by a central nucleus of small dimensions ($10^{-13}-10^{-12}$ cm) around which gravitate a certain number Z of electrons. Almost all the mass of the atom is concentrated in the nucleus. The latter carries a positive electric charge Ze which exactly offsets the total charge $-Ze$ of the electrons, so as to form an electrically neutral entity. The Rutherford atom thus resembles a miniature solar system where the forces of gravity are replaced by electrical forces. Under the action of the latter, namely Coulomb attraction of the nucleus and mutual Coulomb repulsions, the electrons describe stable orbits around the nucleus, orbits whose extension is of the order of atomic dimensions, namely 10^{-8} cm.

While the corpuscular character of matter seems to be confirmed as our knowledge of atomic phenomena progresses, the spectrum of

[1] Cf. J. Perrin, *Les Atomes* (Presses Universitaires, Paris, 1948).

known electromagnetic waves nears completion and extends toward the short wavelengths with the discovery of X-rays (Röntgen, 1895) whose wave nature is established by the experiments on diffraction by crystals (von Laue, 1912). For the sake of completeness one should mention gamma radiation from radioactive bodies as well, whose electromagnetic nature is to be recognized only much later. Figure 1

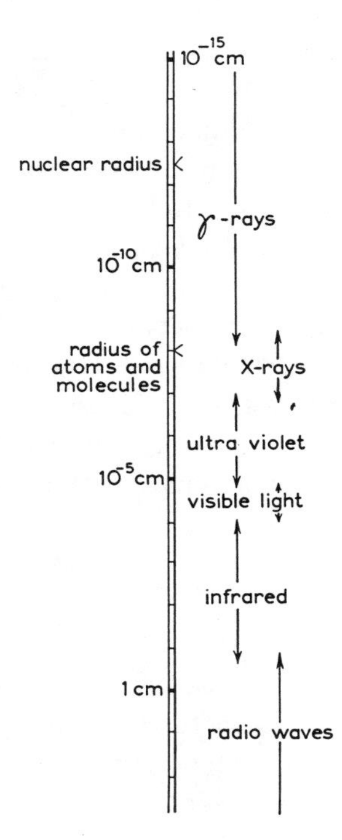

Fig. I.1. Wavelength scale of electromagnetic radiation.

gives the scale of wavelengths of electromagnetic radiation which was thus identified. While at the same time enlarging its domain, spectral analysis of radiation becomes more and more precise and allows the accumulation of a considerable body of information on the problems of emission, scattering and absorption of light by matter, in other words on the interaction of matter and radiation on the microscopic scale. The Lorentz theory of the electron which was already mentioned, a theory of charged particles interacting with the electromagnetic field, makes definite predictions concerning all these phenomena. It is precisely in the comparison of the predictions of this theory with this body of experimental results that the first disagreements between classical theory and experiment became apparent.

The first difficulties appear when one studies the spectral distribution of electromagnetic radiation in thermodynamic equilibrium with matter. The typical case is the one of the *black body*; by definition, it is a body which absorbs all the radiation it receives. Some very general thermodynamic reasoning shows that the radiation emitted by a black body is a function of temperature only. The spectral distribution of intensity of the radiation emitted by the black body is consequently a fundamental expression which must be derivable, by the methods of Statistical Thermodynamics, from the general laws of interaction between matter and radiation. The expression deduced from the Classical Theory is in violent disagreement with experiment. In 1900, Planck succeeds in removing the difficulty by renouncing the classical law of interaction between matter and radiation [1]). He postulates that the energy exchanges between matter and radiation do not take place in a continuous manner, but by discrete and indivisible quantities or *quanta* of energy; he then shows that the quantum of energy must be proportional to the frequency ν of the radiation:

$$\varepsilon_\nu = h\nu,$$

and he obtains an expression for the spectrum in accord with the experimental distribution by properly adjusting the constant of proportionality. This constant h is known henceforth under the name of *Planck's constant.* It has the dimensions of an action (energy $\times$ time, or momentum $\times$ length). In what follows, we shall rather use the constant

$$\hbar = \frac{h}{2\pi} = 1.054 \times 10^{-27} \text{ erg-sec.}$$

Upon its publication, Planck's hypothesis seemed unacceptable; physicists almost unanimously refused to see therein more than a lucky mathematical artifice which could, some day, be explained within the framework of classical doctrine. The very success of Planck's theory could not be considered as irrefutable proof that the energy

---

[1]) For a detailed study of the theory of black-body radiation, cf. M. Born, *loc. cit.*

exchanges between matter and radiation on the microscopic scale actually take place by quanta; Planck's distribution law is a macroscopic law deduced from this hypothesis by statistical methods. It constitutes but an indirect confirmation. One could likewise doubt the validity of the quantum hypothesis itself, in the same manner as one had doubted for a long time the validity of the atomic hypothesis for lack of being able to verify it directly on the microscopic level. Planck's hypothesis, however, was to be confirmed by a whole array of experimental facts which allowed the direct analysis of the elementary processes, and the direct revelation of the existence of discontinuities in the evolution of physical systems on the microscopic scale, where classical theory predicts a continuous evolution.

II. LIGHT QUANTA OR PHOTONS

A first series of experimental facts forces a radical revision of the radiation theory of Maxwell-Lorentz, and a partial return to the old corpuscular theory. This is mainly the case with the *photoelectric effect* and the *Compton effect*.

4. The Photoelectric Effect

The first step in this direction was taken by Einstein in his celebrated note of 1905 on the photoelectric effect. The general attitude toward Planck's theory was to state that "everything behaves as if" the energy exchanges between radiation and the black body occur by quanta, and to try to reconcile this *ad hoc* hypothesis with the wave theory. Taking the opposite view, and going even further than Planck who limits himself to the introduction of the discontinuity in the absorption or emission mechanism, Einstein postulates that light radiation itself consists of a beam of corpuscles, the photons, of energy $h\nu$ and velocity c ($=$velocity of light in vacuo $= 3 \times 10^{10}$ cm/sec). He then shows how this surprising hypothesis may account for a certain number of hitherto unexplained phenomena, the photoelectric effect in particular.

One designates by this name the emission of electrons observed when one irradiates an alkali metal under vacuum with ultraviolet light. The electric current intensity thus produced is proportional to the intensity of the radiation striking the metal. On the other hand, the speed of the electrons does not depend upon the radiation intensity, but only upon its frequency (Lenard, 1902), no matter how large the distance from the light source; only the number of electrons emitted per second is proportional to the intensity, hence inversely proportional to the square of the distance from the source.

Einstein's explanation is very simple. Whatever the distance covered by the light since its emission, it occurs in the form of corpuscles of energy $h\nu$. When one of these photons encounters an electron of the metal, it is entirely absorbed and the electron receives the energy $h\nu$; in leaving the metal, the electron must do an amount of work equal to its binding energy in the metal W, so that the observed electrons have a well-defined kinetic energy:

$$\tfrac{1}{2}mv^2 = h\nu - W. \tag{I.1}$$

This quantitative theory is completely verified by experiment. As expected, the constant W is a characteristic constant of the irradiated metal. As for the constant h, it has the same numerical value as the constant which occurs in the expression for the black-body spectrum of radiation.

In view of the success of the corpuscular theory, one must examine whether the classical wave theory is also capable of explaining the photoelectric effect. This is not *a priori* inconceivable. In fact, a light wave transports a certain quantity of energy proportional to its intensity and can give up all or part of this energy as it penetrates into the metal: the energy which is gradually accumulated in the metal is eventually concentrated on certain electrons who thus succeed in escaping: one can imagine that by some mechanism yet to be specified, an electron could not escape before having received a quantity of energy equal to $h\nu$. The essential difference between this type of explanation and the corpuscular theory lies in the continuous and progressive character of the energy accumulation in the metal; consequently, photoelectric emission, instead of being instantaneous, cannot take place before the metal has received the energy $h\nu$. If one operates with sufficiently fine metallic particles, this minimum delay between the start of the irradiation and the onset of the emission can be made sufficiently long to be detectable experimentally.

Experiments for this purpose were conducted by Meyer and Gerlach (1914) on metallic dusts. Knowing the intensity of the radiation and the dimensions of the dust particles, they could determine the minimum irradiation time for a dust particle to absorb the energy $h\nu$ necessary for the emission of an electron; this amounted to several seconds in the conditions under which they were operating. In every case they observed the emission of electrons right at the start of irradiation. One must therefore conclude that the wave theory of light, at least in its classical form, is entirely incapable of accounting for the photoelectric effect.

5. The Compton Effect

The Compton effect represents another confirmation of the photon theory, and a refutation of the wave theory. One observes it (Compton, 1924) in the scattering of X-rays by free (or weakly bound) electrons. The wavelength of the scattered radiation exceeds that of the incident radiation. The difference $\Delta\lambda$ varies as a function of the angle θ between the direction of propagation of the incident radiation and the direction along which one observes the scattered light, according to Compton's formula:

$$\Delta\lambda = 2 \frac{h}{mc} \sin^2 \frac{\theta}{2}, \tag{I.2}$$

where m is the rest mass of the electron [1]). One notes that $\Delta\lambda$ is independent of the incident wavelength. Compton and Debye have shown that the Compton effect is a simple elastic collision between a photon of the incident light and one of the electrons of the irradiated target.

In order to discuss this corpuscular interpretation it is convenient to state a few properties of photons which derive directly from Einstein's hypothesis. Since they possess the velocity c, photons are particles of zero mass [2]). The momentum $\boldsymbol{p}$ and the energy ε of a photon are thus connected by the relation

$$\varepsilon = pc \tag{I.3}$$

---

[1]) The length $\hbar/mc$, intermediate between the mean radius of atoms and that of atomic nuclei ($\hbar/mc = 3.86 \times 10^{-11}$ cm), plays some role in the quantum theory of the electron. It is called the Compton wavelength of the electron.

[2]) According to the principle of relativity, the (rest) mass m, the energy ε and the momentum p of a particle are connected by the relation: $\varepsilon^2 - p^2c^2 = m^2c^4$; its velocity $v = \partial\varepsilon/\partial p = pc^2/\varepsilon$. If $v = c$, $\varepsilon = pc$ and $m = 0$.

Consider a plane, monochromatic light wave

$$\exp\left[2\pi i\left(\frac{\boldsymbol{u}\cdot\boldsymbol{r}}{\lambda} - \nu t\right)\right].$$

$\boldsymbol{u}$ is a unit vector in the direction of propagation, λ is the wavelength, ν the frequency: $\lambda\nu = c$. In accordance with Einstein's hypothesis, this wave represents a stream of photons of energy $h\nu$. The momentum of these photons is evidently directed along $\boldsymbol{u}$ and its absolute value, according to (I.3), is equal to

$$p = \frac{h\nu}{c} = \frac{h}{\lambda}.$$

This relation is a special case of the relation of L. de Broglie which we shall study in Ch. II. It is often convenient to introduce the angular frequency $\omega = 2\pi\nu$ and the wave vector $\boldsymbol{k} = (2\pi/\lambda)\boldsymbol{u}$ of the plane wave. The connecting relations are then written:

$$\varepsilon = \hbar\omega, \qquad \boldsymbol{p} = \hbar\boldsymbol{k}. \tag{I.4}$$

The corpuscular theory of the Compton effect consists in writing down that the total energy and momentum are conserved in the elastic collision between the incident photon and the electron. Let $\boldsymbol{p}$, $\boldsymbol{p}'$ be the initial and final momenta of the photon, respectively, $\boldsymbol{P}'$ the recoil momentum of the electron after collision (Fig. I.2).

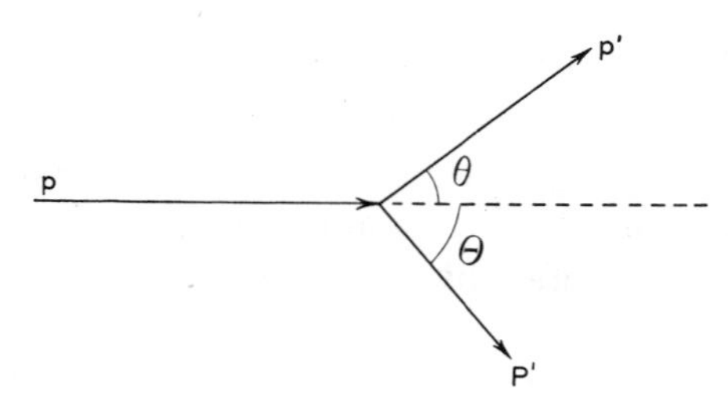

Fig. I.2. Compton collision of a photon with an electron at rest.

The conservation equations are written

$$\begin{aligned} \boldsymbol{p} &= \boldsymbol{p}' + \boldsymbol{P}', \\ mc^2 + pc &= \sqrt{P'^2c^2 + m^2c^4} + p'c. \end{aligned} \tag{I}$$

According to these equations the collision is completely defined once the initial conditions and the direction of emission of the scattered photon are known. Taking into account the relations (I.4), one can easily deduce the Compton formula which is thus explained theoretically (cf. Problem I.1). Since the first work of Compton, all the other predictions of this theory have been confirmed experimentally. The recoil electrons have been observed and the law of their energy variation as a function of the angle of emission Θ is just the one which one derives from equations (I). Coincidence experiments have shown that the scattered photon and electron are emitted simultaneously, and that the correlation between the emission angles θ and Θ agrees with the theory.

It is instructive to compare these results with the predictions of Classical Theory. The Maxwell-Lorentz theory predicts that part of the incident electromagnetic energy is absorbed by each irradiated electron and then re-emitted in the form of radiation of the same frequency. In contrast to the momentum of the absorbed radiation, the total momentum of the emitted radiation is zero. The light-scattering process is thus accompanied by a continuous transfer of momentum (radiation pressure) from the electromagnetic radiation to the irradiated electron; the electron is thereby continuously accelerated in the direction of propagation of the incident wave. The law of absorption and emission at the same frequency holds in the reference system where the electron is at rest. As soon as the electron is set in motion, the frequencies observed in the laboratory system are different because of the Doppler effect. The wavelength shift $\Delta\lambda$ depends upon the angle of observation θ of the scattered radiation. A simple calculation yields

$$\Delta\lambda = 2\lambda \frac{P_{\text{cl.}}\, c}{E_{\text{cl.}} - P_{\text{cl.}}\, c} \sin^2 \frac{\theta}{2}, \tag{I.5}$$

where λ is the incident wavelength, $P_{\text{cl.}}$ is the momentum of the electron (assumed in motion along the direction of propagation), $E_{\text{cl.}} = \sqrt{m^2c^4 + P_{\text{cl.}}^2\, c^2}$ is its energy. $\Delta\lambda$ is an increasing function of $P_{\text{cl.}}$ and increases regularly during the irradiation.

Thus the classical predictions do not agree with the experimental facts. The main defect of the classical theory of the Compton effect is to predict a *continuous* transfer of momentum and energy of the radiation to *all* the electrons exposed to the radiation, while the experimentally observed effect is a *discontinuous* and instantaneous transfer to *certain* electrons among them. Here is a difficulty of the same type as the one we encounter in the case of the photoelectric effect. The two effects are in fact related: Compton scattering may be considered as light absorption followed by re-emission, while the photoelectric effect is a mere absorption.

The introduction of light quanta is unavoidable if one wants to account for the discontinuous transfer of momentum and energy to the electrons. Nevertheless, the similarity between the classical formula (I.5) and the correct formula (I.2) of the Compton effect suggests that the classical theory retains some features of the real phenomenon. This matter merits closer examination.

The Compton effect was calculated above assuming the electron to be initially at rest. Of course, the theory remains valid when its initial velocity is not zero. One may easily generalize the equations (I) and the Compton formula. In the particular case where the electron travels parallel to the incident light wave with momentum $\boldsymbol{P}$ and energy $E=\sqrt{m^2c^4+P^2c^2}$ we obtain (Problem I.1)

$$\Delta\lambda = 2\lambda \frac{(P+p)c}{E-Pc} \sin^2 \frac{\theta}{2}. \tag{I.6}$$

Note the great similarity between this expression and the classical expression (I.5) for the shift $\Delta\lambda$. To pass from one to the other, one must replace the momentum $P_{\text{cl.}}$ in the numerator by the quantity $P+p$ (of the order of the momentum after the photon-electron collision), and in the denominator (in the expression $E_{\text{cl.}}-P_{\text{cl.}}c$) by the momentum before impact P. The mechanism to which expression (I.6) refers, however, is very different from the classical mechanism. Under the effect of prolonged irradiation, each electron receives a first momentum transfer which sets it in motion, then a second, and so on. The momentum transfer varies from one impact to the next around an average value approximately equal to the momentum $\boldsymbol{p}$ of the incident photons. We wish to compare this variation of the momentum by *discontinuous* quanta of the order $\boldsymbol{p}$ and the resulting variation of the shift $\Delta\lambda$ with the *continuous* variations predicted by the classical theory (Fig. I.3).

Such a comparison makes sense only in the limit where the quanta can be considered to be infinitely small and infinite in number, and where one considers the average effect of a very large number of successive impacts. Since the electron gains, on the average, a momentum of the order $\boldsymbol{p}$ per collision and since, after a very large

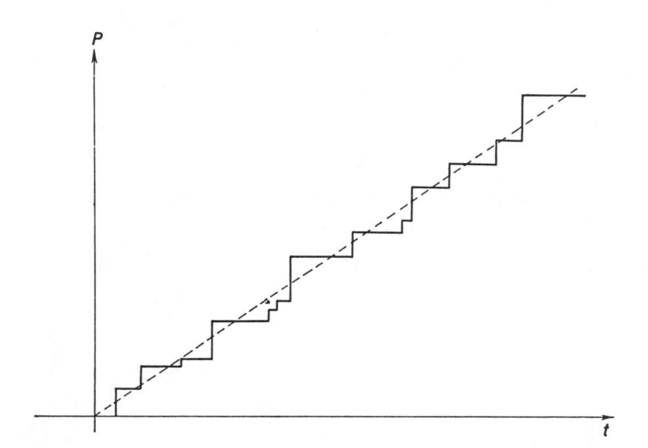

Fig. I.3. Variation as function of time of the momentum P of an electron exposed to monochromatic radiation, by reason of successive Compton collisions (this is a very schematic picture of the phenomenon whose limitations will be discussed in Ch. IV in connection with the uncertainty relations). Dashed line represents the variation $P_{\text{cl}}(t)$ predicted by classical theory.

number of impacts, the fluctuations compensate each other, the net resulting effect is practically the same as if the electron received at each impact just this average momentum. The momentum $\boldsymbol{P}$ of the electron therefore grows by successive jumps, in the direction of the incident radiation. These jumps are of the order of magnitude of the quantum $p = h\nu/c$ and the momentum gain may be likened to a continuous gain to the extent that the quantity p may be regarded as infinitely small. Within these limits of approximation one may define an average momentum $\langle \boldsymbol{P} \rangle$ growing in continuous fashion in the course of time. A thorough experimental study – which we shall not elaborate here – shows that the variation of this average momentum as a function of time is precisely the one predicted by classical theory; in other words, the vectors $\langle \boldsymbol{P} \rangle$ and $\boldsymbol{P}_{\text{cl.}}$ remain equal. Furthermore, since the classical value $P_{\text{cl.}}$, defined to within p, is equal at each instant to the average value of P, the Compton shift predicted by the classical theory [eq. (I.5)] is equal at each instant to the average value of the actually observed Compton shift [eq. (I.6)].

6. Light Quanta and Interference Phenomena

If we have some indications that classical wave theory is macroscopically correct, it is nevertheless clear that on the microscopic scale only the corpuscular theory of light is able to account for typical absorption and scattering phenomena such as the photoelectric effect and the Compton effect, respectively. One must still ascertain how the photon hypothesis may be reconciled with the essentially wave-like phenomena of interference and diffraction.

To be definite, let us consider the scattering of a beam of monochromatic light by a parallel grating (Fig. I.4). A screen conveniently placed allows us to view the interference pattern. The quantitative observation of the phenomenon may occur in various ways, for instance by replacing the screen by a photographic plate which one develops after a given irradiation time. The interference pattern appears as a negative upon development, the blackening of each element of surface of the plate being proportional to the quantity of light received. In fact, the absorption of light by the plate occurs in quanta: each photon penetrating into the plate excites a photosensitive microcrystal which yields a black spot upon development [1]). It is not possible to resolve with the naked eye the spots from each other, and one observes in practice a continuous blackening which is more or less pronounced according to whether the density of impacts is high or low. However the existence of separate impacts may be effectively detected if one observes the plate with a sufficiently powerful microscope. Under normal experimental conditions the number of photons received by the plate is very great, and the nearly continuous distribution of impacts forms the interference pattern predicted by the wave theory.

---

[1]) The description of the phenomenon given here is greatly oversimplified. In reality, the impact of a single photon suffices to sensitize the microcrystal only if the energy of the photon is high enough (far-ultraviolet or X-rays). Even in that case one needs special experimental conditions to be able to assert with near certainty that each photon received by the plate excites one and only one microcrystal (microcrystals of appropriate size, sufficiently thick photographic plate, etc.). All these complications may be passed over without affecting the essential features of the argument given here.

Now, just on the basis of experimental observations, we can *a priori* eliminate all explanations of the phenomenon within the framework of a purely corpuscular theory. Note first of all that photons travel independently of each other, and that their mutual interaction is entirely negligible. Indeed, the interference pattern remains unchanged when one reduces the intensity of the source and increases the time of irradiation, keeping constant the quantity of light received by the grating. In other words, when one sends a certain (very large) number N of photons onto the grating, the distribution of impacts over the photographic plate is the same whether the incident photons be bunched or not; this remains true even in the limit of very weak intensities where the photons "fall upon the grating one by one". One would still obtain the same distribution upon sending a *single photon* onto the grating and repeating the experiment N times.

Let us therefore examine the problem of the scattering of one photon by the grating. Under the conditions of the experiment the initial state of the system (photon + grating) is not known exactly; consequently, the corpuscular picture does not lead to a unique trajectory for the photon, nor, *a fortiori*, to a unique impact of the scattered photon upon the plate, but only to a statistical distribution of possible trajectories and a statistical distribution of impacts. Experimentally, one actually observes a statistical distribution of the

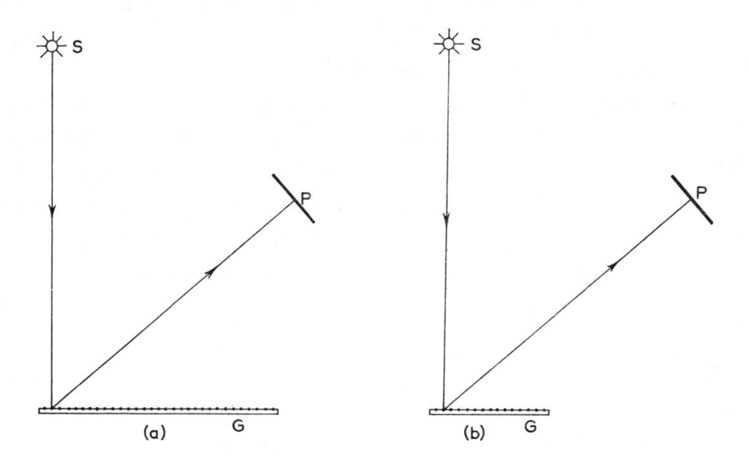

Fig. I.4. Scattering of light by a grating: S = light source; G = grating; P = photographic plate. In the case (*b*), the right-hand half of G has been omitted.

impacts: the latter is, to within a constant, the distribution law of the light intensities making up the interference pattern on the plate. Now, as is well known, the resolving power of a grating depends upon the number of rulings (assumed to be equidistant) and the light fringes are the narrower the larger this number; thus one modifies the interference pattern noticeably when one eliminates one half of the grating. On this very point the corpuscular theory is in clear contradiction with experiment. Indeed, no matter what the equations of motion of each photon interacting with the particles constituting the grating might be, the distribution of trajectories of the photons scattered by the left-hand half of the grating cannot depend upon the presence (Fig. I.4*a*) or the absence (Fig. I.4*b*) of the right-hand half unless one were to attribute to the photon dimensions of the order of those of the grating. If, furthermore, the light source and photographic plate are sufficiently far from each other, the distributions of impacts of the photons scattered by each half are essentially the same. Upon suppressing the right-hand half of the grating, the light intensity received at each point of the plate must thus be reduced without any other modification of the interference pattern. Experiment contradicts these predictions and forces us to assume that the entire grating takes part in the scattering process.

The hypothesis of light particles encounters difficulties of the same kind in all cases of interference or diffraction (Problem I.2). One can effectively detect the arrival of photons, one by one, at the detecting apparatus (screen, photographic plate, or any other more elaborate arrangement), but it is not possible to attribute a precise trajectory to each photon without running into contradictions. The classical doctrine, according to which a corpuscle travels through space in a continuous manner as a function of time, is found to fail. On its way up to the detecting apparatus, everything happens as if light were propagated as a wave; its corpuscular aspect manifests itself only at the instant of detection.

7. Conclusions

We can thus draw a certain number of preliminary conclusions from the experimental results bearing on the interaction between matter and light on the microscopic scale.

Even though the observed discontinuities can properly be explained only with the picture of light corpuscles, it is nevertheless inconceivable to renounce the wave concept. Light presents itself in two forms: wave and corpuscle, each of these aspects appearing in more or less clear-cut fashion depending upon the phenomenon under consideration. The connecting relations (I.4) allow us to pass from one mode of description to the other. The very close bond between these two modes is a *statistical bond* as the discussion of the experiment on scattering by a grating shows: the probability of localizing the photon at a given point is equal to the intensity of the light wave at that point, calculated by the methods of wave optics. *The existence of the wave-corpuscle duality is incompatible with classical doctrine.* It is impossible to consider light either as a stream of classical corpuscles, or as a superposition of classical waves, without entering into contradiction with the experimental facts.

Faced by the prospect of a revision of the classical doctrine, it is particularly important to stress the results of classical wave theory which remain valid. In the first place, *the laws of conservation of momentum and of energy remain rigorously verified.* Furthermore, as we have seen in connection with the Compton effect, *classical theory correctly predicts the average evolution of phenomena in the "macroscopic limit"* where the quantum discontinuities can be treated as infinitely small.

III. QUANTIZATION OF MATERIAL SYSTEMS

8. Atomic Spectroscopy and Difficulties of Rutherford's Classical Model

We have seen in the preceding section how the existence of discontinuities in the interaction mechanism between matter and radiation upsets the Classical Theory of Light. But the upheaval is not limited to light, and equally affects the classical corpuscular theory of matter. This becomes clear when one seeks to reconcile the facts of atomic spectroscopy with the results bearing on the structure of atoms obtained by Rutherford [1]).

[1]) Historically, the first argument showing the necessity of "quantizing" material systems was presented by Einstein in the theory of the specific heat of solids (1907). This theory contains rather crude approximations, which are inevitable in the treatment of a material system which is as complex as a solid body. Furthermore, this theory brings into play results of statistical thermodynamics, similarly to the theory of black-body radiation to which it is closely related. For all these reasons, we shall not enter into detail here, and refer the reader to works such as M. Born, *Atomic Physics, loc. cit.*

One of the most salient features brought to light by the refinement of the study of emission and absorption spectra of light by matter, is the existence of narrow spectral lines. The frequencies of emitted or absorbed radiation vary from one atom to the next; the absorption and emission spectra of a given atom, however, are the same; every line of one spectrum can be observed in the other if one operates under suitable conditions. Each atom may be identified by this spectrum, which therefore constitutes an essential element of information on the structure of the atom and its interaction mechanism with radiation.

The case of hydrogen deserves special mention here since it is the simplest of all atoms (one proton + one electron); all the observed frequencies obey Balmer's empirical formula:

$$\nu = R\left(\frac{1}{n^2} - \frac{1}{m^2}\right),$$

where n and m are positive integers ($m > n$) and R is a numerical constant characteristic of hydrogen (the Rydberg constant).

For more complex atoms there are no equally simple formulae, but there is always a certain correlation among the various observed frequencies: when two frequencies belong to the same spectrum, it often happens that their sum or their difference also occurs in the spectrum. More precisely, one can set up for each atom a table of numbers or spectral terms chosen in such a manner that any frequency of its spectrum is equal to the difference of two of the term values. This rule, of which Balmer's formula is a special case, is named the *Rydberg-Ritz combination principle* (1905). Conversely, all the differences thus formed are not necessarily frequencies of the spectrum, but it is possible to formulate relatively simple *selection rules* which allow us to distinguish those differences which do occur in the spectrum from those which do not.

These experimental facts are in clear-cut disagreement with the classical radiation theory of the Rutherford atom; indeed, the Rutherford model itself encounters serious contradictions when, instead of limiting oneself to the Coulomb interaction, one takes rigorously into account the interaction of the atomic electrons with the electromagnetic field in accordance with the electron theory of Lorentz. Indeed, the electrons radiate while describing their orbits, progressively lose energy, and finally spiral into the nucleus. At each instant, the observed frequencies in the emitted radiation are equal to the frequency of the orbital motion or to one of its harmonics; as the latter varies during the slowing-down process, a *continuous* light spectrum is emitted. The classical theory of the Rutherford atom therefore explains neither the stability of atoms, nor the existence of line spectra. We are confronted with a new manifestation of discontinuities in the interaction of matter and light, where the classical theory predicts a continuous variation.

9. Quantization of Atomic Energy Levels

In 1913, Bohr obtains a general scheme for the explanation of spectra by completing the quantum hypothesis of light by a new postulate incompatible with classical notions: the quantization of the energy levels of atoms.

According to Bohr, the atom does not behave as a classical system capable of exchanging energy in a continuous manner. It can exist only in a certain number of *stationary states* or quantum states each having a well-defined energy. One says that the energy of the atom is quantized. It can vary only by jumps, each jump corresponding to a transition from one state to another.

This postulate allows us to specify the mechanism of absorption or emission of light through quanta. In the presence of light an atom of energy E_i may undergo a transition to a state of higher energy $E_j(>E_i)$ by absorbing a photon $h\nu$ provided that the total energy is conserved, namely

$$h\nu = E_j - E_i.$$

Similarly, the atom can undergo a transition to a state of lower energy $E_k(<E_i)$ by emitting a photon $h\nu$ whose frequency satisfies the relation

$$h\nu = E_i - E_k.$$

If the atom finds itself in its lowest energy state (ground state) it cannot radiate and remains stable.

In this way an explanation is found for the existence of spectral lines characteristic of each atom and satisfying the Rydberg-Ritz combination principle: the spectral terms are equal, to within a factor of h, to the energies of the quantum states of the atom. In particular, for the case of the hydrogen atom, one rediscovers the

Balmer formula by assuming that the energy levels are given by the formula

$$E_n = -h\frac{R}{n^2} \qquad (n=1, 2, 3, \ldots, \infty). \qquad \text{(I.7)}$$

Another confirmation of the quantization of atomic energy levels is furnished by the experiment of Franck and Hertz on the inelastic collisions between electrons and atoms (1914). The experiment consists in bombarding atoms by monoërgic electrons and in measuring the kinetic energy of the scattered electrons. From this one deduces by subtraction the quantity of energy absorbed in the collision by the atoms. Let $E_0, E_1, E_2, \ldots$ be the sequence of quantized energy levels of the atoms, T the kinetic energy of the incident electrons. Under the conditions of the experiment, the atoms of the target are practically all in their ground state. As long as T lies below the difference E_1-E_0 between the energy of the ground state and that of the first excited state, the atom cannot absorb energy and all collisions are elastic. As soon as $T>E_1-E_0$, inelastic collisions can occur in which the electron loses a quantity of energy equal to E_1-E_0 and the atom goes into its first excited state. This is exactly what is found experimentally. One similarly observes collisions with excitation of the second excited state as soon as $T>E_2-E_0$, and so on.

Hence the quantization of atomic energy levels appears as experimental fact. This property is not peculiar to atoms. Progress of experimentation, especially in the field of spectroscopy, has shown that quantization is found in the case of molecules and of more complex systems of particles as well. We thus face a very general property of matter which classical corpuscular theory is unable to explain.

10. Other Examples of Quantization: Space Quantization

Another type of experimentally observed quantization is that of the quantization of orientation, or "space quantization" of atomic systems. One observes it every time the atom is situated in an external field possessing a preferred direction. The orientation of the atomic system is not arbitrary, but limited to certain discrete values.

The most direct confirmation of this type of quantization is furnished by the *Stern-Gerlach experiment* (1922) on the deviation of paramagnetic atomic beams (or molecular beams) in an inhomogeneous magnetic field. Paramagnetic atoms are by hypothesis endowed with a permanent magnetic moment $\boldsymbol{\mu}$, and can be considered as little elementary gyroscopes of angular momentum $\boldsymbol{l}$ proportional to $\boldsymbol{\mu}$:

$$\boldsymbol{\mu} = M\boldsymbol{l}.$$

The orientation of $\boldsymbol{\mu}$ and $\boldsymbol{l}$ defines the orientation of the atom itself. In a magnetic field $\boldsymbol{H}$, the angular momentum executes a precessional motion about $\boldsymbol{H}$ (Larmor precession, cf. Problem I.3). If $\boldsymbol{H}$ is constant, the magnetic energy $-\boldsymbol{\mu}\cdot\boldsymbol{H}$ remains constant and independent of the position of the center of mass of the atom, and the latter remains in uniform rectilinear motion. If $\boldsymbol{H}$ is not constant, the center of mass of the atom is subject to the force $\boldsymbol{F} = \text{grad}\,(\boldsymbol{\mu}\cdot\boldsymbol{H})$ and suffers a certain deflection. This is exactly what one observes in the Stern-Gerlach experiment which is shown schematically in Fig. I.5. Because of the precessional motion about the field $\boldsymbol{H}$, the component μ_z of $\boldsymbol{\mu}$ along the field remains constant while the other components oscillate about zero. Everything takes place as if the atom were subject to the value of the force averaged over several oscillations: $\mu_z\,\text{grad}\,\boldsymbol{H}_z$. Under normal experimental circumstances this average force is directed along Oz and equals $\mu_z(\partial \boldsymbol{H}_z/\partial z)$. Let $2l$ be the distance in the magnetic field traversed by the atom, T the kinetic energy of the atoms in the incident beam; a simple calculation shows that the velocity of each atom is deflected from its initial direction Ox through an angle $\approx \mu_z(\partial \boldsymbol{H}_z/\partial z)(l/T)$. The deflection is thus proportional to the component of $\boldsymbol{\mu}$ in the direction of the field. If the atoms were randomly oriented, μ_z could take on all values between $-\mu$ and $+\mu$ and the deflection angle all values between the corresponding two extreme values. The impacts of the atoms on the screen should then form a single spot elongated in the direction Oz. What one actually observes is a series of small spots, equidistant and aligned parallel to Oz. If one varies the field [and hence $(\partial \boldsymbol{H}_z/\partial z)$] the mutual distance of the spots varies correspondingly, without any other modification of the pattern; in particular the number of spots λ remains constant. Each of the spots corresponds to a definite value of μ_z. Consequently, μ_z is a quantized quantity taking on λ different values. The component l_z of the angular momentum evidently possesses the same property.

Against this interpretation of the Stern-Gerlach experiment one may raise the objection that it is based on a very particular hypothesis concerning the origin of atomic paramagnetism, namely the existence of a permanent magnetic moment proportional to the angular momentum. We shall not elaborate here on the facts and arguments

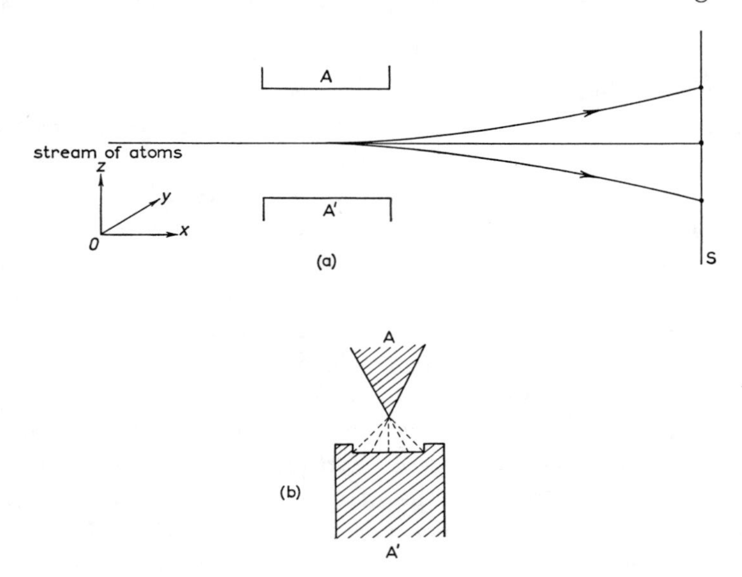

Fig. I.5. The Stern–Gerlach experiment.

(*a*) General scheme of the experiment: the atomic beam passes between the pole pieces AA′ of the magnet where an inhomogeneous magnetic field exists (directed vertically in the figure); the impacts of the atoms are observed on the screen S.

(*b*) Section through the pole pieces of the magnet; dashed lines represent the lines of force of the magnetic field.

which justify such a hypothesis (gyromagnetic effects, Langevin's theory of paramagnetic susceptibility, etc.) which later developments of Quantum Mechanics have amply confirmed. Even if one doubts the details of the explanation given above, one can hardly account for the existence of λ distinct spots on the screen without assuming that certain quantities characteristic of the internal motions are quantized. Indeed, to the extent that the center-of-mass motion follows the laws of Classical Mechanics, its trajectory is entirely determined by the dynamical state of the atom at the entrance of the magnet. The appearance on the screen of a more or less spread-out distribution of impacts indicates that the atoms are not all in the same initial conditions and that the dynamical variables defining the initial state are statistically distributed over a somewhat extended domain. The existence of λ distinct spots attests to the fact that this statistical distribution presents at least **e** discontinuities; in other words, certain dynamical variables of the atom are quantized. Since the atoms are practically all in their ground state (otherwise they would radiate) one cannot be dealing here with a quantization of the energy and, since the effect observed on the screen is directional, the dynamical variable of the atom whose quantization is thus made evident depends upon the orientation of the atom.

In addition to the Stern-Gerlach experiment, there exist quite a few other less direct manifestations of space quantization. In particular we mention the effect of a constant magnetic field on the structure of spectra, or Zeeman effect (1896), to which we shall return later on. All these phenomena have a common origin, namely the quantization of angular momentum; later developments of Quantum Mechanics will bring this out quite clearly.

IV. CORRESPONDENCE PRINCIPLE AND THE OLD QUANTUM THEORY

11. Inadequacy of Classical Corpuscular Theory

The quantization of certain physical quantities – and we have to insist on that point – is an experimental fact incompatible with the classical corpuscular theory of matter. Thus the energy of a system of classical particles is an essentially continuous quantity; no matter how one modifies the force law, not even by introducing additional dynamical variables, can one change this situation: the fact that the energy of a system of particles is limited to a discrete set of allowed values is a result which falls outside Classical Mechanics. The same remarks apply to all other quantized quantities.

Correlated with this, *the evolution in time of a quantized quantity is impossible to describe in strictly classical terms.* Let us take the example of an atom finding itself initially in its first excited state E_1 and falling back to the ground state with emission of a photon. If we adopt the language of classical physics and try to define the evolution in time of the energy of the atom, one must assume that this energy undergoes a discontinuous jump from E_1 to E_0 at a certain instant, since any continuous evolution of the energy between these two values is excluded. Yet it is not possible to predict at what precise instant this jump will occur. Indeed, if *the dynamical state of the atom remains identically the same* during the entire period of time preceding the jump, there is no reason why the latter should occur at any one given instant rather than at another. At best one can speak of the probability per unit time of the occurrence of the jump. In reality, classical physics is unable to describe this situation, and the very picture of a jump occurring at a precise instant is incorrect. One must give up imagining an exact evolution of the energy as a function of time. The only thing one may define is the probability that an initially excited atom be found in its ground state at a given later time. We shall see later on that the probability of de-excitation – as well as the law of disintegration of a radioactive nucleus – follows a decreasing exponential law whose characteristic constant is the de-excitation probability per unit time, otherwise stated, the reciprocal of the mean life of the excited state.

We must therefore, at the price of renouncing certain classical concepts, incorporate this phenomenon of quantization into a coherent theory of matter, from which one might deduce the precise numerical values of the quantized quantities, and the quantities pertaining to the various quantum transitions which are possible, as for instance, the mean lives of excited states which were mentioned above. This program will not be fully realized until the establishment of Quantum Mechanics in its modern form. Previously, Bohr and his school (Kramers, Sommerfeld) formulated a first draft of a quantum theory from which the spectral terms of hydrogen-like systems could be predicted correctly. In spite of the difficulties of principle, and the limitations of this Old Quantum Theory, it is useful to know its main features in order to properly appreciate the later development of the theory. Furthermore, this older theory represents a first example of the application of a heuristic principle which played an essential role in the development of Quantum Mechanics: the correspondence principle. It is on this point that we shall especially focus our attention in the following presentation of the Old Quantum Theory. The latter was completed by a semi-classical theory of the interaction between matter and radiation also based on the correspondence principle; we shall not discuss it in this book.

12. Correspondence Principle

The correspondence principle was not clearly formulated by Bohr until 1923 [1]), but it inspired all previous work. It consists in stating precisely to what extent the notions and the results of Classical Mechanics can serve as guides in the elaboration and interpretation of the correct theory.

We have already discussed the domain of validity of the Classical Theory of radiation in connection with light quanta. What has been said is valid for Classical Theory in general. The latter predicts correctly a vast array of phenomena, from the macroscopic scale down to and including certain phenomena of the microscopic domain; among the latter let us mention the motion of electrons in a static electromagnetic field, the thermal motion of the atoms or molecules of a gas, etc. The major difficulty of Classical Theory in explaining phenomena on the microscopic scale stems from the appearance of discontinuities on that scale.

One may therefore assert that *Classical Theory is "macroscopically correct"*, that is to say, it accounts for phenomena in the limit where quantum discontinuities may be considered infinitely small; in all these limiting cases, the predictions of the exact theory must coincide with those of Classical Theory. We have here a very restrictive condition imposed upon the Quantum Theory. One often expresses it in abbreviated form by saying that: *Quantum Theory must approach Classical Theory asymptotically in the limit of large quantum numbers.*

In order that this condition might be fulfilled, one establishes in principle *that there exists a formal analogy between Quantum Theory and Classical Theory*; this "correspondence" between the two theories persists down to the smallest details and must serve as guide in the interpretation of the results of the new theory.

13. Application of the Correspondence Principle to the Calculation of the Rydberg Constant

Let us verify that expression (I.7), giving the energy levels of the hydrogen atom as a function of the quantum number n, is compatible with the correspondence principle, and let us show that the application of this principle unambiguously determines the numerical value of the constant R which occurs in that expression.

---

[1]) N. Bohr, Z. Physik 13 (1923) 117.

According to the classical theory of Rutherford, the hydrogen atom is composed of an electron and a proton in Coulomb interaction [potential $-(e^2/r)$]. In accordance with Kepler's laws, which we shall assume to be known to the reader, the electron describes an elliptical orbit about the proton, the latter (assumed infinitely heavy) being at its focus. To each orbit there corresponds a certain value of the energy $E(<0)$ and a frequency $\nu_{\mathrm{cl.}}$ of the electron motion along that orbit. These quantities in fact depend only upon the length of the major axis of the ellipse; they are related by the expression:

$$\nu_{\mathrm{cl.}}(E) = \frac{1}{\pi e^2}\left(\frac{2|E|^3}{m}\right)^{\frac{1}{2}} \tag{I.8}$$

(m = mass of the electron).

During the course of this motion, the electron emits a certain radiation made up of a superposition of monochromatic waves whose frequencies are equal to $\nu_{\mathrm{cl.}}$ or one of its harmonics; this radiation is the richer in high harmonics the more eccentric the elliptical orbit. It is emitted in continuous fashion and is accompanied by a steady decrease of the energy E.

This should be compared with the energy degradation by discrete jumps in the Bohr theory. When n is very large, the distance of the level E_n to each of its nearest neighbors is a certain integral multiple of $\mathrm{d}E/\mathrm{d}n = 2Rh/n^3$; for all optical transitions where the relative variation $\Delta n/n$ of the quantum number is very small, the emitted frequency is, just as in classical theory, the harmonic (of order $\Delta n - 1$) of a certain fundamental frequency

$$\nu_{\mathrm{qu.}} \approx 2\frac{R}{n^3} = 2\left(\frac{|E_n|^3}{Rh^3}\right)^{\frac{1}{2}}. \tag{I.9}$$

In the limit where n is very large, the energy E_n is degraded on the average by a succession of small and numerous quantum jumps, and the spectrum of emitted frequencies (more precisely, the low-frequency portion of that spectrum which is associated with quanta of the lowest energy) must be identical to the classical spectrum, in accordance with the correspondence principle. In other words,

$$\nu_{\mathrm{qu.}} \underset{n\to\infty}{\sim} \nu_{\mathrm{cl.}}(E). \tag{I.10}$$

Upon examining expressions (I.8) and (I.9) one sees that this condition may be actually fulfilled if one takes

$$R = \frac{2\pi^2 m e^4}{h^3}. \tag{I.11}$$

The experimental value of R is known with extreme precision ($\approx 10^{-6}$). The theoretical value (I.11) agrees with the former to within less than one part in 10^4 [1]). We have here one of the most spectacular successes of the Bohr theory.

The latter is easily extended to hydrogen-like atoms formed by an electron and a nucleus of charge Ze, in particular to the singly ionized helium atom ($Z = 2$). One has merely to replace e^2 by Ze^2 in all formulae. The spectral terms of He^+ thus obtained coincide, with the same extraordinary precision of 1 in 10^4, with those observed experimentally.

14. Lagrange's and Hamilton's Forms of the Equations of Classical Mechanics

In view of later discussions of the formal correspondence between Quantum Theory and Classical Theory, it is well to recall some points of classical analytical mechanics.

Quite generally, the dynamical state of a classical system is defined by its position – in terms of its coordinates $q_1, q_2, \ldots, q_R$, – and its velocity – defined by the derivatives $\dot{q}_1, \dot{q}_2, \ldots, \dot{q}_R$ of its coordinates with respect to time. R is the number of degrees of freedom of the system [2]). If we are dealing with a system of n particles, one may choose as position coordinates the $3n$ cartesian coordinates of these particles, but the considerations which follow apply equally well to other choices of coordinates. The position of the system can be represented at each instant in a space with R dimensions, *configuration space*, by a point M having $q_1, q_2, \ldots, q_R$ as coordinates in that space.

---

[1]) In order to claim such a precision in the determination of R, it is necessary to take into account the fact that the mass of the proton M is finite. To do this one merely replaces in formula (I.11) the mass m by the reduced mass $m' = mM/(m + M)$. Having taken this correction into account ($\approx 5 \times 10^{-4}$), the theoretical value of R lies slightly below the experimental value. The difference is due essentially to relativistic effects, which amount to a slight increase of the mass m'.

[2]) We only consider here systems without constraints; in other words, the q's may vary independently of each other without any limitation.

Classical Mechanics has as its objective the formulation of the laws of evolution of the system as a function of time or, if one prefers, the laws of motion of its representative point M in configuration space.

For a large number of dynamical systems – the only ones we shall have to consider here – one can write down the laws of motion by introducing a certain characteristic function of the system, the Lagrange function (or Lagrangian):

$$L \equiv L(q_1, q_2, \dots, q_R;\ \dot{q}_1, \dot{q}_2, \dots, \dot{q}_R;\ t).$$

The coordinates q will satisfy R second-order differential equations (Lagrange's equations):

$$\frac{\mathrm{d}}{\mathrm{d}t}\left(\frac{\partial L}{\partial \dot{q}_r}\right) - \frac{\partial L}{\partial q_r} = 0 \qquad (r=1, 2, \dots, R).$$

The quantities

$$p_r \equiv \frac{\partial L}{\partial \dot{q}_r} \qquad (r=1, 2, \dots, R)$$

which occur in these equations, are called the Lagrange *conjugate momenta.* In the case where q_r is one of the cartesian coordinates of a particle of mass m and the forces are derivable from a static potential, p_r is the corresponding coordinate for the momentum of that particle: $p_r = m\dot{q}_r$.

The laws of motion may be expressed equally well in the form of a variational principle. The system of Lagrange's equations is, in fact, equivalent to the *principle of least action* (Maupertuis-Hamilton):

$$\delta \int_{t_1}^{t_2} L\,\mathrm{d}t = 0, \qquad \delta M(t_1) = \delta M(t_2) = 0 \qquad \text{(I.12)}$$

whose significance is as follows: of all the laws $M(t)$ allowing the system to pass from position M_1 at time t_1, to position M_2 at time t_2, the law of motion which is actually realized is the one making the integral $\int_{t_1}^{t_2} L\,\mathrm{d}t$ stationary.

Another, particularly useful form of the laws of Classical Mechanics is *Hamilton's canonical form.* We note that the dynamical state of classical system at a given instant is completely defined by givir R position coordinates $q_1, q_2, \dots, q_R$ and the R corresponding cor momenta $p_1, p_2, \dots, p_R$. It is convenient to introduce a sr

$2R$ dimensions, the *phase space*, where such a dynamical state is represented by a point P having the q's and p's as coordinates. Now let us define the classical Hamiltonian:

$$H \equiv H(q_1, \ldots, q_R;\ p_1, \ldots, p_R;\ t) = \sum_{r=1}^{R} \dot{q}_r \frac{\partial L}{\partial \dot{q}_r} - L. \qquad \text{(I.13)}$$

The equations of motion may be written in the following canonical form

$$\dot{q}_r \equiv \frac{\partial H}{\partial p_r}, \qquad \dot{p}_r \equiv -\frac{\partial H}{\partial q_r} \qquad (r = 1, 2, \ldots, R). \qquad \text{(I.14)}$$

These are first-order differential equations. Knowledge of the coordinates and momenta at the initial instant suffices to determine their values at all later instants. Hence when H does not depend upon time, one and only one trajectory passes through each point of phase space, representing a possible motion of the system.

In the most common cases, L is the difference between the kinetic energy T, a homogeneous quadratic function of the $\dot{q}$, and the potential energy V; $H = T + V$ is the total energy of the system expressed as a function of the q's and p's. However, the formalisms of Lagrange and of Hamilton apply to more general dynamical systems (cf. Problem I.4). By extension, one agrees in all cases to consider H as the total energy of the system. From Hamilton's equations one deduces that $\dot{H} \equiv (\mathrm{d}H/\mathrm{d}t) = (\partial H/\partial t)$; in other words, if the Hamiltonian does not explicitly depend upon the time, the total energy of the system is a constant of the motion. One says that the system is conservative.

In order to illustrate these points, let us consider an electron in the Coulomb field of a proton (considered infinitely heavy). Let $\mathbf{r}(x, y, z)$ be its position in a system of cartesian axes centered at the proton, $\mathbf{v} = \mathrm{d}\mathbf{r}/\mathrm{d}t$ its velocity, and $\mathbf{p}(p_x, p_y, p_z)$ its momentum. The Lagrangian is

$$L = \tfrac{1}{2} m v^2 + \frac{e^2}{r}.$$

The conjugate momenta of the coordinates are the respective components of the momentum $(p_x = \partial L/\partial v_x,\ p_y = \partial L/\partial v_y,\ p_z = \partial L/\partial v_z)$. From the explicit form of the Hamiltonian function:

$$H \equiv \frac{p^2}{2m} - \frac{e^2}{r},$$

one obtains Hamilton's equations:

$$\frac{\mathrm{d}\boldsymbol{r}}{\mathrm{d}t} = \frac{\boldsymbol{p}}{m}, \qquad \frac{\mathrm{d}\boldsymbol{p}}{\mathrm{d}t} = \operatorname{grad}\frac{e^2}{r} = -e^2\frac{\boldsymbol{r}}{r^3}.$$

With these equations one verifies directly that the angular momentum, $\boldsymbol{l} = \boldsymbol{r} \times \boldsymbol{p}$, is a constant of the motion: $\mathrm{d}\boldsymbol{l}/\mathrm{d}t = 0$, a consequence of the central character of the potential $-e^2/r$, and that the electron trajectory is entirely situated in the plane containing the origin and perpendicular to this constant vector $\boldsymbol{l}$.

In the same way one obtains the equations of motion using any other system of coordinates. For the trajectories located in the xy plane ($z = \dot{z} = 0$), one obtains in polar coordinates (r, φ):

$$x = r\cos\varphi, \qquad y = r\sin\varphi,$$

$$L = \tfrac{1}{2}m[\dot{r}^2 + (r\dot{\varphi})^2] + \frac{e^2}{r}, \qquad p_r = m\dot{r}, \qquad p_\varphi = mr^2\dot{\varphi},$$

$$H = \frac{1}{2m}\left(p_r{}^2 + \frac{p_\varphi{}^2}{r^2}\right) - \frac{e^2}{r},$$

whence Hamilton's equations:

$$\begin{aligned} \dot{p}_\varphi &= 0, & \dot{\varphi} &= \frac{p_\varphi}{mr^2}, \\ \dot{p}_r &= \frac{p_\varphi{}^2}{mr^3} - \frac{e^2}{r^2}, & \dot{r} &= \frac{p_r}{m}. \end{aligned} \tag{I.15}$$

p_φ is the magnitude of the angular momentum: it is, indeed, a constant of the motion.

15. Bohr-Sommerfeld Quantization Rules

The Old Quantum Theory is essentially a general method of calculating quantized quantities, based upon the hypotheses of Bohr and the correspondence principle. The procedure is the following. One assumes that the systems of material particles follow the laws of Classical Mechanics. One further postulates that, of all the solutions of the equations of motion, one must only retain those which satisfy certain *ad hoc* quantization rules. One thus selects a discontinuous family of motions; these are, by hypothesis, the only motions which can actually be realized. To each of these motions there corresponds a certain value of the energy. The discontinuous sequence of energy values thus obtained is the spectrum of quantized energy levels. One likewise obtains a discrete spectrum of allowed values for all other constants of the motion.

The determination of the "quantization rules" constitutes the central problem of this Old Quantum Theory. It is above all a matter of intuition: one postulates certain rules and compares the spectra of quantized quantities derived therefrom with the experimental results. In this search, the correspondence principle serves as a valuable guide.

There exists a very simple case where this principle leads to the result in a very natural way, namely the case where the classical motions are periodic motions whose frequency is a function of the energy alone,

$$\nu_{\text{cl.}} = \nu_{\text{cl.}}(E).$$

In particular, this is the case in the hydrogen atom [cf. eq. (I.8)]. Let $E_1, E_2, \ldots, E_n, \ldots,$ be the sequence of quantized energies. One may always consider the energy of the system to be a continuous function $E(n)$ of the quantum number n, the quantization of energy then resulting from the fact that n can take on only integral values. Upon repeating the reasoning of § 13 for the calculation of the Rydberg constant, we obtain the relation for the correspondence between classical and quantum frequency [cf. eq. (I.10)]:

$$\frac{1}{h}\frac{\mathrm{d}E}{\mathrm{d}n} \underset{n\to\infty}{\sim} \nu_{\text{cl.}}(E),$$

from which we obtain the quantization rule

$$\int^{E} \frac{\mathrm{d}E}{\nu_{\text{cl.}}(E)} = nh + \text{constant},$$

for large values of the integer n. It is natural to extend this rule to all values of n and to put

$$\int_{E_{\min}}^{E} \frac{\mathrm{d}E}{\nu_{\text{cl.}}(E)} = nh \qquad (n = 1, 2, \ldots, \infty). \tag{I.16}$$

($E_{\min}$ is the minimum value of the energy of the classical system.) In the case of the hydrogen atom, this quantization rule yields just the terms of the Balmer formula.

This rule applies equally to periodic systems with a single degree of freedom. In that case it is possible to put it into a form more suited to generalizations. Let q be the position coordinate of such a system, p its canonically conjugate momentum, and $H(q, p)=E$ the total energy. The two-dimensional phase space and the periodic motions are represented by closed curves $H(q, p)=\text{const.}$ in this space [1]). Using Hamilton's equations, it is possible to show that

$$\int_{E_{\min}}^{E} \frac{\mathrm{d}E}{\nu(E)} = \oint_{H=E} p \, \mathrm{d}q,$$

where the symbol $\oint_{H=E}$ implies that one must integrate over a complete period of the motion corresponding to the energy E (the integral $\oint p \, \mathrm{d}q$ is called the action integral). One thus obtains the quantization rule equivalent to the rule (I.16):

$$\oint p \, \mathrm{d}q = nh \qquad (n=1, 2, \ldots, \infty). \tag{I.17}$$

It determines the allowed trajectories of phase space and the corresponding quantized energies. This rule is known as the *Bohr-Sommerfeld quantization rule.*

Wilson and Sommerfeld have generalized this rule to multiply-periodic systems. These are systems with several degrees of freedom whose motions can be represented by a sequence of functions $p_1(q_1), p_2(q_2), \ldots, p_R(q_R)$, for a suitable choice of coordinates $q_1, q_2, \ldots, q_R$ and of conjugate momenta $p_1, p_2, \ldots, p_R$; in other words, the trajectories in phase space are such that each momentum p_r is a function of the coordinate q_r only, to the exclusion of all other variables. Each function $p_r(q_r)$ represents a periodic motion of frequency ν_r; the motion of the system results from the combination of periodic motions of respective frequencies $\nu_1, \nu_2, \ldots, \nu_R$. In that case, the quantization rules are the R relations

$$\oint p_r \, \mathrm{d}q_r = n_r h \qquad (r=1, 2, \ldots, R). \tag{I.18}$$

The R (integral) quantum numbers $n_1, n_2, \ldots, n_R$ define the quantized trajectories of the system, and the quantized values of the various

[1]) If q is a cyclic variable (for instance an angular variable), in other words if the values of q differing by an integral multiple of a certain period Q represent the same configuration of the system, the periodic motion in phase space is not represented by a closed curve but by a curve of period Q.

constants of the motion such as the energy, angular momentum, etc. In particular, the energy $E(n_1, n_2, \ldots, n_R)$, considered as a function of the variables $n_1, n_2, \ldots, n_R$, fulfills the conditions of correspondence

$$\frac{\partial E}{\partial n_r}\underset{n_r \to \infty}{\sim} h\nu_r \qquad (r = 1, 2, \ldots, R).$$

By way of an application we shall briefly consider the quantization of the hydrogen atom. Once we have chosen the plane of the electron orbit, this is a two-dimensional problem whose equations in polar coordinates we have already written down [eq. (I.15)]. The angular momentum and the energy are constants of the motion. If one fixes the respective values $L(\geqslant 0)$ and $E(<0)$ of these two quantities, one obtains a possible trajectory of the classical motion: it is an ellipse of eccentricity

$$\sqrt{1 + \frac{2L^2E}{me^4}}.$$

The momenta p_φ and p_r are functions of their respective conjugate coordinates. Indeed

$$p_\varphi = L, \qquad \frac{1}{2m}\left(p_r{}^2 + \frac{L^2}{r^2}\right) - \frac{e^2}{r} = E.$$

We can thus apply the Bohr-Sommerfeld quantization rules:

$$\oint p_\varphi \, \mathrm{d}\varphi = lh, \qquad \oint p_r \, \mathrm{d}r = kh,$$

where l, the azimuthal quantum number, and k, the radial quantum number, are positive (or zero) integers. The first rule yields the quantized value of the angular momentum:

$$L = l\hbar.$$

The second rule gives after a somewhat lengthy but straightforward calculation,

$$\sqrt{\frac{2\pi^2\, me^4}{-E}} - 2\pi L = kh,$$

from which, introducing the "principal quantum number" $n = l + k$, one extracts the Balmer formula:

$$E_n = -\frac{me^4}{2\hbar^2\, n^2},$$

with the same value for the Rydberg constant as the one we calculated previously [eq. (I.11)].

The quantized energy depends only upon the sum of the two quantum numbers l and k. This property, which is characteristic of the Coulomb potential, derives from the fact that the azimuthal and radial frequencies are equal: $\nu_\varphi = \nu_r$. To the energy E_n there correspond n quantized orbits, defined respectively by the following values of l: $l = 1, 2, \ldots, n$ (one eliminates the value $l = 0$ for reasons we shall not discuss here); they are ellipses of excentricity $\sqrt{1-(l^2/n^2)}$. The value $l = n$ corresponds to the circular orbit [1]).

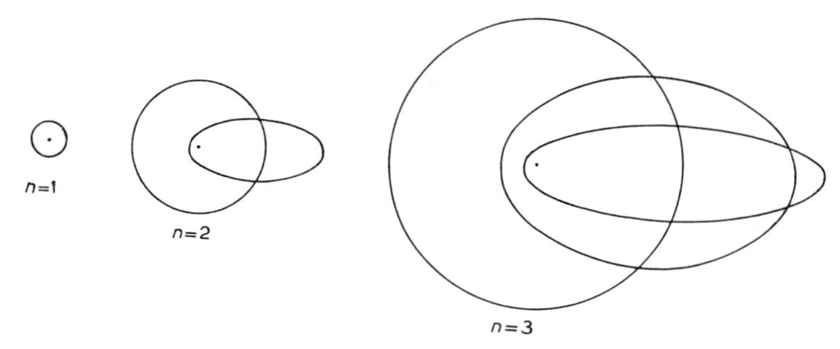

Fig. I.6. The Bohr orbits of the ground state ($n = 1$) and of the first two excited states ($n = 2, 3$) of the hydrogen atom. The relative dimensions of the orbits have been drawn to scale.

One can apply the same quantization rules to the relativistic equations of motion and thus determine the relativistic corrections to this theory of the hydrogen atom. One thus obtains a value of the Rydberg constant in better accord with experiment (cf. footnote p. 31). Furthermore, the "degeneracy" of the energy is lifted: to each value of n there correspond n very closely spaced but distinct values of the quantized energy, each corresponding to a precise value of the angular momentum $l\hbar$. Experimentally, one actually observes a fine structure of the hydrogen atom in fair agreement with the predictions of this theory.

Let us now examine the problem of space quantization. The above treatment, where one quantizes the orbits in the same plane, contains

---

[1]) The quantization of circular orbits led Bohr to find the Balmer formula as early as 1913. The quantization of elliptical orbits is due to Sommerfeld, who extended the theory to the relativistic case (see further on).

no privileged direction and only permits the determination of quantized spectra for scalar quantities such as the energy or the *magnitude* of the angular momentum ($L = l\hbar$). There exists a very general rule for space quantization for systems possessing axial symmetry (atom in constant magnetic field, for instance). In that case, the component L_z of the total angular momentum along the axis of symmetry Oz is a constant of the motion: it can be shown in Classical Mechanics that the latter can always be taken as the momentum conjugate to an angle φ fixing the overall orientation of the system about the axis Oz. Consequently

$$\oint L_z \,\mathrm{d}\varphi = mh \qquad (m \text{ integer})$$

and since L_z is constant,

$$L_z = m\hbar. \tag{I.19}$$

The component of the angular momentum along the axis of symmetry is a (positive or negative) integral multiple of $\hbar$. The quantum number m is known as the *magnetic quantum number*.

16. Successes and Limitations of the Old Quantum Theory

We shall not pursue the discussion of the Old Quantum Theory any further. It has led to considerable progress in the knowledge of spectra, by furnishing a correct evaluation of the spectral terms of a large number of atomic and molecular systems. The treatment of the hydrogen atom is easily extended to hydrogen-like atoms (He^+, Li^{++}) and to the alkali atoms. The theory also applies to the vibration and rotation spectra of molecules, to the X-ray spectra of atoms, and to the normal Zeeman effect. Complemented by the semi-classical theory of the interaction between matter and radiation, it yields the various selection rules and probabilities of the different possible quantum transitions. In all cases, the predictions of the theory agree remarkably well with the experimental results, aside from occasional disagreements for very small quantum numbers, disagreements which one can remedy by imposing some empirical modifications on the quantization (the suppression of the value $l = 0$ of the azimuthal quantum number is an example).

Nevertheless it is an incomplete theory. The Bohr-Sommerfeld rules apply only to periodic or multiply-periodic systems. There exists no method of quantization for aperiodic motions. Thus, the mechanism of an experiment as fundamental as that of Franck and Hertz remains completely unexplained: the Bohr-Sommerfeld theory gives the quantized energy levels of the atomic target, but there is no theoretical treatment whatsoever which would indicate the trajectory actually followed by the electron, and which would describe in detail the inelastic collision between electron and target atom. As a general rule, all collision phenomena lie outside the scope of the theory. Even in the calculation of spectral terms, the success of the theory is limited to the simplest systems; numerous difficulties arise as soon as one tries to set up the problem of the quantization of complex atoms in a rigorous manner. Moreover, there are cases of violent disagreement, such as the one of the (un-ionized) helium atom or the anomalous Zeeman effect.

Furthermore, the theory is by no means free of ambiguity and contradictions. We have an example in the prescriptions concerning space quantization. The quantization rule of the component L_z of the angular momentum of a system possessing axial symmetry about the z axis should apply just as well to systems with spherical symmetry, since the latter possess axial symmetry about any axis passing through the origin. In this last case we are led to the absurd prescription that the component of angular momentum along any axis passing through the origin must be an integral multiple of $\hbar$.

But the difficulties of principle of the Old Quantum Theory are considerably more serious than all these faults and contradictions. The quantization rules are purely formal restrictions imposed upon the solutions of the classical equations of motion; they were determined in an entirely empirical manner. The profound justification of this quantization of classical trajectories is completely absent. In fact, the very notion of trajectory is hard to reconcile with the quantization phenomenon. It implies that the particle possesses at each instant a well-defined position and momentum, and that these quantities vary in a continuous manner in the course of time. Under these conditions what would be the trajectory of a particle such as the electron in the Franck and Hertz experiment? If this electron has a precise trajectory, its energy varies in a continuous manner, and one must consequently give up the idea of an exchange of energy by quanta with the target atom; in other words, one must renounce the quantization of the energy levels of the target atom. Conversely, to postulate this quantization amounts to giving up the (classical) idea of a precise trajectory of the electron and, quite logically, the idea of trajectory in general. We shall see in Chapters II and IV that this renunciation is fully justified, and we shall analyze its physical significance and its consequences in detail. In any event, we are led to renounce the classical equations of motion; under these conditions, one may well question the physical meaning of the solutions of these equations of motion such as those leading to the quantized trajectories of atoms.

The Old Quantum Theory undoubtedly represented a great step forward. Predicting a considerable body of experimental results from a few simple rules, it constitutes a general scheme for the phenomenological explanation of spectra and has, by virtue of this fact, played a large clarifying role in the history of contemporary physics. But this rather haphazard mixture of classical mechanics and *ad hoc* prescriptions can in no way be considered as a definitive theory.

17. Conclusions

In this chapter we have analyzed the main difficulties encountered by Classical Theory in the domain of microscopic physics. These difficulties appear as soon as one makes a detailed study of the interaction between matter and radiation. The essential fact is the appearance of discontinuity on this scale, connected with the existence of an indivisible quantum of action $\hbar$.

This *atomism of action* seems to be a fundamental characteristic of natural phenomena. On the macroscopic scale, the quantum $\hbar$ may be regarded as an infinitesimal quantity and one may be content with a classical description of phenomena, in which the evolution of physical systems is represented by means of dynamical variables, defined at each instant in a precise manner, and varying in continuous fashion with time. In atomic and subatomic physics, on the other hand, $\hbar$ ceases to be negligible and characteristic quantum effects appear. The upheaval affects the entire classical construct.

The classical wave treatment of electromagnetic radiation is irreconcilable with the experimental fact that the transfers of energy and momentum between matter and radiation take place through discrete and indivisible quanta. The photoelectric effect, and the Compton effect, in particular, cannot be understood unless one imagines light as a stream of corpuscles; in contrast, the photon hypothesis cannot be reconciled with the existence of interference or diffraction phenomena, in which the behavior of light is characteristic of that of a superposition of waves. If one keeps to the language of classical physics, it is not possible to account in a completely coherent manner for the totality of light phenomena. One must appeal, depending upon the case, to one or the other of two contradictory representations – stream of corpuscles or superposition of waves. The connection between these two representations is given by the fundamental relations (I.4), into which the quantum $\hbar$ enters. The simplest way to interpret this wave-corpuscle duality is to postulate a statistical bond between waves and corpuscles, where the intensity of the wave at a point gives the probability of finding the photon which is associated with it, at that point.

As far as material systems are concerned, the quantization of certain quantities sets a limit to the conception according to which matter is made up of corpuscles obeying Newtonian Mechanics. The quantization of atomic energy levels, the quantization of the orientation of atoms or molecules placed in suitable conditions, are experimental facts incompatible with classical corpuscular theory.

In order to have a guide in the search for a coherent theory, it is well to list the elements of Classical Theory which can be maintained. The first realization which imposes itself concerns the fundamental laws of the conservation of energy and momentum: none of the experimental facts discussed in this chapter invalidates them. It therefore seems that these laws remain valid on the level of microscopic phenomena. In the second place, the failure of Classical Theory seems to have as sole origin the atomism of action. This theory retains its entire validity on the macroscopic scale, and more generally, whenever the quantum discontinuities may be considered negligible. This second statement lies at the base of the correspondence principle, stated in the last part of this chapter. The successes of the Old Quantum Theory, surprising for a theory whose foundations are so fragile, illustrate the fruitful character of this fundamental heuristic principle.

EXERCISES AND PROBLEMS

1. Consider the scattering of monochromatic photons by free electrons (Compton effect). Determine the wavelength shift, the magnitude and the direction of the recoil momentum of the electron as a function of the deflection angle of the photon, assuming: a) that the electron is initially at rest; b) that the electron has an initial momentum P in the direction of the incident photon.

What is the maximum value of the momentum transferred to the electron in the two cases?

2. Discuss, in connection with the experiment of Young's interference slits, the double aspect — wave and corpuscular — of light. Show that it is not possible to attribute to each photon a precise trajectory passing through one of the two slits.

3. A gyroscope has a magnetic moment $\boldsymbol{\mu}$ proportional to its angular momentum: $\boldsymbol{\mu} = M\boldsymbol{l}$. From the expression $-\boldsymbol{\mu}\cdot\boldsymbol{H}$ of the magnetic energy, derive the equation of motion of $\boldsymbol{l}$ in a constant magnetic field $\boldsymbol{H}$, and show that the gyroscope carries out a precessional motion of circular frequency $\omega_L = MH$ (Larmor frequency).

4. In the non-relativistic limit, the classical equations of motion of an electron in an electromagnetic field are

$$\frac{\mathrm{d}}{\mathrm{d}t}(m\boldsymbol{v}) = e\left(\boldsymbol{E} + \frac{v}{c}\times\boldsymbol{H}\right). \qquad (a)$$

Let $\boldsymbol{A}$ and φ be the vector and scalar potentials of this field, respectively

$$\left(\boldsymbol{E} = -\operatorname{grad}\varphi - \frac{1}{c}\frac{\partial\boldsymbol{A}}{\partial t}; \qquad \boldsymbol{H} = \operatorname{rot}\boldsymbol{A}\right).$$

Show that these equations can be derived from the Lagrangian

$$L = \tfrac{1}{2}mv^2 + e\left(\frac{v\cdot\boldsymbol{A}}{c} - \varphi\right).$$

Calculate the conjugate Lagrange momenta and form the classical Hamiltonian.

5. The equations (a) of Problem (I.4) remain true in the relativistic domain provided that one replaces the rest mass m by the "relativistic mass" $M = m[1 - (v^2/c^2)]^{-\frac{1}{2}}$. Verify that the formalisms of Lagrange and Hamilton remain valid, the equations being deduced from the Lagrangian

$$L = -mc^2\sqrt{1 - \frac{v^2}{c^2}} + e\left(\frac{\boldsymbol{v}\cdot\boldsymbol{A}}{c} - \varphi\right).$$

Show that the Hamiltonian in that case is written

$$H = \left[m^2c^4 + \left(\boldsymbol{p} - \frac{e}{c}\boldsymbol{A}\right)^2 c^2\right]^{\frac{1}{2}} + e\varphi$$

where $\boldsymbol{p}$ stands for the momentum, i.e. the vector whose three components p_x, p_y, p_z are the respective conjugate momenta of x, y, z.

N.B.: $\boldsymbol{p} = M\boldsymbol{v} + (e/c)\boldsymbol{A}$ and $H = Mc^2 + e\varphi\cdot H$ and $\boldsymbol{p}c$ form a four-vector of space-time, just like φ and $\boldsymbol{A}$, and one has: $(H - e\varphi)^2 - (\boldsymbol{p}c - e\boldsymbol{A})^2 = m^2c^4$.

6. A material point of mass m is constrained to move on the x axis where it is subject to a restoring force $(-Kx)$ proportional to its distance from the origin (*harmonic oscillator*). Apply the Bohr-Sommerfeld quantization rule to this system; calculate the energy, the period, and the amplitude of the quantized trajectories.

7. Quantize the circular electronic orbits of the hydrogen atom by applying the Bohr-Sommerfeld rule. Determine the energy, the period, and the radius of the quantized orbits. Calculate specifically the numerical values of the energy, the period, and the radius of the lowest orbit $[mc^2 = 0.51$ MeV; $(\hbar c/e^2) \approx 137]$ and determine in that particular case the relativity correction. Verify that in the limit of large quantum numbers, the Bohr frequencies tend toward those predicted by classical electrodynamics (*correspondence principle*).

CHAPTER II

MATTER WAVES AND THE SCHRÖDINGER EQUATION

1. Historical Survey and General Plan of the Succeeding Chapters

The founding of Quantum Mechanics can be placed between the years of 1923 and 1927. Two equivalent formulations thereof have been proposed almost simultaneously: Matrix Mechanics and Wave Mechanics.

The starting point of *Matrix Mechanics* [1] is a critical analysis of the Old Quantum Theory. The point of view developed by Heisenberg is the following. In any physical theory one must distinguish the concepts and quantities which are physically observable from those which are not. The former must of necessity play a role in the theory, the latter can be modified or abandoned without impairment. In establishing a satisfactory theory of microscopic phenomena, one must as far as possible start only from the former. Now the Old Quantum Theory calls upon a whole set of notions without experimental foundation; it is put in jeopardy to the extent that these notions are erroneous.

The notion of electronic orbit is an example of a concept without experimental foundation. Indeed, let us ask if it is conceivable [2] to follow experimentally the motion of the electron in a Bohr orbit of the hydrogen atom. In order to observe this motion one has to perform successive measurements of the position of the electron, where the tolerated margin of error is very much smaller than the mean radius a of the orbit. Such measurements are conceivable with X-rays of sufficiently short wavelength: $\lambda \ll a$. However, according to the laws of the Compton effect, the collision of each X-ray photon with an electron is accompanied by a momentum transfer of the order of $\hbar/\lambda$ ($\gg \hbar/a$) and consequently by a finite perturbation of the electron motion which one wishes to observe. One can readily show (cf. Problem

[1] W. Heisenberg, Z. Physik **33** (1925) 879; M. Born and P. Jordan, Z. Physik **34** (1925) 858; M. Born, W. Heisenberg, and P. Jordan, Z. Physik **35** (1926) 557; P. A. M. Dirac, Proc. Roy. Soc. London **A 109** (1925) 642.

[2] The limits of possibility of observation at which we shall arrive are imposed by the very nature of things; the difficulties of practical realization of the measurements are of no concern here.

II.1) that this perturbation is the more appreciable the smaller the quantum number n; in particular, when the electron is on the lowest orbit ($n=1$), the mean energy transferred per collision is at least equal to the ionization energy. This uncontrollable perturbation of the observed system by the measuring apparatus limits the precision with which one might hope to know the electronic orbits. In the limit of small quantum numbers, this perturbation is so strong that any observation of the orbit, no matter how coarse, is doomed to failure. Since no experiment allows us to assert that the electron actually describes a precise orbit in the hydrogen atom, nothing prevents us from abandoning the very notion of an orbit; in other words, the fact that the atom is in a well-defined energy state does not necessarily imply that the electron has at each instant a well-defined position and momentum [1]).

The Matrix Mechanics of Heisenberg, Born and Jordan abandons the notion of an electron orbit. Starting exclusively from physically observable quantities such as the frequencies and the intensities of the radiation emitted by atoms, the theory associates with each physical quantity a certain matrix; in contrast to the quantities of ordinary algebra, these matrices obey a non-commutative algebra [2]); it is on this essential point that the new mechanics differs from the Classical Mechanics. The equations of motion of the dynamical variables of a quantized system are thus equations between matrices. Following the correspondence principle, one assumes that these equations are formally identical to the equations (between quantities of ordinary algebra) of the corresponding classical system.

The *Wave Mechanics* of Schrödinger [3]) starts from an entirely different point of view. It originates in the works of L. de Broglie

---

1) This is to be compared with the discussion at the end of the first chapter on the Franck and Hertz experiment, showing that the existence of a continuous trajectory for the bombarding electron is incompatible with the quantization of the energy levels of the target atom. Postulating such a quantization forces us to renounce the notion of a trajectory.

2) The definition of matrices and the presentation of their principal properties will be given in Chapters VII and VIII, where we shall see the close relationship existing between these matrices and the linear operators with which we shall be dealing later on (§ 11).

3) E. Schrödinger, Ann. Physik (4) **79** (1925) 361 and 489; **80** (1926) 437; **81** (1926) 109.

on matter waves [1]). Seeking to establish the bases of a unified theory of matter and radiation, de Broglie had stated the hypothesis that the wave-corpuscle duality is a general property of microscopic objects, and that matter, as well as light, exhibits both wave and corpuscular aspects. Having established the correspondence between the dynamical variables of the corpuscle and the characteristic quantities of the associated wave, he was able to deduce the quantization rules of Bohr-Sommerfeld by a semi-quantitative argument. The speculations of L. de Broglie on the wave nature of matter were to be confirmed very directly a few years later by the discovery of diffraction phenomena analogous to those of wave optics. Meanwhile Schrödinger, pursuing and generalizing this notion of matter waves, discovered the equation of propagation of the wave function representing a given quantum system; a very simple correspondence rule enables us to deduce that fundamental equation from the Hamiltonian of the corresponding classical system. The Schrödinger equation constitutes the essential element of Wave Mechanics.

As Schrödinger has shown [2]), Wave Mechanics and Matrix Mechanics are equivalent. They are two particular formulations of a theory which can be presented in very general terms. The setting up of this general formalism of the Quantum Theory is essentially due to Dirac [3]). The quantum theory thus obtained is a non-relativistic theory of material particles. It was complemented by a quantum theory of the electromagnetic field [4]) to form a coherent whole suitable for the treatment of all problems dealing with systems of non-relativistic, material particles interacting with the electromagnetic field. We might add that the interpretation and internal consistency of the theory were not fully understood until after the works of Born, Heisenberg and Bohr appeared [5]). The major part of this book deals with this theoretical framework and its applications. The problems

---

[1]) L. de Broglie, Nature **112** (1923) 540; *Thesis*, Paris, 1924, Ann. Physique (10) 2 (1925).

[2]) E. Schrödinger, Ann. Physik (4) **79** (1926) 734.

[3]) P. A. M. Dirac, *The Principles of Quantum Mechanics*, (Oxford, Clarendon Press), 1st ed. (1930), 4th ed. (1958).

[4]) P. A. M. Dirac, Proc. Roy. Soc. London **A 114** (1927) 243 and 710; P. Jordan and W. Pauli, Z. Physik **45** (1928) 151.

[5]) M. Born, Z. Physik **38** (1926) 803; W. Heisenberg, Z. Physik 43 (1927) 172; N. Bohr, Naturwiss. **16** (1928) 245; **17** (1929) 483 and **18** (1930) 73. One

of Relativistic Quantum Mechanics will be attacked only in the last part; we shall essentially restrict ourselves to an outline of Dirac's relativistic theory of the electron [1]), and to an elementary introduction to the theory of quantized fields.

Of the various ways of introducing the Quantum Theory, the one which uses the general formalism is undoubtedly the most elegant and the most satisfactory. However, it requires the handling of a mathematical symbolism whose abstract character runs the risk of masking the underlying physical reality. Wave Mechanics, which utilizes the more familiar language of waves and partial differential equations, lends itself better to a first encounter. Furthermore, it is in that form that the Quantum Theory is most frequently used in elementary applications. That is why we shall begin with a general outline of Wave Mechanics. We start this chapter with a discussion of the matter-wave concept; we then set up the Schrödinger equation and discuss its principal properties; we show in particular how this equation allows us to determine the energy levels of stationary states. In order to acquire a certain familiarity with the Schrödinger equation, we devote Chapter III to the handling of simple problems relating to one-dimensional quantum systems, and to the derivation of several important properties of such systems. We shall then be in a position

---

can also find a thorough discussion of the physical interpretation of the theory in W. Heisenberg, *The Physical Principles of the Quantum Theory* (Chicago, University of Chicago Press, 1930); also (New York, Dover); Niels Bohr, *Atomic Theory and the Description of Nature* (Cambridge University Press, Cambridge, 1934) (translated from the Danish).

The interpretation under discussion here is the statistical interpretation of the Copenhagen school. It is the one we are developing in this book. After violent controversies, it has finally received the support of the great majority of physicists. However, it had (and still has) a number of die-hard opponents, among which one should notably list Einstein, Schrödinger, and de Broglie. The controversy has finally reached a point where it can no longer be decided by any further experimental observations; it henceforth belongs to the philosophy of science rather than to the domain of physical science proper. The main arguments can be found in *Albert Einstein, Philosopher-Scientist,* P. A. Schilpp, editor, (New York, Tudor Publishing Company, 1949 and 1951) (see especially the articles of Bohr and Einstein), and in the book of L. de Broglie, *La Théorie de la Mesure en Mécanique Ondulatoire* (Paris, Gauthier–Villars, 1957).

[1]) P. A. M. Dirac, Proc. Roy. Soc. London **A 117** (1928) 610 and **A 118** (1928) 351.

to attack the general problems of interpretation of the Quantum Theory; the latter form the subject of Chapter IV. Chapter V is devoted to the development of the formalism of Wave Mechanics and its statistical interpretation in accordance with the principles defined in Chapter IV. Chapter VI deals with the classical approximation of Wave Mechanics. Only after this general survey of the theory expressed in the language of Wave Mechanics shall we present (Chs. VII–VIII) the general formalism of the Quantum Theory.

I. MATTER WAVES

2. Introduction

The double aspect of light – wave-like and corpuscular – is one of the most striking features connected with the appearance of quanta in physics. Let us suppose that matter too possesses this dual character; just as an electromagnetic wave is associated with each photon, so we associate with each material particle a wave whose angular frequency ω is connected with the energy of the particle E by the relation $E=\hbar\omega$. If one adopts this point of view, the atom behaves as a resonant cavity having a discrete series of proper frequencies; in this way the quantization of its energy levels is explained.

At the same time the possibility exists of establishing a unified theory in which matter and radiation are different varieties of the same type of object, having both wave-like and corpuscular character. These suppositions, which have guided de Broglie in his theory of matter waves, were found to be entirely justified, as we shall see.

The main properties of matter waves are obtained by analogy with optics. Just as for photons, we assume that the value at each point of the intensity of the wave associated with a particle gives the probablity of finding the particle at that point. The particle is better localized, the more restricted the domain occupied by the wave. The conditions for the validity of Classical Mechanics are fulfilled when the wave maintains in the course of time a sufficiently small extension so that it may be approximated by a point and one may attribute a precise motion to the particle. An analogous situation is encountered in optics whenever the wavelength λ may be considered negligibly small; this is the approximation used in *geometrical optics*, where no typically wave-like phenomenon can be detected. This approximation is valid when the optical properties of the medium through which the light travels, remain essentially constant over a distance of several wavelengths ($|\operatorname{grad} \lambda| \ll 1$). This suggests that classical corpuscular theory is correct in the absence of a field or in a slowly varying field, and to the extent that one does not seek to localize the particle with too great precision. These predictions are in accord with the most common observations concerning the motion of the atomic and subatomic particles in the presence of quasi-static and near-uniform fields: trajectories of charged particles in a static electromagnetic field, deflections of paramagnetic atoms in the Stern-Gerlach magnet, etc. In these limiting cases the theory of matter waves must be equivalent to the classical theory (correspondence principle).

3. Free Wave Packet. Phase Velocity and Group Velocity

Consider the propagation of matter waves in a homogeneous, isotropic medium. The simplest type of wave is a plane, monochromatic wave

$$e^{i(\mathbf{k}\cdot\mathbf{r}-\omega t)} \tag{II.1}$$

which represents a vibration of wavelength $\lambda = 2\pi/k$ travelling in the direction of its wave vector $\mathbf{k}$ with constant velocity. The velocity considered here is the velocity of propagation of planes of equal phase, or *phase velocity:*

$$v_\varphi = \frac{\omega}{k}.$$

The frequency ω is independent of the direction of $\mathbf{k}$ but may eventually depend upon the length of that vector. As any wave may be considered as a superposition of plane monochromatic waves, knowledge of the "dispersion law" $\omega(k)$ is sufficient to determine the behavior of any wave in the course of time.

By hypothesis, each frequency ω corresponds to a well-defined energy E of the particle

$$E = \hbar\omega. \tag{II.2}$$

It is therefore natural to associate the wave (II.1) with uniform rectilinear motion of energy E directed parallel to $\mathbf{k}$.

By examining the classical approximation we shall be able to relate $\mathbf{k}$ to the momentum $\mathbf{p}$ of the particle. In order to realize this approximation, one must associate with the particle a wave of limited extension. The wave (II.1) evidently does not satisfy this condition; one may realize it, however, by the superposition of plane waves with neighboring wave vectors. One thus forms a *wave packet*:

$$\psi(\mathbf{r}, t) = \int f(\mathbf{k}') \, \mathrm{e}^{\mathrm{i}(\mathbf{k}'\cdot\mathbf{r} - \omega' t)} \, \mathrm{d}\mathbf{k}'.$$

We designate by A and α the absolute value and the phase of f, respectively. By hypothesis, A has appreciable values only in a small region surrounding $\mathbf{k}$. Our purpose is to examine to what extent and under what conditions the "motion" of this wave packet may be likened to the motion of a classical particle.

To simplify matters, we shall first take up the problem of the one-dimensional wave packet:

$$\psi(x, t) = \int_{-\infty}^{+\infty} f(k') \, \mathrm{e}^{\mathrm{i}(k'x - \omega' t)} \, \mathrm{d}k'.$$

Let us set

$$\varphi = k'x - \omega' t + \alpha,$$

$\psi(x, t)$ is the integral of the product of a function A exhibiting a pronounced peak in a region S of extension Δk surrounding the point $k' = k$, by an oscillating function $\exp(\mathrm{i}\varphi)$. If the number of oscillations of $\exp(\mathrm{i}\varphi)$ in this region is large, the contributions to the integral from the different portions of the region interfere destructively and ψ remains negligible. The largest (absolute) values of ψ obtain when the phase φ remains practically constant in S, i.e. $\mathrm{d}\varphi/\mathrm{d}k \approx 0$ (the symbol $\mathrm{d}/\mathrm{d}k$ denotes the value taken by the derivative of the function with respect to k' when $k' = k$). Roughly speaking the only appreciable values of ψ are those for which $\exp(\mathrm{i}\varphi)$ carries out only one or a fraction of one oscillation, namely

$$\Delta k \times \left| \frac{\mathrm{d}\varphi}{\mathrm{d}k} \right| \lesssim 1.$$

Since

$$\frac{\mathrm{d}\varphi}{\mathrm{d}k} = x - t \frac{\mathrm{d}\omega}{\mathrm{d}k} + \frac{\mathrm{d}\alpha}{\mathrm{d}k},$$

the wave $\psi(x, t)$ is practically concentrated in a region of extension

$$\Delta x \simeq \frac{1}{\Delta k}$$

surrounding the "center of the wave packet", defined by the condition $\mathrm{d}\varphi/\mathrm{d}k=0$, namely

$$x=t\frac{\mathrm{d}\omega}{\mathrm{d}k}-\frac{\mathrm{d}\alpha}{\mathrm{d}k}.$$

This point travels with uniform motion, whose velocity

$$v_{\mathrm{g}}=\frac{\mathrm{d}\omega}{\mathrm{d}k} \tag{II.3}$$

is called the *group velocity* of the wave $\exp\,[\mathrm{i}(kx-\omega t)]$. It is this velocity v_{g}, and not the phase velocity v_{φ} which, in the classical approximation where one considers the extension of the wave packet to be negligible, must be identified with the particle velocity

$$v=\frac{\mathrm{d}E}{\mathrm{d}p} \qquad (\approx p/m \text{ in non-relativistic approximation}).$$

From the condition $v=v_{\mathrm{g}}$ and from relation (I.2) one obtains [1] the de Broglie relation:

$$p=\hbar k=\frac{h}{\lambda}. \tag{II.4}$$

The above treatment is easily generalized to the three-dimensional wave packet: the center of the packet travels with uniform motion at the velocity

$$\boldsymbol{v}_{\mathrm{g}}=\mathrm{grad}_{\boldsymbol{k}}\,\omega. \tag{II.3'}$$

This group velocity must be identified with the particle velocity:

$$\boldsymbol{v}=\mathrm{grad}_{\boldsymbol{p}}\,E.$$

This condition, combined with relation (II.2), leads to the following relations [2] between dynamical variables of the particle and characteristic quantities of the associated wave:

$$E=\hbar\omega, \qquad \boldsymbol{p}=\hbar\boldsymbol{k}. \tag{II.5}$$

---

[1] Rigorously, these two conditions define k as a function of p only to within an additive constant. One fixes the constant by requiring the relation between p and k to be independent of the direction of travel chosen along the coordinate axis.

[2] One obtains the second relation (II.5) to within a constant vector, which one chooses equal to zero by imposing upon that relation the condition to be invariant under a rotation of axes (cf. preceding footnote).

These relations are identical to the relations (I.4) found in the case of photons.

To conclude, we shall examine the preceding results from the point of view of the principle of relativity.

In the non-relativistic approximation, the energy E is defined only to within a constant; a modification of the zero of the energy scale has the effect of adding to the frequency $\omega(k)$ a constant frequency ω_0 [eq. (II.2)] and of multiplying the function $\psi(\mathbf{r}, t)$ by the phase factor $\exp(-i\omega_0 t)$. This in no way affects the preceding results concerning the motion of the wave packet, nor the relations (II.5) derived therefrom.

However, the preceding treatment does not depend at all on the non-relativistic approximation. The relativity principle allows us to define unambiguously the energy $E=\sqrt{m^2c^4+p^2c^2}$, and the frequency ω which corresponds to it. The energy E and the momentum $\mathbf{p}$ are respectively the time and space components of the same four-vector (with the convention $c=1$). The same holds true for the frequency ω and the wave vector $\mathbf{k}$. The relations (II.5) satisfy the principle of relativity: they state that the four-vectors $(E, \mathbf{p})$ and $(\omega, \mathbf{k})$ are proportional.

4. Wave Packet in a Slowly Varying Field

The above results, and especially relations (II.5) may be extended to the motion of particles in a slowly varying field; the conditions for the classical approximation are realized as long as the field variations are negligible on the wavelength scale.

The laws of propagation are those of geometrical optics. In particular, a wave packet of limited extension, analogous to those which were studied in the preceding section, follows the trajectory of a ray with a velocity equal to the group velocity. In order to compare the motion of the wave packet to the motion of a classical particle, it is necessary:

(*a*) that the rays corresponding to the (angular) frequency ω be identical to the classical trajectories of energy $E=\hbar\omega$;

(*b*) that the group velocity along each ray be equal to the velocity of the corresponding classical particle.

The trajectories of the classical particle are given by the principle of least action (I.12); for a fixed energy E, the Lagrangian is, according to equation (I.13), equal to $\boldsymbol{p}\cdot(\mathrm{d}\boldsymbol{r}/\mathrm{d}t)-E$ and the principle may be written:

$$\delta I_{12} \equiv \delta \int_{M_1}^{M_2} \boldsymbol{p}\cdot\mathrm{d}\boldsymbol{r}=0.$$

It expresses the fact that the integral I_{12} calculated along a curve joining the points M_1 and M_2 is stationary when this curve is the actual trajectory of the particle from M_1 to M_2. The momentum $\boldsymbol{p}$ in the most general case is a function of the position $\boldsymbol{r}$ and the velocity $\boldsymbol{v}=(\mathrm{d}\boldsymbol{r}/\mathrm{d}t)$, in other words, a function of the position on the curve and of the direction of its tangent. In the case of a non-relativistic particle in a scalar potential $V(\boldsymbol{r})$

$$E=\frac{p^2}{2m}+V(\boldsymbol{r}) \quad \text{and} \quad \boldsymbol{p}=m\boldsymbol{v}, \tag{II.6}$$

momenta and velocities are parallel ($\boldsymbol{p}\,||\,\mathrm{d}\boldsymbol{r}$). However, the principle is equally valid in more general cases, such as that of a particle in a magnetic field where this condition does not obtain [1]).

At a given fixed frequency ω, the rays of geometrical optics are determined by another variational principle, Fermat's principle, which may be written

$$\delta J_{12} \equiv \delta \int_{M_1}^{M_2} \boldsymbol{k}\cdot\mathrm{d}\boldsymbol{r}=0;$$

$\boldsymbol{k}$ is the wave vector. The integral J_{12} calculated along a given curve joining the points M_1 and M_2 represents the optical path along that curve. Fermat's principle expresses the fact that the ray joining M_1 and M_2 is the curve along which the optical path is stationary. As a general rule $\boldsymbol{k}$ (perpendicular to the surfaces of equal phase) is a function of the position on the curve and of the slope of that curve. In an isotropic medium, where the phase velocity is independent of the direction, the vector $\boldsymbol{k}$ is directed along the tangent to the curve and its length $k=2\pi/\lambda$ depends only upon the position along the curve and not upon the direction of propagation. However, Fermat's principle applies equally well to anisotropic media.

---

[1]) We recall that the momentum (i.e. linear momentum) of a particle is the vector whose three components are the Lagrange-conjugate momenta of the three position coordinates. It is sometimes, but not always, equal to the product of the mass and the velocity of the particle.

Note the perfect formal analogy of the two variational principles. In order that the rays relative to the frequency ω merge with the classical trajectories of energy E [condition (a)], it is sufficient that $\boldsymbol{k}$ and $\boldsymbol{p}$ be proportional:

$$\boldsymbol{p} = \alpha \boldsymbol{k}.$$

The constant of proportionality α is determined by condition (b). The group velocity $\boldsymbol{v}_g$ is the gradient with respect to $\boldsymbol{k}$ of the frequency ω; therefore

$$\boldsymbol{v}_g = \frac{1}{\hbar} \operatorname{grad}_{\boldsymbol{k}} E = \frac{\alpha}{\hbar} \operatorname{grad}_{\boldsymbol{p}} E .$$

As for the velocity of the particle, it is given by the formula

$$\boldsymbol{v} = \operatorname{grad}_{\boldsymbol{p}} E.$$

These two velocities are equal if $\alpha = \hbar$.

We thus come back to the relations (II.5) in a completely general way.

In the case of a non-relativistic particle in a scalar potential $V(\boldsymbol{r})$ which is slowly changing, the wave travels in an isotropic medium and the wavelength λ is given [cf. eq. (II.6)] by the relation

$$\lambda = \frac{h}{p} = \frac{h}{\sqrt{2m(E - V(\boldsymbol{r}))}} . \tag{II.7}$$

5. Quantization of Atomic Energy Levels

The theory of matter waves leads very simply to the quantization of the energy levels of atoms.

To be specific, let us return to the hydrogen atom. Consider an elliptical orbit of energy E. The value of the wave vector $\boldsymbol{k}$ is given at each point of the orbit by relations (II.5). The phase of the wave at a given point of the orbit increases by $\oint \boldsymbol{k} \cdot \mathrm{d}\boldsymbol{r}$ in each revolution. In order that a standing wave be set up, it is necessary that this phase be an integral multiple of 2π. This yields the quantization condition

$$\oint \boldsymbol{p} \cdot \mathrm{d}\boldsymbol{r} = \hbar \oint \boldsymbol{k} \cdot \mathrm{d}\boldsymbol{r} = nh \qquad (n \text{ integer} > 0)$$

which may also be written, adopting the notation of the first chapter

$$\oint p_r \, \mathrm{d}r + \int p_\varphi \, \mathrm{d}\varphi = nh.$$

Similar arguments lead to the Bohr-Sommerfeld quantization rules in all cases of periodic and multiply-periodic motions.

Of course, the preceding results are valid only in the geometrical optics approximation, where the notions of wavelength and of wave vector retain their meaning. In particular, we cannot assert that the quantization conditions remain the same for small quantum numbers. The only certain fact is the energy quantization which is tied to the establishment of a standing wave.

To deal with the more general cases, one must extend this approximate theory as one does classical optics when one goes from geometrical optics to wave optics [1]): having postulated the existence of matter waves, one has yet to discover their equation of propagation. Before attacking this problem, we shall examine how the wave nature of matter may be – and actually is – revealed experimentally, although the concept of corpuscles of matter can by no means be abandoned.

6. Diffraction of Matter Waves

The practical possibilities of observation evidently depend on the wavelength of the object under study. With macroscopic objects, the commonly realized wavelengths are so tiny that typical wave effects cannot be detected in practice. With objects of atomic dimensions, on the other hand, it is possible to form beams of wavelength comparable to that of X-rays and capable of giving rise to similar diffraction effects by crystals [2]).

The first diffraction experiments with matter waves were performed

---

[1]) Hence the name Wave Mechanics given to the theory.

[2]) To be specific, let us consider a particle used to demonstrate Brownian motion. The smallest particles of this kind have a diameter of the order of 1 micron and a mass

$$M \approx 10^{-12} \text{ g}.$$

In thermal equilibrium at ordinary temperatures, their mean kinetic energy $\frac{3}{2}kT$ is about 0.4×10^{-13} erg, from which we obtain a mean wavelength

$$\lambda = \frac{h}{p} = \frac{h}{\sqrt{3MkT}} \simeq 5 \times 10^{-6}\,\text{Å}.$$

At the same energy a helium atom has a wavelength $\lambda \approx 0.9$ Å, a neutron $\lambda \approx 1.9$ Å, and an electron $\lambda \approx 77$ Å.

with electrons by Davisson and Germer (1927), G. P. Thomson (1928) and Rupp (1928).

Davisson and Germer worked with reflection by a single crystal and observed spots of the von Laue type; G. P. Thomson and Rupp studied the Debye-Scherrer rings obtained by diffraction through a thin, polycrystalline target.

In these experiments the incident beam is obtained by the acceleration of electrons in an electrical potential. If E is the energy of the electrons in electron-volts, the de Broglie wavelength of the electrons measured in angstrom units is

$$\lambda = \frac{12.2}{\sqrt{E_{(\mathrm{eV})}}}\ \text{Å}.$$

With energies in the range from 1 to 100 keV, one finds oneself in the domain of ordinary crystal spectrography. Knowing the parameters of the crystal lattice, it is possible to deduce an experimental value for the electron wavelength from the interference pattern; the theoretical value of de Broglie is in perfect accord with this experimental value.

Analogous crystal-diffraction experiments were performed with monoërgic beams of helium atoms and hydrogen molecules (Stern, 1932), furnishing a new verification of the de Broglie relation. The wavelength entering here is the wavelength associated with the motion of the center of gravity of each atom or molecule of the beam. The same observations can be made with beams of slow neutrons from nuclear reactors. All these experiments clearly show that the wave-like structure is not peculiar to electrons, but that one is actually dealing with a very general property of material objects.

7. Corpuscular Structure of Matter

Pursuing the analogy between matter waves and classical wave optics, one may ask if it is possible to renounce the notion of corpuscles of matter once and for all, and to replace the classical theory by a wave theory where the wave $\psi(\boldsymbol{r}, t)$ would play the role played by the electromagnetic field in the theory of radiation.

The picture of corpuscles of matter, localized grains of energy and momentum, would be replaced by that of a continuous wave with a continuous distribution of energy and momentum. The particles of classical mechanics would actually be wave packets of finite but negligibly small size. As shown above such packets obey the laws of motion of classical particles under certain limiting conditions which are precisely those where classical mechanics turns out to be correct. Nevertheless, even in the absence of a field, a wave packet cannot maintain this appearance of a particle indefinitely because it spreads out little by little in the course of its displacement, and finally, after a sufficiently long time, it occupies as large a region of space as one wishes (cf. Problem II.6) [1]). It would be difficult to explain under these conditions why matter appears so often in the form of well-localized particles.

However, the inadequacy of such a pure wave theory is clearly exhibited by a careful analysis of any diffraction experiment with matter waves. Consider, for instance, a beam of monoërgic electrons traversing a polycrystalline foil; on a screen suitably placed on the other side of the foil, one observes a central spot due to the transmitted wave, surrounded by concentric rings due to the diffracted wave. Suppose, to be specific, that the incident wave is a wave packet $\psi(\mathbf{r}, t)$ restricted in space; it is obtained by placing (Fig. II.1) an

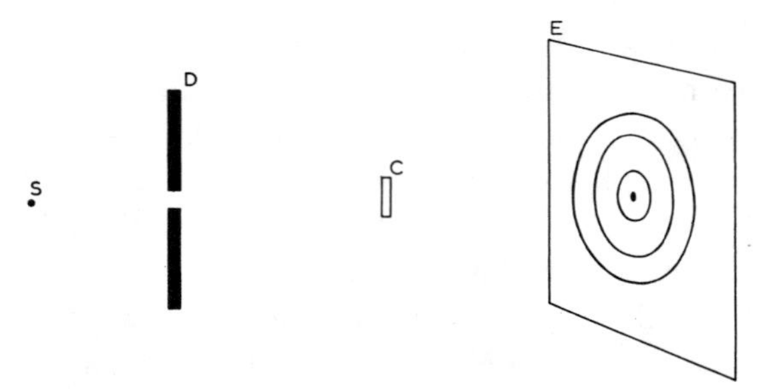

Fig. II.1. Diffraction of electrons by a polycrystalline foil. The electron beam, originating from the source S is collimated by the diaphragm D and then diffracted by the polycrystalline foil C. One observes the diffraction pattern on screen E.

intense source S of monoërgic cathode rays behind a diaphragm D equipped with a shutter whose opening is fixed once and for all. When this wave propagates through the foil C, it splits into a transmitted and a diffracted wave and forms the interference pattern described above, on the screen. Since it is assumed to be a continuous, classical

[1]) The case of the harmonic oscillator forms an exception to this very general rule concerning the indefinite spreading of a wave packet (cf. Ch. VI).

wave, the observed interference pattern must be continuous. If one diminishes the intensity of the incident wave, everything else being equal (for instance, by increasing the distance between the source S and the diaphragm D), the intensity of the interference spots decreases accordingly, but the interference pattern must remain continuous. Experiment invalidates these predictions. The observed pattern is actually made up of a succession of well-localized impacts. If one decreases the intensity of the wave, the number of these impacts decreases proportionately. In the limit of very low intensity, one eventually observes just a single impact, either on the central spot, or on one of the diffraction rings. The simplest explanation is to attribute each impact to the passage of a corpuscle of matter: an electron.

Note the perfect parallelism between the situation described here and the diffraction experiment of light by a grating, discussed in the first chapter. One may pursue the parallelism to the end and conclude that the simplest interpretation one might give of the wave-corpuscle duality is a statistical interpretation: namely that the intensity of the wave at each point of the screen gives the probability of occurrence of an impact at that point.

8. Universal Character of the Wave-Corpuscle Duality

We conclude from all this that microscopic objects have a very general property: they appear under two apparently irreconcilable aspects, the wave aspect on the one hand, exhibiting the superposition property characteristic of waves, and the corpuscular aspect on the other hand, namely localized grains of energy and momentum, There exists a universal relationship between these two aspects, given by equations (II.5). Furthermore, the bond between corpuscles and associated waves is a statistical one, on which we shall have more to say later.

II. THE SCHRÖDINGER EQUATION

9. Conservation Law of the Number of Particles of Matter

All we have said thus far reveals a remarkable similarity o
perties of light and matter. Nevertheless, an important dif
to be noted. Even in the simplest situations, the numb
may vary in the course of time through emission o

contrast, the number of electrons, and more generally the number of elementary particles of matter remains constant. This is indicated by an analysis of the most common phenomena of atomic physics and the very success of the quantum mechanics of systems of particles will confirm the validity of this important conservation law.

In reality we are not dealing here with an absolute conservation law, and the disparity between matter and light is not as pronounced as we have just stated. Since the discovery of the positron (Anderson, 1932), a particle of the same mass m as that of the electron and of opposite charge, one knows that it is possible, under certain circumstances, to create electron-positron pairs (emission of matter) and conversely, that a positron and an electron undergoing collision can annihilate (absorption of matter) giving off energy in the form of radiation. In accordance with the law of equivalence between mass and energy, the energy necessary to create an electron-positron pair is at least equal to $2mc^2$ ($\approx$ 1 MeV). One has another case of emission of electrons (or positrons) in the beta decay of atomic nuclei. If one restricts oneself to phenomena of atomic physics, the positrons are absent, nuclei are stable and the energy transfers lie below the threshold for electron-positron pair creation; the conservation law stated above is then obeyed. We shall assume this in what follows.

This law greatly facilitates the development and interpretation of the quantum theory of matter. The various quantum systems which we consider are made up of a well-defined number of material particles. The simplest system is that of a particle, for instance an electron, in an external force field. The wave which is associated with it at each instant t is a function $\Psi(\mathbf{r}, t)$ of the position coordinates of that particle. The hydrogen atom is a system of two particles, an electron and a proton, interacting with each other. The associated wave $\Psi(\mathbf{r}_\mathrm{e}, \mathbf{r}_\mathrm{p}; t)$ depends upon the position coordinates $\mathbf{r}_\mathrm{e}$ and $\mathbf{r}_\mathrm{p}$ of these two particles. A complex atom is formed of a nucleus of charge Ze defined by its position $\mathbf{R}$, and of Z electrons whose positions are determined by the vectors $\mathbf{r}_1, \mathbf{r}_2, \ldots, \mathbf{r}_Z$, respectively. The associated wave is a certain function $\Psi(\mathbf{R}, \mathbf{r}_1, \mathbf{r}_2, \ldots, \mathbf{r}_Z; t)$. One can similarly define the wave function of more complex systems.

10. Necessity for a Wave Equation and Conditions Imposed upon this Equation

We have seen that the intensity at a given point and at a given instant of the wave associated with a particle gives the probability of finding the particle at that point and at that instant. More generally, we *postulate* that the wave function Ψ of a quantum system completely defines its dynamical state; otherwise stated, all the predictions which can be made concerning the dynamical properties of the system at a given instant of time t can be deduced from a knowledge of Ψ at that instant. Hence the central problem of the theory is the following: knowing the wave function at a given initial instant t_0, determine that function at all later instants. In order to do this, one must know the equation of propagation of the wave Ψ.

It is quite clear that no deductive reasoning can lead us to that equation. Like all equations of mathematical physics, it must be postulated and its only justification lies in the success of the comparison of its predictions with the experimental results. Nevertheless, the choice of a wave equation is restricted by a certain number of *a priori* conditions if one wishes to maintain the previously defined interpretation for Ψ:

(A) The equation must be *linear* and *homogeneous;* the wave thus possesses the property of superposition, characteristic of waves in general: namely if Ψ_1 and Ψ_2 are solutions of the equation, any linear combination $\lambda_1\Psi_1+\lambda_2\Psi_2$ of these functions is also a solution thereof.

(B) It must be a differential equation of the *first order with respect to time*; thus, specifying Ψ at a given initial instant uniquely defines its entire later evolution, in accord with the hypothesis that the dynamical state of the physical system is entirely determined once Ψ is given.

On the other hand, the predictions of the theory must coincide with those of Classical Mechanics in the domain where the latter is valid. In other words, the equation must lead to the same laws of motion of wave packets as the de Broglie theory, in the limit where the geometrical optics approximation is valid. This suggests that the equation bears a formal analogy to certain equations of Classical Mechanics (*correspondence principle*).

All these considerations will lead us to the Schrödinger equation in a very natural way. Before carrying out this program, we shall introduce a mathematical concept which turns out to be very useful later on, namely the operator concept.

11. The Operator Concept

Consider the function $\partial\Psi/\partial t$, the derivative of Ψ with respect to time; one can say that the operator $\partial/\partial t$ acting on the function Ψ yields the function $\partial\Psi/\partial t$. More generally, if a certain operation allows us to bring into correspondence with each function Ψ of a certain function space, one and only one well-defined function Ψ' of that same space, one says that Ψ' is obtained through the action of a given *operator* A on the function Ψ, and one writes

$$\Psi' = A\Psi.$$

By definition A is a *linear operator* if its action on the function $\lambda_1\Psi_1 + \lambda_2\Psi_2$, a linear combination with constant (complex) coefficients, of two functions of this function space, is given by

$$A(\lambda_1\Psi_1 + \lambda_2\Psi_2) = \lambda_1(A\Psi_1) + \lambda_2(A\Psi_2).$$

Among the linear operators acting on the wave functions

$$\Psi \equiv \Psi(\mathbf{r}, t) \equiv \Psi(x, y, z, t)$$

associated with a particle, let us mention:

1) the differential operators $\partial/\partial x$, $\partial/\partial y$, $\partial/\partial z$, $\partial/\partial t$, such as the one which was considered above;

2) the operators of the form $f(\mathbf{r}, t)$ whose action consists in multiplying the function Ψ by the function $f(\mathbf{r}, t)$.

Starting from certain linear operators, one can form new linear operators by the following algebraic operations:

a) multiplication of an operator A by a constant c:

$$(cA)\Psi \equiv c(A\Psi);$$

b) the sum $S = A + B$ of two operators A and B:

$$S\Psi \equiv A\Psi + B\Psi;$$

c) the product $P = AB$ of an operator B by the operator A:

$$P\Psi \equiv AB\Psi \equiv A(B\Psi).$$

Note that in contrast to the sum, *the product of two operators is not commutative*. Therein lies a very important difference between the algebra of linear operators and ordinary algebra. The product AB is not necessarily identical to the product BA; in the first case, B first acts on the function Ψ, then A acts upon the function $(B\Psi)$ to give the final result; in the second case, the roles of A and B are inverted. The difference $AB-BA$ of these two quantities is called the *commutator* of A and B; it is represented by the symbol $[A, B]$:

$$[A, B] \equiv AB - BA. \tag{II.8}$$

If this difference vanishes, one says that the two operators commute:

$$AB = BA.$$

As an example of operators which do not commute, we mention the operator $f(x)$, multiplication by function $f(x)$, and the differential operator $\partial/\partial x$. Indeed we have, for any Ψ,

$$\frac{\partial}{\partial x} f(x)\Psi = \frac{\partial}{\partial x}(f\Psi) = \frac{\partial f}{\partial x}\Psi + f\frac{\partial \Psi}{\partial x} = \left(\frac{\partial f}{\partial x} + f\frac{\partial}{\partial x}\right)\Psi.$$

In other words

$$\left[\frac{\partial}{\partial x}, f(x)\right] = \frac{\partial f}{\partial x} \tag{II.9}$$

and, in particular

$$\left[\frac{\partial}{\partial x}, x\right] = 1. \tag{II.10}$$

However, any pair of derivative operators such as $\partial/\partial x$, $\partial/\partial y$, $\partial/\partial z$, $\partial/\partial t$, commute.

A typical example of a linear operator formed by sum and product of linear operators is the Laplacian operator

$$\triangle \equiv \text{div grad} \equiv (\nabla \cdot \nabla) \equiv \frac{\partial^2}{\partial x^2} + \frac{\partial^2}{\partial y^2} + \frac{\partial^2}{\partial z^2}$$

which one may consider as the scalar product of the vector operator gradient $\nabla \equiv (\partial/\partial x, \partial/\partial y, \partial/\partial z)$ by itself.

12. Wave Equation of a Free Particle

The theory of matter waves leads unambiguously to the wave equation of a free particle (in non-relativistic approximation). Indeed, the wave $\Psi(\mathbf{r}, t)$ is a superposition:

$$\Psi(\mathbf{r}, t) = \int F(\mathbf{p})\, e^{i(\mathbf{p}\cdot\mathbf{r}-Et)/\hbar}\, d\mathbf{p} \tag{II.11}$$

of monochromatic plane waves $\exp[i(\mathbf{p}\cdot\mathbf{r}-Et)/\hbar]$ whose frequency $E/\hbar$ is connected with the wave vector $\mathbf{p}/\hbar$ by the relation connecting momentum and energy for a particle of mass m

$$E = \frac{p^2}{2m}. \tag{II.12}$$

Taking the partial derivatives of the two sides of equation (II.11), — we omit questions of convergence since mathematical rigor is of no concern to us in this argument — one obtains successively:

$$i\hbar \frac{\partial}{\partial t} \Psi(\mathbf{r}, t) = \int E F(\mathbf{p})\, e^{i(\mathbf{p}\cdot\mathbf{r}-Et)/\hbar}\, d\mathbf{p} \tag{II.13}$$

$$\frac{\hbar}{i} \nabla \Psi(\mathbf{r}, t) = \int \mathbf{p} F(\mathbf{p})\, e^{i(\mathbf{p}\cdot\mathbf{r}-Et)/\hbar}\, d\mathbf{p} \tag{II.14}$$

$$-\hbar^2 \triangle \Psi(\mathbf{r}, t) = \int p^2 F(\mathbf{p})\, e^{i(\mathbf{p}\cdot\mathbf{r}-Et)/\hbar}\, d\mathbf{p}. \tag{II.15}$$

According to relation (II.12), the expressions under the integral signs of equations (II.13) and (II.15) are proportional; therefore the integrals themselves differ by the same proportionality factor. Consequently

$$i\hbar \frac{\partial}{\partial t} \Psi(\mathbf{r}, t) = -\frac{\hbar^2}{2m} \triangle \Psi(\mathbf{r}, t). \tag{II.16}$$

This is the Schrödinger equation for a free particle; it satisfies conditions (A) and (B); from the very manner in which it was obtained it also satisfies the requirements of the correspondence principle. Indeed the formal analogy with Classical Mechanics is actually realized: equation (II.16) is in a sense the quantum-mechanical translation of the classical equation (II.12), the energy and momentum being represented in this quantum language by differential operators acting on the wave function according to the correspondence rule

$$E \to i\hbar \frac{\partial}{\partial t}, \qquad \mathbf{p} \to \frac{\hbar}{i} \nabla. \tag{II.17}$$

Thus the quantity $\mathbf{p}^2 = p_x^2 + p_y^2 + p_z^2$ is represented by the operator

$$-\hbar^2 \triangle = \left(\frac{\hbar}{i}\right)^2 \left(\frac{\partial^2}{\partial x^2} + \frac{\partial^2}{\partial y^2} + \frac{\partial^2}{\partial z^2}\right).$$

Just like relation (II.12) from which it originated, equation (II.16) obviously does not satisfy the principle of relativity. On the other hand, the de Broglie theory does not suffer from this limitation. To obtain a relativistic equation of the free particle, one may try to repeat the preceding argument, replacing relation (II.12) by a relation between energy and momentum in conformity with the theory of relativity. The correct relation $E=\sqrt{p^2c^2+m^2c^4}$ is not suitable because of the presence of the square root. To avoid that difficulty, one can use the relation

$$E^2=p^2c^2+m^2c^4 \tag{II.18}$$

from which one deduces the equation

$$-\hbar^2\frac{\partial^2}{\partial t^2}\Psi=-\hbar^2c^2\triangle\Psi+m^2c^4\Psi,$$

which may also be written

$$\left[\Box+\left(\frac{mc}{\hbar}\right)^2\right]\Psi(\mathbf{r},t)=0 \tag{II.19}$$

making use of the Dalembertian operator

$$\Box\equiv\frac{1}{c^2}\frac{\partial^2}{\partial t^2}-\triangle.$$

One again finds the same formal correspondence between equations (II.18) and (II.19), as the one which exists between equations (II.12) and (II.16).

Equation (II.19), the so-called *Klein-Gordon equation*, plays an important role in Relativistic Quantum Theory. As it does not satisfy criterion (B), it cannot be adopted as wave equation without a physical reinterpretation of the wave Ψ. Actually, the fact that a wave can represent the dynamical state of one and only one particle is fully justified only in the non-relativistic limit, i.e. when the law of conservation of the number of particles is satisfied. Hence we shall restrict ourselves from now on to the search for a non-relativistic wave equation.

13. Particle in a Scalar Potential

In order to form the wave equation of a particle in a potential $V(\mathbf{r})$, we operate at first under the conditions of the "geometrical optics approximation" and seek to form an equation of propagation for a wave packet $\Psi(\mathbf{r}, t)$ moving in accordance with the de Broglie theory.

The center of the packet travels like a classical particle whose position, momentum, and energy we shall designate by $\mathbf{r}_{\text{cl.}}$, $\mathbf{p}_{\text{cl.}}$, and $E_{\text{cl.}}$, respectively. These quantities are connected by the relation

$$E_{\text{cl.}} = H(\mathbf{r}_{\text{cl.}}, \mathbf{p}_{\text{cl.}}) \equiv \frac{p^2_{\text{cl.}}}{2m} + V(\mathbf{r}_{\text{cl.}}) \tag{II.20}$$

$H(\mathbf{r}_{\text{cl.}}, \mathbf{p}_{\text{cl.}})$ is the classical Hamiltonian. We suppose that $V(\mathbf{r})$ does not depend upon the time explicitly (conservative system), although this condition is not absolutely necessary for the present argument to hold. Consequently $E_{\text{cl.}}$ remains constant in time, while $\mathbf{r}_{\text{cl.}}$ and $\mathbf{p}_{\text{cl.}}$ are well-defined functions of t. Under the approximate conditions considered here, $V(\mathbf{r})$ remains practically constant over a region of the order of the size of the wave packet; therefore

$$V(\mathbf{r})\, \Psi(\mathbf{r}, t) \approx V(\mathbf{r}_{\text{cl.}})\, \Psi(\mathbf{r}, t). \tag{II.21}$$

On the other hand, if we restrict ourselves to time intervals sufficiently short so that the relative variation of $\mathbf{p}_{\text{cl.}}$ remains negligible, $\Psi(\mathbf{r}, t)$ may be considered as a superposition of plane waves of the type (II.11), whose frequencies are in the neighborhood of $E_{\text{cl.}}/\hbar$ and whose wave vectors lie close to $\mathbf{p}_{\text{cl.}}/\hbar$. Therefore

$$\begin{aligned} i\hbar \frac{\partial}{\partial t} \Psi(\mathbf{r}, t) &\approx E_{\text{cl.}}\, \Psi(\mathbf{r}, t) \\ \frac{\hbar}{i} \nabla \Psi(\mathbf{r}, t) &\approx \mathbf{p}_{\text{cl.}}(t)\, \Psi(\mathbf{r}, t) \end{aligned} \tag{II.22}$$

and taking the divergence of this last expression, one obtains

$$-\hbar^2 \triangle \Psi(\mathbf{r}, t) \approx p^2_{\text{cl.}}\, \Psi(\mathbf{r}, t). \tag{II.23}$$

Combining the relations (II.21), (II.22), and (II.23) and making use of equation (II.20), we obtain

$$i\hbar \frac{\partial}{\partial t} \Psi + \frac{\hbar^2}{2m} \triangle \Psi - V\Psi \approx \left(E_{\text{cl.}} - \frac{p^2_{\text{cl.}}}{2m} - V(\mathbf{r}_{\text{cl.}})\right) \Psi \approx 0.$$

The wave packet $\Psi(\mathbf{r}, t)$ satisfies – at least approximately – a wave equation of the type we are looking for. We are very naturally led to adopt this equation as the wave equation of a particle in a potential, and we postulate that in all generality, even when the conditions for the "geometrical optics" approximation are not fulfilled, the wave Ψ satisfies the equation

$$i\hbar \frac{\partial}{\partial t} \Psi(\mathbf{r}, t) = \left(-\frac{\hbar^2}{2m} \triangle + V(\mathbf{r}) \right) \Psi(\mathbf{r}, t). \tag{II.24}$$

It is the Schrödinger equation for a particle in a potential $V(\mathbf{r})$.

14. Charged Particle in an Electromagnetic Field

The above argument may be repeated in more general cases where the potential V is an explicit function of time, or in the case of a particle with charge e in an electromagnetic field derived from a vector potential $\mathbf{A}(\mathbf{r}, t)$ and a scalar potential $\varphi(\mathbf{r}, t)$. In the latter case, the classical relation (II.20) must be replaced (cf. Problem I.4) by the relation

$$E = \frac{1}{2m} \left(\mathbf{p} - \frac{e}{c} \mathbf{A}(\mathbf{r}, t) \right)^2 + e\varphi(\mathbf{r}, t). \tag{II.25}$$

Considerations of the behavior of wave packets on the "geometrical optics" approximation lead us to the wave equation

$$i\hbar \frac{\partial}{\partial t} \Psi(\mathbf{r}, t) = \left[\frac{1}{2m} \left(\frac{\hbar}{i} \nabla - \frac{e}{c} \mathbf{A} \right)^2 + e\varphi \right] \Psi(\mathbf{r}, t). \tag{II.26}$$

It is the Schrödinger equation of a charged particle in an electromagnetic field [1]).

---

[1]) On the right-hand side of equation (II.26), the operator $\left(\frac{\hbar}{i} \nabla - \frac{e}{c} \mathbf{A} \right)^2$ designates the scalar product of the vector operator $\frac{\hbar}{i} \nabla - \frac{e}{c} \mathbf{A}$ by itself; in other words, the function which results from its action on Ψ is the sum of the expression

$$\left(\frac{\hbar}{i} \frac{\partial}{\partial x} - \frac{e}{c} A_x \right) \left(\frac{\hbar}{i} \frac{\partial}{\partial x} - \frac{e}{c} A_x \right) \Psi \equiv -\hbar^2 \frac{\partial^2 \Psi}{\partial x^2} - \frac{e\hbar}{ic} \left(A_x \frac{\partial \Psi}{\partial x} + \frac{\partial}{\partial x} (A_x \Psi) \right) + \frac{e^2}{c^2} A_x{}^2 \Psi$$

and of two other expressions which are obtained from it by substituting y and z for x, namely

$$-\hbar^2 \triangle \Psi - 2 \frac{e\hbar}{ic} (\mathbf{A} \cdot \nabla \Psi) + \left(-\frac{e\hbar}{ic} \operatorname{div} \mathbf{A} + \frac{e^2}{c^2} A^2 \right) \Psi.$$

In all of this one must realize that the components of the operator ∇ and those of the operator $\mathbf{A}$ do not in general commute with each other.

Equations (II.24) and (II.26) are the generalizations of equation (II.16) and the same remarks apply to them. They are indeed linear, homogeneous, partial differential equations of the first order in the time [conditions (A) and (B)]. Furthermore, they can be deduced from the classical relations (II.20) and (II.25), respectively, by the correspondence relation defined by (II.17).

15. General Rule for Forming the Schrödinger Equation by Correspondence

Generalizing this correspondence operation, one can formulate a systematic method for constructing the Schrödinger equation, which can be applied to the most general systems.

Consider a classical dynamical system whose Hamiltonian is $H(q_1, \ldots, q_R;\ p_1, \ldots, p_R;\ t)$. This function depends on the coordinates $q_1, \ldots, q_R$ of the system in configuration space, on their respective momenta $p_1, \ldots, p_R$, and on the time t. The total energy E of the system is

$$E = H(q_1, \ldots, q_R; p_1, \ldots, p_R; t). \tag{II.27}$$

To this classical system corresponds a quantum system whose dynamical state is represented by a wave function $\Psi(q_1, \ldots, q_R; t)$ *defined in configuration space* and whose wave equation can be obtained by performing on both sides of equation (II.27) the substitutions

$$E \rightarrow \mathrm{i}\hbar \frac{\partial}{\partial t} \qquad p_r \rightarrow \frac{\hbar}{\mathrm{i}} \frac{\partial}{\partial q_r} \qquad (r = 1, 2, \ldots, R) \tag{II.28}$$

and by writing down that these two quantities, considered as operators, give identical results when acting on Ψ. The equation thus obtained is the *Schrödinger equation* of the corresponding quantum system:

$$\begin{aligned} \mathrm{i}\hbar \frac{\partial}{\partial t} \Psi(q_1, \ldots, q_R; t) &= \\ &= H\left(q_1, \ldots, q_R; \frac{\hbar}{\mathrm{i}} \frac{\partial}{\partial q_1}, \ldots, \frac{\hbar}{\mathrm{i}} \frac{\partial}{\partial q_R}; t\right) \Psi(q_1, \ldots, q_R; t). \end{aligned} \tag{II.29}$$

The operator $H\left(q_1, \ldots, q_R; \frac{\hbar}{\mathrm{i}} \frac{\partial}{\partial q_1}, \ldots, \frac{\hbar}{\mathrm{i}} \frac{\partial}{\partial q_R}; t\right)$ is called the *Hamiltonian* of the system under consideration.

It is important to note that the correspondence rule stated above does not define the Schrödinger equation uniquely. Indeed, there exist two causes of ambiguity.

The first cause of ambiguity comes from the fact that this rule is not invariant under a change of coordinates of configuration space. To illustrate this point, take the simple case of a free particle in two-dimensional space. From the classical Hamiltonian

$$\frac{p_x{}^2 + p_y{}^2}{2m}$$

in cartesian coordinates, one deduces the equation

$$i\hbar \frac{\partial}{\partial t} \Psi(x, y; t) = -\frac{\hbar^2}{2m}\left(\frac{\partial^2}{\partial x^2} + \frac{\partial^2}{\partial y^2}\right) \Psi(x, y; t).$$

If by carrying out a change of variables one goes over to polar coordinates (r, φ), one obtains, after a straightforward calculation, the following equation for the wave function Ψ considered as a function of these new coordinates:

$$i\hbar \frac{\partial}{\partial t} \Psi(r, \varphi; t) = -\frac{\hbar^2}{2m}\left(\frac{\partial^2}{\partial r^2} + \frac{1}{r}\frac{\partial}{\partial r} + \frac{1}{r^2}\frac{\partial^2}{\partial \varphi^2}\right) \Psi(r, \varphi; t).$$

If, on the other hand, one applies correspondence rule (II.28) directly to the classical Hamiltonian in polar coordinates

$$\frac{1}{2m}\left(p_r{}^2 + \frac{p_\varphi{}^2}{r^2}\right)$$

one obtains a different equation, namely

$$i\hbar \frac{\partial}{\partial t} \Psi(r, \varphi; t) = -\frac{\hbar^2}{2m}\left(\frac{\partial^2}{\partial r^2} + \frac{1}{r^2}\frac{\partial^2}{\partial \varphi^2}\right) \Psi(r, \varphi; t).$$

In order to avoid this type of ambiguity, we adopt the convention not to apply rule (II.28) unless the coordinates q are cartesian coordinates[1]).

---

[1]) This convention is not arbitrary. It automatically ensures the invariance of the form of the Schrödinger equation under a rotation of axes. In fact, one can remove this restriction and formulate the correspondence rule in covariant form by adopting a suitable metric in configuration space, and by replacing the operation $\partial/\partial q$ in (II. 28) by the covariant derivative. In this connection, see Brillouin, L., *Les Tenseurs en Mécanique et en Elasticité* (Paris, Masson, 2nd. ed, 1949), p. 200; also (New York, Dover, 1946).

The second cause of ambiguity stems from the fact that rule (II.28) substitutes for quantities obeying the rules of ordinary algebra, operators which do not all commute with each other. As a consequence different Hamiltonians may correspond to equivalent forms of the classical Hamiltonian. Thus, to the two equivalent expressions for the kinetic energy (in a one-dimensional problem)

$$\frac{p^2}{2m} \quad \text{and} \quad \frac{1}{2m}\frac{1}{\sqrt{q}}\,pqp\,\frac{1}{\sqrt{q}},$$

there respectively correspond the operators

$$-\frac{\hbar^2}{2m}\frac{\partial^2}{\partial q^2} \text{ and } -\frac{\hbar^2}{2m}\left(\frac{1}{\sqrt{q}}\frac{\partial}{\partial q}\cdot q\cdot\frac{\partial}{\partial q}\cdot\frac{1}{\sqrt{q}}\right) \equiv -\frac{\hbar^2}{2m}\left(\frac{\partial^2}{\partial q^2}+\frac{1}{4q^2}\right)$$

which differ by the quantity $\hbar^2/8mq^2$.

No rule based on the correspondence with Classical Mechanics can resolve such ambiguities, since the latter arise from the non-commutability of operators, which in turn is tied to the finite and non-zero character of $\hbar$. One must therefore fix empirically the precise form of the Hamiltonian function to which one applies rule (II.28). In all cases of practical interest, one must conform to the following prescriptions:

In *cartesian* coordinates, the classical Hamiltonian is the sum of a quadratic expression in the p's completely independent of the q's, a function depending exclusively upon the q's, and possibly a linear function of the p's of the form $\sum_i p_i f_i(q_1, \ldots, q_R)$. Having put the Hamiltonian function into this form, one replaces the last term of the summation by the "symmetrized" expression

$$\tfrac{1}{2}\textstyle\sum_i [p_i f_i(q_1, \ldots, q_R) + f_i(q_1, \ldots, q_R) p_i];$$

the correspondence rule (II.28) must be applied to this expression.

The "symmetrization" of the linear terms in p is, as we shall see in Chapter IV, a necessary condition for the statistical interpretation of the wave function to be consistent. The case of a particle in an electromagnetic field [eqs. (II.25) and (II.26)] is an example of a system where this procedure must be carried out.

To conclude this section consider the following application. We propose to form the Schrödinger equation of a complex atom, formed by a nucleus of charge Ze and of mass M, and by Z electrons of charge $-e$ and mass m. The Hamiltonian function is made up of $Z+1$ kinetic energy terms, Z terms of Coulomb attraction of the Z electrons by the nucleus, and $\frac{1}{2}Z(Z-1)$ terms of mutual Coulomb repulsion between each pair of electrons; thus, adopting the notation of § 9,

$$\frac{P^2}{2M}+\sum_{i=1}^{Z}\frac{p_i^2}{2m}-\sum_{i=1}^{Z}\frac{Ze^2}{|\mathbf{R}-\mathbf{r}_i|}+\sum_{i<j}\frac{e^2}{|\mathbf{r}_i-\mathbf{r}_j|}.$$

From it we derive the Schrödinger equation

$$i\hbar\frac{\partial}{\partial t}\Psi(\mathbf{R},\mathbf{r}_1,\ldots,\mathbf{r}_Z;t)=$$
$$=\left[-\hbar^2\left(\frac{\triangle_R}{2M}+\sum_{i=1}^{Z}\frac{\triangle_i}{2m}\right)-\sum_{i=1}^{Z}\frac{Ze^2}{|\mathbf{R}-\mathbf{r}_i|}+\sum_{i<j}\frac{e^2}{|\mathbf{r}_i-\mathbf{r}_j|}\right]\Psi \tag{II.30}$$

where the operator $\triangle_R$ designates the Laplacian with respect to the position vector $\mathbf{R}$, i.e. $\partial^2/\partial X^2+\partial^2/\partial Y^2+\partial^2/\partial Z^2$, and the operator $\triangle_i$ denotes the Laplacian with respect to the position vector of the ith electron $\mathbf{r}_i$.

In the case of the hydrogen atom ($Z=1$) the equation reads

$$i\hbar\frac{\partial}{\partial t}\Psi(\mathbf{r}_p,\mathbf{r}_e;t)=\left(-\frac{\hbar^2}{2M}\triangle_p-\frac{\hbar^2}{2m}\triangle_e-\frac{e^2}{|\mathbf{r}_p-\mathbf{r}_e|}\right)\Psi(\mathbf{r}_p,\mathbf{r}_e;t) \tag{II.31}$$

(here M is the mass of the proton, $\mathbf{r}_p$ its position, and $\mathbf{r}_e$ the position of the electron). As a first approximation, one can consider the proton to be infinitely heavy, and treat the hydrogen atom as an electron in an attractive Coulomb field $-e^2/r$, $\mathbf{r}$ designating the position of the electron in a coordinate system whose origin is located at the proton (assumed to be at rest). The wave function $\Psi(\mathbf{r},t)$ of the electron satisfies the Schrödinger equation:

$$i\hbar\frac{\partial}{\partial t}\Psi(\mathbf{r},t)=\left(-\frac{\hbar^2}{2m}\triangle-\frac{e^2}{r}\right)\Psi(\mathbf{r},t). \tag{II.32}$$

III. THE TIME-INDEPENDENT SCHRÖDINGER EQUATION

16. Search for Stationary Solutions

The Schrödinger equation of a quantum system is formally written

$$i\hbar\frac{\partial}{\partial t}\Psi=H\Psi. \tag{II.33}$$

Let us assume that the Hamiltonian H does not explicitly depend upon the time. This is the case of conservative systems, corresponding to classical systems whose energy is a constant of the motion. We look for a solution Ψ representing a dynamical state of well-defined energy E.

Such a wave Ψ must have a well-defined (angular) frequency ω, namely the one given by the Einstein relation $E=\hbar\omega$. Recall that this relationship between frequency of the wave and energy of the system constitutes the basic postulate of the theory of matter waves. Ψ is thus put into the form

$$\Psi = \psi \mathrm{e}^{-\mathrm{i}Et/\hbar} \tag{II.34}$$

where ψ depends upon the coordinates of configuration space but not upon the time. Substitution of this expression into equation (II.33) leads to the equation

$$H\psi = E\psi. \tag{II.35}$$

It is the so-called *time-independent Schrödinger equation.*

When the system is in a state represented by a wave of type (II.34) it is said to be in a *stationary* state of energy E; the time-independent wave function ψ is usually called the wave function of the stationary state, although the true wave function differs from the latter by the phase factor $\exp(-\mathrm{i}Et/\hbar)$.

17. General Properties of the Equation. Nature of the Energy Spectrum

To simplify our presentation we shall discuss the special case of a particle of mass m in a scalar potential $V(\boldsymbol{r})$. We further assume that $V(\boldsymbol{r}) \to 0$ as $r \to \infty$. The function ψ is a function of the vector $\boldsymbol{r}(x, y, z)$ defining the position of the particle, and the time-independent Schrödinger equation is written

$$H\psi(\boldsymbol{r}) \equiv \left[-\frac{\hbar^2}{2m}\triangle + V(\boldsymbol{r})\right]\psi(\boldsymbol{r}) = E\psi(\boldsymbol{r}). \tag{II.36}$$

In the language of the theory of partial differential equations, an equation of the type (II.36) is known as an *eigenvalue equation.* A solution $\psi_E(r)$ of this equation is called an *eigenfunction* (proper function) corresponding to the eigenvalue E of the operator H.

In fact, this eigenvalue problem is not well defined unless one specifies the conditions of "regularity" and the boundary conditions which the function ψ must satisfy. The conditions to be imposed upon $\psi(\boldsymbol{r})$ must of course be compatible with the interpretation given to the wave function. We shall return to this point in Chapter IV. Let us agree for the time being that $\psi(\boldsymbol{r})$ and its partial derivatives of the first order shall be *continuous*, *uniform*, and *bounded* functions over all space.

One can then show the following results, which we shall accept without proof and verify later on in numerous examples.

a) If $E < 0$, equation (II.36) has solutions only for certain particular values of E forming a *discrete spectrum*. The eigenfunction $\psi(\boldsymbol{r})$ corresponding to it – or each of the eigenfunctions when several exist – vanishes at infinity. More precisely, the integral $\int|\psi(\boldsymbol{r})|^2 \mathrm{d}\boldsymbol{r}$ extended over the entire configuration space is convergent. Following the statistical interpretation, there is a vanishing probability of finding the particle at infinity, and the particle remains practically localized in a finite region. The particle is said to be in a *bound state*.

b) If $E > 0$, equation (II.36) may be solved for any positive value of E: the positive energies form a *continuous spectrum*. However, the corresponding eigenfunctions do not vanish at infinity; their asymptotic behavior is analogous to that of the plane wave $\exp(\mathrm{i}\boldsymbol{k}\cdot\boldsymbol{r})$. More precisely, when $r \to \infty$ the absolute value $|\psi(\boldsymbol{r})|$ approaches a non-zero constant, or oscillates indefinitely between limits, one of which at least is not zero. The particle does not remain localized in a finite region. Wave functions of this type are used in collision problems; one is dealing with a so-called *unbound state*, or stationary state of collision.

We thus obtain a first fundamental result: the quantization of the energy of bound states, one of the most striking facts among those which caused the abandonment of the Classical Theory. The determination of the quantized energies appears here as an eigenvalue problem. To solve this problem as accurately as possible is one of the central problems of Wave Mechanics. For some particularly simple forms of the Hamiltonian, it can be solved rigorously. This is especially the case of the hydrogen atom (we shall treat it in detail in Chapter XI) whose energy levels are the eigenvalues of the operator

$$[-(\hbar^2/2m)\triangle - (e^2/r)].$$

As will be shown in Chapter XI the eigenvalue spectrum is identical to that predicted by the Old Quantum Theory; we have already emphasized its extraordinary agreement with experiment. In more complex situations, one has to have recourse to suitable methods of approximation. In all the cases where the energy spectrum could be calculated with reasonable precision, the agreement with experimental results is as good as might be expected from a non-relativistic theory.

The eigensolution ψ_E itself may be to some extent subject to experimental check. Indeed, the eigenfunctions of the discrete spectrum enter into the calculations of various measurable quantities, such as transition probabilities. As to the eigenfunctions of the continuous spectrum, their asymptotic form is very directly related to the cross sections – characteristic parameters of the collision phenomena, whose precise definition will be given later on (Ch. X).

EXERCISES AND PROBLEMS

1. Let us try to observe the motion of the electron along a circular Bohr orbit of the hydrogen atom by carrying out several successive measurements of the electron position with sufficiently hard X-rays.

Evaluate the order of magnitude of the kinetic energy transfer ΔT to the electron in a collision with an X-ray photon as a function of the wavelength λ of the latter. To observe the motion along an orbit, λ must be much smaller than the radius of this orbit. Compare in that case the perturbation ΔT with the distance between neighboring energy levels. What must one conclude from this concerning the observability of Bohr orbits?

2. In relativistic quantum mechanics, the total energy E and the momentum p of a free particle of (rest) mass m and velocity v are respectively equal to

$$\frac{mc^2}{\sqrt{1-v^2/c^2}} \quad \text{and} \quad \frac{mv}{\sqrt{1-v^2/c^2}}.$$

Verify that the equations of motion may be written in the Hamiltonian form by taking for the Hamiltonian function $H = E = \sqrt{m^2c^4 + p^2c^2}$.

From this, deduce the equality between the velocity of this particle and the group velocity v_g of the associated de Broglie wave. Calculate the phase velocity v_φ of this wave: show that it is superior to the velocity c and that $v_g v_\varphi = c^2$.

3. Examine the soundness of a classical description of the motion of an atom in a diatomic molecule. To do this, assume that the atom executes harmonic oscillations of angular frequency ω with an average kinetic energy of the order of $\frac{1}{2}kT$, and compare the average wavelength of the atom with the amplitude of these oscillations. First, treat the case of the hydrogen molecule at ordinary temperatures: $T = 300°$ K, $kT = 0.025$ eV, $\hbar\omega = 0.5$ eV; then consider the

case of a molecule with heavy atoms of mass 200 times that of hydrogen assuming the restoring force to be the same as for hydrogen, at 300° K, and also at 10° K.

4. An electron follows a circular trajectory in a constant magnetic field $\boldsymbol{H}$. Apply to this rotational motion the de Broglie resonance condition. Show that the kinetic energy of the electron is quantized, that the energy levels are equidistant, and that the distance between levels is equal to $(e\hbar/mc)\boldsymbol{H}$ [this result differs from that of the rigorous quantum theory only by an overall displacement of all the energy levels by an amount $(e\hbar/2mc)\boldsymbol{H}$.] Calculate the radius, the momentum and the kinetic energy of the quantized trajectories for a field of 10^4 gauss. Compare the radius of the orbit of quantum number unity to that of the Bohr orbit of the ground state of the hydrogen atom.
[*N.B.* Be sure to distinguish in this problem between the momentum $\boldsymbol{p}$ and the quantity $m\boldsymbol{v}$. If $\boldsymbol{A}$ is the vector potential from which the magnetic field is derived, $\boldsymbol{p} = m\boldsymbol{v} + (e/c)\boldsymbol{A}$.]

5. Utilizing the fact that any wave can be considered as a superposition of plane waves, show that in the absence of a field, the matter wave $\psi(\boldsymbol{r}_2, t_2)$ at the point $\boldsymbol{r}_2$ at the instant t_2 can be deduced from the values $\psi(\boldsymbol{r}_1, t_1)$ taken by the wave at the instant t_1, by the operation

$$\psi(\boldsymbol{r}_2, t_2) = \int K(\boldsymbol{r}_2 - \boldsymbol{r}_1; t_2 - t_1)\, \psi(\boldsymbol{r}_1, t_1)\, \mathrm{d}\boldsymbol{r}_1 \tag{1}$$

where

$$K(\boldsymbol{\rho}; \tau) = (2\pi\hbar)^{-3} \int \exp\left[\frac{\mathrm{i}}{\hbar}(\boldsymbol{p}\cdot\boldsymbol{\rho} - E\tau)\right] \mathrm{d}\boldsymbol{p},$$

an expression in which E, a function of $\boldsymbol{p}$, is equal to the energy of the particle, corresponding to the momentum $\boldsymbol{p}$. Show that for a non-relativistic particle of mass m,

$$K(\boldsymbol{\rho}; \tau) = \mathrm{e}^{-\frac{3}{4}\pi\mathrm{i}} \left(\frac{m}{2\pi\hbar\tau}\right)^{\frac{3}{2}} \exp\left(\mathrm{i}\,\frac{m\rho^2}{2\hbar\tau}\right).$$

Deduce from this that the main contribution to integral (1) giving the wave function at $\boldsymbol{r}_2$ at time t_2 comes from a region surrounding the point $\boldsymbol{r}_2$ whose radius is of the order of

$$\left[\frac{2\hbar(t_2 - t_1)}{m}\right]^{\frac{1}{2}}.$$

6. How does the method of the preceding problem have to be modified in order to apply to a particle in one dimension? Using this method, determine the wave function at the instant t of a free, non-relativistic particle of mass m whose wave function at the instant $t = 0$ is

$$\psi(x, 0) = (\pi\xi_0^2)^{-\frac{1}{4}} \exp(\mathrm{i}p_0 x/\hbar) \exp(-x^2/2\xi_0^2).$$

The intensity (modulus squared) of that wave at the instant $t = 0$ is a Gaussian of width ξ_0. Show that the shape remains Gaussian at all later instants, but that its width increases according to the law

$$\xi = \xi_0 \left(1 + \frac{\hbar^2 t^2}{m^2 {\xi_0}^4}\right)^{\frac{1}{2}}$$

(spreading of the wave packet, cf. Ch. VI).

CHAPTER III

ONE-DIMENSIONAL QUANTIZED SYSTEMS

1. Introduction

In order to acquire a certain amount of practice in dealing with the Schrödinger equation before we attack problems of interpretation of the Quantum Theory, we shall study the wave mechanics of one-dimensional systems. One-dimensional problems are of interest, not only as simple models allowing us to display a certain number of properties which one encounters again in more complex situations, but also because in a number of problems one is led, after some suitable manipulations, to equations of the same type as the one-dimensional Schrödinger equation.

Let us consider the motion of a particle of mass m constrained to move on the x axis in a certain potential $V(x)$. Its Schrödinger equation is:

$$\mathrm{i}\hbar \frac{\partial}{\partial t} \Psi(x, t) = \left(-\frac{\hbar^2}{2m} \frac{\partial^2}{\partial x^2} + V(x) \right) \Psi(x, t). \tag{III.1}$$

We shall especially be concerned with the search for the stationary states. If E is the energy of the stationary state, we have

$$\Psi(x, t) = \psi(x)\, \mathrm{e}^{-\mathrm{i}Et/\hbar} \tag{III.2}$$

and the function $\psi(x)$ is a solution of the time-independent Schrödinger equation

$$\left(-\frac{\hbar^2}{2m} \frac{\mathrm{d}^2}{\mathrm{d}x^2} + V(x) \right) \psi = E\psi. \tag{III.3}$$

Throughout this chapter we adopt the notation

$$V(x) = \frac{\hbar^2}{2m} U(x), \qquad E = \frac{\hbar^2}{2m} \varepsilon \tag{III.4}$$

which allows us to rewrite the preceding equation in the form

$$\psi'' + [\varepsilon - U(x)]\psi = 0. \tag{III.5}$$

It is a differential equation of the Sturm-Liouville type for which we intend to find solutions which are finite, continuous and differentiable over the entire interval $(-\infty, +\infty)$.

If such a solution exists, all multiples of that solution are solutions as well; we shall not consider two solutions to be distinct if they differ only by a constant multiplicative factor. If two linearly independent solutions are acceptable solutions, any linear combination thereof is a solution as well; the eigenvalue is said to be *degenerate* of the second order. The order of degeneracy is by definition the number of linearly independent eigenfunctions.

Equation (III.5) is real [$V(x)$ is a real function of x]. If ψ is an eigenfunction, so are its real part and its imaginary part (the latter are necessarily multiples of one another if the eigenfunction is not degenerate). As a consequence one merely needs to know the real eigenfunctions to construct all the eigenfunctions belonging to a given eigenvalue. This remark simplifies the calculations considerably.

In the first section, we solve the eigenvalue problem exactly for some simple square potentials. We shall dwell upon the principal points of difference between classical and quantized motions, especially the quantization of the energies of bound states, and the phenomena of wave reflection, resonance and potential-barrier penetration in the motion of unbound "particles". In the second section, we shall make a systematic study of equation (III.5) for an arbitrary potential $U(x)$, and show that many results obtained with square potentials are quite general.

I. SQUARE POTENTIALS

2. General Remarks

In order that typical quantum effects may appear, the potential $V(x)$ must possess an appreciable relative variation over a distance of the order of a wavelength. The simplest type of potential fulfilling these conditions is the *square potential*; it is a potential which exhibits discontinuities of the first kind (that is to say, sudden jumps by a finite amount) at certain points, and remains constant everywhere else. The x axis is thus subdivided into n intervals, say, in each of which the potential remains constant.

The existence of discontinuities of the first kind in the potential

$U(x)$ does not modify the conditions of regularity imposed upon the function ψ. Indeed, according to the Schrödinger equation

$$\psi'' = (U - \varepsilon)\psi.$$

At each discontinuity of the potential, U and consequently ψ'' exhibit a sudden jump by a finite amount, but the integral of ψ'' remains continuous at these points: ψ' and *a fortiori* ψ are therefore continuous everywhere.

Now, let U_i be the (constant) value of $U(x)$ in the ith region ($i = 1, 2, \ldots, n$). The general solution in this region is a linear combination of exponentials. Its behavior is in fact very different according to whether $(\varepsilon - U_i)$ is positive or negative.

If $\varepsilon > U_i$, it is a linear combination of imaginary exponentials

$$\mathrm{e}^{\mathrm{i}k_i x} \quad \text{and} \quad \mathrm{e}^{-\mathrm{i}k_i x} \qquad (k_i = \sqrt{\varepsilon - U_i})$$

or, equivalently, a combination of sine and cosine: it has an "oscillatory" behavior.

If $\varepsilon < U_i$, we have a combination of real exponentials

$$\mathrm{e}^{\varkappa_i x} \quad \text{and} \quad \mathrm{e}^{-\varkappa_i x} \qquad (\varkappa_i = \sqrt{U_i - \varepsilon}).$$

In that case, we shall say that the solution has an "exponential" behavior.

To obtain the general solution of the differential equation, one writes it in the form of a linear combination of (real or imaginary) exponentials for each of the n regions where the potential is constant. The parameters of these combinations ($2n$ in number) are fixed by the conditions of continuity of the function and its derivative at the points of discontinuity of the potential. This yields $2(n-1)$ conditions since there are $(n-1)$ points of discontinuity. Therefore the general solution which one thus obtains depends upon two arbitrary parameters, as expected. In order to be acceptable as eigenfunction, the solution must remain bounded everywhere, i.e. bounded at both of the limits $x = +\infty$ and $x = -\infty$. Note that if the energy remains lower than the potential over the entire interval $(-\infty, +\infty)$, the general solution has an exponential behavior everywhere; its second derivative is always of the same sign as the function itself. From this, one readily deduces that the latter increases exponentially at one of the limits $-\infty$ or $+\infty$, and possibly at both limits. The eigenvalue problem has, therefore, no solution. In classical mechanics too, motion is possible only if the energy exceeds the potential in at least part of the interval.

If ε exceeds at least one of the quantities U_i, the existence and the number of eigenfunctions essentially depend upon the exponential or oscillatory character of the general solution at the two extremities of the x axis $-\infty$ and $+\infty$.

3. Potential Step. Reflection and Transmission of Waves

The simplest example of a square potential is that of a sudden jump of the potential ($n=2$) such as the one shown in Fig. III.1:

$$U(x)=\begin{cases} U_1 & \text{if} \quad x>0 \qquad \text{(region I)} \\ U_2 & \text{if} \quad x<0. \qquad \text{(region II)} \end{cases}$$

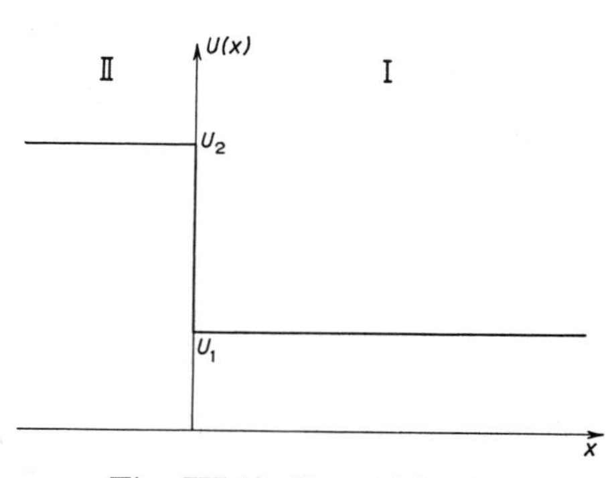

Fig. III.1. Potential step.

For definiteness, we shall assume that $U_2>U_1$.

Two cases may occur:

a) $U_2>\varepsilon>U_1$. The general solution has an oscillatory behavior in region I ($x>0$), and an exponential behavior in region II ($x<0$). To be acceptable as an eigenfunction, it must be exponentially decreasing in region II. There is always one and only one solution fulfilling this condition. Each value of ε contained in this interval is thus a non-degenerate eigenvalue. The energy spectrum is *continuous and non-degenerate*. In each of the two regions this function takes on the general form

$$y=\begin{cases} A_1 \sin(k_1 x+\varphi) & x>0 \\ A_2\, e^{\varkappa_2 x} & x<0. \end{cases} \qquad \text{(III.6)}$$

The conditions of continuity define y to within a constant. Instead of writing down the continuity condition of the function and its derivative, it is more convenient to write down the continuity of the function and of its *logarithmic derivative* y'/y. The continuity of the logarithmic derivative fixes the phase φ:

$$k_1 \cot \varphi = \varkappa_2.$$

φ is determined to within a multiple of π since a change from φ to $\varphi + \pi$ is equivalent to changing the sign of A_1. For the value of φ we take:

$$\varphi = \tan^{-1} \frac{k_1}{\varkappa_2} \tag{III.6a}$$

where $\tan^{-1}$ denotes the value of the arc tangent located in the interval

$$\left(-\frac{\pi}{2}, + \frac{\pi}{2} \right).$$

The continuity of the function determines the ratio A_2/A_1, i.e.

$$\frac{A_2}{A_1} = \sin \varphi = \frac{k_1}{\sqrt{k_1^2 + \varkappa_2^2}} = \sqrt{\frac{\varepsilon - U_1}{U_2 - U_1}}. \tag{III.6b}$$

b) $\varepsilon > U_2$. The general solution has an oscillatory behavior over all space and therefore represents an acceptable eigenfunction. To each value of ε there thus correspond two linearly independent eigenfunctions: the spectrum of eigenvalues is *continuous* and its *degeneracy is of order* 2.

We shall form the eigenfunction whose character in region II is $\exp(-ik_2 x)$. It is defined to within a constant which we fix by taking the coefficient of $\exp(-ik_1 x)$ equal to unity in the expression for that function in region I.

In other words

$$\chi = \begin{cases} e^{-ik_1 x} + R\, e^{ik_1 x} & x > 0 \\ S\, e^{-ik_2 x} & x < 0. \end{cases} \tag{III.7}$$

The (*a priori* complex) constants R and S are determined by the continuity conditions at $x = 0$. The continuity of the logarithmic derivative gives

$$R = \frac{k_1 - k_2}{k_1 + k_2}. \tag{III.7a}$$

Continuity of the function itself yields

$$S = 1 + R = \frac{2k_1}{k_1 + k_2}. \tag{III.7b}$$

The function χ^* is an eigenfunction linearly independent of χ. All the eigenfunctions corresponding to the eigenvalue ε can thus be put in the form of a linear combination of χ and χ^*.

We shall compare the present situation to that which would obtain if the system were classical. The motion of a classical particle in this potential is very different in case (*a*) and in case (*b*).

In case (*a*) the classical motion is one of a particle of energy $(\hbar^2/2m)\varepsilon$ which, coming from $+\infty$ runs along the positive semi-axis with constant velocity $\hbar k_1/m$ in the direction of decreasing x, rebounds elastically at the point $x=0$ and starts out again in the opposite direction with the same velocity traveling to infinity. In order to realize an analogous phenomenon in wave mechanics, one must construct a wave packet by superposing waves of the form $y \exp(-\mathrm{i}Et/\hbar)$ with neighboring energies. Rather than the function given by eq. (III.6), it is more convenient to use the wave

$$\psi_\varepsilon(x) = \begin{cases} \mathrm{e}^{-\mathrm{i}k_1 x} - \mathrm{e}^{\mathrm{i}(k_1 x + 2\varphi)} & x > 0 \\ \dfrac{2A_2}{\mathrm{i}A_1}\,\mathrm{e}^{\mathrm{i}\varphi}\,\mathrm{e}^{\varkappa_2 x} & x < 0 \end{cases} \tag{III.8}$$

obtained by dividing y by $\frac{1}{2}\mathrm{i}A_1 \exp(-\mathrm{i}\varphi)$. We write it with the subscript ε in order to remind us that it is an eigenfunction corresponding to the eigenvalue ε. Consider the wave packet

$$\Psi(x, t) = \int_0^\infty f(k_1' - k_1)\psi_{\varepsilon'}(x)\,\mathrm{e}^{-\mathrm{i}E't/\hbar}\,\mathrm{d}k_1'. \tag{III.9}$$

The function $f(k_1' - k_1)$ is a real, more or less regular function of k_1', with a very pronounced peak at the point $k_1' = k_1$ (the "prime" no longer denotes the derivative with respect to x for the time being; the meaning of the quantities k_1', ε', E' and their relationships are self-evident). For the sake of simplicity, we further assume that $f(k_1' - k_1)$ vanishes when $k_1^2 > U_2 - U_1$. $\Psi(x, t)$ is thus a superposition of eigenfunctions of case (*a*), multiplied by the exponential $\exp(-\mathrm{i}E't/\hbar)$ corresponding to their respective time dependences. By its very construction Ψ is clearly a solution of the time-dependent Schrödinger equation. Its motion is easily exhibited if one refers back to the study of free wave packets of Chapter II.

In region I, $\Psi(x, t)$ is a superposition of two quantities: an "incident wave packet",

$$\Psi_{\rm i}(x, t) = \int_0^\infty f(k_1' - k_1)\, {\rm e}^{-{\rm i}k_1'x}\, {\rm e}^{-{\rm i}E't/\hbar}\, {\rm d}k_1' \tag{III.10a}$$

whose center $x = -(1/\hbar)({\rm d}E/{\rm d}k_1)t = -v_1 t$ travels at the velocity $v_1 = \hbar k_1/m$ in the negative direction and reaches the point $x = 0$ at the time $t = 0$; and a "reflected wave packet",

$$\Psi_{\rm r}(x, t) = -\int_0^\infty f(k_1' - k_1)\, {\rm e}^{{\rm i}[k_1'x + 2\varphi' - (E't/\hbar)]}\, {\rm d}k_1' \tag{III.10b}$$

whose center $x = v_1 t - 2\, {\rm d}\varphi/{\rm d}k_1$ travels with the velocity v_1 in the opposite direction and leaves the origin at the time

$$\tau = \frac{2}{v_1}\frac{{\rm d}\varphi}{{\rm d}k_1} = 2\hbar\frac{{\rm d}\varphi}{{\rm d}E} \tag{III.11}$$

later than the time $t = 0$ at which the center of the "incident wave packet" arrived there. The motion of the center of the packet is thus almost identical to that of the classical particle. The only difference consists in the *delay* τ suffered by the center of the packet upon reflection at the point of discontinuity $x = 0$, whereas the rebound of the classical particle is instantaneous. We note in this connection that the consideration of the motion of the center of the packet makes sense only if the shape of the packet is not too violently modified in the course of its motion. This is actually so for the incident wave packet as long as the center is at a distance from the origin large compared to the width Δx of the packet. For this to hold for the reflected wave packet as well, it is further necessary that the width Δk of the peak of the function f be sufficiently small so that φ does not vary appreciably over the region which contributes most to the integral (III.9), namely $\Delta k({\rm d}\varphi/{\rm d}k_1) \ll 1$. As the spatial extension Δx of the packet is of the order $1/\Delta k$, this condition may also be written

$$\Delta x \gg \frac{{\rm d}\varphi}{{\rm d}k_1}. \tag{III.12}$$

Consequently, $\Delta x/v_1 \gg \tau$. The wave packet is so wide that the time it spends passing a point of the axis is clearly longer than the delay τ caused by the reflection.

Besides the delay τ there exists another difference between the motion of the classical particle and the reflection of the quantum-mechanical wave packet. The wave Ψ does not always vanish in region II. A study analogous to the one above shows that Ψ is equal to the product of $2A_2 \exp(\varkappa_2 x)/A_1$ and a quantity which takes on appreciable values during a certain time interval about the instant $t=\frac{1}{2}\tau$, a time which one can interpret as the collision time with the potential "wall" located at the point $x=0$. Therefore at that moment there exists a non-vanishing probability of finding the particle in region II, while the classical particle never penetrates into that region.

Let us now examine case (*b*). There are two possible classical motions corresponding to the same value of the energy [1]). One is the motion of a particle running along the entire length of the x axis from $+\infty$ to $-\infty$; its velocity which is constant and equal to $v_1=\hbar k_1/m$ in region I, suddenly jumps from v_1 to $v_2=\hbar k_2/m$ at the point of discontinuity of the potential, and continues its motion to $-\infty$ at the velocity v_2. The other is the exactly reversed motion, namely that of a particle travelling along the x axis in the positive direction with velocity v_2 in region II, and velocity v_1 in region I.

Let us compare these classical motions to those of wave packets with similar initial conditions. We shall carry out that comparison for the first of these motions (displacement in the negative direction). Proceeding as in case (*a*), we form a wave packet analogous to the one of eq. (III.9) by superposition of eigenfunctions corresponding to neighboring values of the energy ε. Let us attach the subscript ε to the eigenfunction χ of the type (III.7) to remind ourselves that it depends upon the energy. *A priori*, the packet has to be formed by the superposition of functions χ_ε and χ_ε^*. But in order to realize the desired initial conditions, it must only contain the functions χ_ε as this investigation will show. Let us write therefore

$$\Psi(x, t)=\int_0^\infty f(k_1'-k_1)\chi_{\varepsilon'}(x)\, \mathrm{e}^{-\mathrm{i}E't/\hbar}\, \mathrm{d}k_1'.$$

The only difference with expression (III.9) is that the peak $k_1=\sqrt{\varepsilon-U_1}$

---

[1]) This fact is to be compared with the existence of a degeneracy of order 2 in the corresponding quantum-mechanical problem.

of the function f lies in the energy region (b) instead of the energy region (a). The motion of the above wave packet is found in similar fashion to that of the wave packet (III.9). One readily verifies that the desired initial conditions are actually fulfilled, namely that when $t \ll 0$, $\Psi(x, t)$ remains practically zero in region II, and that in region I the only appreciable contribution comes from the term $\exp(-ik_1x)$. This yields a wave packet whose center $x = -v_1t$ moves like the classical particle at velocity v_1 in the direction of decreasing x and reaches the origin at time $t = 0$; later, $\Psi(x, t)$ splits into two packets, a "transmitted wave packet"

$$\Psi_t(x, t) = \int_0^\infty f(k_1' - k_1)\, S' e^{-ik_2'x}\, e^{-iE't/\hbar}\, dk_1'$$

whose center $x = -v_2t$ rigorously continues the motion of the classical particle, and a "reflected wave packet",

$$\Psi_r(x, t) = \int_0^\infty f(k_1' - k_1)\, R' e^{ik_1'x}\, e^{-iE't/\hbar}\, dk_1'$$

whose center $x = v_1t$ moves as the classical particle would have if it had suffered an elastic collision at $x = 0$. Thus there is a very important difference with the classical motion: *the quantum "particle" has a non-vanishing probability of being "reflected" at the potential discontinuity.*

In order to pursue this analysis, one must define the probabilities in a precise manner; this will be done in Ch. IV. We merely indicate here without proof that the probability of finding the particle in the reflected wave is equal to $|R|^2$, that of finding it in the transmitted wave is equal to $(k_2/k_1)|S|^2$ (cf. Problem IV.2). These results are quite consistent, since the sum of these two quantities is equal to unity:

$$|R|^2 + \frac{k_2}{k_1}|S|^2 = 1 \qquad \text{(III.13)}$$

as one may easily verify by substituting in that equation the values given by eqs. (III.7a) and (III.7b).

The quantity

$$T = \frac{k_2}{k_1}|S|^2 = \frac{4k_1k_2}{(k_1 + k_2)^2} \qquad \text{(III.14)}$$

is called the *transmission coefficient* and measures the relative importance of the phenomenon of transmission. It increases with energy and approaches unity as $\varepsilon \rightarrow \infty$. One may say that in that limit the result of classical mechanics holds.

Note that T is a symmetric function of k_1 and k_2. A wave of the same energy but propagating in the opposite direction (from II toward I) therefore has the same transmission coefficient: the transmission coefficient is independent of the direction of travel.

All these results are not surprising when we realize the great analogy which they bear to problems of lightwave propagation. The problem treated here is exactly the same as that of the propagation of a light signal in a non-absorbing medium of variable index of refraction. In the case (*a*), the index of refraction changes suddenly from a real value (medium I) to an imaginary value (medium II) at the point $x=0$: we have total reflection. In the case (*b*), the index remains real, and the media I and II have different indices of refraction: this sudden change of index causes a partial reflection of the signal.

4. Infinitely High Potential Barrier

A limiting case of the preceding one is that of a particle encountering an infinitely high potential barrier. For definiteness, let us suppose that $U(x) = +\infty$ when $x < 0$. We are in a situation which is analogous to case (*a*) of the preceding problem as $U_2 \rightarrow +\infty$. Inspection of the wave function y as given by eqs. (III.6), (III.6*a*) and (III.6*b*), shows that it vanishes at the point $x=0$ in this limiting case ($\varkappa_2 \rightarrow \infty$).

We are dealing here with a general result which is valid no matter what the form of the function $V(x)$ in the region $x > 0$. Indeed, the wave function necessarily takes the form $A \exp(\varkappa_2 x)$ in the region $x < 0$; its logarithmic derivative is therefore $\varkappa_2$. In the limit where the potential V_2 tends toward infinity, $\varkappa_2$ also becomes infinite. Consequently, the function must have an infinite logarithmic derivative at the point $x=0$: thus, the function itself must vanish there.

In conclusion, *the wave function must vanish at the edge of an infinitely high potential barrier.*

5. Infinitely Deep Square Potential Well. Discrete Spectrum

As a second simple example, consider an infinitely deep square potential well. The value of the potential at the bottom of the well is taken as the origin for the energy scale. This region of zero potential occupies a certain interval $(-L/2, +L/2)$ of the x axis; it is bounded on both sides by infinite potential barriers (Fig. III.2).

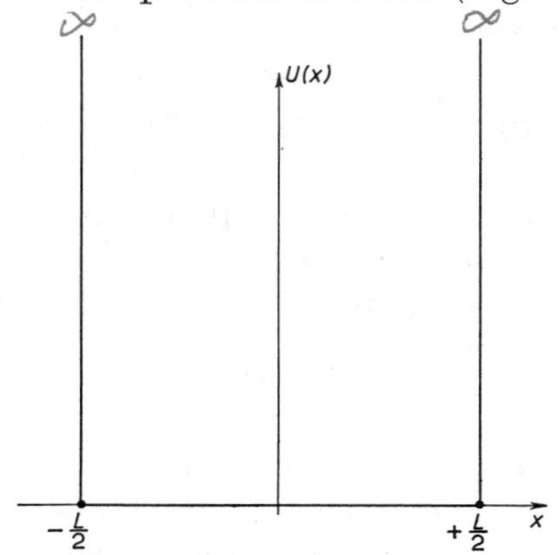

Fig. III.2. Infinitely deep square well.

The eigenvalue problem reduces to the search for a function ψ which vanishes at the points $+L/2$ and $-L/2$ and which satisfies the Schrödinger equation in the interval $(-L/2, +L/2)$

$$\psi'' + \varepsilon\psi = 0.$$

The general solution is a linear combination of $\sin kx$ and $\cos kx$ $(k = \sqrt{\varepsilon})$. Solutions satisfying simultaneously both boundary conditions exist for certain discrete values of ε, namely

$$\varepsilon_n = \frac{n^2\pi^2}{L^2} \qquad (n = 1, 2, \ldots, \infty) \tag{III.15}$$

(solutions for which $kL = n\pi$). To each of these eigenvalues there corresponds one and only one eigenfunction (no degeneracy), namely

if n is odd $$\psi_n = \cos \frac{n\pi}{L} x, \tag{III.16a}$$

if n is even $$\psi_n = \sin \frac{n\pi}{L} x. \tag{III.16b}$$

This very simple result calls for a certain number of remarks of a general nature.

In the first place, the present result differs profoundly from that obtained when dealing with a classical system. In the same potential, a classical particle can actually move with any energy as long as it is positive; it is a (periodic) back-and-forth motion between the two potential walls located at the ends of the interval $(-L/2, +L/2)$. In wave mechanics, the motion takes place only for certain discrete values of the energy [1]: *the energy is quantized.*

The second remark concerns the *parity* of the eigenfunctions [2]. They are even if n is odd [eq. (III.16a)], odd if n is even [eq. (III.16b)]. The fact that the eigenfunctions have a definite parity is due to the potential $U(x)$ being invariant under reflection through the origin:

$$U(x) = U(-x).$$

The question of parity will be examined in all generality in § 14.

The last remark concerns the *number of nodes* of the eigenfunctions. By definition, the nodes are the zeros of the eigenfunction (except for those which lie at the ends $x = L/2$ and $x = -L/2$). The number of nodes increases regularly with the eigenvalue of the energy, increasing by unity when going from one eigenvalue to the one lying immediately above: the eigenfunction of the ground state ψ_1 has no nodes, . . . that of the $(n-1)$th excited state ψ_n has $(n-1)$ nodes, etc. It is instructive to stress the analogy of this result with the one concerning the number of nodes of the stationary states of vibrating strings. Indeed, the analogy may be carried all the way since the two problems are mathematically identical.

6. Study of a Finite Square Well. Resonances

The results we have obtained with the potential step and the infinitely deep square potential well reappear in more complicated cases. As a new example, we consider the potential of Fig. III.3. Here the function $U(x)$ takes the form

$$U(x) = \begin{cases} U_1 & x > a \qquad \text{(region I)} \\ U_2 & a > x > b \qquad \text{(region II)} \\ U_3 & b > x \qquad \text{(region III)} \end{cases}$$

$$(U_2 < U_1 < U_3).$$

---

[1] The period of the classical motion is in fact equal to $h/\Delta E$, where ΔE is the distance between neighboring energy levels, in accord with the correspondence principle.

[2] A function $f(x)$ is even if $f(x) = f(-x)$, and odd if $f(x) = -f(-x)$.

The eigenvalue problem appears in a different manner depending on the value of ε compared to the constants U_1, U_2, U_3.

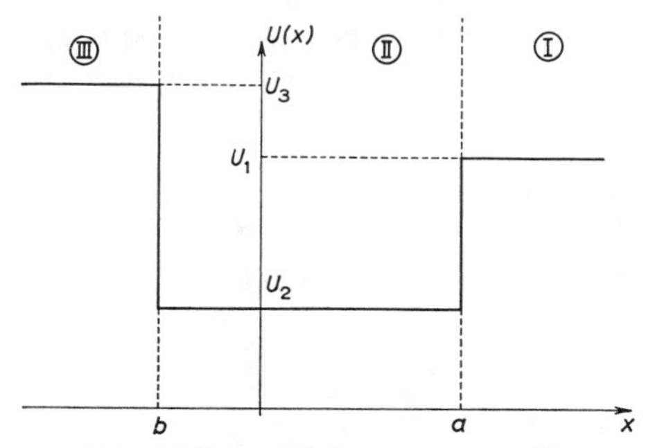

Fig. III.3. Finite square well.

a) $U_2 < \varepsilon < U_1$. DISCRETE SPECTRUM AND BOUND STATES

The general solution has an exponential behavior in the two external regions I and III and an oscillatory behavior in the interior region. In order to be acceptable as an eigenfunction, it must be exponentially decreasing in the two external regions. There exists one and only one exponentially decreasing solution in region I, and one and only one exponentially decreasing solution in region III; these two solutions join smoothly only for certain discrete values of ε. From this we conclude that the energy spectrum is certainly discrete and non-degenerate.

The function ψ is assumed real (cf. Ch. III, § 1) and assumes the following forms in each of the three regions

$$\psi = \begin{cases} A_1 \, \mathrm{e}^{-\varkappa_1 x} & x > a \\ A_2 \sin (k_2 x + \varphi) & a > x > b \\ A_3 \, \mathrm{e}^{\varkappa_3 x} & b > x. \end{cases} \tag{III.17}$$

If the phase φ is known, the two continuity conditions of the function determine the constants A_1, A_2, A_3 (to within a factor). As for φ, it must simultaneously satisfy both continuity conditions of the logarithmic derivative

$$k_2 \cot (k_2 a + \varphi) = -\varkappa_1 \qquad k_2 \cot (k_2 b + \varphi) = \varkappa_3 \tag{III.18}$$

which can also be written

$$\begin{aligned} \varphi &= -k_2 a - \tan^{-1} \frac{k_2}{\varkappa_1} + n\pi \quad (n \text{ positive integer}) \\ \varphi &= -k_2 b + \tan^{-1} \frac{k_2}{\varkappa_3} \end{aligned} \tag{III.19}$$

(φ must be determined to within a multiple of π) we have fixed this arbitrariness by forcing $k_2b+\varphi$ to lie within the interval $(-\pi/2, +\pi/2)$. This is possible if and only if the right-hand sides of these last two equations are equal. This equality can be achieved only for certain discrete values ε_n of ε, namely those which satisfy the equation

$$n\pi - k_2(a-b) = \tan^{-1}\frac{k_2}{\varkappa_1} + \tan^{-1}\frac{k_2}{\varkappa_3}. \tag{III.20}$$

Let us introduce the following notations:

$$K=\sqrt{U_1-U_2}, \quad L=b-a, \quad \cos\gamma=\sqrt{\frac{U_1-U_2}{U_3-U_2}} \qquad \left(0<\gamma<\frac{\pi}{2}\right)$$

and the new variable

$$\xi=\frac{k_2}{K}=\sqrt{\frac{\varepsilon-U_2}{U_1-U_2}}.$$

Equation (III.20) may also be written in the form of a condition on ξ:

$$n\pi-\xi KL=\sin^{-1}\xi+\sin^{-1}(\xi\cos\gamma).$$

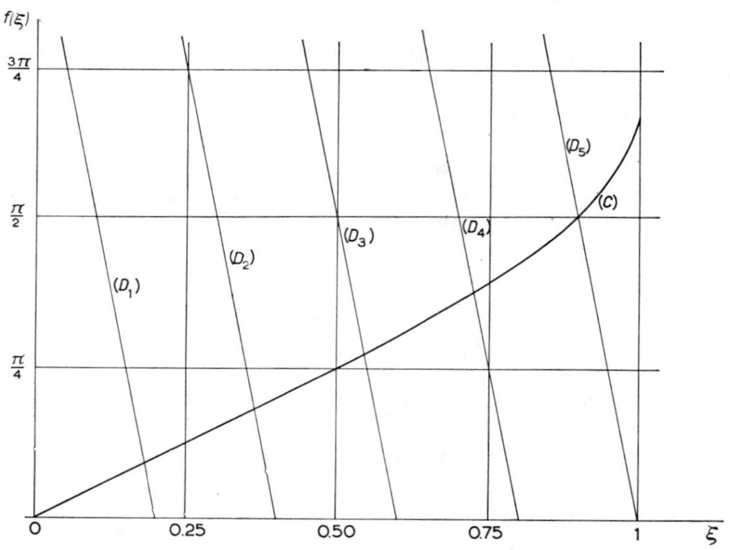

Fig. III.4. Graphical determination of the discrete eigenvalues: $\xi=[(\varepsilon-U_2)/(U_1-U_2)]^{\frac{1}{2}}$. The eigenvalues are the points of intersection of the curve (C) whose equation is:

$$f(\xi) = \sin^{-1}\xi + \sin^{-1}(\xi\cos\gamma)$$

with each of the straight lines (D_n) with equations: $y(\xi)=n\pi-\xi KL$ (we have chosen: $\gamma=\pi/3$, $KL/\pi=5$).

This last equation has been solved graphically in Fig. III.4. When ε increases from U_2 to U_1, ξ increases from 0 to 1, the right-hand side increases from 0 to $\pi-\gamma$ following curve (C) (which depends only upon parameter γ), and the left-hand side decreases from $n\pi$ to $n\pi-KL$ along the straight-line segment (D_n). In order that (C) and (D_n) intersect, it is necessary and sufficient that the integer n be sufficiently small:

$$KL \geqslant (n-1)\pi + \gamma.$$

If $KL<\gamma$, there are no eigenvalues; if $\gamma \leqslant KL \leqslant \pi+\gamma$ there is just one eigenvalue ε_1; if $\pi+\gamma \leqslant KL < 2\pi+\gamma$, there are two eigenvalues ε_1 and ε_2 $(\varepsilon_1<\varepsilon_2)$; and so on. It is easy to see that the eigenvalues are always arranged in increasing order of n. They form a *discrete* and *finite* sequence, from the ground state eigenvalue ε_1 to a maximum eigenvalue corresponding to the largest integer contained in the number $1+(KL-\gamma)/\pi$.

The quantum number n has a quite precise mathematical meaning. Inspection of eqs. (III.19) shows that the function $\sin(k_2x+\varphi)$ vanishes $(n-1)$ times as x crosses the interval (a, b). But, according to eq. (III.17), the zeros of that function are those of the function ψ. Consequently, the *number of nodes* of the eigenfunction corresponding to the nth eigenvalue ε_n is $n-1$.

To conclude, let us make a comparison with the classical situation as we did for the infinitely deep well. In the present case, there exists a further point of difference, besides the quantization of the energy: since the wave function has values different from zero in regions I and III, there is a non-vanishing probability of finding the particle in regions whose access is forbidden for the corresponding classical particle.

b) $U_1<\varepsilon<U_3$. CONTINUOUS, NON-DEGENERATE SPECTRUM. WAVE REFLECTION

The situation here is analogous to that of case (*a*) in the problem of the potential step. To each value of ε corresponds one and only one solution bounded everywhere: the one which is exponentially decreasing in region III. In the interval (U_1, U_3) the spectrum of eigenvalues is *continuous* and *non-degenerate*.

We seek a solution of the form

$$\psi = \begin{cases} e^{-ik_1x} + e^{i(k_1x+2\varphi_1)} & x>a \\ 2A\, e^{i\varphi_1} \sin(k_2x+\varphi_2) & a>x>b \\ 2B\, e^{i\varphi_1}\, e^{\varkappa_3 x} & b>x. \end{cases} \qquad \text{(III.21)}$$

As in the preceding problems, the continuity conditions of the logarithmic derivative determine the phases φ_1 and φ_2. One finds

$$\varphi_2 = -k_2 b + \tan^{-1}\frac{k_2}{\varkappa_3}, \quad \varphi_1 = -k_1 a - \frac{\pi}{2} + \tan^{-1}\left[\frac{k_1}{k_2}\tan\left(k_2 L + \tan^{-1}\frac{k_2}{\varkappa_3}\right)\right].$$

Then the value of A and B follow from the continuity of the function.

We assume below that $U_3 - \varepsilon \gg \varepsilon - U_2$; hence $k_2 \ll \varkappa_3$ and *a fortiori* $k_1 \ll \varkappa_3$. Everything behaves as if region III were occupied by an infinitely repulsive potential so that $B=0$. The interesting quantities to be considered here are φ_1 and A^2.

Let us take $a=0$, $b=-L$, and set

$$\eta = \frac{k_1}{K} = \sqrt{\xi^2 - 1}.$$

After a straightforward calculation, one obtains

$$\varphi_1 = \tan^{-1}\left(\frac{\eta}{\xi}\tan \xi KL\right) - \frac{\pi}{2}$$

$$A^2 = \frac{\eta^2}{\eta^2 + \cos^2 \xi KL}.$$

As the energy increases, the phase φ_1 grows in more or less regular fashion, while the quantity A^2 which measures the relative intensity of the wave in region II oscillates between the extreme values $\eta^2/(1+\eta^2)$ and 1, These oscillations are more pronounced the larger KL and the smaller η. Thus let us assume that

$$KL \gg \pi, \qquad \eta \ll 1.$$

Then A^2, considered as a function of η^2 (i.e. of the energy) exhibits a series of sharp peaks of width $4\eta/KL$ separated from each other by $2\pi/KL$. Figure III.5 illustrates this remarkable behavior of A^2 as well as that of φ_1.

We are faced with a typical wave phenomenon, the *resonance* phenomenon. Over certain restricted energy regions (of width $4\eta/KL$) the intensity of the wave in the interior region is of order unity: the resonance energies are those for which $\varphi_2 = (n+\frac{1}{2})\pi$, and hence where region II contains $(n+\frac{1}{2})$ "half wavelengths". Outside of these resonance regions, the intensity is comparatively very weak.

As in the problem of the potential step, we may compare the motion of a wave packet of the type given by eq. (III.9) with the motion of a classical particle in the same potential. Coming from $+\infty$ at constant velocity $v_1 = (\hbar K/m)\eta$, the classical particle suffers a sudden acceleration at $x = 0$, crosses the region II at velocity $v_2 = (\hbar K/m)\sqrt{1+\eta^2}$,

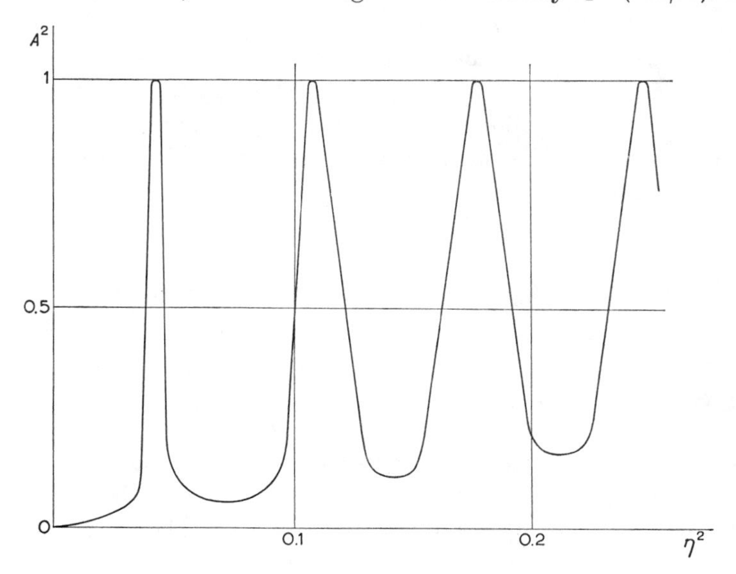

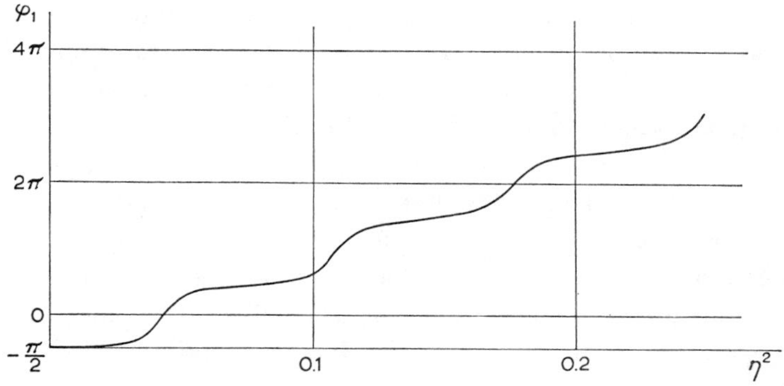

Fig. III.5. Resonances of reflection. Variation of A^2 and of φ_1 [cf. eq. (III.21)] as function of the energy. The curves correspond to $KL = (b - a)\sqrt{U_1 - U_2} = 100$. The energy scale is expressed in units of the variable

$$\eta^2 = (\varepsilon - U_1)/(U_1 - U_2).$$

bounces back at $x = -L$; it starts back in the opposite direction at velocity v_2 in region II, then at velocity v_1 in region I. The time spent in region II is $\tau_{\text{cl.}} = 2L/v_2$. The center of the wave packet carries out an analogous motion, at least in the region of very large x, where the shape of the wave packet remains practically unchanged so that the motion of a center remains meaningful. Everything behaves as if it had carried out the classical motion except for the following slight difference: the "time spent in region II" is no longer $\tau_{\text{cl.}}$ but $\tau=(2/v_1)\,\mathrm{d}\varphi_1/\mathrm{d}k_1=(2/v_1K)\,\mathrm{d}\varphi_1/\mathrm{d}\eta$. We shall not dwell upon the details of this study which is in all respects similar to the one of § 3. The behavior of the various quantities entering the discussion is summarized in the following table:

Incident energy	φ_1	$\dfrac{\mathrm{d}\varphi_1}{\mathrm{d}k_1}$	$\tau/\tau_{\text{cl.}}$	A^2
At resonance	$n\pi$	L	$1/\eta$	1
Half-way between two resonances .	$(n+\frac{1}{2})\pi$	$L\eta^2$	η	η^2

Between the resonances, A^2 remains very small, and the time of passage τ in region II is very short compared to $\tau_{\text{cl.}}$: the wave packet practically does not penetrate into region II. The wave is almost entirely reflected upon arriving at the point $x=0$. One meets an identical situation in optics where a sudden and appreciable change of index of refraction almost always entails total reflection. At resonance on the other hand, $A^2=1$, the wave penetrates entirely into region II, and remains concentrated there during a relatively long time, in fact much longer than $\tau_{\text{cl.}}$. In conformity with condition (III.12), the picture obtained here is applicable only for wave trains whose extent is considerably greater than that of region II itself ($\mathrm{d}\varphi_1/\mathrm{d}k_1=L$ at resonance); consequently, the front of the wave packet reaches the point of reflection $x=-L$ long before the wave has finished crossing the potential step at $x=0$. The effect is typical of waves, namely one of interference between incident and reflected wave in region II.

c) $\varepsilon>U_3$. Continuous and Degenerate Spectrum. Reflection and Transmission of Waves

The situation here is analogous to that of case (b) in the problem of the potential step. To each value of ε correspond two linearly independent eigenfunctions: in the interval (U_3, ∞) the eigenvalue spectrum is continuous and degenerate of order 2.

As in the problem of the potential step we build the eigenfunction of the form

$$\chi = \begin{cases} e^{-ik_1 x} + R\, e^{ik_1 x} & x > a \\ P\, e^{-ik_2 x} + Q\, e^{ik_2 x} & a > x > b \\ S\, e^{-ik_3 x} & b > x. \end{cases} \qquad \text{(III.22)}$$

The continuity conditions at points a and b give the values of R, Q, P, and S. Without entering into calculation details, we simply list the results concerning the quantities R and S. We use the following notation and conventions:

$$a = 0, \qquad b = -L, \qquad K = \sqrt{U_1 - U_2},$$

$$\xi = \frac{k_2}{K}, \qquad \eta = \frac{k_1}{K}, \qquad \zeta = \frac{k_3}{K}.$$

One obtains

$$R = \frac{\xi(\eta - \zeta) \cos \xi KL + i(\xi^2 - \eta\zeta) \sin \xi KL}{\xi(\eta + \zeta) \cos \xi KL - i(\xi^2 + \eta\zeta) \sin \xi KL}$$

$$S = e^{-i\zeta KL} \frac{2\eta\xi}{\xi(\eta + \zeta) \cos \xi KL - i(\xi^2 + \eta\xi) \sin \xi KL}$$

With the help of these expressions, one is able to compare the motion of a wave packet formed with waves of the type (III.22) and neighboring energies, with the motion of a classical particle of the same energy in the same potential.

The incident wave packet [made up of waves $\exp(-ik_1 x)$ of region I] moves in region I with constant velocity $v_1 = \hbar k_1/m$ and enters region II; once the collision has taken place, it splits into a reflected wave packet [built of waves $R \exp(ik_1 x)$ of region I] moving with velocity v_1 toward $+\infty$, and a transmitted wave packet [built of waves $S \exp(-ik_3 x)$ of region III] travelling with velocity v_3 toward $-\infty$. In contrast to the classical particle, *the wave is in general only partially transmitted*, and we can define a transmission coefficient

$$T = \frac{k_3}{k_1} |S|^2 = \frac{4\eta\zeta\xi^2}{\xi^2(\eta + \zeta)^2 \cos^2 \xi KL + (\xi^2 + \eta\zeta)^2 \sin^2 \xi KL} \qquad \text{(III.23)}$$

as we did in the case of the potential step.

Here too, we note that the transmission coefficient at a given energy is independent of the direction of travel (η and ζ occur symmetrically in T).

One further verifies the equality

$$|R|^2 + \frac{k_3}{k_1} |S|^2 = 1. \qquad \text{(III.24)}$$

The ratio of reflected to transmitted wave amplitudes varies with the energy, and we can observe *resonance phenomena* of the same kind as those of case (*b*). The latter are particularly sharp when

$$KL \gg \pi \qquad \zeta < \eta \ll 1. \qquad \text{(thus } \xi = 1\text{)}.$$

In that case, inspection of equation (III.23) shows that the transmission coefficient considered as a function of η^2 (i.e. of the energy) remains very small, of the order of $4\eta\zeta$, almost everywhere, but exhibits a series of sharp maxima equal to $4\eta\zeta/(\eta+\zeta)^2$; the width of these peaks is about $4(\eta+\zeta)/KL$. Their positions correspond to the energies for which region II contains an integral number n of "half wavelengths", namely $\xi KL = n\pi$ (their separation is approximately $2\pi/KL$).

One may pursue this study by considering the phases of the amplitudes R and S, and defining a "transit time" of the transmitted wave, or a "reflection time" of the reflected wave; one then compares these times with the time necessary for the classical particle to cross region II. One is led to the following qualitative picture: at resonance, the wave remains concentrated in region II during a time which is very much greater [$(\eta+\zeta)^{-1}$ times greater] than the classical time before splitting up into a transmitted and a reflected wave; off resonance, the wave practically does not penetrate into region II, and is almost totally reflected at the boundary of regions I and II, in near-instantaneous fashion (cf. Problem III.1).

7. Penetration of a Square Potential Barrier. The "Tunnel" Effect

As a last example, we examine the problem of the penetration of a square potential barrier (Fig. III.6).

$$U(x) = \begin{cases} 0 & x > L \qquad \text{(region I)} \\ U_0(>0) & 0 < x < L \qquad \text{(region II)} \\ 0 & x < 0. \qquad \text{(region III)} \end{cases}$$

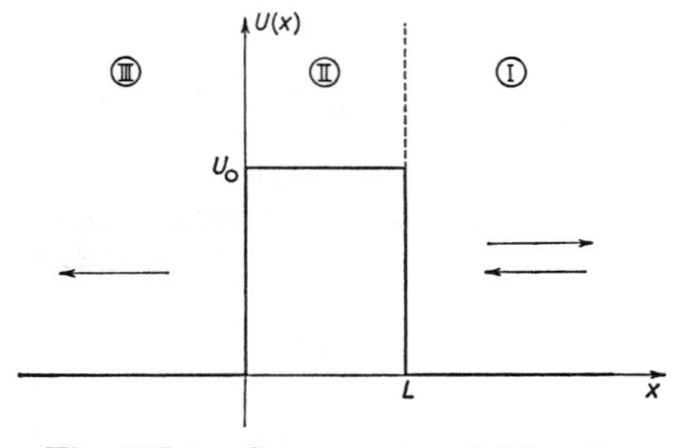

Fig. III.6. Square potential barrier.

In this case, all positive values of ε are doubly degenerate eigenvalues. Two cases occur according to whether ε is larger or smaller than U_0. In both cases we construct the solution representing a wave traveling in the negative direction in region III, i.e. the solution of the form

$$\begin{array}{ll} e^{-i\sqrt{\varepsilon}x} + R\, e^{+i\sqrt{\varepsilon}x} & \text{for} \quad x > L \\ S\, e^{-i\sqrt{\varepsilon}x} & \text{for} \quad x < 0. \end{array}$$

Its behavior in region II is

$$\begin{array}{llll} \text{exponential:} & A\, e^{\varkappa x} + B\, e^{-\varkappa x}, & \text{if} \quad \varepsilon < U_0 & (\varkappa = \sqrt{U_0 - \varepsilon}) \\ \text{sinusoidal:} & C\, e^{ikx} + D\, e^{-ikx}, & \text{if} \quad \varepsilon > U_0 & (k = \sqrt{\varepsilon - U_0}). \end{array}$$

Let us merely give the result of the calculation of the transmission coefficient (Fig. III.7):

$$T = |S|^2 = \begin{cases} \dfrac{4\varepsilon(\varepsilon - U_0)}{4\varepsilon(\varepsilon - U_0) + U_0^2 \sin^2 kL}, & \text{if} \quad \varepsilon > U_0 \\[2ex] \dfrac{4\varepsilon(U_0 - \varepsilon)}{4\varepsilon(U_0 - \varepsilon) + U_0^2 \sinh^2 \varkappa L}, & \text{if} \quad \varepsilon < U_0. \end{cases}$$

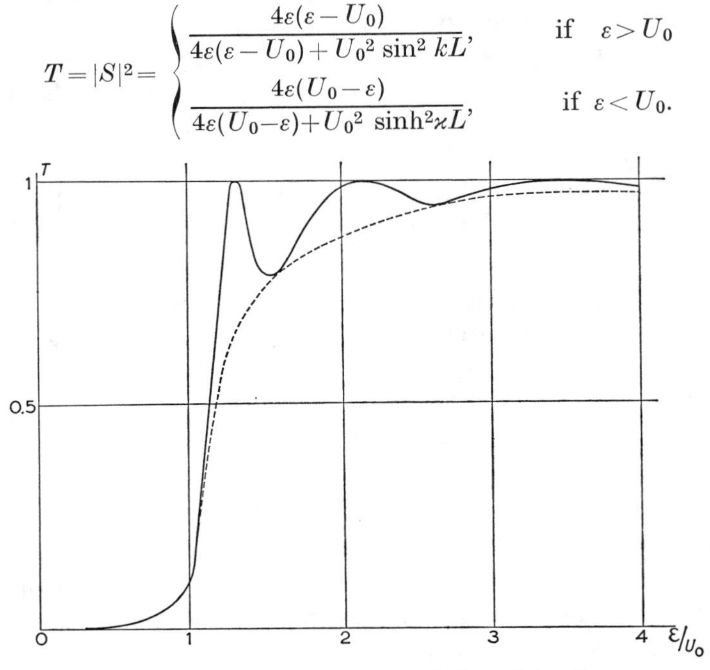

Fig. III.7. Variation of the transmission coefficient through a square potential barrier as a function of the energy (barrier of Fig. III.6). We have used:

$$U_0 L^2 = 40.$$

As in the preceding sections, one can compare the motion of a wave packet of the type (III.9) built of wave functions of that form, with the motion of a classical particle of the same energy coming from $+\infty$.

The most spectacular difference occurs when $\varepsilon < U_0$. The classical particle rebounds from the barrier without being able to penetrate it. The wave packet splits into a reflected packet, and a transmitted packet whose intensity never vanishes: as ε increases from 0 to U_0, the transmission coefficient increases regularly from 0 to the value $[1+U_0L^2/4]^{-1}$. This effect is known as the *tunnel effect* and plays an important role in the theory of alpha radioactivity. It is more important the lower and thinner the barrier.

When $\varepsilon > U_0$, the classical particle is slowed down in region II but traverses it nevertheless and continues its journey in region III toward $-\infty$. The wave packet, on the other hand, is almost always partially reflected. Complete transmission ($T=1$) occurs only for certain values of the energy, namely those for which kL is a multiple of π. As the energy increases, the transmission coefficient oscillates between this maximum value and a minimum value of the order of $4\varepsilon(\varepsilon-U_0)/(2\varepsilon-U_0)^2$. The effect is particularly marked when the barrier is very high or very thick, and when the kinetic energy $\varepsilon - U_0$ in region II is small. Note the similarity with the resonance phenomena discussed in the preceding sections (cf. Problem III.2).

II. GENERAL PROPERTIES OF THE ONE-DIMENSIONAL SCHRÖDINGER EQUATION

8. Property of the Wronskian

Let us return to the equation

$$y'' + [\varepsilon - U(x)]y = 0. \tag{III.25}$$

We propose to show some very general properties of this eigenvalue equation. In all that follows the only restrictions imposed upon the real function $U(x)$ are: to be bounded from below, and to be piecewise continuous over the entire interval $(-\infty, +\infty)$.

A large number of these properties derive directly from an important theorem concerning the Wronskian of two solutions, which we shall henceforth call the *Wronskian Theorem*.

By definition, the Wronskian of two functions y_1, y_2 is

$$W(y_1, y_2) \equiv y_1y_2' - y_2y_1'.$$

It is a bilinear expression in y_1 and y_2, antisymmetric in the exchange of these two functions. If it vanishes at a point of the x axis, the functions y_1 and y_2 have the same logarithmic derivative at that point; if it vanishes over the entire interval $(-\infty, +\infty)$, the two functions are multiples of each other.

WRONSKIAN THEOREM. – *If z_1 and z_2 are solutions of the equations*

$$z_1'' + F_1(x) z_1 = 0 \tag{III.25'}$$

$$z_2'' + F_2(x) z_2 = 0 \tag{III.25''}$$

respectively, in an interval (a, b) where the functions $F_1(x)$ and $F_2(x)$ are piecewise continuous, the overall variation of their Wronskian in this interval is given by the expression

$$W(z_1, z_2)\Big|_a^b = \int_a^b [F_1(x) - F_2(x)] z_1 z_2 \, \mathrm{d}x. \tag{III.26}$$

To prove this theorem, we multiply eq. (III.25′) by z_2, eq. (III.25″) by z_1, and subtract term by term. We obtain

$$[z_2 z_1'' - z_1 z_2''] + (F_1 - F_2) z_1 z_2 = 0.$$

The first term is, except for a sign, the derivative with respect to x of the Wronskian $W(z_1, z_2)$. Upon integrating term by term over the interval (a, b), one obtains relation (III.26), Q.E.D.

This theorem if of particular interest to us when the equations (III.25′) and (III.25″) are equations of the type (III.25) with the same potential $U(x)$. One then obtains the three following important corollaries:

Corollary I. – If y_1 and y_2 are solutions of the equation (III.25) corresponding to the values ε_1, ε_2 of the constant ε, respectively, one has for any pair of values a, b of the variable x, located in the interval where these solutions are defined:

$$W(y_1, y_2)\Big|_a^b = (\varepsilon_1 - \varepsilon_2) \int_a^b y_1 y_2 \, \mathrm{d}x. \tag{III.27}$$

Corollary II. – If y and z are two solutions of eq. (III.25) corresponding to the same value of ε, their Wronskian is independent of x:

$$W(y, z) = \text{constant}.$$

Corollary III. – Let $Y(x;\varepsilon)$ be the solution of eq. (III.25) whose logarithmic derivative (with respect to x) has a fixed value f_a at point a of the x axis, and let $f(x;\varepsilon)$ be its logarithmic derivative at a point x of the axis. When considered as a function of ε, $f(x;\varepsilon)$ is a *monotonic* function of that variable, increasing if $x<a$, decreasing if $x>a$, whose derivative is

$$\frac{\partial f}{\partial \varepsilon} = -\frac{1}{Y^2(x;\varepsilon)}\int_a^x Y^2(\xi;\varepsilon)\,\mathrm{d}\xi. \tag{III.28}$$

[Considered as a function of ε, $f(x;\varepsilon)$ has a behavior analogous to the tangent or the cotangent, with a vertical asymptote at each point where $Y(x)$ vanishes.]

Corollaries I and II are direct applications of the Wronskian Theorem. The proof of Corollary III is as follows. ε being fixed, a solution of eq. (III.25) is completely determined if one gives its value and that of its derivative at a given point $x=a$ of the x axis. Let $Y(x;\varepsilon)$ be that particular solution

$$Y(a;\varepsilon)=y_a, \qquad Y'(a;\varepsilon)=y_a'.$$

If one varies ε while maintaining these boundary conditions constant, $Y(x;\varepsilon)$ is a certain continuous function of ε (and of x). To two infinitely close values ε, $\varepsilon+\delta\varepsilon$ correspond two expressions Y, $Y+\delta Y$ lying infinitely close together. Apply Corollary I to the latter in the interval (a, b):

$$W(Y, Y+\delta Y)\Big|_a^b = -\delta\varepsilon\int_a^b Y^2\,\mathrm{d}x.$$

For $x=a$, $W(Y, Y+\delta Y)=0$ by hypothesis. For all other values of x,

$$W(Y, Y+\delta Y)= W(Y, \delta Y)=Y\,\delta Y'-Y'\,\delta Y=Y^2\,\delta\left(\frac{Y'}{Y}\right)=Y^2\,\delta f.$$

We have introduced $f=Y'/Y$. The logarithmic derivative f is, like Y, a continuous function of ε and of x. We have

$$-Y^2\,\delta f\Big|_{x=b}=\delta\varepsilon\int_a^b Y^2\,\mathrm{d}x.$$

In other words

$$\left.\frac{\partial f}{\partial \varepsilon}\right|_{x=b} = -\frac{1}{Y^2(b)}\int_a^b Y^2(x)\,\mathrm{d}x. \qquad \text{Q.E.D.}$$

The properties of the solutions of the Schrödinger equation contained in these three corollaries derive their interest from the fact that they are independent of the particular shape of the potential $U(x)$.

9. Asymptotic Behavior of the Solutions

The asymptotic form of the general solution of eq. (III.25) at the boundaries of the interval $(-\infty, +\infty)$ is very different depending on the sign of $\varepsilon - U$ as x approaches these limits. We look for the asymptotic form in the limit $x = +\infty$. Similar conclusions are applicable at the limit $x = -\infty$. Suppose thus that $\varepsilon - U(x)$ keeps the same sign when x exceeds a certain value x_0. Two cases can then occur:

First Case: $\varepsilon > U(x)$ *when* $x > x_0$.

We shall assume — as is always the case in practice — that $U(x)$ tends monotonically toward a finite limit U_+ as $x \to \infty$. Let us put

$$k = \sqrt{\varepsilon - U_+}.$$

We shall show that when $x \to \infty$:

(a) the real solutions of equation (III.25) remain bounded and oscillate indefinitely between two opposite values;

(b) if, furthermore, $U(x)$ tends toward U_+ faster than $1/x$,

$$y \underset{x \to \infty}{\sim} A_+ \sin(kx + \varphi_+) \tag{III.29}$$

where A_+ and φ_+ are two suitable real constants.

For this purpose, one notes that equation (III.25) "tends asymptotically" toward the equation $z'' + k^2 z = 0$ whose general solution is $A \sin(kx + \varphi)$ and depends on two arbitrary constants A and φ. In order to determine the asymptotic form of y, let us introduce (method of variation of constants) the functions $A(x)$ and $\varphi(x)$ defined by

$$y = A \sin(kx + \varphi), \qquad y' = Ak \cos(kx + \varphi). \tag{III.30}$$

Equation (III.25) is equivalent to the two first-order differential equations

$$\frac{A'}{A} = \frac{U - U_+}{2k} \sin 2(kx + \varphi), \qquad \varphi' = -\frac{U - U_+}{k} \sin^2(kx + \varphi).$$

Upon integration they yield

$$A(x) = A(x_0) \exp \left\{ \int_{x_0}^{x} \frac{U(\xi) - U_+}{2k} \sin 2[k\xi + \varphi(\xi)] \, d\xi \right\} \qquad \text{(III.31)}$$

$$\varphi(x) = \varphi(x_0) - \int_{x_0}^{x} \frac{U(\xi) - U_+}{k} \sin^2 [k\xi + \varphi(\xi)] \, d\xi. \qquad \text{(III.32)}$$

The integral of the right-hand side of (III.31) certainly converges: $A(x)$ approaches a finite limit A_+ as $x \to \infty$; furthermore, since $\varphi' \to 0$, the function $\sin (kx + \varphi)$ in expression (III.30) for y oscillates with a period which tends asymptotically to $2\pi/k$. This proves the first stated result. If in addition $U(x)$ approaches U_+ more rapidly than $1/x$, the integral of the right-hand side of eq. (III.32) also converges: A and φ both approach finite limits A_+ and φ_+, respectively, whence the asymptotic form (III.29). Q.E.D.

Second Case: $\varepsilon < U(x)$ *when* $x > x_0$.

The results we shall obtain are independent of the behavior of $U(x)$ at infinity. We merely assume that

$$U(x) - \varepsilon \geqslant M^2 > 0 \quad \text{when} \quad x > x_0.$$

This case corresponds to the exponential solutions in square well potential problems.

We shall show that when $x \to \infty$:

(a) there exists *one* particular solution (defined to within a constant) of equation (III.25) which approaches 0 at least as rapidly as $\exp(-Mx)$;

(b) all other solutions tend toward ∞ at least as rapidly as $\exp(Mx)$.

Since the solutions are defined to within a constant, let us fix this constant by the condition $y(x_0) = 1$, and let us look at the behavior of the solutions $y(x)$ satisfying this normalization condition. Some of them are represented in the graph of Fig. III.8.

If we designate by $Y(x)$ and $Z(x)$ the particular solutions defined by the boundary conditions

$$Y(x_0) = 1, \qquad Y'(x_0) = 0 \qquad \text{and} \qquad Z(x_0) = 0, \qquad Z'(x_0) = 1$$

respectively, the solutions considered above are of the form

$$y(x) = Y(x) + fZ(x). \qquad \text{(III.33)}$$

The parameter $f = y'(x_0)$ may take on all values between $-\infty$ and $+\infty$.

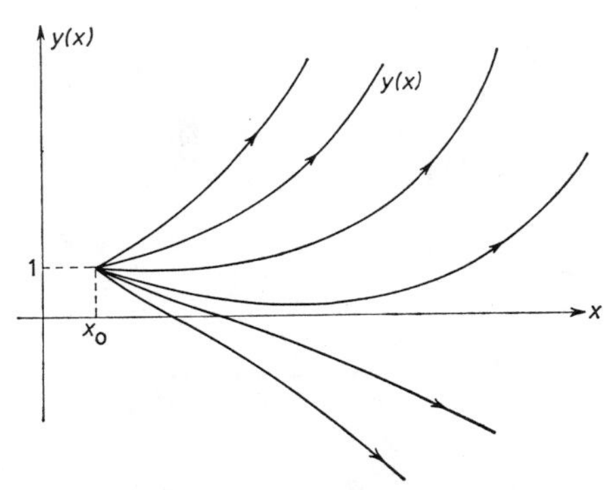

Fig. III.8. Diagram representing several solutions of eq. (III.25) satisfying the normalization condition $y(x_0) = 1$ for the case where $U(x) - \varepsilon \geqslant M^2 > 0$ for $x > x_0$.

$Y(x)$ and $Z(x)$ remain positive over the entire interval (x_0, ∞) and tend toward infinity at least as fast as $\exp(Mx)$. Indeed, like all solutions of eq. (III.25), these functions keep the same sign as their second derivatives. From this one derives by simple inspection of their boundary conditions that these functions necessarily increase indefinitely since they are concave upward. In order to evaluate the rapidity of their growth, note that $Y'' \geqslant M^2 Y$ and $Z'' \geqslant M^2 Z$ and compare these functions with the solutions of the differential equation $u'' - M^2 u = 0$ having the same initial conditions at x_0, namely $\cosh M(x - x_0)$ and $\sinh M(x - x_0)$, respectively. Y and Z remain everywhere superior (or equal) to these comparison functions.

By applying the Wronskian Theorem one has in fact

$$W[Y, \cosh M(x - x_0)] \leqslant 0,$$

therefore

$$Y'/Y \geqslant M \tanh M(x - x_0)$$

and, by integration

$$Y \geqslant \cosh M(x - x_0);$$

one shows similarly that $Z \geqslant \sinh M(x - x_0)$. Note that

$$Y' \geqslant MY \tanh M(x - x_0),$$

thus at infinity $Y' \geqslant MY$; similarly, at infinity $Z' \geqslant MZ$.

On the other hand (Corollary II)

$$Z'Y - Y'Z = 1 \quad \text{for any } x. \qquad \text{(III.34)}$$

Let us introduce the functions

$$u(x) \equiv \frac{Y}{Z}, \qquad v(x) \equiv \frac{Y'}{Z'}.$$

From the property (III.34) and from the fact that Y and Z are solutions of equation (III.25), one obtains:

$$\begin{aligned} u - v &\equiv \frac{Y}{Z} - \frac{Y'}{Z'} = \frac{1}{ZZ'} \\ u' &= \frac{Y'Z - YZ'}{Z^2} = -\frac{1}{Z^2} \\ v' &= \frac{Y''Z' - Y'Z''}{Z'^2} = \frac{U - \varepsilon}{Z'^2}. \end{aligned} \qquad \text{(III.35)}$$

In the interval (x_0, ∞), u is a decreasing function, v is an increasing function and their difference vanishes at infinity. They therefore have a common (positive) limit C as $x \to \infty$, and we have

$$v(x) < C < u(x),$$

an inequality which can be rewritten with the help of eq. (III.35)

$$-\frac{1}{ZZ'} < v - C < 0 < u - C < \frac{1}{ZZ'}. \qquad \text{(III.36)}$$

The particular solution

$$\hat{y}(x) \equiv Y - CZ = [u(x) - C]\, Z(x)$$

and its derivative

$$\hat{y}'(x) \equiv Y' - CZ' = [v(x) - C]\, Z'(x)$$

always satisfy the inequalities

$$-\frac{1}{Z} < \hat{y}' < 0 < \hat{y} < \frac{1}{Z'}.$$

Thus, $\hat{y}$ is positive everywhere, and tends toward 0 at least as fast as $1/Z'$, and *a fortiori* at least as fast as $\exp(-Mx)$. Similarly, $\hat{y}'$, a negative function everywhere, tends toward 0 at least as rapidly as $\exp(-Mx)$. The solution $\hat{y}$ is the solution vanishing at infinity which we were seeking.

Obviously, there exist no others, because if $f \neq -C$, the solution $y(x)$ can also be written

$$y = \hat{y} + (f + C)Z,$$

and its asymptotic behavior is the same as that of Z, to within the non-vanishing factor $f+C$. Q.E.D.

10. Nature of the Eigenvalue Spectrum

Let U_+ and U_- be the respective limits of $U(x)$ when x tends toward $+\infty$ and $-\infty$. (The following conclusions hold when either one of these limits is replaced by $+\infty$.) U_+ and U_- divide the domain of variation of ε into three regions in which the eigenvalue spectra have different properties. To be specific, let us assume that $U_+ < U_-$.

When $\varepsilon > U_-$, $\varepsilon - U(x)$ remains positive at the two ends of the interval $(-\infty, +\infty)$; since any solution of eq. (III.25) which remains bounded as $x \to \pm\infty$, is acceptable as eigenfunction, ε is a degenerate eigenvalue of order 2. The eigenvalue spectrum is *continuous* and *degenerate*. On the other hand, in the two asymptotic regions the eigenfunctions oscillate indefinitely between two finite and opposite limits: they represent *unbound states*.

When $U_- > \varepsilon > U_+$, since $\varepsilon - U(x)$ is negative in the limit $x = -\infty$, only one solution remains bounded (exponentially decreasing) in this asymptotic region. This solution remains bounded and oscillates indefinitely in the other asymptotic region since $\varepsilon - U(x)$ is positive there; it is therefore an acceptable eigenfunction and represents an *unbound state*. The eigenvalue spectrum is *continuous* and *non-degenerate*.

When $U_+ > \varepsilon$, $\varepsilon - U(x)$ is negative in both asymptotic regions. The bounded solution, if it exists, vanishes (exponentially) at the two extremities of the interval and represents a *bound state*. However, it exists only for discrete values of ε. Indeed, let $\hat{y}_-$ be the solution vanishing at the limit $-\infty$, $\hat{y}_+$ the solution vanishing at the limit $+\infty$, and f_- and f_+ their respective logarithmic derivatives at a definite point of the x axis. ε is an eigenvalue if, and only if, $\hat{y}_-$ and $\hat{y}_+$ are equal (to within a constant factor), i.e. if $f_- = f_+$. Considered as functions of ε, f_- is a monotonically decreasing function, f_+ a monotonically increasing function (Corollary III [1]). Therefore, the values of ε for which these two functions are equal, are necessarily

[1] In fact, we are dealing with an extension of Corollary III to the case where $a = +\infty$ or $-\infty$. It is easy to see that this corollary remains valid with a slight change in the definition of the function $Y(x;\varepsilon)$. The latter is the solution of eq. (III.5) which vanishes (and whose derivative vanishes as well) at the point a $(= \pm\infty)$. Note that f_- and f_+ can exhibit vertical asymptotes for some values of ε.

isolated from each other. The spectrum is *discrete* and *non-degenerate.*

The *number of eigenvalues of the discrete spectrum* depends upon the particular function $U(x)$ considered. It may go from 0 to ∞. It is certainly zero if $U(x)$ everywhere exceeds the smaller of its two asymptotic values, U_+. If not, one can show – and we state it here without proof – that it is of the order of magnitude of

$$\frac{1}{\pi}\int\sqrt{U_+-U(x)}\,\mathrm{d}x=\frac{1}{\pi}\int\frac{\sqrt{2m[V_+-V(x)]}}{\hbar}\,\mathrm{d}x,$$

this integral being extended over the entire region of the x axis where $U(x)<U_+$. In particular, if this integral diverges, the eigenvalues are infinite in number.

11. Unbound States: Reflection and Transmission of Waves

With the eigenfunctions of the continuous, doubly-degenerate part of the spectrum, one may construct wave packets which are partly transmitted and partly reflected by the potential $U(x)$. To simplify matters we assume that $U(x)$ reaches its two asymptotic limits, U_+ and U_- more rapidly than $1/|x|$; we can then use the asymptotic expressions given by eq. (III.29).

To form the wave packets under consideration, two types of eigenfunctions may be used. The functions of the first type are the functions $u(x)$ whose behavior in the two asymptotic zones is

$$\begin{aligned} u &\underset{x\to-\infty}{\sim} \mathrm{e}^{\mathrm{i}k_-x}+R_u\,\mathrm{e}^{-\mathrm{i}k_-x} \\ &\underset{x\to+\infty}{\sim} S_u\,\mathrm{e}^{\mathrm{i}k_+x} \end{aligned} \qquad (k_\pm=\sqrt{\varepsilon-U_\pm}).$$

The wave packet one obtains by superposing functions of neighboring energies in a way analogous to that given by eq. (III.9) represents an incident wave exp $(\mathrm{i}k_-x)$ traveling from $-\infty$ in the positive direction; it enters the potential $U(x)$ and splits up into a reflected wave R_u exp $(-\mathrm{i}k_-x)$ moving in the opposite sense, and a transmitted wave S_u exp $(\mathrm{i}k_+x)$ propagating toward $+\infty$ (Fig. III.9*a*).

The functions of the second type are the functions $v(x)$ whose asymptotic behavior is

$$\begin{aligned} v &\underset{x\to-\infty}{\sim} S_v\,\mathrm{e}^{-\mathrm{i}k_-x} \\ &\underset{x\to+\infty}{\sim} \mathrm{e}^{-\mathrm{i}k_+x}+R_v\,\mathrm{e}^{\mathrm{i}k_+x} \end{aligned}$$

and which can represent an analogous wave packet traveling in the opposite sense (Fig. III. 9*b*).

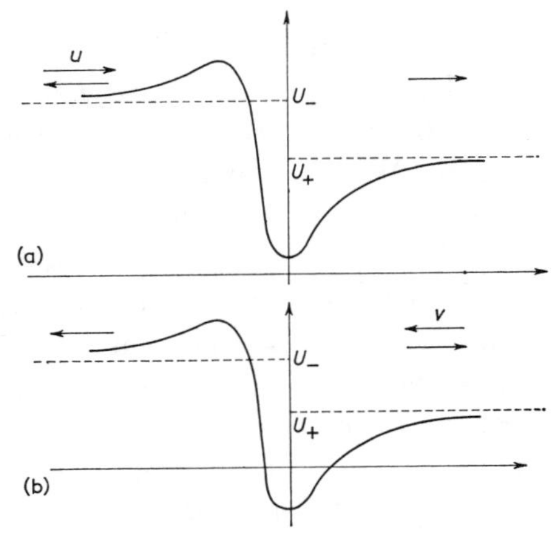

Fig. III.9. Reflection and transmission of waves by a potential:
(*a*) solution of type u: wave coming from negative x;
(*b*) solution of type v: wave coming from positive x.

The functions u and v and their complex conjugates u^* and v^* are solutions of the same Schrödinger equation. The Wronskian of any two such functions is independent of x (Corollary II); in particular it takes on the same value in the two asymptotic regions; equating these two values we obtain a relation between the coefficients R_u, S_u, R_v, and S_v, or their complex conjugates. Six such relations can be formed with the four functions u, v, u^* and v^*. They are very general relations, which hold whatever the shape of the potential function $U(x)$.

We obtain [1])

$$\frac{\mathrm{i}}{2} W(u, u^*) = k_-[1-|R_u|^2] = k_+|S_u|^2 \tag{III.37}$$

$$\frac{\mathrm{i}}{2} W(v, v^*) = -k_-|S_v|^2 = -k_+[1-|R_v|^2] \tag{III.38}$$

$$\frac{\mathrm{i}}{2} W(u, v) = k_- S_v = k_+ S_u \tag{III.39}$$

$$\frac{\mathrm{i}}{2} W(u, v^*) = -k_- R_u S_v^* = k_+ S_u R_v^* \tag{III.40}$$

---

[1]) This calculation is quite simple if one makes use of the fact that the Wronskian of two functions is bilinear and antisymmetric, and if one notes that $W(\mathrm{e}^{\mathrm{i}kx}, \mathrm{e}^{\mathrm{i}kx}) = W(\mathrm{e}^{-\mathrm{i}kx}, \mathrm{e}^{-\mathrm{i}kx}) = 0$ and that $W(\mathrm{e}^{-\mathrm{i}kx}, \mathrm{e}^{+\mathrm{i}kx}) = 2\mathrm{i}k$.

and two relations deduced from the two last ones by complex conjugation.

The equations (III.37) and (III.38) are called the *relations of conservation of flux*; we have already verified them in special cases [eqs. (III.13) and (III.24)]. The origin of this name comes from the following interpretation of the wave function ψ of an unbound state in the asymptotic region. We shall examine the justification of that interpretation when we shall discuss collision problems (Ch. X). Let $A \exp(ikx) + B \exp(-ikx)$ be the expression of the wave function ψ in one of the asymptotic regions, $-\infty$ for instance. If one builds with this wave function a wave packet analogous to the one given by expression (III.9), it is made up of two terms; the first one formed with $A \exp(ikx)$ is a term of relative intensity $|A|^2$ which is traveling in the direction of increasing x with velocity $\hbar k/m$; the other, formed with $B \exp(-ikx)$, has an intensity $|B|^2$ and travels at the same velocity but in the opposite direction. Sufficiently far in the asymptotic region these two terms do not interfere, the first leaving that region long before the other reaches it. However, instead of forming with this wave function a packet representing the motion of a particle, one can equally well consider – and this is precisely the interpretation mentioned above – that the wave itself represents a statistical ensemble of a very large number of particles. This wave is composed of a beam of density $|A|^2$ in the asymptotic region under consideration, traveling in the direction of increasing x with velocity $\hbar k/m$, and of a beam of density $|B|^2$ moving at the same speed but in the opposite direction. The total flux (counted positively in the direction of increasing x) of particles passing a given point is the difference between the flux $(\hbar k/m)|A|^2$ of particles traveling in the positive sense, and the flux $(\hbar k/m)|B|^2$ of particles traveling in the negative sense. This flux is equal, to within a constant, to the Wronskian $W(\psi, \psi^*)$:

$$\frac{\hbar k}{m}[|A|^2 - |B|^2] = \frac{i}{2}\frac{\hbar k}{m} W(\psi, \psi^*).$$

The equality of the Wronskian $W(\psi, \psi^*)$ at both ends of the interval $(-\infty, +\infty)$ means that the number of particles entering the interaction region per unit time is equal to the number which leave it.

According to this interpretation, one or the other of eqs. (III.37) and (III.38) may just as well be written:

$$\text{incident flux} - \text{reflected flux} = \text{transmitted flux}.$$

Following this same interpretation, the *transmission coefficient* T may be defined as follows:

$$T \equiv \frac{\text{transmitted flux}}{\text{incident flux}}.$$

We have in particular

$$T_u = \frac{k_+}{k_-}\,|S_u|^2, \qquad T_v = \frac{k_-}{k_+}\,|S_v|^2.$$

Writing that the absolute values of the two sides of equation (III.39) are equal, one obtains the equality

$$T_u = T_v. \tag{III.41}$$

At a given energy, the transmission coefficient of a wave is independent of the direction of travel. This is the *reciprocity property of the transmission coefficient,* which was already pointed out in special cases in § 3 and § 6. A potential barrier is as difficult to traverse in one direction as in the other.

The equality of the absolute values of the two sides of eq. (III.40), coupled with the conservation relations (III.37) and (III.38), again yields the reciprocity relation (III.41).

From eqs. (III.39) and (III.40) we also obtain relations between the phases of the reflection and transmission amplitudes:

$$\text{phase }(S_u) = \text{phase }(S_v)$$

$$\text{phase}\left(\frac{R_u}{S_u}\right) = \pi - \text{phase}\left(\frac{R_v}{S_v}\right).$$

These relations are of interest due to the fact that the phases are related to "retardation" effects in the propagation of the wave packets. We have seen several times in Section I (§§ 3 and 9) that the quantity $\hbar\partial(\text{phase})/\partial E$, the product of $\hbar$ and the derivative with respect to the energy of the phase of the reflection or transmission amplitude, a quantity with the dimension of a time, may be interpreted as the "retardation" of the wave during reflection or transmission. This interpretation proves to be quite general.

12. Number of Nodes of Bound States

Let us now consider the eigenfunctions (if they exist) of the discrete spectrum. There is no degeneracy in that case; consequently these functions are certainly real to within a phase.

Let us return to the notation of § 8. We assume the functions y_1 and y_2 to be real, $\varepsilon_2 > \varepsilon_1$, and we apply relation (III.27) by taking as limits of the integration interval two consecutive zeros of y_1. We obtain

$$y_2 y_1' \Big|_a^b = (\varepsilon_2 - \varepsilon_1) \int_a^b y_1 y_2 \, \mathrm{d}x.$$

In the interval (a, b), y_1 keeps the same sign. Let us suppose for instance, that $y_1 > 0$. In that case $y_1'(a) > 0$, $y_1'(b) < 0$. Hence y_2 certainly changes sign in the interval (a, b); if not, the right-hand side of the equation would have the same sign as y_2, the left-hand side would have the opposite sign. Hence y_2 certainly has at least one zero within the interval (a, b). Between two nodes of y_1, there is always at least one node of y_2.

Assume now that y_1 and y_2 are eigenfunctions of the discrete spectrum. They both vanish ("exponentially") at the two boundaries of the interval $(-\infty, +\infty)$. The n_1 nodes of y_1 subdivide that interval into (n_1+1) partial intervals, in each of which y_2 has at least one zero: thus, the function y_2 has at least (n_1+1) nodes. Consequently an eigenfunction has the more nodes the higher its eigenvalue.

By repeating the argument used in § 10 to build up the successive eigenfunctions, and following the increase of the number of nodes of the functions $\hat{y}_-$ and $\hat{y}_+$ as the energy ε increases continuously, one obtains the more precise statement which follows (Problems III.4 and III.5):

If one arranges the eigenstates in the order of increasing energies $\varepsilon_1, \varepsilon_2, \ldots, \varepsilon_n, \ldots,$ *the eigenfunctions likewise fall in the order of increasing number of nodes; the nth eigenfunction has* $(n-1)$ *nodes between each of which the following eigenfunctions all have at least one node.*

13. Orthogonality Relations

One obtains another very important consequence of the Wronskian Theorem (Corollary I) by letting the limits of integration a and b in eq. (III.27) go to $-\infty$ and $+\infty$, respectively.

Assume that y_1 and y_2 are two eigenfunctions belonging to two distinct eigenvalues of the *discrete spectrum*. They both vanish at infinity, and so does their Wronskian; and since $\varepsilon_2 - \varepsilon_1 \neq 0$,

$$\int_{-\infty}^{+\infty} y_1 y_2 \, \mathrm{d}x = 0. \tag{III.42}$$

When the integral of the product $y_1 y_2$ of two (real) functions extended over all space is zero, these two functions are said to be orthogonal. More generally, two complex functions y_1 and y_2 are said to be *orthogonal* if

$$\int_{-\infty}^{+\infty} y_1{}^* y_2 \,\mathrm{d}x = 0.$$

Thus, the eigenfunctions of the discrete spectrum are orthogonal.

Clearly, this result remains valid if only one of the two eigenfunctions belongs to the discrete spectrum.

The limiting procedure is more delicate when both eigenfunctions belong to the continuous spectrum. The Wronskian $W(y_1, y_2)$ exhibits infinite oscillations at least at one of the limits of integration; the integral $\int y_1 y_2 \,\mathrm{d}x$ therefore possesses the same property. However, if one replaces at least one of the eigenfunctions in the integral, say y_2, by a wave packet made up of the eigenfunctions of a small domain of energy $\delta\varepsilon$ surrounding ε_2, the orthogonality relation is verified in the limit where $\delta\varepsilon \ll |\varepsilon_1 - \varepsilon_2|$. Indeed, let us write y_2 in the form $y(x; \varepsilon)$ in order to recall that it is a function of the energy ε [1]). Consider also the wave packet

$$Y_2(x; \delta\varepsilon) = \frac{1}{\sqrt{\delta\varepsilon}} \int_{\varepsilon_2}^{\varepsilon_2+\delta\varepsilon} y(x; \varepsilon) \,\mathrm{d}\varepsilon. \tag{III.43}$$

As the Wronskian $W(y_1, y_2)$ depends linearly upon the function y_2, one obtains upon integrating equation (III.27) term by term

$$W(Y_2, y_1)\Big|_a^b = (\varepsilon_2 - \varepsilon_1) \int_a^b y_1 Y_2 \,\mathrm{d}x + \int_a^b y_1 \left[\frac{1}{\sqrt{\delta\varepsilon}} \int_{\varepsilon_2}^{\varepsilon_2+\delta\varepsilon} (\varepsilon - \varepsilon_2)\, y(x; \varepsilon) \,\mathrm{d}\varepsilon\right] \mathrm{d}x.$$

The interest in this procedure lies in the fact that Y_2 goes to zero (as $1/x$) in the asymptotic regions where y_2 has an oscillatory behavior. When one let a and b go to $-\infty$ and $+\infty$, respectively, the left-hand side vanishes; thus the two converging integrals on the right-hand

---

[1]) Knowing ε is not sufficient to define the solution $y(x; \varepsilon)$ since the latter further depends upon one or two arbitrary constants, according to whether the eigenvalue is simple or degenerate. One resolves this arbitrariness by a suitable condition on the asymptotic form of $y(x; \varepsilon)$ at one of the limits, $-\infty$ or $+\infty$, of the range of integration.

side add up to zero. In the limit where $\delta\varepsilon \ll |\varepsilon_2 - \varepsilon_1|$, the second integral becomes negligible. We can therefore write

$$\lim_{\delta\varepsilon\to 0} \int_{-\infty}^{+\infty} y_1(x)\, Y_2(x;\, \delta\varepsilon)\, \mathrm{d}x = 0. \tag{III.42'}$$

One calls "eigendifferential" of the function $y_2(x)$ the wave packet $Y_2(x;\, \delta\varepsilon)$ defined by eq. (III.43), in which $\delta\varepsilon$ is a very small quantity which will eventually be made to vanish at the end of the calculations.

In conclusion, *two eigenfunctions belonging to two distinct eigenvalues are orthogonal,* with the convention that at least one of the eigenfunctions must be replaced by its eigendifferential in the orthogonality relation [eq. (III.42')] when the two eigenfunctions belong to the continuous spectrum.

We have mentioned the concept of eigendifferential in a rather summary fashion. It is never used in practice. We shall describe later a rather elegant mathematical device which will enable us to formulate the orthogonality relations in a quite general way, without having recourse to that concept.

14. Remark on Parity

Let us return to the concept of parity, encountered for the first time in connection with the infinitely deep square well. The property is entirely general.

If the potential is even, that is to say if

$$U(x) = U(-x),$$

the Schrödinger Hamiltonian does not change when we replace x by $-x$: *it is invariant under reflection through the origin.* Consequently, if $\psi(x)$ is an eigenfunction of the eigenvalue E,

$$H\psi(x) = E\psi(x),$$

this equation continues to hold true if one changes x to $-x$, hence

$$H\psi(-x) = E\psi(-x).$$

Moreover, the even function $\psi(x) + \psi(-x)$ and the odd function $\psi(x) - \psi(-x)$ are also eigenfunctions for the same eigenvalue E; at least one of these two functions certainly does not vanish identically. Two cases may then arise:

1. The eigenvalue E is not degenerate. The four functions above are multiples of each other: $\psi(x)$ is a multiple of that one of the two functions $\psi(x)+\psi(-x)$ or $\psi(x)-\psi(-x)$ which is not identically equal to zero (the other one is necessarily identically zero). In other words, the eigenfunctions of the non-degenerate portion of the spectrum have a well-defined parity: some are even, others are odd. In fact, an even function necessarily has an even number of nodes, an odd function has an odd number of nodes. Thus, if the eigenfunctions are arranged in the order of increasing eigenvalues of the energy, these functions are alternately even and odd, the eigenfunction of the ground state is always even. The results of § 5 agree with these predictions quite well.

2. The eigenvalue E is degenerate. In that case all functions can be put into the form $\lambda\psi+\mu\varphi$ where ψ and φ are two linearly independent eigenfunctions. Let us assume that at least one of these two functions, ψ say, has no well-defined parity: under these conditions, neither of the functions $\psi_+=\psi(x)+\psi(-x)$ and $\psi_-=\psi(x)-\psi(-x)$ is identically zero. These two functions, being of opposite parity, are of necessity linearly independent; as noted above, they are eigenfunctions of the same eigenvalue E. One can thus express ψ, φ, and consequently $\lambda\psi+\mu\varphi$ in the form of linear combinations of these two functions. Therefore, the eigenfunctions of a degenerate eigenvalue may be written as linear combination of two functions, each having a well-defined parity.

One readily sees by inspection that the eigenvalues of the continuous spectrum are all doubly degenerate, and that to each of them corresponds an even eigenfunction (whose derivative vanishes at the origin and an odd function (which vanishes at the origin).

It often happens in Quantum Mechanics that the Hamiltonian of the system under study is invariant under some specific transformations. From that invariance property there result certain symmetry properties characteristic of the eigensolutions of the Schrödinger equation. Parity represents a very simple example of this situation.

EXERCISES AND PROBLEMS

1. In the square well problem of § 6, calculate the constants P, Q, R, S occurring in the expression (III.22) of the solution χ as a function of the parameters of the well; verify expression (III.23) giving the transmission coefficient,

and the conservation relation (III.24). Assuming that $KL \gg \pi$ and $\zeta < \eta \ll 1$, define and calculate the "transit time" of the transmitted wave, and the "reflection time" of the reflected wave. Show the existence of resonances and compare the motion of the transmitted wave to that of the corresponding classical particle.

2. Calculate the transmission coefficient of the square barrier defined in § 7. Calculate the "transit time" of the transmitted wave and compare the motion of this wave to that of the classical particle.

3. Investigate the motion of a particle in a square potential consisting of an infinite barrier for $x < 0$, and in the region of positive x

$$\begin{cases} V(x) = V_{\mathrm{I}}, & 0 < x < a \\ V(x) = V_{\mathrm{II}}, & a < x < b \\ V(x) = 0, & x > b. \end{cases}$$

One assumes that $V_{\mathrm{I}} < 0 < V_{\mathrm{II}}$. Compare the motion of a wave packet undergoing reflection at $x = 0$ with that of the coresponding classical particle. Investigate the "reflection delay" when the incident energy E is less than V_{II}.

Show the existence of resonances and discuss the connection between the width of these resonances and the "reflection delay" when $E \ll V_{\mathrm{II}}$ and $(b-a)\sqrt{2mV_{\mathrm{II}}} \gg \hbar$.

4. The energy spectrum of a particle in one dimension in an arbitrary potential may occasionally include a discrete portion. Show that if one arranges the eigenstates of the discrete spectrum in the order of increasing energy, the eigenfunctions are found to lie also in the order of increasing number of nodes, the nth eigenfunction having $(n-1)$ nodes, between any two of which the following eigenfunctions all have at least one node.

5. A particle in one dimension in the interval $(0, \infty)$ lies in a potential $V(x)$ tending asymptotically to 0, and bounded at $x = 0$ by an infinitely repulsive barrier. Show that the number of bound states is equal to the number of nodes of the solution of the Schrödinger equation which vanishes at the origin, corresponding to an infinitely small negative energy.

CHAPTER IV

STATISTICAL INTERPRETATION OF THE WAVE-CORPUSCLE DUALITY AND THE UNCERTAINTY RELATIONS

1. Introduction

If one attempts to locate the underlying causes of discrepancy between the experimental facts concerning microscopic phenomena, and the predictions of classical theory, one arrives at two conclusions of entirely general applicability.

The first is what we have called the *atomism of action*, responsible for the appearance of discontinuity in microscopic physics: the fact that the change of action of a physical system, and the exchanges of action between physical systems can occur only by *discrete* and *indivisible quanta*. We have discussed this new concept at length in the first chapter, and we have seen how the description of phenomena within the framework of Classical Theory can be made only in the limit where the quantum of action $\hbar$ can be treated as negligibly small.

The second conclusion is the wave-corpuscle duality, that very general property of microscopic objects of appearing under one of two contradictory aspects, as the case may be: that of waves or that of corpuscles. This character of duality is in fact very intimately connected with the atomism of action, as is illustrated by the appearance of the constant $\hbar$ in the general correspondence relations (II.5) between waves and corpuscles. We have seen how the very general character of this duality property was recognized only rather late, and what a decisive role the recognition of this fact had played in the development of the Quantum Theory of natural systems.

A systematic examination of the actual course of diffraction experiments (cf. Ch. I, §§ 5 and 6, Ch. II, §§ 7 and 8) lends support to a very simple interpretation of the wave-corpuscle duality of microscopic objects, namely that there exists a statistical bond between their wave aspect and their corpuscular aspect. The intensity of the wave in a region of space gives the probability of finding the particle in this region. It is upon this statistical interpretation that we shall elaborate in this chapter, as well as check its internal consistency and compatibility with the experimental facts.

In Section I, we develop the statistical interpretation of the wave function of Wave Mechanics for systems of matter. The discussions and results of that section will lead us in Section II to formulate a very general consequence of the statistical interpretation of the wave-corpuscle duality: the uncertainty relations of Heisenberg. In Section III, we show how these uncertainty relations can in no way be invalidated by experiment, no matter how they might appear; we must only take into account the fact that the measuring instruments are themselves quantum objects obeying the same relations. Consequently, when a system is subjected to a given measurement, the perturbation caused in its evolution by the intervention of the measuring device can never be made arbitrarily small or perfectly controllable.

In fact, at the quantum level of accuracy, it is impossible to separate the object to be measured from the measuring instrument. Now, when one describes a measuring operation in *ordinary language*, one implicitly assumes a clear-cut separation between the object to be measured and everything that might serve to carry out that measurement. The intervention of the measuring instrument then appears as an uncontrollable perturbation, which cannot be made to vanish because of the atomism of action. The existence of such an uncontrollable disturbance in fact sets a limit to that necessary distinction between subject and object, and entails a revision of the classical ideas concerning the description of phenomena. This question is treated in Section IV of this chapter.

I. STATISTICAL INTERPRETATION OF THE WAVE FUNCTIONS OF WAVE MECHANICS

2. Probabilities of the Results of Measurement of the Position and the Momentum of a Particle

Let us first treat the case of a quantized system consisting of a single particle. Let us designate its wave function by $\Psi(\mathbf{r}, t)$. It obeys the Schrödinger equation and is fully determined at each instant if one knows its value $\Psi(\mathbf{r}, t_0)$ at the initial time t_0. For the time being we analyze the situation at a given time t and simply denote by $\Psi(\mathbf{r})$ the wave function of the particle at time t.

The dynamical state of a classical particle is defined at every instant upon specifying precisely its position $\mathbf{r}(x, y, z)$ and its momentum $\mathbf{p}(p_x, p_y, p_z)$. On the other hand, since the wave function $\Psi(\mathbf{r})$ has a certain spatial extension one cannot attribute to a quantum particle a precise position; one can only define the probability of finding the particle in a given region of space *when one carries out a measurement of position*. Denote by $P(\mathbf{r})\,\mathrm{d}\mathbf{r}$ the probability of finding the particle in the volume element $(\mathbf{r}, \mathbf{r}+\mathrm{d}\mathbf{r})$. The probability $P(V)$ of finding it in a finite volume V is obtained by integrating the "probability density" $P(\mathbf{r})$ over that volume: $P(V)=\int_V P(\mathbf{r})\,\mathrm{d}\mathbf{r}$. Similarly, one cannot in general attribute a precise momentum to a quantum particle. If the associated wave is a plane wave $\exp(i\mathbf{k}\cdot\mathbf{r})$, it actually represents a particle with momentum $\mathbf{p}=\hbar\mathbf{k}$, according to de Broglie's law of correspondence; however, Ψ is in general a *superposition* of plane waves of variable wave vector $\mathbf{k}$. One can only define the probability of finding the momentum in a given region of momentum space *when one carries out a measurement of momentum*; we shall denote by $\Pi(\mathbf{p})\,\mathrm{d}\mathbf{p}$ the probability of finding the momentum of the particle in the interval $(\mathbf{p}, \mathbf{p}+\mathrm{d}\mathbf{p})$. The probability $\Pi(D)$ of finding the momentum in the finite region D of momentum space derives from it by integration $\Pi(D)=\int_D \Pi(\mathbf{p})\,\mathrm{d}\mathbf{p}$. The probability densities $P(\mathbf{r})$ and $\Pi(\mathbf{p})$ are necessarily positive quantities, satisfying the conditions

$$\int P(\mathbf{r})\,\mathrm{d}\mathbf{r}=1, \qquad \int \Pi(\mathbf{p})\,\mathrm{d}\mathbf{p}=1, \tag{IV.1}$$

with the integrals of the left-hand side to be taken over the whole configuration space and the whole momentum space, respectively.

The distributions $P(\mathbf{r})$ and $\Pi(\mathbf{p})$ are completely determined once the wave function $\Psi(\mathbf{r})$ is known. We define $P(\mathbf{r})$ by the relation

$$P(\mathbf{r})=\Psi^*(\mathbf{r})\,\Psi(\mathbf{r})=|\Psi(\mathbf{r})|^2. \tag{IV.2}$$

This definition agrees well with the idea that the larger the probability of presence of a particle the more intense the wave at that point. Relation (IV.1) leads to the *normalization condition*

$$N \equiv \int |\Psi(\mathbf{r})|^2\,\mathrm{d}\mathbf{r}=1. \tag{IV.3}$$

Thus, the wave function $\Psi(\mathbf{r})$ must be square-integrable; furthermore, its normalization must remain constant in the course of time. We shall prove in § 3 that this second condition of consistency of the statistical interpretation is actually fulfilled.

To define $\Pi(\boldsymbol{p})$, let us consider the process of momentum measurement of the particle associated with the wave Ψ. It looks analogous to that of the spectral analysis of a composite light wave. The analogy is particularly striking if the momentum measurement is carried out with the aid of a diffraction arrangement, but the present discussion is independent of the particular nature of the measuring device used. We introduce the Fourier transform $\Phi(\boldsymbol{p})$ of the wave function defined as follows [1])

$$\Phi(\boldsymbol{p}) = (2\pi\hbar)^{-3/2} \int \Psi(\boldsymbol{r})\, \mathrm{e}^{-\mathrm{i}(\boldsymbol{p}\cdot\boldsymbol{r})/\hbar}\, \mathrm{d}\boldsymbol{r} \tag{IV.4}$$

$$\Psi(\boldsymbol{r}) = (2\pi\hbar)^{-3/2} \int \Phi(\boldsymbol{p})\, \mathrm{e}^{+\mathrm{i}(\boldsymbol{p}\cdot\boldsymbol{r})/\hbar}\, \mathrm{d}\boldsymbol{p}. \tag{IV.5}$$

According to eq. (IV.5), $\Psi(\boldsymbol{r})$ may be viewed as a linear combination of elementary waves $\exp(\mathrm{i}\boldsymbol{p}\cdot\boldsymbol{r}/\hbar)$ of well-defined momentum $\boldsymbol{p}$, each elementary wave having a coefficient $(2\pi\hbar)^{-3/2}\Phi(\boldsymbol{p})$. If there was just one single term $\exp(\mathrm{i}\boldsymbol{p}_0\cdot\boldsymbol{r}/\hbar)$, the result of the measurement would with certainty be $\boldsymbol{p}_0$. If $\Phi(\boldsymbol{p})$ has appreciable values only in a small region surrounding $\boldsymbol{p}_0$ as is the case for the wave packets studied in Ch. II, experiment shows that the value found is almost certainly a value lying in the neighborhood of $\boldsymbol{p}_0$. More generally speaking, the probability $\Pi(\boldsymbol{p})\,\mathrm{d}\boldsymbol{p}$ of finding a value of the momentum in the volume element $(\boldsymbol{p}, \boldsymbol{p}+\mathrm{d}\boldsymbol{p})$ is large when $|\Phi(\boldsymbol{p})|$ is large. We are thus led to the definition

$$\Pi(\boldsymbol{p}) = \Phi^*(\boldsymbol{p})\Phi(\boldsymbol{p}) = |\Phi(\boldsymbol{p})|^2. \tag{IV.6}$$

Since the scalar product is conserved by Fourier transformation (Theorem IV, § A.16),

$$\int |\Phi(\boldsymbol{p})|^2\, \mathrm{d}\boldsymbol{p} = \int |\Psi(\boldsymbol{r})|^2\, \mathrm{d}\boldsymbol{r}$$

and the normalization condition (IV.1) is automatically satisfied if the function $\Psi(\boldsymbol{r})$ is normalized to unity.

The Fourier transformation establishes a one-to-one correspondence between the square-integrable functions $\Psi(\boldsymbol{r})$ and $\Phi(\boldsymbol{p})$. The knowledge of $\Phi(\boldsymbol{p})$ is sufficient, as is a knowledge of $\Psi(\boldsymbol{r})$ to define the dynamical state of the particle; hence one calls $\Phi(\boldsymbol{p})$ the *wave function in mo-*

---

[1]) If the integrals of the right-hand sides of equations (IV.4) and (IV.5) do not converge, the latter must be modified in conformity with the prescriptions of Theorem I of Appendix A (§ 16) (recall that the function $\Psi(\boldsymbol{r})$ is square-integrable). The results below remain valid with this slight modification. In the following, these points of mathematical rigor will be omitted.

mentum space, a terminology which is justified inasmuch as the functions Ψ and Φ play entirely symmetrical roles in the definitions (IV.2) and (IV.6). The functions Ψ and Φ are also said to be equivalent *representations* of the same dynamical state.

The physical meaning of the quantities $P(\boldsymbol{r})$ and $\Pi(\boldsymbol{p})$ should be clearly understood. The particle associated with the wave generally possesses neither a precise position, nor a precise momentum. When carrying out a measurement on either one of these dynamical variables on an individual system represented by Ψ, no definite prediction can be made about the result. The predictions defined here apply to a very large number $\mathscr{N}$ of equivalent systems independent of each other, each system being represented by the same wave function Ψ. If one carries out a position measurement on each of them, $P(\boldsymbol{r})$ gives the distribution of the $\mathscr{N}$ results of measurement in the limit where the number $\mathscr{N}$ of members of this statistical ensemble approaches infinity. Similarly, if one measures the momentum, $\Pi(\boldsymbol{p})$ gives the distribution of the results of measurement of the momentum.

The above definitions of $P(\boldsymbol{r})$ and $\Pi(\boldsymbol{p})$ have been based on plausibility and self-consistency arguments. It is not obvious that expressions (IV.2) and (IV.6) are the only ones which follow from basing oneself on arguments of this type. In fact, the quantities $P(\boldsymbol{r})$ and $\Pi(\boldsymbol{p})$ may in principle be directly compared with the experimental results; the justification of definitions (IV.2) and (IV.6) definitely rests on the success of this comparison.

3. Conservation in Time of the Norm

In order that the definition of the probabilities of the preceding section be consistent, the norm N of the wave function must remain constant in time. Now the functions Ψ and Ψ^* satisfy respectively the Schrödinger equation (II.33) and its complex conjugate equation, namely

$$i\hbar \frac{\partial}{\partial t} \Psi = H\Psi,$$

$$i\hbar \frac{\partial}{\partial t} \Psi^* = -(H\Psi)^*.$$

Hence

$$\frac{\partial}{\partial t} |\Psi|^2 = \Psi^* \left(\frac{\partial}{\partial t} \Psi\right) + \left(\frac{\partial}{\partial t} \Psi^*\right) \Psi = \frac{1}{i\hbar} [\Psi^*(H\Psi) - (H\Psi)^* \Psi]. \qquad \text{(IV.7)}$$

Upon integrating the two sides of eq. (IV.7) over all of configuration space, we obtain

$$\frac{\mathrm{d}N}{\mathrm{d}t}=\frac{1}{\mathrm{i}\hbar}\int[\Psi^*(H\Psi)-(H\Psi)^*\,\Psi]\,\mathrm{d}\mathbf{r}.$$

In order that the norm remain constant in time, it is necessary and sufficient that

$$\int\Psi^*(H\Psi)\,\mathrm{d}\mathbf{r}=\int(H\Psi)^*\Psi\,\mathrm{d}\mathbf{r}. \tag{IV.8}$$

This property must hold no matter what the dynamical state of the particle, hence for any square-integrable function Ψ in configuration space.

In the language of operators, one says that an operator such as H is *Hermitean* if it satisfies the property (IV.8) for all functions Ψ of function space in which this operator was defined. The main properties of Hermitean operators will be studied in Chapter V.

The Schrödinger Hamiltonian actually possesses this hermiticity property. Let us verify it in the simple case of a particle in a scalar potential (the case of a charged particle in an electromagnetic field is the subject of Problem IV.1):

$$H\equiv-\frac{\hbar^2}{2m}\triangle+V(\mathbf{r}).$$

Since $V(\mathbf{r})$ is a real quantity, eq. (IV.8) can in this case be written

$$\int[\Psi^*(\triangle\Psi)-(\triangle\Psi)^*\Psi]\,\mathrm{d}\mathbf{r}=0.$$

If the integral of the left-hand side were extended over a certain volume bounded by a surface S, it would be, according to Green's theorem, equal to the surface integral

$$\int_S\left(\Psi^*\frac{\mathrm{d}\Psi}{\mathrm{d}n}-\frac{\mathrm{d}\Psi^*}{\mathrm{d}n}\Psi\right)\mathrm{d}S,$$

where $\mathrm{d}/\mathrm{d}n$ is the external normal derivative. One obtains the integral of the left-hand side by letting the volume of integration extend over all of space. In that limit all elements of the surface S are removed to infinity. However, since the wave Ψ represents the dynamical state of a physical system, it is necessarily square-integrable; consequently the surface integral goes to zero. Q.E.D.

Thus, it is sufficient that the normalization condition (IV.3) be fulfilled at the initial instant for it to remain satisfied at all later times. Since the Schrödinger equation is homogeneous, its solutions are defined to within a constant. The normalization constant at the initial time fixes the absolute value of that constant; its phase remains arbitrary.

4. Concept of Current

The property of conservation of the norm has a simple interpretation if one introduces the notion of current. The right-hand side of eq. (IV.7) can always be put in the form of a divergence of a suitably defined vector, the vector probability current density or *vector current*. Let us restrict ourselves here to the case of a particle in a scalar potential (cf. Problem IV.1). We define the current $\boldsymbol{J}(\boldsymbol{r}, t)$ at point $\boldsymbol{r}$ and at time t by

$$\boldsymbol{J}(\boldsymbol{r}, t) = \operatorname{Re}\left[\Psi^* \frac{\hbar}{\mathrm{i}m} \nabla \Psi\right]. \tag{IV.9}$$

It is easily verified that

$$\operatorname{div} \boldsymbol{J} = \frac{\mathrm{i}}{\hbar} [\Psi^*(H\Psi) - (H\Psi)^* \Psi]; \tag{IV.10}$$

thus we can write eq. (IV.7) in the form

$$\frac{\partial}{\partial t} P + \operatorname{div} \boldsymbol{J} = 0. \tag{IV.11}$$

Relation (IV.11) commonly occurs in hydrodynamics. It is the conservation law for a fluid of density P and current $\boldsymbol{J}$ in a medium without source or sink. We are thus led to liken the motion of the quantum particle to that of a classical fluid [1]). The mass of fluid contained in a given volume $\mathscr{V}$ is equal to the integral of the density P extended over that volume. From eq. (IV.11) comes the well-known result that the fluid contained in $\mathscr{V}$ is equal to

$$-\int_{\mathscr{V}} \operatorname{div} \boldsymbol{J} \, \mathrm{d}\boldsymbol{r} = -\int_S \boldsymbol{J} \cdot \mathrm{d}\boldsymbol{S},$$

---

[1]) Of course the analogy between this probability fluid and a classical fluid should not be pushed too far. All pictures based on this analogy contain no more than the property (IV.11) (cf. Ch. VI).

i.e. to the flux of the vector current crossing the surface S bounding that volume. The total mass remains constant (conservation of the norm) if the flux across S goes to zero in the limit where $\mathscr{V}$ extends over all space.

The definition of $\boldsymbol{J}$ inherently has some arbitrariness: the property (IV.11) remains valid when adding to vector $\boldsymbol{J}$ a vector with vanishing divergence. However, the definition (IV.9) has the advantage of simplicity. Besides, it can be deduced by correspondence from the classical definition of current. Indeed, according to the correspondence principle, the operator $(\hbar/im)\nabla$ represents the quantity $\boldsymbol{p}/m$, i.e. the velocity of the particle; the quantity $\boldsymbol{J}$ corresponds to the product of the velocity and the density, i.e. to the current. In particular if Ψ is a plane wave $A \exp[(i/\hbar)(\boldsymbol{p}\cdot\boldsymbol{r}-Et)]$, $\boldsymbol{J}(\boldsymbol{r}, t)=A^2(\boldsymbol{p}/m)$ is actually equal to the product of the (probability) density and the velocity.

Property (IV.11) is much more general than the conservation of the norm. If the function Ψ is a stationary solution of the Schrödinger equation

$$\Psi(\boldsymbol{r}, t) = \psi(\boldsymbol{r})\, e^{-iEt/\hbar},$$

the conservation of the norm is either obvious, or meaningless. It is obvious if the state is bound; it is meaningless if the state is unbound since the function ψ is not square-integrable in this case. In both cases, however, eq. (IV.11) remains valid, and as the density $|\Psi|^2$ is time-independent, it may be written

$$\operatorname{div} \boldsymbol{J} = 0. \tag{IV.12}$$

This property of the eigenfunction ψ is of interest because it does not depend upon the specific form of the potential term in the Hamiltonian [1]).

5. Mean Values of Functions of $\boldsymbol{r}$ or of $\boldsymbol{p}$

Having checked the consistency of our definitions for the probability densities P and Π, we may use them to calculate the mean values of functions of $\boldsymbol{r}$ or of $\boldsymbol{p}$.

Knowing the distribution $P(\boldsymbol{r})$ of the results of position measurements at a given time, one can define the mean value assumed at

---

[1]) In three-dimensional problems its role is analogous to the property of conservation of the Wronskian $W(y^*, y)$ in one-dimensional problems (cf. Ch. III, § 11).

that time by some arbitrary function $F(\mathbf{r}) \equiv F(x, y, z)$ of the coordinates of the particle. The physical meaning to be attributed to this mean value is the same as we have stated in the definition of $P(\mathbf{r})$: it is the average value of the measurements of $F(\mathbf{r})$ carried out on a very large number $\mathcal{N}$ of equivalent systems, independent of each other and represented by the same wave function Ψ.

We shall designate this quantity by the notation $\langle F(\mathbf{r})\rangle$. It is evident that

$$\langle F(\mathbf{r})\rangle = \int P(\mathbf{r})\, F(\mathbf{r})\, \mathrm{d}\mathbf{r}.$$

Similarly, we obtain for the average value of a function of the momentum $G(p) \equiv G(p_x, p_y, p_z)$:

$$\langle G(\mathbf{p})\rangle = \int \Pi(\mathbf{p})\, G(\mathbf{p})\, \mathrm{d}\mathbf{p}.$$

Using the definitions of probability densities of § 2, we obtain the expressions (provided that the integrals converge)

$$\langle F(\mathbf{r})\rangle = \int \Psi^*(\mathbf{r})\, F(\mathbf{r})\, \Psi(\mathbf{r})\, \mathrm{d}\mathbf{r} \tag{IV.13}$$

$$\langle G(\mathbf{p})\rangle = \int \Phi^*(\mathbf{p})\, G(\mathbf{p})\, \Phi(\mathbf{p})\, \mathrm{d}\mathbf{p}. \tag{IV.14}$$

Thus the average value of the coordinate x of the particle is

$$\langle x\rangle = \int \Psi^*(\mathbf{r})\, x\, \Psi(\mathbf{r})\, \mathrm{d}\mathbf{r} \tag{IV.15}$$

and that of the component p_x of the momentum

$$\langle p_x\rangle = \int \Phi^*(\mathbf{p})\, p_x\, \Phi(\mathbf{p})\, \mathrm{d}\mathbf{p}. \tag{IV.16}$$

We shall obtain another form of expression (IV.16) by using the properties of Fourier transforms stated in appendix A. If the function $p_x\Phi(\mathbf{p})$ is square-integrable – and we shall always assume this to be so – its Fourier transform is $(\hbar/\mathrm{i})(\partial/\partial x)\Psi(\mathbf{r})$ (Theorem A.III). Applying the conservation property of the scalar product (Theorem A.IV) to the functions Φ and $p_x\Phi$, we have

$$\langle p_x\rangle = \int \Psi^*(\mathbf{r}) \left(\frac{\hbar}{\mathrm{i}}\frac{\partial}{\partial x}\right) \Psi(\mathbf{r})\, \mathrm{d}\mathbf{r}. \tag{IV.17}$$

Note the formal analogy between the right-hand sides of eqs. (IV.16) and (IV.17): one goes over from the one to the other by replacing the integration over $\mathbf{p}$ by an integration over $\mathbf{r}$, $\Phi(\mathbf{p})$ by its Fourier transform, $\Phi^*(\mathbf{p})$ by its complex conjugate function, and p_x by $(\hbar/\mathrm{i})(\partial/\partial x)$, where $(\partial/\partial x)$ stands for partial derivation with respect to x applied to the function on its right.

In analogous fashion, one goes over from equation (IV.15) to the expression

$$\langle x\rangle = \int \Phi^*(\mathbf{p}) \left(\mathrm{i}\hbar \frac{\partial}{\partial p_x}\right) \Phi(\mathbf{p})\, \mathrm{d}\mathbf{p}. \tag{IV.18}$$

These results can be generalized to more complex types of functions. Thus, from the fact that $p_x{}^2\Phi(\mathbf{p})$ (assumed to be square-integrable) is the Fourier transform of

$$\left(\frac{\hbar}{\mathrm{i}}\frac{\partial}{\partial x}\right)^2 \Psi(\mathbf{r}) = -\,\hbar^2 \frac{\partial^2\Psi}{\partial x^2}$$

(repeated application of Theorem A.III), one deduces

$$\langle p_x{}^2\rangle = \int \Phi^*\, p_x{}^2\, \Phi\, \mathrm{d}\mathbf{p} = -\,\hbar^2 \int \Psi^* \frac{\partial^2\Psi}{\partial x^2}\, \mathrm{d}\mathbf{r}. \tag{IV.19}$$

More generally, if $G(\mathbf{p})$ is a polynomial or an absolutely converging series in p_x, p_y, p_z, one has

$$\langle G(\mathbf{p})\rangle = \int \Psi^*(\mathbf{r})\, G\!\left(\frac{\hbar}{\mathrm{i}}\nabla\right) \Psi(\mathbf{r})\, \mathrm{d}\mathbf{r} \tag{IV.20}$$

provided that some straightforward conditions of convergence hold. Under the same conditions one finds for the average value of $F(\mathbf{r})$:

$$\langle F(\mathbf{r})\rangle = \int \Phi^*(\mathbf{p})\, F(\mathrm{i}\hbar\nabla_p)\, \Phi(\mathbf{p})\, \mathrm{d}\mathbf{p}. \tag{IV.21}$$

Since the results obtained above are sufficient to undertake the main discussion of this chapter, we shall not pursue further the development of the statistical interpretation of the function Ψ. In addition to the statistics of the measurements of position and momentum, and the results derived therefrom concerning the mean values of quantities such as $F(\mathbf{r})$ and $G(\mathbf{p})$, the specification of Ψ must yield the statistics of the measurement of any measurable physical quantity. This question will be treated in Chapter V. We shall limit ourselves here to a few remarks which will guide us in its study.

The quantities $\langle x\rangle$ and $\langle p_x\rangle$ are real as a consequence of their definition. The right-hand sides of eqs. (IV.15) and (IV.17) are real as well. In other words – and this is precisely the definition of hermiticity [cf. eq. (IV.8)] – the operators x and $(\hbar/\mathrm{i})(\partial/\partial x)$ are *Hermitean*. Similarly, the two other components of $\boldsymbol{r}$, and the two other components of $-\mathrm{i}\hbar\nabla$ are Hermitean operators, and so are the operators $F(\boldsymbol{r})$, $G(-\mathrm{i}\hbar\nabla)$ if the functions F and G are real.

Consider the expressions of the mean values formed with the function Ψ [eqs. (IV.13), (IV.20), (IV.15), and (IV.17)]. They are all of the same form. With the quantity whose average value is to be taken is associated some linear (Hermitean) operator A and the value sought is given by the expression

$$\int \Psi^* A \Psi \,\mathrm{d}\boldsymbol{r} \tag{IV.22}$$

in which, according to usual convention, the operator acts on the function of $\boldsymbol{r}$ on its right. Operator A is obtained by a very simple correspondence rule: if we are dealing with a function $F(\boldsymbol{r})$ of the coordinates of its position, the corresponding operator is the function itself. If it is a function $G(\boldsymbol{p})$, the corresponding operator is obtained by substituting in G the respective components of $-\mathrm{i}\hbar\nabla$ for those of $\boldsymbol{p}$. This is just the correspondence rule (II.17) between the momentum $\boldsymbol{p}$ and the operator $-\mathrm{i}\hbar\nabla$ which led to the Schrödinger equation.

Each of these mean values may be calculated indifferently with either one of the wave functions Ψ and Φ: expressions (IV.21), (IV.14), (IV.18), and (IV.16) constructed with function Φ are respectively equivalent to expressions (IV.13), (IV.20), (IV.15), and (IV.17) constructed with function Ψ. They are even formally alike. To the quantity whose average value is to be taken there corresponds some linear (Hermitean) operator B – acting this time on functions of $\boldsymbol{p}$ – and the value sought is given by the expression

$$\int \Phi^* B \Phi \,\mathrm{d}\boldsymbol{p}. \tag{IV.23}$$

The operator B is found by a correspondence rule similar to that which led to A: if one is dealing with a function $G(\boldsymbol{p})$, the operator is the function itself; if it is a function $F(\boldsymbol{r})$, it is obtained by substituting in F the components of $\mathrm{i}\hbar\nabla_p$ $[\nabla_p \equiv (\partial/\partial p_x, \partial/\partial p_y, \partial/\partial p_z)]$ for those of $\boldsymbol{r}$, respectively.

Just as the wave functions Φ and Ψ are equivalent *representations* of the same dynamical state, the operators B and A are equivalent representations of one and the same physical entity. The calculation of physically measurable quantities such as the mean values considered here may be carried out in a formally identical manner in either of these representations. This suggests that Quantum Theory might be formulated in a general way, independent of any representation. This general formulation will be given in Chapters VII and VIII.

6. Generalization to Systems of Several Particles

The preceding definitions and results may be easily generalized to quantum systems of several particles.

Most generally, let $\Psi(q_1, \ldots, q_R; t)$ be the wave function of an R-dimensional quantum system, whose dynamical variables are its R coordinates $q_1, \ldots, q_R$ and the R conjugate momenta $p_1, \ldots, p_R$. We assume the coordinates to be cartesian, and denote by $\mathrm{d}\tau \equiv \mathrm{d}q_1 \ldots \mathrm{d}q_R$ and by $\mathrm{d}\omega \equiv \mathrm{d}p_1 \ldots \mathrm{d}p_R$ the volume elements in q space and p space, respectively. The wave function in p space is

$$\Phi(p_1, \ldots, p_R; t) = (2\pi\hbar)^{-\frac{1}{2}R} \int \Psi(q_1, \ldots, q_R; t)\, \mathrm{e}^{-(\mathrm{i}/\hbar) \sum_{p_i} p_i q_i}\, \mathrm{d}\tau.$$

$|\Psi|^2\, \mathrm{d}\tau$ is the probability of finding the coordinates q in region $(\tau, \tau + \mathrm{d}\tau)$, $|\Phi|^2\, \mathrm{d}\omega$ that of finding the momenta p in region $(\omega, \omega + \mathrm{d}\omega)$. Taking this as a starting point, one can repeat the entire procedure of the preceding sections.

Consider, for example, a two-particle system, and denote by $\boldsymbol{r}_1(x_1, y_1, z_1)$, $\boldsymbol{r}_2(x_2, y_2, z_2)$ their respective positions, and by $\boldsymbol{p}_1(p_{x1}, p_{y1}, p_{z1})$, $\boldsymbol{p}_2(p_{x2}, p_{y2}, p_{z2})$ their respective momenta. $P(\boldsymbol{r}_1, \boldsymbol{r}_2)\, \mathrm{d}\boldsymbol{r}_1 \mathrm{d}\boldsymbol{r}_2$ is the probability of finding particle 1 in volume element $(\boldsymbol{r}_1, \boldsymbol{r}_1 + \mathrm{d}\boldsymbol{r}_1)$ and particle 2 in volume element $(\boldsymbol{r}_2, \boldsymbol{r}_2 + \mathrm{d}\boldsymbol{r}_2)$; $\Pi(\boldsymbol{p}_1, \boldsymbol{p}_2)\mathrm{d}\boldsymbol{p}_1\mathrm{d}\boldsymbol{p}_2$ is the probability of finding the momentum of particle 1 in the interval $(\boldsymbol{p}_1, \boldsymbol{p}_1 + \mathrm{d}\boldsymbol{p}_1)$ and that of particle 2 in interval $(\boldsymbol{p}_2, \boldsymbol{p}_2 + \mathrm{d}\boldsymbol{p}_2)$. One may also introduce the probability density of presence $P_1(\boldsymbol{r}_1)$ of particle 1 at the point $\boldsymbol{r}_1$, the position of particle 2 being unspecified. This quantity is the statistical distribution obtained when measuring the position of particle 1 without caring about the position of particle 2. Obviously:

$$P_1(\boldsymbol{r}_1) = \int P(\boldsymbol{r}_1, \boldsymbol{r}_2)\, \mathrm{d}\boldsymbol{r}_2.$$

Similarly one may introduce the probability densities $P_2(\boldsymbol{r}_2)$, $\Pi_1(\boldsymbol{p}_1)$, and $\Pi_2(\boldsymbol{p}_2)$. All these statistical distributions are essentially positive quantities satisfying the normalization conditions

$$\iint P(\boldsymbol{r}_1, \boldsymbol{r}_2)\, \mathrm{d}\boldsymbol{r}_1\, \mathrm{d}\boldsymbol{r}_2 = 1, \ldots, \int \Pi_2(\boldsymbol{p}_2)\, \mathrm{d}\boldsymbol{p}_2 = 1$$

[the symbol $\iint \mathrm{d}\boldsymbol{r}_1\, \mathrm{d}\boldsymbol{r}_2$ designates the six-fold integral extended over all configuration space, the symbol $\int \mathrm{d}\boldsymbol{p}_2$ denotes a triple integral extended over all momentum space of particle 2, etc.]

The dynamical state of the system of two particles at a given time is defined by the wave function $\Psi(\boldsymbol{r}_1, \boldsymbol{r}_2)$ at that time. Its Fourier transform is the wave function in momentum space:

$$\Phi(\boldsymbol{p}_1, \boldsymbol{p}_2) = (2\pi\hbar)^{-3} \iint \mathrm{e}^{-\mathrm{i}(\boldsymbol{p}_1\cdot\boldsymbol{r}_1 + \boldsymbol{p}_2\cdot\boldsymbol{r}_2)/\hbar}\, \Psi(\boldsymbol{r}_1, \boldsymbol{r}_2)\, \mathrm{d}\boldsymbol{r}_1\, \mathrm{d}\boldsymbol{r}_2$$

$$\Psi(\boldsymbol{r}_1, \boldsymbol{r}_2) = (2\pi\hbar)^{-3} \iint \mathrm{e}^{\mathrm{i}(\boldsymbol{p}_1\cdot\boldsymbol{r}_1 + \boldsymbol{p}_2\cdot\boldsymbol{r}_2)/\hbar}\, \Phi(\boldsymbol{p}_1, \boldsymbol{p}_2)\, \mathrm{d}\boldsymbol{p}_1\, \mathrm{d}\boldsymbol{p}_2.$$

The generalization of definitions (IV.2) and (IV.6) is clearly

$$P(\boldsymbol{r}_1, \boldsymbol{r}_2) = |\Psi(\boldsymbol{r}_1, \boldsymbol{r}_2)|^2, \qquad \Pi(\boldsymbol{p}_1, \boldsymbol{p}_2) = |\Phi(\boldsymbol{p}_1, \boldsymbol{p}_2)|^2 \tag{IV.24}$$

and the normalization conditions of the probabilities imply the normalization condition of the wave functions:

$$\iint |\Psi(\boldsymbol{r}_1, \boldsymbol{r}_2)|^2\, \mathrm{d}\boldsymbol{r}_1\, \mathrm{d}\boldsymbol{r}_2 = \iint |\Phi(\boldsymbol{p}_1, \boldsymbol{p}_2)|^2\, \mathrm{d}\boldsymbol{p}_1\, \mathrm{d}\boldsymbol{p}_2 = 1.$$

This normalization condition can actually be fulfilled at *any instant* if the Hamiltonian in the Schrödinger equation of the system is a Hermitean operator. This hermiticity property is easily verified. The wave functions Ψ and Φ are then determined to within an arbitrary phase constant.

From the preceding definitions one derives the definitions of various other distributions introduced above. Thus

$$P_1(\boldsymbol{r}_1) = \int |\Psi(\boldsymbol{r}_1, \boldsymbol{r}_2)|^2\, \mathrm{d}\boldsymbol{r}_2.$$

Similarly, one also obtains the average values of functions $F(\boldsymbol{r}_1, \boldsymbol{r}_2)$ and $G(\boldsymbol{p}_1, \boldsymbol{p}_2)$ of the positions and the momenta. Thus for instance

$$\langle x_1 \rangle = \iint \Psi^*\, x_1 \Psi\, \mathrm{d}\boldsymbol{r}_1\, \mathrm{d}\boldsymbol{r}_2 = -\frac{\hbar}{\mathrm{i}} \iint \Phi^* \frac{\partial \Phi}{\partial p_{x_1}}\, \mathrm{d}\boldsymbol{p}_1\, \mathrm{d}\boldsymbol{p}_2,$$

$$\langle p_{x_2} \rangle = \iint \Phi^*\, p_{x_2} \Phi\, \mathrm{d}\boldsymbol{p}_1\, \mathrm{d}\boldsymbol{p}_2 = \frac{\hbar}{\mathrm{i}} \iint \Psi^* \frac{\partial \Psi}{\partial x_2}\, \mathrm{d}\boldsymbol{r}_1\, \mathrm{d}\boldsymbol{r}_2.$$

All the remarks made at the end of § 5 remain entirely valid. If the wave function $\Psi(\boldsymbol{r}_1, \boldsymbol{r}_2)$ can be factored:

$$\Psi(\boldsymbol{r}_1, \boldsymbol{r}_2) = \Psi_1(\boldsymbol{r}_1)\, \Psi_2(\boldsymbol{r}_2)$$

[$\Psi_1(\mathbf{r}_1)$ and $\Psi_2(\mathbf{r}_2)$ are assumed normalized to unity], the same is true for the wave function in momentum space, and for the distributions P and Π:

$$P(\mathbf{r}_1, \mathbf{r}_2) = P_1(\mathbf{r}_1)\, P_2(\mathbf{r}_2), \qquad \Pi(\mathbf{p}_1, \mathbf{p}_2) = \Pi_1(\mathbf{p}_1)\, \Pi_2(\mathbf{p}_2),$$

as can easily be seen from the definitions of these quantities. In other words, there is no correlation in the statistics of measurements carried out on each of these particles. The statistical predictions concerning the results of measurement on one of them, particle 1, say, are the same as if it were in the dynamical state represented by the wave function $\Psi_1(\mathbf{r}_1)$. Indeed, one easily verifies that $P_1(\mathbf{r}_1) = |\Psi_1(\mathbf{r}_1)|^2$ and that $\Pi_1(\mathbf{p}_1) = |\Phi_1(\mathbf{p}_1)|^2$, $\Phi_1(\mathbf{p}_1)$ denoting the wave function of momentum space associated with Ψ_1. In all calculations bearing on the measurements carried out on this particle (average values, fluctuations, etc.) one may simply ignore the other particle and treat the problem as if one were dealing with a single particle whose wave function is $\Psi_1(\mathbf{r}_1)$.

If the two particles do not interact or if for some reason their interaction may be considered negligible, this property of factorization of the wave function persists in the course of time. Indeed the Hamiltonian of the system may then be put in the form of a sum of two terms:

$$H = H_1 + H_2,$$

of which one, H_1, acts only on the functions of the variable $\mathbf{r}_1$, and the other, H_2, on those of the variable $\mathbf{r}_2$. Assume that at the initial instant

$$\Psi(\mathbf{r}_1, \mathbf{r}_2; t_0) = \Psi_1(\mathbf{r}_1, t_0)\Psi_2(\mathbf{r}_2, t_0)$$

and let $\Psi_1(\mathbf{r}_1, t)$ and $\Psi_2(\mathbf{r}_2, t)$ be the solutions of the Schrödinger equation:

$$\left[i\hbar \frac{\partial}{\partial t} - H_1\right] \Psi_1 = 0, \qquad \left[i\hbar \frac{\partial}{\partial t} - H_2\right] \Psi_2 = 0,$$

with the respective initial conditions $\Psi_1(\mathbf{r}_1, t_0)$ and $\Psi_2(\mathbf{r}_2, t_0)$. The factored wave function

$$\Psi(\mathbf{r}_1, \mathbf{r}_2; t) = \Psi_1(\mathbf{r}_1, t)\Psi_2(\mathbf{r}_2, t)$$

satisfies the Schrödinger equation of the system, since

$$
\begin{aligned}
i\hbar \frac{\partial}{\partial t} \Psi &= i\hbar \left(\frac{\partial \Psi_1}{\partial t} \Psi_2 + \Psi_1 \frac{\partial \Psi_2}{\partial t} \right) = (H_1 \Psi_1) \Psi_2 + \Psi_1 (H_2 \Psi_2) \\
&= H_1 \Psi_1 \Psi_2 + H_2 \Psi_1 \Psi_2 = (H_1 + H_2) \Psi_1 \Psi_2 \\
&= H\Psi.
\end{aligned}
$$

The motions of each particle remain completely independent, as required by logic, and there is no correlation at any moment between the statistics of measurements carried out on either of them.

II. HEISENBERG'S UNCERTAINTY RELATIONS

7. Position-Momentum Uncertainty Relations of a Quantized Particle

Let us return to the definitions of probabilities of § 2. The distributions $P(\mathbf{r})$ and $\Pi(\mathbf{p})$ being defined in terms of the same wave function $\Psi(\mathbf{r})$, are not independent of each other, although the function $\Psi(\mathbf{r})$ could *a priori* be any square-integrable function. One of them can always be fixed arbitrarily by means of a suitable choice of Ψ; if, for instance, one picks $P(\mathbf{r})$ it suffices to take a function with absolute value $\sqrt{P}$: $\Psi(\mathbf{r}) = \sqrt{P(\mathbf{r})} \exp [i\alpha(\mathbf{r})]$, the phase $\alpha(\mathbf{r})$ remaining completely indeterminate. However, it is not possible to obtain by a suitable choice of $\alpha(\mathbf{r})$ any distribution $\Pi(\mathbf{p})$ specified in advance, although the distribution $\Pi(\mathbf{p})$ when considered as a function of $\alpha(\mathbf{r})$ can vary over a rather wide range. The fact that there always exists some correlation between the distributions $P(\mathbf{r})$ and $\Pi(\mathbf{p})$ is characteristic of Quantum Theory [1]). It is expressed quantitatively by *Heisenberg's uncertainty relations.*

[1]) Herein lies an essential difference between the statistical distributions $P(\mathbf{r})$ and $\Pi(\mathbf{p})$ on the one hand, and the corresponding distributions $P_{\text{cl.}}(\mathbf{r})$ and $\Pi_{\text{cl.}}(\mathbf{p})$ of classical *statistical* mechanics with which one would be tempted to make a comparison. The latter are obtained by means of the density in phase space $\varrho(\mathbf{r}, \mathbf{p})$:

$$P_{\text{cl.}}(\mathbf{r}) = \int \varrho(\mathbf{r}, \mathbf{p}) \, d\mathbf{p}, \qquad \Pi_{\text{cl.}}(\mathbf{p}) = \int \varrho(\mathbf{r}, \mathbf{p}) \, d\mathbf{r};$$

$\varrho(\mathbf{r}, \mathbf{p})$ is a positive function, subject to the condition $\iint \varrho(\mathbf{r}, \mathbf{p}) \, d\mathbf{r} \, d\mathbf{p} = 1$, but otherwise arbitrary. Now one may fix arbitrarily and simultaneously the distributions $P_{\text{cl.}}(\mathbf{r})$ and $\Pi_{\text{cl.}}(\mathbf{p})$; indeed there exists at least one density in phase space, $\varrho(\mathbf{r}, \mathbf{p}) = P_{\text{cl.}}(\mathbf{r}) \Pi_{\text{cl.}}(\mathbf{p})$ which leads to these distributions.

Consider at first a particle in one dimension. Denote its position by x and its momentum by p; let $\psi(x)$ and

$$\varphi(p) = (2\pi\hbar)^{-\frac{1}{2}} \int_{-\infty}^{+\infty} \psi(x)\, \mathrm{e}^{-(\mathrm{i}px)/\hbar}\, \mathrm{d}x,$$

be the wave functions representing its dynamical state in x-space and p-space, respectively.

Heisenberg's result essentially expresses the mathematical fact that the extension of the wave ψ and that of its Fourier transform φ cannot simultaneously be made arbitrarily small. *If the wave ψ occupies a region of order Δx in x-space, and the wave φ occupies a region of order Δp in p-space, the product $\Delta x \cdot \Delta p$ always remains larger than a quantity of order $\hbar$:*

$$\Delta x \cdot \Delta p \gtrsim \hbar. \tag{IV.25}$$

We already met this result in connection with the construction of wave packets in the theory of matter waves. It also entered, although less directly, in the discussion of Chapter III on the motion of wave packets formed by superposition of unbound states.

One may convince oneself of its validity by the following semi-quantitative argument (which merely reproduces the analysis of Ch. II, § 3 in different terms). Any wave $\psi(x)$ is a superposition of plane waves $\exp(\mathrm{i}kx)$ of wavelength $2\pi/k$. Let Δk be the extent of the domain of variation of parameter k. In order that the wave $\psi(x)$ be restricted to a certain region Δx, it is necessary that constructive interference among these different waves occurs only in that region, and that they interfere destructively everywhere else. The number of wavelengths $2\pi/k$ contained in Δx is $k\Delta x/2\pi$. In order that the various plane waves forming $\psi(x)$ may interfere destructively at the limits of the interval Δx, this number must change at least by one unit when k runs over its domain of variation: $\Delta x \cdot \Delta k \gtrsim 2\pi$. We are dealing here with an order of magnitude argument; we shall therefore not bother about the factor 2π and simply write

$$\Delta x \cdot \Delta k \gtrsim 1.$$

Since the momentum p is connected to k by the relation $p = \hbar k$, we arrive immediately at relation (IV.25).

One usually calls the quantities Δx and Δp the uncertainties in position and in momentum, respectively, and often states the result of Heisenberg in the following manner:

The product of the uncertainty in position and the uncertainty in momentum is necessarily greater than a quantity of order $\hbar$.

Let us illustrate this result by some examples.

The Gaussian wave packet (not normalized to unity)

$$\psi(x) = \exp\left[\frac{\mathrm{i}}{\hbar}\, p_0 x - \frac{(x-x_0)^2}{2\xi^2}\right]$$

occupies a region of dimension ξ about the point x_0. The wave

$$\varphi(p) = \frac{\xi}{\sqrt{\hbar}} \exp\left[\frac{\mathrm{i}}{\hbar}\, x_0(p_0-p) - \tfrac{1}{2}\frac{\xi^2}{\hbar^2}(p-p_0)^2\right]$$

of momentum space corresponding to it (cf. table at the end of Appendix A) occupies in that space a region of extension $\hbar/\xi$

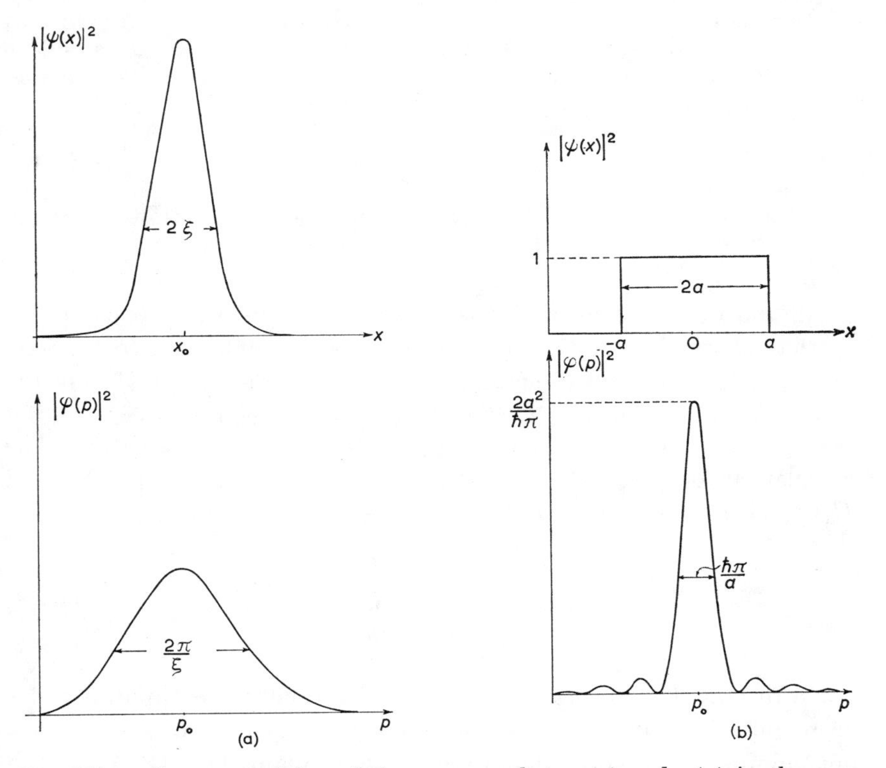

Fig. IV.1. Square modulus of the wave packets $\psi(x)$ and $\varphi(p)$ in the case: (*a*) of a Gaussian wave packet; (*b*) of a square pulse.

about the point p_0. Upon diminishing ξ, one decreases Δx, but one increases Δp by a corresponding amount so that their product remains of order $\hbar$.

As another example, let us consider the "square pulse"

$$\psi(x) = \begin{cases} e^{ip_0x/\hbar} & \text{for} \quad |x| < a \\ 0 & \text{for} \quad |x| > a \end{cases}$$

which extends over a region $2a$ surrounding the point $x=0$. We have in this case (cf. Appendix A)

$$\varphi(p) = \frac{\sqrt{(2\hbar/\pi)}}{p-p_0} \sin \frac{(p-p_0)a}{\hbar}.$$

The function $|\varphi(p)|^2$ exhibits a very sharp peak at the point p_0 surrounded on either side by a succession of vanishing minima [reached when $p=p_0+(n\pi\hbar/a)$] separated by maxima whose height decreases as $[1/(p-p_0)]^2$. One can say that the wave $\varphi(p)$ is mainly concentrated between the first zeros of $|\varphi(p)|^2$ located on either side of the main peak, i.e. an extension $\Delta p \approx 2\pi\hbar/a$. The smaller the extension $\Delta x \approx 2a$, the larger Δp:

$$\Delta x \cdot \Delta p \approx 4\pi\hbar.$$

Figure IV.1 shows the curves $|\psi(x)|^2$ as a function of x and $|\varphi(p)|^2$ as a function of p for these two wave packets.

All the considerations on the extension of the wave ψ compared to that of its Fourier transform may be easily extended to the case of a particle in three-dimensional space. Denote by $\Delta x, \Delta y, \Delta z$ the respective uncertainties in the three position coordinates, and by $\Delta p_x, \Delta p_y, \Delta p_z$ the corresponding quantities for the momentum. The correlation between the statistical distributions $P(\mathbf{r}) = |\Psi(\mathbf{r})|^2$ and $\Pi(\mathbf{p}) = |\Phi(\mathbf{p})|^2$ is expressed by the uncertainty relations

$$\begin{aligned} \Delta x \cdot \Delta p_x &\gtrsim \hbar, \\ \Delta y \cdot \Delta p_y &\gtrsim \hbar, \\ \Delta z \cdot \Delta p_z &\gtrsim \hbar. \end{aligned} \qquad \text{(IV.26)}$$

Up to now, the uncertainty relations have been presented as order of magnitude relations. Of course this is inevitable as long as one has not adopted a precise definition of the quantities $\Delta x, \Delta p_x$, etc., measuring the various uncertainties. By taking suitable definitions for these quantities, we shall arrive at a precise statement. However, without underestimating the virtues of a precise statement might have, one must insist on the fact that the essential interest of the uncertainty relations is already contained in the order of magnitude result: one may not, under any circumstances, attribute simultaneously to the quantum particle a rigorously precise position and a rigorously precise momentum. To think of the particle as endowed with precise position and momentum is justified only to the extent that the quantum of action $\hbar$ may be considered negligible, that is, in the domain of validity of Classical Theory.

8. Precise Statement of the Position-Momentum Uncertainty Relations

Let us first treat in detail the case of a particle in one dimension. We take for the uncertainties in x and p the following precise definitions [1]: Δx and Δp are the root-mean-square deviations of the distributions $|\psi(x)|^2$ and $|\varphi(p)|^2$. Adopting the notation of Chapter IV, § 5, we therefore have

$$\begin{aligned} \Delta x &\equiv \sqrt{\langle x^2 \rangle - \langle x \rangle^2} \\ \Delta p &\equiv \sqrt{\langle p^2 \rangle - \langle p \rangle^2}. \end{aligned} \tag{IV.27}$$

The quantity Δx is thus directly related to the position measurement: it is the statistical fluctuation of the result of measurement around the average value $\langle x \rangle$; the same remark applies to Δp relative to the momentum measurement. We proceed to show the following very general result:

$$\Delta x \cdot \Delta p \geqslant \tfrac{1}{2}\hbar. \tag{IV.28}$$

Let us consider the positive-definite expression

$$I(\lambda) = \int_{-\infty}^{+\infty} \left| x\psi + \lambda\hbar \frac{\partial \psi}{\partial x} \right|^2 \mathrm{d}x \qquad [I(\lambda) \geqslant 0 \text{ for any } \lambda].$$

[1] This is probably the most convenient definition one might imagine. In most of the cases, the quantities Δx and Δp thus defined give a good idea of the uncertainty in x and p. Nevertheless, it sometimes happens that this mathematical definition of the uncertainty differs rather substantially from the rough estimates one might make. The "square pulse" is an example of this type of accident: the quantity Δp defined by equation (IV.27) is infinite whereas a rough estimate of this quantity gave $2\pi\hbar/a$.

Developing and integrating by parts, we obtain successively

$$I(\lambda) = \int_{-\infty}^{+\infty} |x\psi|^2 \,\mathrm{d}x + \lambda\hbar \int_{-\infty}^{+\infty} \left(\frac{\partial\psi^*}{\partial x}\, x\psi + x\psi^* \frac{\partial\psi}{\partial x}\right) \mathrm{d}x + \lambda^2\hbar^2 \int_{-\infty}^{+\infty} \left|\frac{\partial\psi}{\partial x}\right|^2 \mathrm{d}x$$

$$= \int_{-\infty}^{+\infty} \psi^* x^2 \psi \,\mathrm{d}x - \lambda\hbar \int_{-\infty}^{+\infty} \psi^*\psi \,\mathrm{d}x - \lambda^2\hbar^2 \int_{-\infty}^{+\infty} \psi^* \frac{\partial^2\psi}{\partial x^2} \,\mathrm{d}x,$$

which yields, assuming ψ to be normalized to unity and using the results of § 5:

$$I(\lambda) = \langle x^2 \rangle - \lambda\hbar + \lambda^2 \langle p^2 \rangle. \qquad \text{(IV.29)}$$

Since the polynomial $I(\lambda)$ of the second degree in λ is positive-definite (or zero) its discriminant $\hbar^2 - 4\langle p^2\rangle\langle x^2\rangle$ is necessarily negative (or zero), thus:

$$\langle x^2\rangle\langle p^2\rangle \geqslant \tfrac{1}{4}\hbar^2. \qquad \text{(IV.30)}$$

Condition (IV.30) is less restrictive than condition (IV.28). But one can make an identical calculation starting from a slightly different expression $I(\lambda)$, i.e. by replacing in the definition of $I(\lambda)$, x by $x - \langle x\rangle$ and $\hbar(\partial/\partial x)$ by $\hbar(\partial/\partial x) - \mathrm{i}\langle p\rangle$ [or, what amounts to the same thing, replacing $\psi(x)$ by $\exp(-\mathrm{i}\langle p\rangle x/\hbar)\ \psi(x + \langle x\rangle)$]. The result is analogous to eq. (IV.29):

$$I(\lambda) = (\Delta x)^2 - \lambda\hbar + \lambda^2 (\Delta p)^2 \geqslant 0.$$

Condition (IV.28) expresses the fact that the discriminant of this second-degree expression in λ is never positive. Q.E.D.

The foregoing proof applies equally well to a particle in three-dimensional space. The wave function $\Psi(\boldsymbol{r})$ is a function of three-dimensional space and the integrals introduced in the proof extend over the three dimensions of configuration space instead of extending only along the x axis. The reader will easily verify that all the operations carried out on these integrals remain valid. Similarly, the definition (IV.27) of the root-mean-square deviations is easily generalized, and one obtains the *Heisenberg uncertainty relations*:

$$\begin{aligned} \Delta x \cdot \Delta p_x &\geqslant \tfrac{1}{2}\hbar \\ \Delta y \cdot \Delta p_y &\geqslant \tfrac{1}{2}\hbar \\ \Delta z \cdot \Delta p_z &\geqslant \tfrac{1}{2}\hbar. \end{aligned} \qquad \text{(IV.31)}$$

9. Generalization: Uncertainty Relations Between Conjugate Variables

One finds the same relations in the general case of R-dimensional quantum systems. We use the notation of § 6. By extension of the definition (IV.27) we measure the uncertainty in q_i and p_i, respectively, by the root-mean-square deviations of their statistical distribution:

$$\begin{aligned} \Delta q_i &= \sqrt{\langle q_i^2 \rangle - \langle q_i \rangle^2}, \\ \Delta p_i &= \sqrt{\langle p_i^2 \rangle - \langle p_i \rangle^2}. \end{aligned} \tag{IV.32}$$

The proof of the foregoing section can be repeated without modification leading to the uncertainty relations between conjugate (cartesian) variables

$$\Delta q_i \cdot \Delta p_i \geqslant \tfrac{1}{2}\hbar \qquad (i = 1, 2, \ldots, R).$$

10. Time-Energy Uncertainty Relation

Position-momentum uncertainty relations originate from the fact that the momentum is defined, to within a constant, as the characteristic wave number of a plane wave, and that, rigorously speaking, a plane wave extends over all space; to localize the momentum at an exact point of space has no more meaning than to localize a plane wave.

Just as momentum is a wave number and cannot be localized in space, so energy is a frequency and cannot be localized in time. Thus there exists, as suggested by the principle of relativity, a time-energy uncertainty relation, analogous to the position-momentum uncertainty relations discussed previously:

$$\Delta t \cdot \Delta E \gtrsim \hbar. \tag{IV.33}$$

However, its physical interpretation is quite different. In the position-momentum uncertainty relations, the position and momentum variables play exactly symmetrical roles; they can both be measured at *a given time t*. The statistical distributions of the results of measurement and consequently the uncertainties Δq_i, Δp_i are all derivable from the value of the wave function at that time. In relation (IV.33), on the other hand, energy and time play fundamentally different roles: the energy is a dynamical variable of the system, whereas the time t is a parameter. Relation (IV.33) *connects the uncertainty ΔE* in the value taken by this dynamical variable *to a time interval Δt characteristic of the rate of change of the system.*

At first, let us consider the free particle case. The monochromatic plane wave $\exp[i(\mathbf{k}\cdot\mathbf{r}-\omega t)]$ represents a particle of well-defined momentum $\hbar\mathbf{k}$ and energy $\hbar\omega$. By superposition of plane waves, one can form a wave packet such as the one of eq. (II.11). For simplicity we consider a wave packet in one dimension and take a wave train such as the square pulse studied above. Denote its length by Δx and its group velocity by v. It travels with velocity v along the x axis; however, the instant at which it passes a given point of the axis is not determined precisely, but carries an uncertainty

$$\Delta t \approx \frac{\Delta x}{v}.$$

Furthermore, this wave train has a certain spread in momentum space, hence a certain uncertainty ΔE in the value of the energy of the particle

$$\Delta E \approx \frac{\partial E}{\partial p}\,\Delta p = v\Delta p.$$

From the two foregoing equations we deduce

$$\Delta t \cdot \Delta E \approx \Delta x \cdot \Delta p$$

and, applying the momentum-position uncertainty relation, we obtain relation (IV.33) which sets a lower limit to the product of the spread ΔE of the energy spectrum of the particle, and the precision Δt with which the instant of passage of the particle at a given point can be predicted.

The preceding proof can be extended without difficulty to the case of a wave packet in a slowly varying field, but ceases to be valid in more general cases. In order to obtain a relation such as (IV.33) one must consider the time dependence of the wave function in these more general cases.

The simplest situation obtains when the quantum system has a well-defined energy. We know (cf. Ch. II, § 16) that the wave function of a quantum system of given energy $E=\hbar\omega$ varies in time proportionally to the exponential factor $\exp(-i\omega t)$. Consider, for example, the case of a particle in a force field. If its quantum state corresponds to a well-defined value of the energy $E=\hbar\omega$, its wave function is written $\Psi(\mathbf{r}, t)=\psi(\mathbf{r})\exp(-i\omega t)$. Consequently, the distribution of position of this particle $P(\mathbf{r})=|\psi(\mathbf{r})|^2$ is independent of time. The momentum distribution is easily seen to possess this same property. Consequently, the result of a measurement (of position or momentum) carried out on this system is independent of the moment at which the measurement is made. This is expressed in short by saying that the physical properties of such a system are time-independent, or else that the system is in a *stationary state.*

Assume now that the quantum state of the particle is the superposition of two stationary states of energy E_1 and E_2, respectively. Its wave function is of the form

$$\psi_1(\mathbf{r})\,\mathrm{e}^{-\mathrm{i}E_1t/\hbar}+\psi_2(\mathbf{r})\,\mathrm{e}^{-\mathrm{i}E_2t/\hbar}$$

and the distribution

$$P(\mathbf{r},t)=|\psi_1(\mathbf{r})|^2+|\psi_2(\mathbf{r})|^2+2\mathrm{Re}\,\psi_1{}^*\psi_2\,\mathrm{e}^{\mathrm{i}(E_1-E_2)t/\hbar}$$

oscillates in time [between two extreme values $(|\psi_1|-|\psi_2|)^2$ and $(|\psi_1|+|\psi_2|)^2$] with the period

$$\tau=\frac{h}{|E_1-E_2|}.$$

The momentum distribution has the same property.

τ thus appears as a characteristic time for the rate of change of the physical properties of the system. The result of measurement – more precisely, the statistical distribution of the results of measurement – made at two different times t_1 and t_2 will be practically identical if the difference $\Delta t=|t_1-t_2|$ is small compared to τ. In other words, in order that the physical properties of the system be notably modified over a time interval Δt, the product of Δt and the energy uncertainty $\Delta E=|E_1-E_2|$ must at least be equal to a quantity of the order of $\hbar$: $\Delta t\cdot\Delta E\gtrsim\hbar$. Expressed in this fashion, this result remains valid when the state of the system is any superposition of any number of stationary states. It is thus entirely general. A precise proof thereof will be given later on (Ch. VIII, § 13).

An important application of relation (IV.33) is the *lifetime-width relation* for radioactive systems (radioactive nucleus, excited state of an atom, unstable elementary particle, etc.). A radioactive system is not stationary; it does not correspond to a well-defined value of the energy, but to an energy spectrum with a certain spread ΔE, usually called the level width. The mean lifetime τ here plays the role of the characteristic time considered above. One must wait for a time of order τ to observe an appreciable change in the properties of the system. Consequently

$$\tau \cdot \Delta E \approx \hbar.$$

Another consequence of (IV.33) has to do with energy measurements in general. *The precision ΔE of the measurement is connected with the time Δt required for the measurement by relation* (IV.33). Thus, one may consider the measurement of the excitation energy of the first-excited state of the hydrogen atom by bombarding it with monoërgic electrons and measuring the energy lost by the electrons in the corresponding inelastic collision. The duration of the measurement is at least equal to the collision time, that is to say to the time of passage Δt of the wave packet representing the electron at the location of the hydrogen atom. The measurement error is at least equal to the uncertainty ΔE in the energy of the incident electron, and we have in fact $\Delta t \cdot \Delta E \gtrsim \hbar$.

11. Uncertainty Relations for Photons

The uncertainty relations for systems of material particles follow from the wave-corpuscle duality of matter. For the same reason, there exist analogous relations for photons. But when formulating them one must be aware that the number of photons contained in a physical system is not in general a well-defined quantity, and that one cannot, strictly speaking, consider the motion of a particular photon unless it is free from any interaction.

With these restrictions in mind we may nevertheless represent a free photon by a wave packet formed by superposition of plane, monochromatic waves propagating at velocity c [1]).

Thus, by illuminating a screen equipped with a diaphragm, which one keeps open during a certain time Δt, one obtains a light signal which eventually may contain but a single light quantum. This photon is represented by a wave packet whose extension along the three directions of the axes, Δx, Δy, Δz, depends upon the dimensions of the diaphragm, and upon its open time. This wave packet is a super-

---

[1]) To simplify matters we ignore the existence of the polarization of light. To take it into account, one must attribute to the photon an internal degree of freedom.

position of monochromatic waves. It has all the properties of the wave packet we studied previously. It does not represent a photon of well-defined momentum and energy, but the components of its momentum and its energy cover a certain finite region, namely Δp_x, Δp_y, Δp_z, and ΔE, respectively.

All these quantities satisfy the relations

$$\Delta x \cdot \Delta p_x \gtrsim \hbar, \qquad \Delta y \cdot \Delta p_y \gtrsim \hbar, \qquad \Delta z \cdot \Delta p_z \gtrsim \hbar, \qquad \Delta t \cdot \Delta E \gtrsim \hbar.$$

From these relations, some interesting conclusions may be drawn concerning the mechanism of interaction between such a photon and matter. Thus, in the absorption of a photon by an atom (photoelectric effect), the product of the uncertainty ΔE in the energy transfer to the atom, and the uncertainty Δt concerning the instant at which this energy is transferred, is at least of order $\hbar$. Conversely, if an atom initially in an excited state, decays to the ground state by emitting a photon, the moment at which this transition occurs cannot be defined with a precision greater than the mean life τ of this excited state. The emitted photon is represented by a wave packet of spatial extension $c\tau$ and, consequently, its spread in energy ΔE is such that $\tau \cdot \Delta E \approx \hbar$. This result is in good agreement with experiment. One could also have obtained it by starting from the relation between mean life and width discussed earlier, and noting that, owing to the conservation of energy, the spread in energy of the emitted photon (final state) must be equal to the spread in energy of the excited atom (initial state).

III. UNCERTAINTY RELATIONS AND THE MEASUREMENT PROCESS

12. Uncontrollable Disturbance During the Operation of Measurement

In the following, we shall concentrate particularly on the position-momentum uncertainty relations. The spreads Δx, Δp which enter into these relations refer to measurements carried out under the conditions stated earlier when we defined the probabilities. They must no be confused with the ordinary errors of measurement caused by the fact that the measuring instrument is never perfect and does not permit the determination of the measured quantities with infinite precision. In all previous reasoning this type of error is assumed negligible.

We proceed to discuss more closely the mechanism of a measurement and its consistency with the statistical interpretation outlined above. Consider, for instance, the measurement of the position coordinate of a particle in a quantized system whose dynamical state is represented by the wave function Ψ (we shall henceforth say, more briefly, that the system is in the state Ψ). We assume that the measuring device at our disposal is *infinitely precise*. Thus, the impossibility to predict with certainty the result of observation is not due to any imperfection of the measuring device; the state Ψ does not in general correspond to a precise value of x; rather, it is a superposition of dynamical states, each corresponding to a given value of x. Immediately after the operation of measurement, we can assert that the system is in a dynamical state where the x coordinate has the precise value x' indicated by the measuring apparatus. Such a state can definitely not be represented by the wave function Ψ: *the intervention of the measuring device has modified the dynamical state of the measured system. Moreover, this perturbation of the system during the measuring operation is to a certain extent uncontrollable* in the sense that one cannot predict exactly what the state of the system will be after the measurement, but only the probability that it is in a dynamical state corresponding to one of several values x' of the x coordinate.

It is not too surprising that there should be a disturbance of the system in the course of measurement. In order to obtain a measurement, it is necessary to let the system interact in some way with an appropriate measuring apparatus; the latter undergoes some change during the process. Such a change is necessary, since it determines the response of the apparatus, that is to say, the value taken by the physical quantity one wishes to measure. Conversely, one cannot avoid some perturbation of the system itself during the process of measurement.

On the scale of macroscopic phenomena, one can always proceed in such a way that this disturbing action of the measuring device on the system is negligible, or at least known to a good precision. Let us illustrate this point with an example. One can determine the position of a macroscopic object by forming its image with the aid of a set of lenses on a photographic plate. The dynamical state of the object is unavoidably modified during the measurement owing to the very fact that it is exposed to light (radiation pressure). In the classical approximation where the incident light can be treated as a continuous wave (numerous photons present), this modification may in principle be calculated exactly; provided that the photographic plate is sensitive enough, one can in fact decrease the illumination of the object indefinitely and thus make the modification arbitrarily small.

The above argument is evidently valid only in the limit of the classical approximation. In fact, the action of the measuring instrument on the object cannot be decreased indefinitely since this mutual interaction takes place via discrete quanta. The deflection angle of a light quantum by the object is not defined in a precise manner, in other words, the momentum transfer during the deflection may be considered uncontrollable. The position measurement is thus accompanied by an uncontrollable change in the momentum of the object (the semi-quantitative analysis of this effect will be given in the following section). Since modifications of this type are entirely negligible on the macroscopic scale, classical theory postulates that *all* the dynamical variables of a physical system can be measured simultaneously with an arbitrarily small error; the dynamical state of the system is then defined at any given time by the precise specification of the values assumed by all these variables at that instant. On the level of precision of microscopic phenomena, this postulate lacks experimental support and must be abandoned. *The quantum theory, however, assumes that the unpredictable and uncontrollable disturbance suffered by the physical system during a measurement is always sufficiently strong so that the uncertainty relations always holds true.* Thus, in the measurement of x contemplated above, the system has gone over from state Ψ to state Ψ'. As was already pointed out, state Ψ in general corresponds neither to a precise value of x, nor to a precise value of p, but to a probability distribution $P(x)$ of finding some value of x when a precise measurement of this quantity is undertaken; or to a probability distribution $\Pi(p)$ of finding some value of p in the case of a precise measurement of p. To the new state Ψ' correspond new distributions $P'(x)$ and $\Pi'(p)$; in particular the root-mean-square deviations $\Delta' x$ and $\Delta' p$ of these distributions necessarily satisfy the relation $\Delta' x \cdot \Delta' p \geqslant \frac{1}{2}\hbar$. In particular, if the measurement of x is infinitely precise ($\Delta' x = 0$), the quantity $\Delta' p$ must be infinite. One often expresses this result by stating that one cannot diminish the uncertainty in the position variable x without increasing the uncertainty in the momentum variable p by at least a corresponding amount, and *vice versa*.

We shall examine a few specific examples of measurements and show that the perturbation suffered by the measured system is actually always sufficiently large so that the uncertainty relations of Heisenberg are never contradicted [1]).

13. Position Measurements

a) *Use of a Diaphragm.* – Consider a beam of monoërgic electrons moving parallel to Oz. Let us measure the component of the electron position along some direction Ox perpendicular to Oz. To this effect, one inserts into the path of the beam a screen with a slit (Fig. IV.2). If d is the width of the slit, the x-component of the position of any electron crossing this diaphragm is defined with a precision $\Delta x = d$.

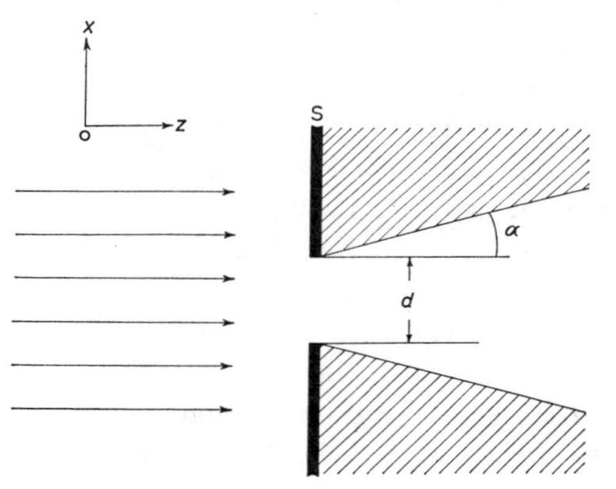

Fig. IV.2. Measurement of position by means of a diaphragm.

However, this electron is represented by a de Broglie wave of wavelength $\lambda = \hbar/p$. The crossing of the diaphragm is accompanied by a

[1]) This especially happens in all experimental arrangements which permit the "trajectory" of a particle to be displayed, as is the case for photographic plates or Wilson cloud chambers for the observation of charged particles. Thus in the cloud chamber, the particle ionizes a certain number of atoms along its path, and these ions become centers of condensation on which visible droplets form. The position of the particle is thus obtained with an uncertainty Δq at least equal to the radius of the ionized atom (in practice it is much larger). However, in the interaction with the measuring device, that is to say in the process of ionization of the atom by the particle, the latter suffers an unpredictable and uncontrollable momentum transfer Δp of the order of $\hbar/\Delta q$. The "trajectory" of the particle can thus be observed only within the limits of precision $\Delta q \cdot \Delta p \approx \hbar$.

diffraction effect; the beam therefore diverges by a certain angle α of the order

$$\sin \alpha \approx \frac{\lambda}{d} = \frac{h}{p \cdot \Delta x}.$$

This amounts to a spread $\Delta p_x = p \sin \alpha$ of the electron momentum in the direction Ox, and we have

$$\Delta x \cdot \Delta p_x \approx h.$$

The x-component of the electron momentum, assumed perfectly well-known before the measuring operation ($p_x = 0$) has been shifted by an uncontrollable amount of order $\hbar/\Delta x$ during the measuring process (crossing of the diaphragm).

It is important to convince oneself thoroughly here that the momentum transferred from the diaphragm to the electron during the measurement cannot actually be determined to better than within $\hbar/\Delta x$; otherwise the preceding argument would be vitiated. Indeed, to carry out the position measurement, the diaphragm must be kept stationary and its position (along Ox) must be known to within δx, δx being very small compared to the slit width: $\delta x \ll \Delta x$. But the diaphragm is a quantum object, just like the electron. Its momentum is therefore not defined to better than δp, and we have

$$\delta p \approx \frac{\hbar}{\delta x} \gg \frac{h}{\Delta x}.$$

It can nevertheless remain practically motionless in the course of the measurement provided it is sufficiently heavy; this limitation therefore does not interfere with the measuring operation itself. But it is obviously not possible to determine the momentum variation of the diaphragm with a greater precision than δp and *a fortiori* with a greater precision than $h/\Delta x$.

This discussion stresses an important point. One must postulate that *the measuring apparatus is a quantized object which also obeys the uncertainty relations. This supposes that the uncertainty relations are of an entirely universal nature.* Otherwise, the physical interpretation of the Quantum Theory would have to be profoundly revised.

b) *Use of the Microscope.* – The use of a diaphragm is undoubtedly the most direct way to measure the position of an object. A less direct but equally valid method consists in illuminating that object and observing the image through a microscope. That is, we are considering the determination of the position x of an electron by observing it through a microscope (Fig. IV.3). The precision of the measurement is limited by the fact that the image of each point is a

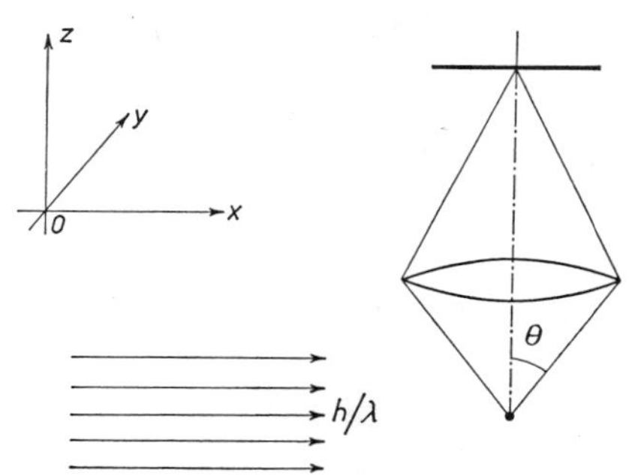

Fig. IV.3. Measurement of position by means of a microscope.

diffraction spot of finite extension. From the extension of that spot one can deduce the limit of precision Δx of the measurement:

$$\Delta x \approx \frac{\lambda}{\sin \theta}$$

where λ is the wavelength of the light, and θ is the half-angle of divergence of the beam scattered by the electron and focused in the microscope. However, the scattering of light proceeds by discrete quanta and is accompanied by a partly uncontrollable momentum transfer (Compton effect). This effect is at a minimum when the scattered light contains but one single photon; the photon momentum has a well-defined magnitude, $p = h/\lambda$; its direction of propagation, however, is defined only to within the angle θ. The momentum transferred to the electron is therefore not exactly known; it has an inherent uncertainty

$$\Delta p \approx \frac{h}{\lambda} \sin \theta.$$

The more precise the position measurement, the more important is this effect, and one always has: $\Delta x \cdot \Delta p \approx \hbar$.

One might object that the precision in x is not given by the size of the diffraction spot, but by the precision to which the center of that spot can be determined. This precision is greater the larger the number N of photons participating in the formation of the spot. According to the laws of statistics, the error in x calculated in this way is $\sqrt{N}$ times smaller than the error computed above:

$$\Delta x \approx \frac{1}{\sqrt{N}} \frac{\lambda}{\sin \theta}.$$

However, if the momentum transferred by each photon has an inherent uncertainty $(h \sin \theta)/\lambda$, the uncertainty in the momentum transferred by the N photons is $\sqrt{N}$ times greater (addition of the square of the errors), hence

$$\Delta p \approx \sqrt{N} \frac{h}{\lambda} \sin \theta.$$

Consequently: $\Delta x \cdot \Delta p \approx h$.

14. Momentum Measurements

The same conclusions hold for momentum measurements. The momentum of a particle may *a priori* be measured with arbitrary precision; however, the measuring process is necessarily accompanied by a perturbation which increases the uncertainty in the corresponding position coordinate, in such a way that the uncertainty relations continue to be satisfied. This will be shown in the following two examples.

a) Deflection in a Magnetic Field. – The momentum of a charged particle is usually measured by deflection in a constant magnetic field. The momentum p is related to the radius of curvature R of the particle trajectory by the well-known relation

$$p = \frac{e}{c} \mathscr{H} R,$$

where $\mathscr{H}$ is the magnitude of the magnetic field and e is the charge of the particle.

We shall examine the measurement of the momentum of an electron by this method. Figure IV.4 shows the schematic diagram of the measuring set-up. The electron enters the magnet after passing through the diaphragm A and leaves it through diaphragm B after having suffered a 180-degree deflection (this particular angle has been chosen for simplicity). At the instant immediately *preceding* the beginning of the measurement (i.e. just before the crossing of diaphragm A), we assume the direction of propagation perfectly defined (direction Oy) and the y coordinate of the electron in this direction

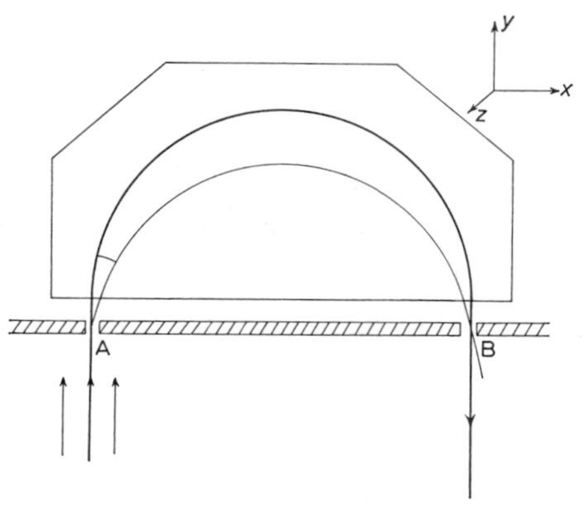

Fig. IV.4. Measurement of momentum by deflection in a magnetic field.

perfectly known ($\Delta y = 0$). These initial conditions ($p_x = p_z = 0$, $y = y_A$) can always be realized in principle by means of a collimator equipped with a shutter whose "exposure time" is sufficiently short. The radius of curvature R is equal to half the distance between the two diaphragms; if $2d$ and $2d'$ are the respective widths of these diaphragms, R is measured to within $d + d'$. The momentum of the electron is therefore known to a precision

$$\Delta p = \frac{e}{c}\,\mathscr{H}(d + d') = \frac{p}{R}\,(d + d').$$

The quantity measured here is the momentum component along the axis Oy. Let us show that once the measurement is carried out, the y coordinate of the electron has an uncertainty Δy such that $\Delta y \cdot \Delta p \gtrsim h$. The quantum effect which enters here is the diffraction of the electron wave upon crossing the diaphragm A (the reader can convince himself that diffraction by diaphragm B plays no role whatsoever in this problem). If this effect did not exist, the momentum would be exactly parallel to Oy when entering the magnet; the electron would then very accurately describe a semi-circle from A to B and the time of flight from A to B would be independent of the value of p and equal to $\pi mc/e\mathscr{H}$ (m = mass of the electron). Because of this effect, the angle between the momentum and the Oy axis at the entrance of the magnet has an inherent indeterminacy $\alpha \approx \lambda/d = h/pd$; the trajectory of the electron (i.e. the ray in the geometrical optics approximation) is a circular segment defined to within 2α. The moment when the electron reaches B has an indeterminacy $\Delta t = 2\alpha mc/e\mathscr{H}$, rather than being defined in a precise manner; the uncertainty Δy is p/m times larger, namely

$$\Delta y \approx 2\alpha \frac{pc}{e\mathscr{H}} \approx 2h \frac{c}{e\mathscr{H}d}.$$

Therefore

$$\Delta y \cdot \Delta p \approx 2h \left(1 + \frac{d'}{d}\right).$$

b) *Collision with a Photon.* – Another method of measuring momentum consists in letting the particle under consideration collide with another particle, a photon, say, whose initial momentum is perfectly well-known and then to measure the momentum transferred to that second particle in the collision. Once again let us take the electron of the foregoing problem characterized by the initial conditions $p_x = p_z = 0$, $y = y_A$. To measure its momentum p_y, we illuminate it with perfectly monochromatic light of frequency ν and direction of propagation parallel to the y axis. One of the photons of this radiation might suffer a Compton collision and one measures its final momentum. For simplicity, we shall assume the latter to be also directed along Oy (in the opposite sense) (Fig. IV.5). Let ν' be the

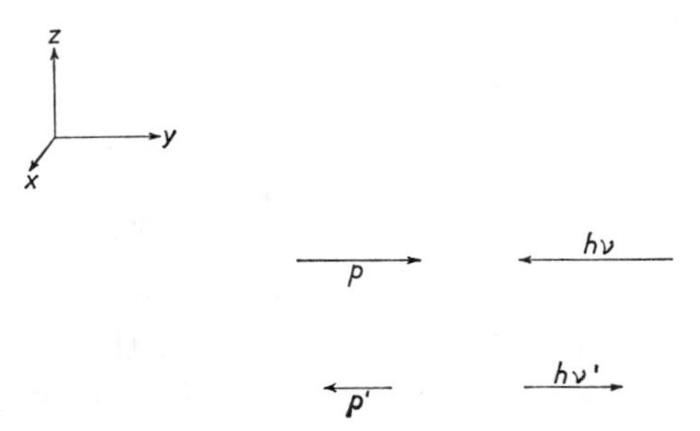

Fig. IV.5. Measurement of momentum of an electron by Compton collision with a photon. Scheme of the collision.

frequency of the photon after collision; the theory of the Compton effect yields the initial momentum p and final momentum p' of the electron as a function of the frequencies ν and ν'. We shall assume that the conditions of non-relativistic approximation hold ($p, p' \ll mc$ and $\nu, \nu' \ll mc^2/h$). One obtains after some calculations

$$p = mc\,\frac{\nu'-\nu}{\nu'+\nu} + \frac{h}{2c}\,(\nu'+\nu), \qquad p' = mc\,\frac{\nu'-\nu}{\nu'+\nu} - \frac{h}{2c}\,(\nu'+\nu),$$

and the precision in the determination of these quantities is connected with the precision in the determination of ν' by the relation

$$\Delta p \approx \Delta p' \approx mc\,\frac{\Delta\nu'}{\nu'+\nu}.$$

The position y of the electron after measurement can be calculated knowing its initial value y_A and that the electron travels with velocity p/m before the Compton collision, and with velocity p'/m afterwards. If the momentum and the position of the scattered photon could be measured simultaneously with extreme precision, the moment when the collision takes place could be determined in a precise manner, and the indeterminacies in p and y could be simultaneously made very small. The reader can easily verify this fact. Indeed, the measurement of ν' is a frequency measurement; the more precise it is, the larger the indeterminacy in the time when the photon passes a given point, and in particular the larger the uncertainty Δt in the time of collision: $\Delta\nu' \cdot \Delta t \gtrsim 1$. Now, the uncertainty in the position of the electron after collision Δy is at least equal to the product of Δt and the change in velocity, hence

$$\Delta y \gtrsim \frac{|p-p'|}{m\Delta\nu'} = \frac{h}{mc}\,\frac{\nu+\nu'}{\Delta\nu'},$$

which yields

$$\Delta y \cdot \Delta p \quad \text{and} \quad \Delta y \cdot \Delta p' \gtrsim h.$$

In this measuring operation as in the foregoing one, the measured quantity p_y is itself modified in the course of the measuring process. *This modification must not be confused with the unpredictable and uncontrollable perturbation suffered by the system during the measurement.* Indeed, this modification is known exactly or, at any rate, the values p and p' of p_y before and after collision are known with equal precision; this precision may be made as great as desired. On the other hand, short of being able to predict and control the perturbation suffered by the particle during the measuring operation (uncertainty in the time at which the momentum and energy are being transferred), one cannot know y once the measurement is carried out, without some uncertainty Δy; the latter is the larger the more precise the measurement of p.

Let us stress again that the uncertainty relations must be of a universal character. Let us assume, for the sake of argument, that the quantum of action for photons is a quantity $\hbar'$ very much smaller than $\hbar$. Clearly, all our arguments concerning measurement (*b*) could be taken over with $\hbar'$ instead of $\hbar$ in all formulae; this would lead us to a value of order $\hbar'$ for the product of the uncertainties $\Delta y \cdot \Delta p'$ and consequently $\Delta y \cdot \Delta p' \approx \hbar' \ll \hbar$. The uncertainty relations would be violated and the entire statistical interpretation which we have developed would be in contradiction with experiment.

The two measurements of momentum discussed here require a certain delay. This delay can actually be reduced to some extent (by increasing $\mathscr{H}$ in the first case; by increasing ν in the second case; cf. Problem IV.7) without affecting the precision of the measurement of p_y. The probabilities defined in §§ 2 and 5 are calculated from the wave function at a given time t, which ought to be specified. If the measurement is instantaneous, t is the time when it is performed. If the measurement requires some delay, t is the time when it begins, that is, when the system starts interacting with the measuring device. The probability law defined in § 2 [eq. (IV.6)] refers to the momentum p_y, i.e. to the momentum before measurement.

IV. DESCRIPTION OF PHENOMENA IN QUANTUM THEORY. COMPLEMENTARITY AND CAUSALITY

15. Problems Raised by the Statistical Interpretation

Clearly, the representation of a quantum system by its wave function is unfamiliar because of its abstract character, and the statistical interpretation discussed above is difficult to grasp intuitively. However, when one seeks to picture microscopic phenomena in a more concrete fashion, one always runs into contradictions.

For instance, let us consider a helium atom; for simplicity, we assume the helium nucleus to be infinitely heavy and motionless, and we treat the atom as a system of two electrons represented by the wave function $\Psi(\mathbf{r}_1, \mathbf{r}_2; t)$. The simplest picture one can draw is that of two corpuscles, the electrons, performing a somewhat complicated motion around the nucleus. This implies that each of them has a well-defined trajectory, whereas the function Ψ does not give more than the statistics of positions in the case of position measurement, and the statistics of momenta in the case of momentum measurement. Since we assume that the dynamical state of the atom is completely defined by the wave function Ψ, this corpuscular representation is partly incorrect. At the other extreme, one can imagine these two electrons to be a continuous distribution of electricity in the space surrounding the nucleus, or preferably, as a continuous wave charged with electricity. But this picture also gives rise to difficulties. In the first place, the wave function Ψ is defined in configuration space and not in ordinary space; it can therefore not be identified with the concrete wave which we are discussing, in fact, no satisfactory definition of such a concrete wave has, as yet, been proposed. In the second place, the picture of a continuous wave cannot be reconciled with such phenomena as ionization, where the existence of individual and, to some extent, localizable corpuscles is exhibited.

Of course, there is no existing proof that a consistent and concrete representation of microscopic phenomena is impossible to formulate. The fact remains, however, that nobody has ever succeeded. In fact logical necessity does not require that the more or less abstract concepts of a physical theory be expressible in concrete language. All our intuition, all our sense of what constitutes concreteness are based upon our everyday experience, and the terms used to describe a phenomenon concretely are necessarily drawn from that experience. There is no indication that such a language could be used without contradictions for phenomena which are as far removed from it as those of microscopic physics. This is not the first example of difficulties of this kind. Some results of the theory of relativity similarly offend our common sense [1]), such as for instance the contraction of lengths and the dilatation of time when the relative velocity of the reference frames is close to c. It is therefore hardly surprising that the concrete pictures one tries to draw of microscopic phenomena cannot be pushed all the way without giving rise to some contradiction or some paradox.

In this connection, it is instructive to note the parallelism between

[1]) The occasions to come to grips between human common sense and the discoveries of physics by no means date from the time of relativity theory ("Eppur' si muove . . ." etc.).

the roles played by the constant $\hbar$ in Quantum Theory, and by the constant c in Relativity Theory. The fact that c is finite rather than infinite imposes a revision of the concept of simultaneity and henceforth sets a limit to the domain of validity of Newtonian Mechanics. In the same way, the fact that $\hbar$ is finite rather than zero imposes a revision of the concept of simultaneous measurements and limits the domain of validity of Classical Theory. The concrete pictures suggested by our everyday experience are those of a world where c seems infinite and $\hbar$ seems to be zero; they cannot bodily be brought over into a domain where one or the other of these approximations ceases to be valid.

The absence of a concrete and coherent representation of phenomena in Quantum Theory may therefore in no way be considered as a shortcoming of the theory. Nevertheless it is open to criticism from another point of view.

The question arises whether the description of phenomena in Quantum Theory actually fulfills all the requirements one should expect from a completely satisfactory theory. The first thing one requires of a theory is, of course, that its predictions be in accord with experimental observations. It is quite certain that Quantum Theory meets this condition, at least in the domain of atomic and molecular physics. But a physical theory cannot claim to be *complete* if it does not go further than to state what one observes when one does a given experiment. At the outset of any scientific endeavor one establishes as a fundamental postulate that nature possesses an objective reality independent of our sensory perceptions or of our means of investigation; the object of physical theory is to give an intelligible account of this objective reality.

Now, all the conclusions of the Quantum Theory can always be put into the following form: "One obtains this or that result if one makes this or that observation". One may therefore question whether Quantum Theory actually furnishes a complete description of objective reality.

The question is even more legitimate since the predictions of the theory are of a statistical nature. In Classical Theory, one has recourse to the language and methods of statistics when the information on the physical systems under study is incomplete. The concepts introduced in Classical Statistics cannot lead to a complete description of objective reality; they only allow us to obtain certain average properties and some results concerning the physical systems under study, despite incomplete information. These results do not apply to one particular system, but to a very large number $\mathcal{N}$ of identical and independent systems. In the same way, Quantum Theory does not generally yield with certainty the result of a given measurement performed on an individually selected system, but the statistical distribution of the results obtained when one repeats the same measurement on a very large number $\mathcal{N}$ of independent systems represented by the same wave function.

One would thus be tempted to conclude that Quantum Theory furnishes a correct description of statistical distributions of systems of microscopic objects, but that it cannot claim to describe completely each system when taken individually. According to this view, the knowledge of the wave function would not suffice to define completely the dynamical state of an individual physical system. In order to do so, one should have a certain number of additional data which it is impossible to obtain because of the insufficiency of our means of observation. In other words, the dynamical state of the physical system should be defined at each instant by a certain number of *hidden variables* whose evolution would be governed by some specific laws. The impossibility to predict with certainty the results of a given measurement would simply come from our inability to know the precise values of these hidden parameters. The wave function would not represent the objective state of the system under study; rather it would be a mathematical object containing the totality of information which one possesses on an incompletely known system.

Although this opinion is perfectly tenable, the current view holds [1]) that Quantum Theory furnishes a complete description of natural phenomena. This is based upon the analysis, due to Bohr (1927), of the very special conditions of observation on the microscopic scale, and on a general principle which evolves from Bohr's analysis – the complementarity principle.

16. Description of Microscopic Phenomena and Complementarity

Any description of natural phenomena – whose objective reality is by no means questioned here – must inevitably involve at some stage the results of our observations bearing on these phenomena.

---

[1]) Cf. footnote Ch. II, p. 48.

Now – and this is the first point of Bohr's analysis – *no matter how far the phenomena transcend the scope of Classical Physics, their account must be expressed in classical terms.* Indeed, to account for an experiment means to give an unambiguous description of the circumstances of the experiment and of the observed results; it means to state, for instance, that "this pointer has stopped on this dial at that point and at that moment". The point we wish to emphasize here is the necessity of using unambiguous language, in which no element of uncertainty on the part of the observer may enter. This is absolutely indispensable, since the experiment must be reproducible, and its progress must remain completely independent of the observer who performs it.

However, *on the microscopic level, one cannot make the sharp separation required by the ordinary concept of observation, between the natural phenomenon and the instrument with which it is observed.* To describe the object and the observing instrument as separate entities is justified only to the extent where the quantum $\hbar$ may be considered negligible. This sets a limit to the analysis of phenomena, when carried out in classical language; any attempt to push the analysis beyond this limit requires a modification of the experimental arrangement which introduces new possibilities of interaction between object and measuring instruments.

As a consequence, *evidences obtained under different experimental conditions cannot be comprehended within a single picture.* However, they must be regarded as *complementary* in the sense that only the totality of the observational results exhausts the possible types of information about the objects of microscopic physics. Such is the content of the complementarity principle.

17. Complementary Variables. Compatible Variables

The description of the phenomena in microscopic physics is thus made up of complementary elements; these elements which are simultaneously needed in any attempt to build a classical picture, are respectively defined by means of mutually exclusive experimental arrangements.

A given component of the position x and the corresponding momentum component p form a pair of complementary elements in the above sense. The precise measurement of x and that of p require incompatible experimental arrangements, so that the simultaneous measurement of these two quantities cannot be carried beyond the limit of precision $\Delta x \cdot \Delta p \approx \hbar$. This is what follows from the discussions of §§ 13 and 14. The mutually exclusive character of the experimental arrangements designed for measuring x and p is very simply exhibited when observing an electron by means of a diaphragm (§ 13*a*). The diaphragm used for the position measurement can serve equally well for the momentum measurement. The latter is obtained by measuring the momentum transferred to the screen through its interaction with the electron. But since electron + screen form an indivisible quantum system, and since it is impossible to make a sharp separation between its two parts, one cannot use ordinary classical language and speak of a separate evolution of the electron (measured system) and the screen (measuring instrument) unless one assumes that their mutual interaction is to some extent uncontrollable. If one wishes to know x with precision Δx, one cannot control the momentum transferred to the screen with a precision greater than $h/\Delta x$. If one wants to know p with precision δp, adding to the screen all the experimental devices necessary for a precise determination of the momentum transfer, one cannot control the position of the screen with a precision greater than $h/\delta p$, and this limits our knowledge of x correspondingly (cf. Problem IV.10).

The variables x and p are said to form a pair of *complementary variables*. One often states the complementarity principle in the following more restrictive form:

The description of the physical properties of microscopic objects in classical language requires pairs of complementary variables; the accuracy in one member of the pair cannot be improved without a corresponding loss in the accuracy of the other member.

This statement stresses the essential point of difference between Quantum Mechanics and Classical Mechanics, namely that the dynamical variables of a quantum system cannot all be defined simultaneously with infinite accuracy.

By definition, two dynamical variables are said to be *compatible* if they can simultaneously be defined with infinite precision. As will be shown in Chapter V, two compatible variables are represented by commuting linear operators. Thus the coordinates x and y of a particle are compatible variables. Of particular importance are the *complete sets of compatible variables* $a, b, c, \ldots$, composed of pairwise compatible variables, and such that any other variable compatible with each of them is a function $f(a, b, c, \ldots)$ of these variables. Consider, for example, a one-particle quantum system. The three position variables x, y, z form a complete set of compatible variables. Indeed, all three can be defined simultaneously with infinite precision. On the other hand, any dynamical variable is a function of x, y, z, and p_x, p_y, p_z; the only ones compatible with x, y, and z are those which are independent of p_x, p_y, and p_z, since each of the latter quantities is incompatible with x, y, and z, respectively; hence they are the functions $F(x, y, z)$. In the same way p_x, p_y, and p_z form a complete set; similarly p_x, y, and p_z (the specification of these three variables completely defined the dynamical state of the system before measurement in the examples of § 14). A precise measurement carried out on a complete set of compatible variables of a system represents the *maximum information* one can obtain on that system. It therefore defines the dynamical state completely, and a definite wave function corresponds to it. We shall come back to this point in Chapter V (§§ 15 and 16).

18. Wave-Corpuscle Duality and Complementarity

If one adopts the principle of complementarity, the wave-corpuscle duality ceases to be paradoxical: the wave aspect and the corpuscular aspect are two complementary aspects which are exhibited only in mutually exclusive experimental arrangements. Any attempt to reveal one of the two aspects requires a modification of the experimental set-up which destroys any possibility of observing the other aspect.

Consider for instance, Young's diffraction experiment. The monochromatic radiation originating from the source S crosses the screen Y through two openings separated by d, and then forms the interference pattern on the screen E placed at a distance D from Y (Fig. IV.6). The distance between fringes is $\lambda D/d$, λ being the wavelength of the radiation. In this experiment, the source S and the screens Y and E are rigidly mounted on a common support. In fact the fringes cannot be observed unless the position of Y with respect to S and E is controlled to a tolerance δx smaller than the fringe spacing:

$$\delta x < \frac{\lambda D}{d}.$$

(For the sake of simplicity the source S is assumed to be at infinity.) The corpuscular aspect of the radiation is displayed if one can find out through which hole of the screen Y the particle has passed. This

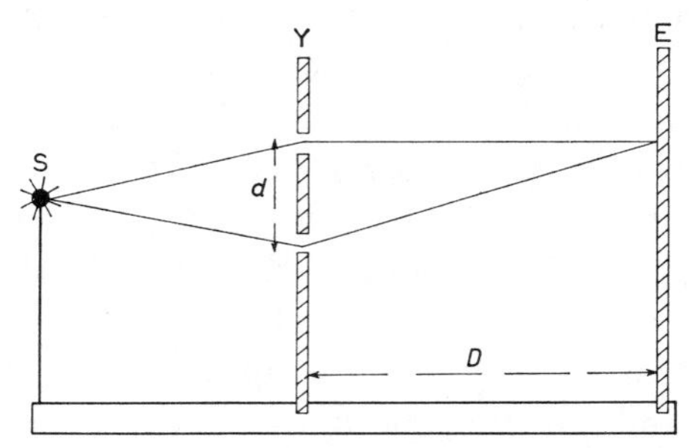

Fig. IV.6. Complementarity between the corpuscular aspect and the wave aspect in the Young double-slit experiment. S = source; Y = slotted screen; E = detection screen. In the figure, S, Y and E are rigidly attached to a common support; the experimental arrangement allows the display of the interference pattern, but all information concerning the trajectory followed by each photon escapes detection by the same token.

may be done most simply by measuring the momentum transferred to the screen Y; it differs by $\approx pd/D$ according to whether the corpuscle passed through one or the other opening, $p=h/\lambda$ denoting the momentum of the corpuscle. The measurement must therefore be made to a precision such that

$$\delta p < \frac{hd}{\lambda D}.$$

In order to reveal simultaneously the wave aspect and the corpuscular aspect, both of these inequalities must be satisfied simultaneously, i.e. one must have

$$\delta x \cdot \delta p < h,$$

which is clearly impossible since the observing instrument is a quantum object and satisfies the uncertainty relations. Any other attempt to reveal the corpuscular aspect would similarly result in a suppression of the interference fringes (cf. Problem IV.11).

19. Complementarity and Causality

The discussion of phenomena in the framework of Quantum Theory implies some definite restriction on the ordinary principle of causality.

Causality, rigorously, applies only to isolated systems. The dynamical state of such a system is represented at a given instant of time by its wave function at that instant. The causal relationship between the wave function $\psi(t_0)$ at an initial time t_0, and the wave function $\psi(t)$ at any later time, is expressed through the Schrödinger equation. However, as soon as it is subjected to observation, the system experiences some reaction from the observing instrument. Moreover, the above reaction is to some extent unpredictable and uncontrollable since there is no sharp separation between the observed system and the observing instrument. They must be treated as an indivisible quantum system whose wave function $\Psi(t)$ depends upon the coordinates of the measuring device as well as upon those of the observed system. During the process of observation, the measured system can no longer be considered *separately* and the very notion of a dynamical state defined by the simpler wave function $\psi(t)$ loses its meaning. Thus the intervention of the observing instrument destroys all causal connection between the state of the system before and after the measurement; this explains why one cannot in general predict with certainty in what state the system will be found after the measurement; one can only make predictions of a statistical nature [1]).

In order to show how the notions which enter into the description of a phenomenon are inseparable from the manner in which it is observed, and what limitations are thereby imposed on causality, we shall take the example of an atom which de-excites through emission of a photon. This example will at the same time illustrate complementarity between wave and corpuscular aspects.

The atom is assumed to be in its excited state at time $t=0$; call τ its mean life, and $\hbar\omega$ the energy of the emitted photon. We wish to

---

[1]) The statistical predictions concerning the results of measurement are derived very naturally from the study of the mechanism of the measuring operation itself, a study in which the measuring instrument is treated as a quantized object and the complex (system + measuring instrument) evolves in causal fashion in accordance with the Schrödinger equation. A very concise and simple presentation of the measuring process in Quantum Mechanics is given in F. London and E. Bauer, *La Théorie de l'Observation en Mécanique Quantique* (Paris, Hermann, 1939). More detailed discussions of this problem may be found in J. von Neumann, *Mathematical Foundations of Quantum Mechanics* (Princeton, Princeton University Press, 1955), and in D. Bohm, *Quantum Theory* (New York, Prentice-Hall, 1951).

make a precise measurement of the time of photon emission. To this effect, we completely surround the atom with suitable counters; one of these counters will eventually be struck by the photon. Knowing the time at which the count took place, the distance separating the atom from the counter, and the velocity of propagation of light c, one easily calculates the moment at which the emission occurred. However, the predictions of Quantum Theory are only statistical in nature. Indeed, consider the system (atom+photon) at time $t \gg \tau$: the atom is then with near certainty in its ground state, and the photon is to a very good approximation – we shall not give the proof here – represented by a wave packet $\psi(\mathbf{r}, t)$ whose dependence upon t and r is of the form

$$f(r, t) = \begin{cases} 0 & \text{when} \quad r > ct \\ \dfrac{1}{r} \exp\left[-\left(i\omega + \dfrac{1}{2\tau}\right)\left(t - \dfrac{r}{c}\right)\right] & \text{when} \quad r < ct \end{cases} \qquad \text{(IV.34)}$$

(r is the distance between the photon and the atom).

The probability of detecting a photon at a distance R from the atom is zero as long as $t < R/c$, and proportional to $\exp\,[-(t-R/c)/\tau]$ at any later time; hence, we have the well-known exponential law giving the statistical distribution of the time of decay.

This law actually fits experimental observations on a very large number of decaying atoms. However, Quantum Theory is unable to predict the time of decay of each individual atom. One is therefore tempted to conclude that the Quantum Theory describes correctly the de-excitation of statistical ensembles of excited atoms, but that it does not describe completely the phenomenon of de-excitation of a single atom.

The reply to this objection is as follows. The notion of decay time cannot be separated from the experimental arrangement which serves to reveal it, and must not be regarded as a property characterizing the evolution of the atom under consideration, independently of this arrangement. In fact, there exist other experimental arrangements, complementary to the one just considered, through which aspects of the decay phenomenon are revealed which are totally incompatible with the very existence of a precise time of decay. These are all the arrangements devised for the observation of the interference effects of the emitted light. Through those arrangements, the wave (IV.34) is split into two wave packets which are later made to recombine after having traveled along different optical paths. Obviously, the above wave packets cannot interfere unless the difference between the two optical paths is smaller than the spatial extension of the wave (IV.34) [1]). In such phenomena, the spatial extension of the wave is clearly exhibited and the very notion of a decay time is meaningless.

As far as causality is concerned, it actually applies to the system (atom + photon) as long as the latter has not yet interacted with the observing instrument. During that entire period, the description of the dynamical state of this system by means of a wave function obeying the Schrödinger equation remains possible; in particular, the photon is represented by the packet of outgoing waves (IV.34) whose wave front ($r=ct$) recedes from the atom with velocity c. However, this causal description loses validity as soon as the system (atom + photon) starts to interact with the observing instrument. One can no longer consider the system as a separate entity since the ensemble (system + observing instrument) forms an indivisible whole. The properties commonly attributed to the system are in fact properties of the complex. Therefore, no strict causal relationship can exist between the state of the system before the measurement and its state afterwards.

This situation differs very profoundly from the strict causality of Classical Theory; the dynamical variables of a classical system are all defined at each instant in a precise manner and their evolution in time is strictly determined. Those of a quantum system are only defined in the limits of approximation required by the uncertainty relations between pairs of complementary variables, and their evolution in time is only partially determined. Surprising as it might appear, this limitation on the causality principle does not run the risk of contradicting any experimental fact, since the evolution in time of the ensemble of dynamical variables of the system can be observed experimentally only within the limits of approximation required by the uncertainty relations.

---

[1]) The lifetime of an excited state emitting visible light is usually of the order of 10^{-8} sec. Consequently, the length of the emitted wave train $c\tau$ is of the order of 3 m. Interference patterns corresponding to optical path differences of the order of a meter have actually been observed. This confirms the conclusions which we are drawing here on the spatial extension of the wave train, and on the ambiguity of the notion of decay time.

EXERCISES AND PROBLEMS

1. Show that the Hamiltonian of a particle in an electromagnetic field [eq. (II.25)] is Hermitean and that, consequently, the normalization of an arbitrary solution Ψ of the Schrödinger equation is conserved in the course of time. Show that one can write a continuity equation in that case which is identical to eq. (IV.11) if one takes as definition of the current the one suggested by correspondence with Classical Mechanics, namely

$$\boldsymbol{J} = \mathrm{Re}\left[\Psi^* \left(\frac{\hbar}{\mathrm{i}m}\nabla - \frac{e}{mc}\boldsymbol{A}\right)\Psi\right]$$

($\boldsymbol{A}$ = vector potential of the electromagnetic field).

2. A particle in one dimension of mass m moves in a potential $V(x)$ which goes asymptotically (more rapidly than $1/|x|$) to V_+ and V_- when x approaches $+\infty$ and $-\infty$, respectively. Introduce the wave numbers $k_\pm = \sqrt{2m(E-V_\pm)}/\hbar$ and consider the solutions $u_E(x)$ whose asymptotic behavior at the two ends of the x axis is

$$u_E \underset{x\to-\infty}{\sim} \mathrm{e}^{\mathrm{i}k_- x} + R\,\mathrm{e}^{-\mathrm{i}k_- x}$$

$$u_E \underset{x\to+\infty}{\sim} S\,\mathrm{e}^{\mathrm{i}k_+ x}.$$

By superposition of solutions of this type, form a wave packet (normalized to unity) representing a particle moving in the direction of increasing x. Show that after a sufficiently long time this packet splits into a transmitted wave packet and a reflected wave packet, and that the probabilities of finding the particle in the one or the other of these packets are equal to $(k_+/k_-)|S|^2$ and $|R|^2$, respectively.

3. The wave function associated with a particle constrained to move along the x axis is

$$\psi(x) = (2\pi\xi^2)^{-\frac{1}{4}} \exp\left(\frac{\mathrm{i}}{\hbar}\,p_0 x - \frac{x^2}{4\xi^2}\right).$$

Calculate the wave function $\varphi(p)$ of momentum space. Verify that $\psi(x)$ and $\varphi(p)$ are normalized to unity. Calculate the average values of x and x^2 by successively applying the expressions (IV.13) and (IV.21), and compare the results. Calculate the mean values of p, p^2, and $\exp(\mathrm{i}pX/\hbar)$ (X is a given real constant) by successively applying expressions (IV.14) and (IV.20), and compare the results.

4. Show that the wave packet for which $\Delta x \cdot \Delta p = \hbar/2$ (minimum wave packet) is necessarily of the form

$$(2\pi\xi^2)^{-\frac{1}{4}} \exp\left[\frac{\mathrm{i}}{\hbar}\,p_0 x - \frac{(x-x_0)^2}{4\xi^2}\right]$$

and that one has $\langle x\rangle = x_0$, $\langle p\rangle = p_0$, $\Delta x = \xi$, $\Delta p = \hbar/2\xi$.

5. Let $\langle x \rangle$ and $\langle p \rangle$ be the mean values of x and its conjugate momentum p, for a system in the dynamical state $\psi(x)$. Show that the mean values of x and p for the dynamical state

$$\exp\left(-\frac{\mathrm{i}}{\hbar}\langle p \rangle\, x\right)\psi(x + \langle x \rangle)$$

vanish.

6. Show that in the position measurement of § 13*b*, the quantum nature of the measuring device (light beam + microscope) plays an essential role, and that the uncertainty relations would be violated if the light were not quantized. Show, in this example, that the constant h which enters in the uncertainty relations for the electron is necessarily no greater than the constant which occurs in the definition of light quanta.

7. In the momentum measurement by deflection in a magnetic field (Ch. IV, § 14*a*), show that one can decrease to a certain extent the duration of the measurement $\tau = \pi mc/e\mathscr{H}$ without changing the precision of the measurement, but that the precision $\Delta\tau$ with which this duration of measurement is defined, remains unchanged. Discuss the time-energy uncertainty relation in this example. Elaborate on the same questions bearing on the duration $1/\Delta\nu'$, and the measurement of momentum by collision with a photon (§ 14*b*).

8. The momentum measurement of § 14*b* may be undertaken by using any kind of particle in the place of photons. It consists in making the electron undergo a collision with that particle, and to measure the momentum transferred to the latter in the collision. Discuss the momentum-position uncertainty relations in this case in the non-relativistic approximation. Show that the uncertainty introduced by the measurement of the position of the electron depends upon the constant $\hbar$ occurring in the uncertainty relations belonging to the particle used.

9. A quantized particle is made to pass through an opening of diameter d equipped with a shutter which opens for a time τ. Show that the particle necessarily exchanges with this device (diaphragm + shutter) a momentum of the order of $\hbar/d$ and an energy of the order of $\hbar/\tau$.

10. In the position measurement of an electron by means of a microscope (§ 13*b*) one seeks to determine the momentum imparted to the electron by carrying out a precise measurement of the momentum transferred to the microscope in the course of the operation. Show that one can improve the precision of this momentum measurement only at the expense of a decrease in precision in the position measurement, in accord with the uncertainty relations.

11. One performs the experiment of Young's double slit on a monoërgic beam of charged particles and one seeks to reveal the interference fringes by using a cloud chamber as a detector. This is possible provided the chamber is sufficiently far from Young's screen. Show that the observation of the "trajectory" followed by each particle in the cloud chamber is not accurate enough to allow us to decide through which slit the particle has traveled. (Cf. footnote p. 142.)

CHAPTER V

DEVELOPMENT OF THE FORMALISM OF WAVE MECHANICS AND ITS INTERPRETATION

1. Introduction

In the preceding chapter we have given the basic elements of the statistical interpretation of Quantum Theory; taking a very general point of view, we have examined the internal consistency of that interpretation, its compatibility with the experimental facts and the very special conditions it implied for the description of natural phenomena. Restricting ourselves henceforth to the narrower framework of the Wave Mechanics of systems of particles (in the non-relativistic approximation), we shall complete and refine this interpretation in accordance with the general principles we previously established.

Thus, we postulate that the wave function completely defines the dynamical state of the system under consideration. In contrast to what occurs in Classical Theory, the dynamical variables of the system cannot in general be defined at each instant with infinite precision. However, if one performs the measurement of a given dynamical variable, the results of measurement follow a certain probability law, and that law must be completely determined upon specifying the wave function.

The probability laws for position and momentum measurements were given explicitly in Chapter IV, Section I; we have also derived a general formula for the mean values of any function of the position coordinates, and of any function of the momentum coordinates. However, the rule for deducing the probability law from the wave function has not been given in the most general case, namely in the case where the measured dynamical variable is a function of *both* position *and* momentum coordinates. We shall correct this flaw in the first three sections of this chapter. The fundamental postulates are presented in Section I. With every dynamical variable $\mathscr{A}$ one associates a certain Hermitean operator A acting on the wave functions, and the mean value of the results of measurements of $\mathscr{A}$ is an expression formed in terms of A, generalizing those of Chapter IV

(§ 5). Assuming this postulate to be valid for any dynamical variable and any function of these variables, the statistical distribution sought is exactly determined. Its explicit determination is closely related to the solution of the eigenvalue problem of A. In Section II, we study the properties of the eigenvalues and eigenfunctions of A in the special case where the spectrum is discrete, and the eigenfunctions are square-integrable. The probability law turns out to be closely related to the coefficients of the expansion of the wave function in a series of eigenfunctions of A. In Section III, the same problem is treated for the general case where the eigenvalue spectrum is partly continuous.

In Section IV, we formally treat the problem of deducing the wave function of a quantum system from the results obtained when carrying out simultaneous measurements of a complete set of compatible variables. When this "maximum observation" is not realized, the information concerning the dynamical state of the system is incomplete; the study of the system can nevertheless be pursued by the methods of statistics, the term statistics being taken here in its usual sense.

With two compatible variables there are associated commuting operators. If the commutators of all pairs of operators were to vanish all the dynamical variables of the system could be simultaneously defined with arbitrary precision. In contrast with this situation characteristic of Classical Theory, some pairs of variables are incompatible and the corresponding commutators differ from zero. The commutators therefore play an essential role in Quantum Theory. Section V is devoted to the study of commutators, to the calculation of the most common commutators, and to the establishment and discussion of some equations in which the commutator concept turns out to be particularly useful.

I. HERMITEAN OPERATORS AND PHYSICAL QUANTITIES

All examples needed for illustrative purposes in the following will be taken from the wave mechanics of one-dimensional systems (Ch. III) or three-dimensional systems (that is, one-particle systems). However, one should never lose sight of the fact that these considerations are very generally applicable to quantum systems with any number of dimensions.

2. Wave-Function Space

The wave functions capable of representing a given quantum system belong to a function space which should be specified precisely. In order that the distributions $P(\boldsymbol{r})$ and $\Pi(\boldsymbol{p})$ defined in Ch. IV, § 2 have a meaning, it is necessary and sufficient that the normalization condition (IV.3) could be applied to the wave function $\psi(r)$. We are thus led to the following definition of wave-function space:

The wave functions of wave mechanics are the square-integrable functions of configuration space, that is to say the functions $\psi(q_1, \ldots, q_R)$ such that the integral $\int |\psi(q_1, \ldots, q_R)|^2 \,\mathrm{d}\tau$ converges [1]) [$\mathrm{d}\tau$ denotes the volume element $\mathrm{d}q_1 \,\mathrm{d}q_2 \ldots \mathrm{d}q_R$].

We could restrict the function space somewhat more by requiring the wave functions to be normalized to unity [eq. (IV.3)]. However, it turns out to be more convenient to relax this normalization condition; this can be done, as we shall see below, at the price of a slight modification in the definition of the statistical distributions and probabilities.

In the language of mathematics, the function space defined above is a *Hilbert space*. It possesses indeed the properties characteristic of such a space, as shown below.

In the first place, it is a *linear space*. If ψ_1 and ψ_2 are two square-integrable functions, their sum, the product of each by a complex number and, more generally, any linear combinations

$$\lambda_1 \psi_1 + \lambda_2 \psi_2,$$

where λ_1 and λ_2 are arbitrarily chosen complex numbers, are also square-integrable functions.

In the second place, one can define a *scalar product* in that space. By definition, the scalar product of the function ψ by the function φ is

$$\langle \varphi, \psi \rangle \equiv \int \varphi^*(q_1, \ldots, q_R)\, \psi(q_1, \ldots, q_R) \,\mathrm{d}\tau. \qquad (\mathrm{V}.1)$$

---

[1]) The Fourier transform $\varphi(p_1, \ldots, p_R)$ of such a function always exists: it is a square-integrable function possessing the same normalization as $\psi(q_1, \ldots, q_R)$. In fact (cf. Appendix A) the correspondence between φ and ψ is one-to-one if one adopts the convention not to consider as distinct two functions which differ only by an ensemble of measure zero, which will always be done.

If it is zero, the functions φ and ψ are said to be *orthogonal*. The *norm* N_ψ of a function ψ is the scalar product of this function by itself:

$$N_\psi \equiv \langle \psi, \psi \rangle.$$

The fundamental properties of the scalar product are as follows:

a) the scalar product of φ by ψ is the complex conjugate of the scalar product of ψ by φ, namely

$$\langle \psi, \varphi \rangle = \langle \varphi, \psi \rangle^* ; \tag{V.2}$$

b) the scalar product of ψ by φ is linear with respect to ψ, in other words

$$\langle \varphi, \lambda_1 \psi_1 + \lambda_2 \psi_2 \rangle = \lambda_1 \langle \varphi, \psi_1 \rangle + \lambda_2 \langle \varphi, \psi_2 \rangle ; \tag{V.3}$$

c) the norm of a function ψ is a real, non-negative number:

$$\langle \psi, \psi \rangle \geqslant 0 \tag{V.4}$$

and if $\langle \psi, \psi \rangle = 0$, we have necessarily [1] $\psi = 0$.

All the above properties are easily deduced from the very definition of the scalar product itself. From properties (*a*) and (*b*) one easily shows that the scalar product $\langle \varphi, \psi \rangle$ does not depend linearly, but antilinearly on φ:

$$\langle \lambda_1 \varphi_1 + \lambda_2 \varphi_2, \psi \rangle = \lambda_1^* \langle \varphi_1, \psi \rangle + \lambda_2^* \langle \varphi_2, \psi \rangle. \tag{V.3'}$$

From the properties (*a*), (*b*), and (*c*) follows a very important property of the scalar product, the *Schwarz inequality* (cf. Problem V.1):

$$|\langle \varphi, \psi \rangle| \leqslant \sqrt{\langle \varphi, \varphi \rangle \langle \psi, \psi \rangle}. \tag{V.5}$$

Equality obtains when the functions φ and ψ are multiples of each other, and only in that case. The Schwarz inequality insures that the integral (V.1) defining the scalar product converges when the functions φ and ψ are square-integrable functions.

In addition to the fact that it is linear and that one can define a scalar product there, the space of square-integrable functions possesses the property of being *complete*; this is what allows us to

[1] Rigorously, ψ can take on values different from zero on an ensemble of measure zero. Following the conventions of the preceding footnote, such a function is not different from zero.

identify it as a Hilbert space. To be complete means that any set of square-integrable functions satisfying the Cauchy criterion, converges (in the quadratic mean) toward a square-integrable function. Conversely, any square-integrable function can be considered as the limit (in the quadratic mean) of a converging series (in the sense of Cauchy) of square-integrable functions (*separability*) [1]).

3. Definition of Mean Values

In Chapter IV, § 5 we were led to associate with any dynamical variable of the type $F(\mathbf{r})$ or of the type $G(\mathbf{p})$ some linear operator A equal to one of the expressions $F(\mathbf{r})$ or $G(-i\hbar\nabla)$; moreover – and this resulted from the definitions adopted for $P(\mathbf{r})$ and $\Pi(\mathbf{p})$ – the mean value of this dynamical variable was given by expression (IV.22) which, with our present notation, can just as well be written $\langle\Psi, A\Psi\rangle$. If the wave function is not normalized to unity, the expressions (IV.2) and (IV.6) for $P(\mathbf{r})$ and $\Pi(\mathbf{p})$ must be divided by the norm $\langle\Psi, \Psi\rangle$ of the wave function, and the expression for the mean value considered above must be replaced by the expression $\langle\Psi, A\Psi\rangle/\langle\Psi, \Psi\rangle$.

Generalizing this result to any dynamical variable, we are therefore led to *postulate* that:

a) with the dynamical variable $\mathscr{A} = A(q_1, \ldots, q_R; p_1, \ldots, p_R)$ is associated the linear operator

$$A\left(q_1, \ldots, q_R; \frac{\hbar}{i}\frac{\partial}{\partial q_1}, \ldots, \frac{\hbar}{i}\frac{\partial}{\partial q_R}\right).$$

b) the mean value of this dynamical variable when the system is in the dynamical state defined by the function $\Psi(q_1, \ldots, q_R)$, is

$$\langle A\rangle = \frac{\langle\Psi, A\Psi\rangle}{\langle\Psi, \Psi\rangle}. \qquad \text{(V.6)}$$

The correspondence between the classical Hamiltonian function and the Schrödinger Hamiltonian (Ch. II, § 15) is a special case of the

[1]) For a rigorous and detailed study of Hilbert space, see M. H. Stone, *Linear Transformations in Hilbert Space* (New York, Amer. Math. Soc., 1932). The main properties are given together with their proof in J. von Neumann, *loc. cit.* For the definition of convergence in the quadratic mean, see Appendix A, footnote p. 475.

correspondence (*a*) between dynamical variables and linear operators. The restrictions made in Ch. II, § 15 concerning this correspondence equally apply to the general case. Let us recall that the q's must be cartesian coordinates. Furthermore there exists a certain ambiguity in the definition of the operator A due to the fact that one replaces in the correspondence operation, quantities of ordinary algebra by operators which might not commute. In practice, one removes this ambiguity by confirming to the empirical rules of Chapter II, § 15.

Furthermore, if the dynamical variable $\mathscr{A}$ represents a physical quantity, it is a real function of the q's and p's, and the results of measurements of $\mathscr{A}$, and *a fortiori* the average value $\langle A\rangle$, are real quantities. Hence $\langle \Psi, A\Psi\rangle$ is real

$$\langle \Psi, A\Psi\rangle = \langle A\Psi, \Psi\rangle. \tag{V.7}$$

and this is so for any dynamical state of the system to which the measurement is applied, hence for any function Ψ. To put it differently (cf. Ch. IV, § 3), *the operator A must be Hermitean.* It is easy to see – and will be readily verified for all further examples – that as long as the operators q_i and $(\hbar/\mathrm{i})\partial/\partial q_i$ are Hermitean, one can always associate by the correspondence rule (*a*) a Hermitean operator with any real dynamical variable. It is precisely with this in mind that the prescription for "symmetrization" indicated in Chapter II, § 15 was given.

The properties of Hermitean operators will be studied systematically in § 5 and following sections. Let us anticipate here an important property of these operators. If A is Hermitean, the mean value of $\mathscr{A}$ calculated for a linear combination $\Phi+\lambda\Psi$ of two functions Φ and Ψ of the function space to which A is applied, is a real quantity. Consequently the quantity

$$\langle \Phi+\lambda\Psi, A(\Phi+\lambda\Psi)\rangle = \langle \Phi, A\Phi\rangle + \lambda\langle \Phi, A\Psi\rangle + \lambda^*\langle \Psi, A\Phi\rangle + |\lambda|^2\langle \Psi, A\Psi\rangle$$

is real. This must be true for any value of the complex number λ. Since $\langle \Phi, A\Phi\rangle$ and $\langle \Psi, A\Psi\rangle$ are real quantities, we conclude therefrom, denoting the phase of λ by α, that

$$\mathrm{e}^{\mathrm{i}\alpha}\langle \Phi, A\Psi\rangle + \mathrm{e}^{-\mathrm{i}\alpha}\langle \Psi, A\Phi\rangle$$

is real, or stated differently that

$$\mathrm{e}^{\mathrm{i}\alpha}(\langle \Phi, A\Psi\rangle - \langle A\Phi, \Psi\rangle) = \mathrm{e}^{-\mathrm{i}\alpha}(\langle A\Psi, \Phi\rangle - \langle \Psi, A\Phi\rangle).$$

Since this equation must hold for any value of α, the quantities between parentheses on both sides of the equation must vanish. In other words, if Ψ and Φ are any two functions of the function space in which the Hermitean operator A acts,

$$\langle \Phi, A\Psi \rangle = \langle A\Phi, \Psi \rangle \tag{V.8}$$

or, what is equivalent,

$$\langle \Phi, A\Psi \rangle = \langle \Psi, A\Phi \rangle^{*}. \tag{V.8$'$}$$

Equation (V.8′) is often taken as the definition of hermiticity.

From the double postulate (*a*) and (*b*) one can deduce the statistical distribution of the values of the physical quantity $\mathscr{A}$. The end of this section and the two following sections are devoted to this problem.

4. Absence of Fluctuations and the Eigenvalue Problem

The fluctuations of the statistical distribution we are looking for are expressed by the root-mean-square deviation ΔA:

$$(\Delta A)^2 = \langle (A - \langle A \rangle)^2 \rangle = \langle A^2 \rangle - \langle A \rangle^2 \qquad (\geqslant 0).$$

A^2 being a dynamical variable by the same token as A, its mean value is given by postulate (*b*). When the deviation ΔA is zero, the fluctuations are absent and one may assert with certainty that $\mathscr{A}$ takes on a well-defined value; the latter is obviously equal to $\langle A \rangle$.

Let us see what condition the relation $\Delta A = 0$ imposes upon the function Ψ. By applying the definition (V.6) to the mean values of the operators A and A^2, one can put this relation into the form

$$\langle \Psi, A^2\Psi \rangle \langle \Psi, \Psi \rangle = \langle \Psi, A\Psi \rangle^2.$$

However, the quantity $\langle \Psi, A^2\Psi \rangle \equiv \langle \Psi, A(A\Psi) \rangle$ is equal to $\langle A\Psi, A\Psi \rangle$, as can easily be shown by applying property (V.8) of the Hermitean operator A to the functions Ψ and $A\Psi$. We therefore have

$$\langle \Psi, A\Psi \rangle^2 = \langle \Psi, \Psi \rangle \langle A\Psi, A\Psi \rangle.$$

We have the case where the Schwarz inequality reduces to an equality. Consequently the functions Ψ and $A\Psi$ are proportional. The fluctuations of the statistical distribution of $\mathscr{A}$ vanish for the dynamical states Ψ_a such that

$$A\Psi_a = a\Psi_a, \tag{V.9}$$

where a is a constant.

Equation (V.9) is an eigenvalue equation; the time-independent Schrödinger equation has already furnished an example of equations of this type. We have thus reached a first conclusion:

The physical quantity $\mathscr{A}$ possesses with certainty (that is to say with probability equal to 1) *a well-defined value if, and only if, the dynamical state of the physical system is represented by an eigenfunction Ψ_a of the Hermitean operator associated with $\mathscr{A}$, and the value assumed by this quantity is the eigenvalue a associated with that function.*

What has just been stated must hold true in particular for the energy $H(q_i; p_i)$ of the system. The corresponding operator, the Schrödinger Hamiltonian, was actually associated with the energy of the system when we established the time-independent Schrödinger equation (Ch. II, Sec. III). We assumed then that the energy of the system takes a well-defined value E when the system is in a stationary state, and that its wave function is an eigenfunction of the operator H corresponding to the eigenvalue E. These assumptions agree with the general postulates (*a*) and (*b*) introduced in the previous section.

The arguments which led to equation (V.9) assume that the functions Ψ, $A\Psi$, $A^2\Psi$ which occur in the various scalar products belong to Hilbert space. Eq. (V.9) itself is an equation in which, until stated otherwise, the unknown function Ψ_a must be square-integrable.

Stated in this way, the eigenvalue problem may very well have no solution. Such is the case of operators as common as q_i and $(\hbar/\mathrm{i})\partial/\partial q_i$. Consider more closely the eigenvalue problem for these two operators in the case of one-dimensional systems. For q, we write

$$(q-q')\,\psi(q)=0,$$

which is possible only if $\psi(q)$ is zero everywhere except when $q=q'$. Not only is there no square-integrable solution satisfying such conditions, but the only possible solution is a particularly strange function, zero everywhere except at one point. We shall discuss this matter more fully in § 8. For $(\hbar/\mathrm{i})\mathrm{d}/\mathrm{d}q$ the eigenvalue problem is written

$$\frac{\hbar}{\mathrm{i}}\frac{\mathrm{d}}{\mathrm{d}q}\,\psi(q) = p'\psi(q).$$

It has a solution, defined to within a constant, for any value of p':

the function $\exp(ip'q/\hbar)$. This eigenfunction is not square-integrable.

For the two preceding operators the result stated above is useless. Clearly, in order to arrive at a statement of general interest, one must also consider the solutions of the eigenvalue problem (V.9) which are not square-integrable. We have already dealt with a particular eigenvalue equation, the Schrödinger equation of one-dimensional systems (cf. Ch. III), and we can be guided by the results of that study in arriving at a general statement. The eigenvalue spectrum of the Schrödinger Hamiltonian generally consists of two parts: a series of discrete eigenvalues whose eigenfunctions have a finite norm, and a continuous spectrum of eigenvalues whose eigenfunctions are bounded in all space but have infinite norm. By superposition of eigenvalues of the continuous spectrum corresponding to neighboring values of the energy, it is intuitively clear that one can form square-integrable functions whose energy is determined, if not in precise fashion, at any rate with a root-mean-square deviation which can be made as small as desired.

Let us elaborate on this point by taking up the notation of Chapter III, § 13. Starting from the eigenfunction $y(x;\, \varepsilon)$ corresponding to the eigenvalue ε of the continuous spectrum of H:

$$Hy(x;\, \varepsilon) = \varepsilon y(x;\, \varepsilon) \tag{V.10}$$

one constructs the "eigendifferential" [1]):

$$Y_\varepsilon(x;\, \delta\varepsilon) = (\delta\varepsilon)^{-\frac{1}{2}} \int_\varepsilon^{\varepsilon+\delta\varepsilon} y(x;\, \varepsilon')\, \mathrm{d}\varepsilon'. \tag{V.11}$$

The latter is a square-integrable function and the quantities $\langle H \rangle$ and ΔH calculated with such a wave function have a well-defined meaning. Using eqs. (V.10) and (V.11) it is easy to see that

$$\langle H \rangle = \frac{\langle Y_\varepsilon,\, HY_\varepsilon \rangle}{\langle Y_\varepsilon,\, Y_\varepsilon \rangle} \underset{\delta\varepsilon \to 0}{\sim} \varepsilon + \mathrm{O}(\delta\varepsilon).$$

$$\langle H^2 \rangle = \frac{\langle Y_\varepsilon,\, H^2 Y_\varepsilon \rangle}{\langle Y_\varepsilon,\, Y_\varepsilon \rangle} \underset{\delta\varepsilon \to 0}{\sim} \varepsilon^2 + \varepsilon \cdot \mathrm{O}(\delta\varepsilon).$$

Therefore, the root-mean-square deviation $\Delta H \equiv \sqrt{\langle H^2 \rangle - \langle H \rangle^2}$ tends

---

[1]) The factor $(\delta\varepsilon)^{-\frac{1}{2}}$ was introduced in the definition of the "eigendifferential" in order that its normalization remain finite in the limit where $\delta\varepsilon \to 0$.

to zero like $(\varepsilon\delta\varepsilon)^{\frac{1}{2}}$ when $\delta\varepsilon$ tends to zero. It can be made arbitrarily small. This result can be expressed in the following manner: by superposition of wave functions (of infinite norm) corresponding to eigenvalues located in a restricted region $(\alpha, \alpha+\delta\alpha)$ of the continuous spectrum of A (if it exists), one can form square-integrable functions, and the root-mean-square deviation of the distribution of the values of A about its mean value $[\approx \alpha+\mathrm{O}(\delta\alpha)]$ may be made as small as desired by choosing the size $\delta\alpha$ of the region sufficiently small.

In conclusion, it is clear that the eigenvalue problem defined by equation (V.9) must play a fundamental role not only in the domain of the discrete spectrum, but also in the domain of the continuous spectrum, domain in which the eigenfunctions no longer belong to Hilbert space. We shall now make a systematic study of this eigenvalue problem.

II. STUDY OF THE DISCRETE SPECTRUM

5. Eigenvalues and Eigenfunctions of a Hermitean Operator

Consider the eigenvalue equation

$$A\psi_a = a\psi_a. \tag{V.9}$$

Throughout this section, we consider only eigensolutions ψ_a *located in Hilbert space. This restricts us to the discrete spectrum.* The general study, including the continuous spectrum (if it exists), will be made in Section III.

Since A is a linear operator:

1. If ψ_a is an eigenfunction, $c\psi_a$, c being an arbitrary constant, is also an eigenfunction of the same eigenvalue. To fix this constant, it is customary to *normalize* the eigenfunctions to unity:

$$\langle \psi_a, \psi_a \rangle = 1.$$

By doing so, ψ_a is defined to within an arbitrary phase.

2. If two linearly independent [1]) functions $\psi_a^{(1)}$ and $\psi_a^{(2)}$ correspond to the same eigenvalue, the same holds true for any linear

---

1) Two functions ψ_1 and ψ_2 are linearly independent if it is impossible to find two non-zero constants λ_1, λ_2 such that $\lambda_1\psi_1 + \lambda_2\psi_2 = 0$.

combination of these functions. One says in this case that there is *degeneracy*. The maximum number of linearly independent functions of a given eigenvalue is called the *order* of the degeneracy (we have encountered examples of degeneracy of order 2 in Chapter III while studying the continuous spectrum).

From the definition of hermiticity and the property (V.8) which follows from it, there result two fundamental properties.

1. *Every eigenvalue a is real.* Indeed, taking the scalar product of both sides of eq. (V.9) by ψ_a from the left, one shows that a is equal to the mean value of A with respect to the dynamical state ψ_a:

$$a = \frac{\langle \psi_a, A\psi_a \rangle}{\langle \psi_a, \psi_a \rangle}$$

and it is real, since A is Hermitean.

2. *Two eigenfunctions corresponding to different eigenvalues are orthogonal* (cf. Ch. III, § 13). Indeed, let us assume that

$$A\psi_1 = a_1\psi_1, \qquad A\psi_2 = a_2\psi_2.$$

Taking the scalar product from the left by ψ_2 in the first equation and the scalar product from the right by ψ_1 in the second equation, and subtracting term by term, one obtains, taking into account the property (V.8)

$$0 = \langle \psi_2, A\psi_1 \rangle - \langle A\psi_2, \psi_1 \rangle = (a_1 - a_2)\langle \psi_2, \psi_1 \rangle.$$

Therefore, if $a_1 \neq a_2$, there necessarily follows

$$\langle \psi_2, \psi_1 \rangle = 0.$$

Consequently, *two eigenfunctions ψ_1 and ψ_2 belonging to different eigenvalues are linearly independent.* Indeed, let us assume that one could find two numbers λ_1, λ_2 such that

$$\lambda_1\psi_1 + \lambda_2\psi_2 = 0.$$

Upon taking the scalar product of each term from the left by ψ_1, we have, taking into account the orthogonality,

$$\lambda_1\langle \psi_1, \psi_1 \rangle = 0,$$

hence λ_1 is necessarily zero. One shows in the same manner that $\lambda_2 = 0$.

If a is a degenerate eigenvalue of order n, each of its eigenvalues can always be put in the form of a linear combination of n particular, linearly independent eigenfunctions $\psi^{(1)}, \psi^{(2)}, \ldots, \psi^{(n)}$. There is a large arbitrariness in the choice of these n basis functions. But one can always choose them such that they are normalized to unity, and orthogonal in pairs. Starting from a set $\psi^{(1)}, \ldots, \psi^{(n)}$ which does not possess these properties one can for instance perform the following manipulations (*Schmidt's process of orthogonalization*). One defines $\varphi^{(1)}$ by

$$c_1 \varphi^{(1)} = \psi^{(1)}$$

and one adjusts the constant c_1 so that $\langle \varphi^{(1)}, \varphi^{(1)} \rangle = 1$, namely

$$|c_1|^2 = \langle \psi^{(1)}, \psi^{(1)} \rangle .$$

One defines $\varphi^{(2)}$ by

$$c_2 \varphi^{(2)} = \psi^{(2)} - \varphi^{(1)} \langle \varphi^{(1)}, \psi^{(2)} \rangle .$$

The left-hand side is certainly not zero since $\psi^{(1)}$ and $\psi^{(2)}$ are linearly independent. It is clear that $\langle \varphi^{(2)}, \varphi^{(1)} \rangle = 0$. c_2 is adjusted in such a way that $\langle \varphi^{(2)}, \varphi^{(2)} \rangle = 1$. $\varphi^{(3)}$ is defined by

$$c_3 \varphi^{(3)} = \psi^{(3)} - \varphi^{(1)} \langle \varphi^{(1)}, \psi^{(3)} \rangle - \varphi^{(2)} \langle \varphi^{(2)}, \psi^{(3)} \rangle .$$

This function is certainly not zero; it is evidently orthogonal to $\varphi^{(1)}$ and $\varphi^{(2)}$, and can be normalized by a suitable choice of c_3. One continues in this way. The n functions $\varphi^{(1)}, \ldots, \varphi^{(n)}$ thus formed satisfy the $n(n+1)/2$ relations

$$\langle \varphi^{(l)}, \varphi^{(m)} \rangle = \delta_{lm} \qquad (l, m = 1, 2, \ldots, n),$$

where δ_{lm} is the Kronecker symbol:

$$\delta_{lm} = \begin{cases} 1 & \text{if} \quad l = m \\ 0 & \text{if} \quad l \neq m. \end{cases}$$

One says that they form a set of *orthonormal* functions.

The study of degenerate eigenvalues must be completed by the following property, which we shall state without proof (cf. footnote p. 166). If the order of degeneracy is infinite, i.e. if one can find an arbitrarily large number of linearly independent eigenfunctions, one can always form a (denumerably infinite) set $\varphi^{(1)}, \varphi^{(2)}, \ldots, \varphi^{(r)}, \ldots$

of orthonormal eigenfunctions such that any eigenfunction corresponding to the eigenvalue a can be expanded in a series of these functions.

One can also show (cf. footnote p. 166) that the eigenvalues form a *discrete set* (finite or denumerably infinite) $a_1, a_2, \ldots, a_p, \ldots$. This property is characteristic of the eigenvalues corresponding to eigenfunctions *located in Hilbert space.*

6. Expansion of a Wave Function in a Series of Orthonormal Eigenfunctions

As was shown above, one can associate with each eigenvalue a_p of A a set of orthonormal eigenfunctions

$$\varphi_p^{(1)}, \varphi_p^{(2)}, \ldots, \varphi_p^{(r)}, \ldots$$

containing one element, a finite number of elements, or an infinite number of elements according to whether a_p is non-degenerate, degenerate of finite order, or infinitely degenerate. We shall denote by $\{\varphi_p^{(r)}\}$ the set formed by all these functions. Any function of the set satisfies the relations

$$A\varphi_p^{(r)} = a_p \varphi_p^{(r)} \qquad \text{(V.12)}$$

$$\langle \varphi_p^{(r)}, \varphi_q^{(s)} \rangle = \delta_{pq}\delta_{rs}. \qquad \text{(V.13)}$$

The question arises now whether an arbitrary wave function ψ, belonging to Hilbert space, can be represented by a series of functions of the set $\{\varphi_p^{(r)}\}$. This series certainly exists if ψ is an eigenfunction of A; in that case, the only non-vanishing terms of the expansion are the coefficients of the functions φ corresponding to the same eigenvalue. If any ψ can be expanded in a series of eigenfunctions, one says that the set $\{\varphi_p^{(r)}\}$ is *complete.*

We shall indicate without proof (cf. footnote p. 166) some properties of expansions in series of orthonormal functions.

Let $u_1, u_2, \ldots, u_n, \ldots$ be a set of orthonormal functions:

(*i*) If ψ can be expanded in a series of these functions:

$$\psi = \sum_n c_n u_n,$$

the expansion coefficients are

$$c_n = \langle u_n, \psi \rangle$$

and satisfy the Parseval relation:

$$\sum_n |c_n|^2 = \langle \psi, \psi \rangle.$$

(*ii*) Conversely, if the numerical series $\sum_n |c_n|^2$ converges to a finite value, say N, the expansion $\sum_n c_n u_n$ converges (in quadratic mean) to a function ψ of norm N.

(*iii*) If the expansions $\sum_n c_n u_n$ and $\sum_n d_n u_n$ converge to ψ and φ respectively, the series $\sum_n d_n^* c_n$ converges to the scalar product of ψ by φ:

$$\langle \varphi, \psi \rangle = \sum_n d_n^* c_n.$$

(*iv*) For any square-integrable ψ the series

$$\hat{\psi} = \sum_n u_n \langle u_n, \psi \rangle$$

always converges. The difference $\psi - \hat{\psi}$ is orthogonal to all the functions u_n, and its norm is equal to $\langle \psi, \psi \rangle - \langle \hat{\psi}, \hat{\psi} \rangle$. One therefore always has

$$\langle \psi, \psi \rangle \geqslant \langle \hat{\psi}, \hat{\psi} \rangle$$

and when the equality holds, $\psi = \hat{\psi}$.

All these properties remain valid if the functions u are characterized by several discrete indices instead of a single one. They therefore apply to the set $\{\varphi_p^{(r)}\}$. In particular, *if the set* $\{\varphi_p^{(r)}\}$ *is complete*, any wave function Ψ can be represented by the series

$$\Psi = \sum_{p,r} c_p^{(r)} \varphi_p^{(r)} \tag{V.14}$$

whose coefficients are given by

$$c_p^{(r)} = \langle \varphi_p^{(r)}, \Psi \rangle \tag{V.15}$$

and satisfy the Parseval relation

$$\sum_{p,r} |c_p^{(r)}|^2 = \langle \Psi, \Psi \rangle. \tag{V.16}$$

Moreover, the scalar product of two wave functions Ψ_1 and Ψ_2 is

$$\langle \Psi_1, \Psi_2 \rangle = \sum_{p,r} \langle \Psi_1, \varphi_p^{(r)} \rangle \langle \varphi_p^{(r)}, \Psi_2 \rangle. \tag{V.17}$$

In fact, if one is not too concerned about mathematical rigor, one easily verifies equation (V.15) and (V.16) by substituting for Ψ the expression (V.14) in the right-hand side of these equations and making use of the orthonormality relations (V.13). Equation (V.17) is obtained by an analogous procedure.

The analogy with ordinary vector space is striking. The complete set of orthonormal functions plays the role of a set of basis vectors of unit length being pairwise orthogonal. The function Ψ is a vector of this space (with an infinite number of dimensions), the coefficients $\langle \varphi_p^{(r)}, \Psi \rangle$ being its components along these basis vectors [eq. (V.15)], and the norm of this vector, that is to say the square of its length, is equal to the sum of the squares of the absolute values of its components [eq. (V.16)]. The scalar product of Ψ_2 by Ψ_1 is equal to the sum of the products of each component of Ψ_2 by the complex conjugate of the corresponding component of Ψ_1.

7. Statistical Distribution of the Results of Measurement of a Quantity Associated with an Operator having a Complete Set of Eigenfunctions with Finite Norm

The possibility of representing any wave function Ψ by an expansion of the type (V.14) greatly facilitates the study of all problems involving A. Let us therefore suppose that the operator possesses a complete set of orthonormal eigenfunctions [the Hamiltonian of the harmonic oscillator (Ch. XII) being an example]. This set is certainly not unique; one can always change the phases of the functions, or else replace orthonormal functions corresponding to the same eigenvalue by orthonormal linear combinations of these functions. However, the properties given below are independent of the choice of the system.

A priori, a function such as $A\Psi$ is not necessarily square-integrable. However, according to (V.14),

$$A\Psi = \sum_{p,r} c_p{}^{(r)} A\varphi_p{}^{(r)} = \sum_{p,r} a_p c_p{}^{(r)} \varphi_p{}^{(r)}.$$

This expansion converges [§ 6, properties (*i*) and (*ii*)] if and only if $\sum_{p,r} a_p{}^2 |c_p{}^{(r)}|^2$ converges and, in that case, the sum of this numerical series is equal to the norm of $A\Psi$. In this way we obtain a criterion for deciding if $A\Psi$ belongs to Hilbert space.

One arrives at analogous conclusions for the function $A^2\Psi$. More generally, starting from a function $F(x)$, one can, by means of the expansion (V.14) define the operator $F(A)$, a function of the operator A. Its action on Ψ is given by

$$F(A)\Psi = \sum_{p,r} F(a_p)\, c_p{}^{(r)}\, \varphi_p{}^{(r)}.$$

The operator is well-defined if the series converges, i.e. for all functions Ψ such that the numerical series

$$\sum_{p,r} |F(a_p)|^2\, |c_p{}^{(r)}|^2$$

is convergent. In that case – as can be verified easily by the reader – the function $F(A)\Psi$ thus defined is the same whatever the set $\{\varphi_p{}^{(r)}\}$ which served for its definition.

In particular, the operator exp $(\mathrm{i}\xi A)$, where ξ is a given parameter, is defined for all functions of Hilbert space. Indeed

$$\mathrm{e}^{\mathrm{i}\xi A}\Psi = \sum_{p,r} \mathrm{e}^{\mathrm{i}\xi a_p}\, c_p{}^{(r)}\, \varphi_p{}^{(r)} \tag{V.18}$$

and the convergence criterion for the series, namely the convergence of the numerical series $\sum_{p,r} |c_p{}^{(r)}|^2$, is always fulfilled.

We are now in a position to establish the statistical distribution of the quantity $\mathscr{A}$ for any dynamical state of the system. Indeed, the characteristic function $f(\xi)$ of this distribution [1]), the knowledge of which suffices to determine the latter completely, is by definition

---

[1]) To within a constant, it is the Fourier transform of this distribution. Let X be a random variable able to take on all values contained in the interval $(-\infty, +\infty)$ and let $P(x)$ be the probability of finding X in the interval $(x, x + \mathrm{d}x)$; the characteristic function $f(\xi)$ of its statistical distribution is the average value of exp $(\mathrm{i}\xi x)$, that is to say

$$f(\xi) = \int_{-\infty}^{+\infty} \mathrm{e}^{\mathrm{i}\xi x}\, P(x)\, \mathrm{d}x.$$

If X can only take certain discrete values $x_1, \ldots, x_n, \ldots$ and if $w_1, \ldots, w_n, \ldots$ are the respective probabilities of these values,

$$f(\xi) = \sum_n w_n\, \mathrm{e}^{\mathrm{i}\xi x_n}.$$

More generally, if $F(x)$ is the probability that $X \leqslant x$, one has, upon introducing the Stieltjes integral

$$f(\xi) = \int_{-\infty}^{+\infty} \mathrm{e}^{\mathrm{i}\xi x}\, \mathrm{d}F(x).$$

the mean value of the quantity $\exp(\mathrm{i}\xi A)$ in that state. By a very natural extension of postulate (*b*) of § 3, we define this average value by the expression

$$f(\xi) = \frac{\langle \Psi, \mathrm{e}^{\mathrm{i}\xi A} \Psi \rangle}{\langle \Psi, \Psi \rangle} \tag{V.19}$$

(which is always meaningful, even if the mean value of A is not defined).

Let Ψ be the wave function representing the dynamical state under consideration. Making use of the expansions (V.14) and (V.18) and expression (V.17) for the scalar product, one has

$$f(\xi) = \sum_p w_p \, \mathrm{e}^{\mathrm{i}\xi a_p},$$

with the notation

$$w_p = \frac{\sum_r |c_p^{(r)}|^2}{\langle \Psi, \Psi \rangle} = \frac{\sum_r |\langle \varphi_p^{(r)}, \Psi \rangle|^2}{\langle \Psi, \Psi \rangle}.$$

By inspection of the characteristic function, we conclude that:

1. *The only values the quantity $\mathscr{A}$ can take on are the eigenvalues $a_1, a_2, \ldots, a_p, \ldots$ of the operator associated with it;*

2. *The probability that $\mathscr{A}$ takes one of these values, say a_p, is equal to w_p.*

One easily verifies that $\sum_p w_p = 1$ (Parseval's relation), and that the average value of $\mathscr{A}$ is given, provided that it converges, by

$$\langle A \rangle = \sum_p w_p a_p,$$

also, more generally, the mean value of a function $f(\mathscr{A})$ is given, if it exists, by

$$\langle f(A) \rangle = \sum_p w_p f(a_p).$$

In particular, in order that $\mathscr{A}$ take on a given value with certainty, it is necessary and sufficient that Ψ be an eigenfunction corresponding to that eigenvalue, in accordance with the deductions of § 4.

One can express the general result above in a way which reveals even more clearly the fact that it does not depend upon the choice of the system $\{\varphi_p^{(r)}\}$. The function Ψ_p defined by

$$\Psi_p = \sum_r \varphi_p^{(r)} \langle \varphi_p^{(r)}, \Psi \rangle,$$

is indeed independent of this choice (cf. Problem V.4). The expansion (V.14) can also be written

$$\Psi = \sum_p \Psi_p. \tag{V.20}$$

Stated differently, one can in a unique manner put Ψ in the form of a sum of eigenfunctions of A, each of which corresponds to a different eigenvalue. Then, the probability w_p of finding the particular value a_p is equal to the ratio of the norm of Ψ_p to that of Ψ:

$$w_p = \frac{\langle \Psi_p, \Psi_p \rangle}{\langle \Psi, \Psi \rangle}. \tag{V.21}$$

III. STATISTICS OF MEASUREMENT IN THE GENERAL CASE

8. Difficulties of the Continuous Spectrum. Introduction of the Dirac δ-Functions

All the conclusions which we have reached are vitiated if the set $\{\varphi_p^{(r)}\}$ is not complete. That case is by no means exceptional. However, the discussions of § 4 suggest a possible extension of the foregoing theory, in which one would still start from equation (V.9) but impose on the eigensolutions conditions less restrictive than to belong to Hilbert space. These more general eigensolutions may have an infinite norm. Our first step is to extend to them the concepts of orthogonality and normalization so important in this whole theory.

For purposes of orientation, consider two examples relating to one-dimensional systems, namely the statistical distributions of position and momentum. In that case, the statistical distributions are known; the formal extension of the results of the preceding section is thereby simplified. The position coordinate q may take on all possible values in the interval $(-\infty, +\infty)$ and the probability of finding q in the interval $(q', q'+\mathrm{d}q')$ is equal to

$$P(q')\,\mathrm{d}q' = |\psi(q')|^2\,\mathrm{d}q' \tag{V.22}$$

where $\psi(q)$ is the wave function – assumed normalized to unity – representing the dynamical state of the system. Likewise, the momentum p represented by the operator $(\hbar/\mathrm{i})\mathrm{d}/\mathrm{d}q$ may take on all possible values in the interval $(-\infty, +\infty)$ and the probability of finding p in the interval $(p', p'+\mathrm{d}p')$ is equal to

$$\Pi(p')\,\mathrm{d}p' = |\varphi(p')|^2\,\mathrm{d}p' \tag{V.23}$$

where $\varphi(p)$ is the Fourier transform, suitably normalized, of the wave function $\psi(q)$.

In both cases, the possible values of the quantities considered form a *continuous spectrum*. Therein lies the principal difference from the foregoing situation characterized by a discrete spectrum and by the possibility of representing any wave function ψ by a series [cf. eq. (V.14) or (V.20)] each term of which is associated with one of the possible values of that spectrum. The natural extension to the case of the continuous spectrum consists in representing ψ no longer by a series but by an integral.

Let us recall briefly the theory of the entirely discrete spectrum.

The Hermitean operator A possesses a discrete set of eigenvalues, assumed non-degenerate for the sake of simplicity. To each of these, a_n, corresponds an eigenfunction φ_n (defined to within a phase) and we have

$$A\varphi_n = a_n\varphi_n \tag{V.24}$$

$$\langle \varphi_n, \varphi_{n'} \rangle = \delta_{nn'}. \tag{V.25}$$

The orthonormal set $\{\varphi_n\}$ being complete, any function assumed normalized to unity can be represented by the series

$$\psi = \sum_n c_n \varphi_n \tag{V.26}$$

and we have from the orthonormality relations (V.25),

$$\langle \varphi_n, \psi \rangle = \sum_{n'} c_{n'} \langle \varphi_n, \varphi_{n'} \rangle = c_n. \tag{V.27}$$

By applying the same relations, one obtains for the expression of the characteristic function

$$\langle \psi, e^{i\xi A}\psi \rangle = \sum_{n,n'} c_n{}^* c_{n'} e^{i\xi a_{n'}} \langle \varphi_n, \varphi_{n'} \rangle = \sum_n |c_n|^2 e^{i\xi a_n}, \tag{V.28}$$

whence one deduces that the probability that A take the value a_n is equal to the square $|c_n|^2$ of the modulus of the coefficient of φ_n in the expansion (V.26).

By analogy, let us denote by $u(p'; q)$ the eigensolution of the operator $p = (\hbar/i)d/dq$ relative to the eigenvalue p',

$$pu(p'; q) \equiv \frac{\hbar}{i} \frac{d}{dq} u(p'; q) = p'u(p'; q). \tag{V.24'}$$

Pursuing the formal development, we represent the wave function $\psi(q)$ by an integral of the form

$$\psi(q) = \int_{-\infty}^{+\infty} c(p')\, u(p'; q)\, \mathrm{d}p'.$$

If the formal analogy can be carried to the end, $|c(p')|^2\, \mathrm{d}p'$ is the probability that p is found in the interval $(p', p'+\mathrm{d}p')$. One must therefore have $|c(p')|^2 = |\varphi(p')|^2$; in other words $c(p')$ is equal to $\varphi(p')$ to within a phase factor. Since the eigensolution $u(p'; q)$ is itself defined to within a constant, one can always choose the latter in such a way that the analogue of eq. (V.26) can be written

$$\psi(q) = \int_{-\infty}^{+\infty} \varphi(p')\, u(p'; q)\, \mathrm{d}p'. \tag{V.26'}$$

The coefficient $\varphi(p')$ corresponding to each eigenvalue p' must be, by generalization of eq. (V.27), equal to

$$\varphi(p') = \langle u(p'; q), \psi(q)\rangle \tag{V.27'}$$

which gives, after substituting for $\psi(q)$ its integral representation (V.26′) in the scalar product,

$$\varphi(p') = \int_{-\infty}^{+\infty} \varphi(p'')\, \langle u_{p'}, u_{p''}\rangle\, \mathrm{d}p'', \tag{V.29}$$

an expression in which we have used the abbreviated notation $u_{p'}$ to denote the eigensolution corresponding to the eigenvalue p'. This property of the scalar product $\langle u_{p'}, u_{p''}\rangle$ must hold for any wave function $\varphi(p')$ of momentum space (the only condition imposed upon φ is to be square-integrable); this property thus generalizes the orthonormality relations (V.25).

In fact, no function of p' and p'' possesses the desired property. Nevertheless, if one does not bother too much about mathematical rigor [1]), one can, following Dirac, make use of the "singular function"

[1]) We have already relaxed the mathematical rigor by writing down equation (V.29). The only correct equation is

$$\varphi(p') = \langle u_{p'}, \int \varphi(p'')\, u_{p''}\, \mathrm{d}p''\rangle$$

from which one obtains (V.29) by inverting the order of the integration over q and p'' This manipulation is certainly not justified since the scalar product $\langle u_{p'}, u_{p''}\rangle$ diverges.

$\delta(x)$ defined by the property

$$\int_a^b f(x)\,\delta(x-x_0)\,\mathrm{d}x = \begin{cases} f(x_0) & \text{for } x_0 \text{ in the interval } (a, b) \\ 0 & \text{for } x_0 \text{ outside the interval } (a, b) \end{cases} \tag{V.30}$$

for any function that is continuous at the point $x=x_0$.

Equation (V.29) is satisfied if

$$\langle u_{p'}, u_{p''}\rangle = \delta(p'-p''). \tag{V.25'}$$

This relation is the generalization to the continuous spectrum of the orthonormality relation (V.25).

One can visualize the "Dirac function" $\delta(x)$ as the limit of a function that is zero everywhere except in a very small interval surrounding the point $x=0$ where it has a very narrow and very high peak such that its integral over the entire interval is equal to 1. In the limit where the width of the peak goes to zero, one has

$$\delta(x) = \begin{cases} 0 & \text{if } x \neq 0 \\ +\infty & \text{if } x = 0 \end{cases} \quad \text{and} \int_{-\infty}^{+\infty} \delta(x)\,\mathrm{d}x = 1. \tag{V.31}$$

$\delta(x)$ is certainly not a function in the usual sense of the word since the integral, if it exists, of a function which is zero everywhere except at one point, necessarily vanishes. We shall not dwell here on the mathematical justification of the use of the "Dirac function". It requires the introduction of a new concept, that of distribution, of which the usual functions (more precisely the locally integrable functions) are particular cases. The mathematicians do not speak of the function $\delta(x-x_0)$ but of the distribution $\delta_{x_0}[f]$, defined as the functional of $f(x)$ equal to $f(x_0)$. In other words, the definition (V.30) must be replaced by the definition:

$$\delta_{x_0}[f] = f(x_0).$$

As the concept of distribution may be unfamiliar to the reader, we shall refer to it as little as possible and we shall continue to make use of the (incorrect) notation $\delta(x-x_0)$ which has in fact some indisputable formal advantages. Its main rules of calculation are listed in Appendix A, where one will also find a brief outline of Distribution Theory.

Let us now take up the problem of the measurement of p. The eigensolution $u(p'; q)$ of equation (V.24′) is the function $c \exp(\mathrm{i}p'q/\hbar)$.

The generalized orthonormality relation (V.25′) is satisfied if one takes $c = 1/\sqrt{2\pi\hbar}$, namely

$$u(p'\,;\,q) = \frac{1}{\sqrt{2\pi\hbar}}\,\mathrm{e}^{\mathrm{i}p'q/\hbar}.$$

With this, and using eq. (A.22), we have

$$\langle u_{p'}, u_{p''}\rangle = \frac{1}{2\pi\hbar}\int_{-\infty}^{+\infty} \mathrm{e}^{\mathrm{i}(p''-p')q/\hbar}\,\mathrm{d}q = \delta(p'-p'').$$

In addition, the eigenfunctions $u_{p'}$ form a complete set since any square-integrable function $\psi(q)$ can be put in the form (properties of the Fourier integral)

$$\psi(q) = \int_{-\infty}^{+\infty} \varphi(p')\,\frac{\mathrm{e}^{\mathrm{i}p'q/\hbar}}{\sqrt{2\pi\hbar}}\,\mathrm{d}p'.$$

The coefficient $\varphi(p')$ of this "expansion in a series of eigenfunctions" is actually equal to the scalar product $\langle u_{p'}, \psi\rangle$. Indeed

$$\langle u_{p'}, \psi\rangle = \langle u_{p'}, \int_{-\infty}^{+\infty} \varphi(p'')\,u_{p''}\,\mathrm{d}p''\rangle = \int_{-\infty}^{+\infty} \varphi(p'')\,\langle u_{p'}, u_{p''}\rangle\,\mathrm{d}p''$$
$$= \int_{-\infty}^{+\infty} \varphi(p'')\,\delta(p'-p'')\,\mathrm{d}p'' = \varphi(p').$$

We recognize the reciprocity property of the Fourier integral, making use of the generalized orthonormality relation.

Pursuing further the analogy with the case of the discrete spectrum, we introduce the operator $\exp(\mathrm{i}\xi p)$. We have

$$\mathrm{e}^{\mathrm{i}\xi p}\,u(p'\,;\,q) = \mathrm{e}^{\mathrm{i}\xi p'}\,u(p'\,;\,q),$$

hence

$$\mathrm{e}^{\mathrm{i}\xi p}\,\psi(q) = \int_{-\infty}^{+\infty} \varphi(p')\,\mathrm{e}^{\mathrm{i}\xi p'}\,u(p'\,;\,q)\,\mathrm{d}p'\,;$$

from this the characteristic function

$$f(\xi) = \langle \psi, \mathrm{e}^{\mathrm{i}\xi p}\,\psi\rangle = \int_{-\infty}^{+\infty} \varphi^*(p'')\,\mathrm{d}p'' \int_{-\infty}^{+\infty} \mathrm{e}^{\mathrm{i}\xi p'}\varphi(p')\,\mathrm{d}p' \times \langle u_{p''}, u_{p'}\rangle$$
$$= \int_{-\infty}^{+\infty} \varphi^*(p'')\,\mathrm{d}p'' \int_{-\infty}^{+\infty} \mathrm{e}^{\mathrm{i}\xi p'}\varphi(p')\,\mathrm{d}p' \times \delta(p'-p'')$$
$$= \int_{-\infty}^{+\infty} |\varphi(p')|^2\,\mathrm{e}^{\mathrm{i}\xi p'}\,\mathrm{d}p'.$$

The statistical distribution which corresponds to it (cf. footnote p. 177) is just the expected distribution (V.23).

The study of the position measurement can be made by the same scheme. The eigensolution, suitably normalized, corresponding to the eigenvalue q' of operator q, is $\delta(q'-q)$. Indeed, according to equation (A.19),

$$q\delta(q'-q)=q'\delta(q'-q).$$

The ensemble of functions $\delta(q'-q)$, where q' can take all possible values from $-\infty$ to $+\infty$, forms an orthonormal set since [eq. (A.21)]

$$\int_{-\infty}^{+\infty} \delta(q-q')\,\delta(q-q'')\,\mathrm{d}q=\delta(q'-q''),$$

and this system is complete since any wave function $\psi(q)$ has the integral representation

$$\psi(q)=\int_{-\infty}^{+\infty} \psi(q')\,\delta(q'-q)\,\mathrm{d}q. \tag{V.32}$$

One easily verifies that the coefficient $\psi(q')$ is equal to the scalar product $\langle\delta(q'-q), \psi(q)\rangle$. Following the same prescription as for the measurement of p, one arrives at the conclusion that the square of the modulus of that coefficient $|\psi(q')|^2$ is equal to the probability density that $q=q'$, in accordance with eq. (V.22).

9. Expansion in a Series of Eigenfunctions in the General Case. Closure Relation

Let us return to the eigenvalue problem

$$A\psi=a\psi. \tag{V.9}$$

We no longer require that the eigensolutions have a finite norm but merely that their scalar product by any wave function (i.e. by any function of finite norm) be convergent.

Most generally, the eigenvalue spectrum consists of:

1. a discrete part, i.e. a finite or denumerably infinite set of values a_n, labelled by the integral index n;

2. a continuous part, i.e. a set of values $a(\nu)$ which can be conveniently defined as the values taken by a uniform, continuous and monotonic function of some index ν.

The eigenfunctions of the discrete part of the spectrum have finite norm. All properties of the discrete spectrum were studied in § 5, and do not need to be repeated here.

Let $\psi(\nu; q_1, \ldots, q_R)$ be an eigenfunction of the continuous spectrum corresponding to the eigenvalue $a(\nu)$. It is a continuous function of the parameter ν whose normalization integral certainly diverges (it is a "vector of infinite length" in function space). However, the eigendifferential

$$(\Delta\nu)^{-\frac{1}{2}} \int_{\nu}^{\nu+\Delta\nu} \psi(\nu'; q_1, \ldots, q_R)\, \mathrm{d}\nu'$$

is supposed to be a function whose norm is finite and tends to a constant value when $\Delta\nu$ goes to zero. We know that the eigenfunctions of the continuous spectrum of the Hamiltonian of a one-dimensional system possess this property (cf. Ch. III); the same holds true, as is easily verified, for the operators q and $(\hbar/\mathrm{i})\mathrm{d}/\mathrm{d}q$ of the preceding paragraph.

One often says that a function is *normalizable* when its norm is finite. By a very natural extension of the terminology, we shall likewise apply this term to fun tions of infinite norm whose eigendifferential has a finite norm. Thus the eigenfunctions of the continuous spectrum are normalizable in this sense, as well as those of the discrete spectrum, although they do not belong to Hilbert space.

Using property (V.8) (which has a meaning only for square-integrable functions Φ and Ψ) suitably applied to eigendifferentials rather than to eigenfunctions, one obtains the fundamental properties concerning the continuous spectrum which are the counterpart to the properties of the discrete spectrum shown in Chapter V, § 5:

1. every eigenvalue $a(\nu)$ is real;

2. two eigenfunctions corresponding to distinct eigenvalues are orthogonal. To be precise the orthogonality property is the generalization of equation (III.42′). It is not correct to write

$$\langle \psi_\nu, \psi_{\nu'} \rangle = 0$$

since the scalar product $\langle \psi_\nu, \psi_{\nu'} \rangle$ diverges in general. But one has

$$\langle \psi_\nu, (\Delta\nu')^{-\frac{1}{2}} \int_{\nu'}^{\nu'+\Delta\nu'} \psi_{\nu''}\, \mathrm{d}\nu'' \rangle = 0$$

when ν lies outside the interval $(\nu', \nu'+\Delta\nu')$. The proof of these properties is not difficult and will be left to the reader.

If the eigenvalues of the continuous spectrum are not degenerate, one can always normalize them in such a way that

$$\langle \psi_\nu, \psi_{\nu'} \rangle = \delta(\nu - \nu').$$

The extension to degenerate eigenvalues is straightforward. We shall merely quote the results. To a given eigenvalue there corresponds, depending upon the nature of the degeneracy, either a finite, or an infinite (denumerable or non-denumerable) number of linearly independent eigenfunctions. One can label such functions either with the aid of an index taking on a finite number of values, or by one or more indices taking an infinite number of discrete values, or else by one or more indices varying continuously; or even by a number of discrete indices and a number of continuous indices. Assume, to be definite, that the labelling is done with a discrete index r and a continuous index ϱ.

One can always arrange matters in such a way that the eigenfunctions $\varphi^{(r)}(\nu, \varrho)$ are orthonormal, in the generalized sense of the term, that is to say, such that

$$\langle \varphi^{(r)}(\nu, \varrho), \varphi^{(r')}(\nu', \varrho') \rangle = \delta_{rr'} \delta(\varrho - \varrho')\, \delta(\nu - \nu'). \tag{V.33}$$

Together with the orthonormal eigenfunctions $\varphi_n^{(r)}$ of the discrete spectrum, these functions form an orthonormal set of eigenfunctions of the Hermitean operator A; any eigenfunction of A can be written as a linear combination of the functions of this set.

Let us assume that a wave function Ψ can be expanded in a series of these functions, that is:

$$\Psi = \sum_{nr} c_n^{(r)} \varphi_n^{(r)} + \sum_r \int \gamma^{(r)}(\nu, \varrho)\, \varphi^{(r)}(\nu, \varrho)\, \mathrm{d}\nu\, \mathrm{d}\varrho. \tag{V.34}$$

The coefficients of each eigenfunction in this expansion are obtained by taking the scalar product of both sides of this equation from the left by the eigenfunction in question. One obtains, taking into account the orthonormality relations

$$c_n^{(r)} = \langle \varphi_n^{(r)}, \Psi \rangle \tag{V.35a}$$

$$\gamma^{(r)}(\nu, \varrho) = \langle \varphi^{(r)}(\nu, \varrho), \Psi \rangle. \tag{V.35b}$$

From these same relations, we arrive at the generalized Parseval relation

$$\langle \Psi, \Psi \rangle = \sum_{n,r} |c_n^{(r)}|^2 + \sum_r \int |\gamma^{(r)}(\nu, \varrho)|^2\, \mathrm{d}\nu\, \mathrm{d}\varrho. \tag{V.36}$$

If every wave function (i.e. every square-integrable function) can be expanded in a series of type (V.34), the set $\{\varphi\}$ is then, by definition, a *complete* orthonormal set.

There is a very simple way of writing down that an orthonormal set is complete: it consists in writing the expansion (V.34) for the function

$$\delta(q-q') \equiv \delta(q_1-q_1')\,\delta(q_2-q_2') \dots \delta(q_R-q_R').$$

The coefficients of this expansion are obtained by substituting in expressions (V.35) and (V.36) the function $\delta(q-q')$ for the function $\Psi(q) \equiv \Psi(q_1, q_2, \dots, q_R)$. One is led to the so-called *closure relation*

$$\delta(q-q') = \sum_{nr} \varphi_n^{*(r)}(q')\,\varphi_n^{(r)}(q) + \sum_r \int \varphi^{*(r)}(\nu, \varrho; q')\,\varphi^{(r)}(\nu, \varrho; q)\,\mathrm{d}\nu\,\mathrm{d}\varrho. \quad \text{(V.37)}$$

Together with the orthonormality relations

$$\langle \varphi_n^{(r)}, \varphi_{n'}^{(r')} \rangle = \delta_{nn'}\,\delta_{rr'} \qquad \text{(V.38a)}$$

$$\langle \varphi_n^{(r)}, \varphi^{(r')}(\nu, \varrho) \rangle = 0 \qquad \text{(V.38b)}$$

$$\langle \varphi^{(r)}(\nu, \varrho), \varphi^{(r')}(\nu', \varrho') \rangle = \delta_{rr'}\,\delta(\varrho-\varrho')\,\delta(\nu-\nu') \qquad \text{(V.38c)}$$

it forms a set of necessary and sufficient conditions for the set $\{\varphi\}$ to be orthonormal and complete.

Expansion (V.34) is obtained, with the correct values (V.35) of the coefficients, by writing that

$$\Psi(q) = \int \delta(q-q')\,\Psi(q')\,\mathrm{d}\tau$$

and substituting for $\delta(q-q')$ the right-hand side of eq. (V.37).

Note that the complete, orthonormal set, if it exists, is certainly not unique. Indeed, one can, just as in the case when the spectrum is entirely discrete,

1) arbitrarily change the phases of each of the eigenfunctions;

2) choose in an infinite number of different ways the set of orthonormal functions corresponding to the same degenerate eigenvalue.

In addition:

3) there is some arbitrariness in the normalization of the eigenfunctions of the continuous spectrum. Indeed, one can replace any continuous index ν by the index $\mu \equiv \mu(\nu)$, where $\mu(\nu)$ is a continuous, differentiable monotonic function of ν, but otherwise arbitrary. The

normalization condition (V.38*c*) is replaced by an analogous condition with index μ instead of ν; the latter is satisfied if one takes as a new eigenfunction

$$\varphi^{(r)}(\mu, \varrho; q) = \left|\frac{\mathrm{d}\mu}{\mathrm{d}\nu}\right|^{-\frac{1}{2}} \varphi^{(r)}(\nu, \varrho; q). \tag{V.39}$$

All Hermitean operators do not possess a complete, orthonormal set of eigenfunctions [1]). However, the Hermitean operators capable of representing physical quantities possess such a set. For this reason, we give the name *observables* to such operators. To prove that a specific Hermitean operator is an observable is, often, a difficult mathematical problem. The proof has actually been given for simple cases such as, for instance, the position or momentum coordinates, the Hamiltonian of one-dimensional quantum systems, angular momentum, etc. In the following, we shall always take for granted that all the operators associated with physical quantities possess a complete orthonormal set of eigenfunctions. In fact, the completeness property is so closely related to the physical interpretation of these operators that the whole theory would have to be profoundly revised if it did not hold true.

10. Statistical Distribution of the Results of Measurement in the General Case

We assume that A is an observable. Since the expansion (V.34) exists for any square-integrable Ψ it is possible to define, except for possible divergence of the series, the action of an operator of the form $F(A)$ on any such function.

Assume for simplicity that the spectrum of A possesses no degeneracy. Thus

$$\Psi = \sum_n c_n \varphi_n + \int \gamma(\nu)\, \varphi(\nu)\, \mathrm{d}\nu. \tag{V.40}$$

---

[1]) Let us consider the operator $\mathrm{i}(\mathrm{d}/\mathrm{d}x)$ acting on the square-integrable functions $\psi(x)$ defined on the semi-axis $(0, +\infty)$. This operator is Hermitean if one restricts it to functions which vanish for $x = 0$. In fact, it is under these conditions that one has

$$\int_0^\infty \psi_1^* \left(\mathrm{i}\frac{\mathrm{d}}{\mathrm{d}x}\psi_2\right) \mathrm{d}x - \int_0^\infty \left(\mathrm{i}\frac{\mathrm{d}}{\mathrm{d}x}\psi_1\right)^* \psi_2\, \mathrm{d}x = \mathrm{i}\, \psi_1^*\, \psi_2 \Big|_0^\infty = 0.$$

However, it possesses no eigensolution at all. The only possible eigensolutions are in fact of the form $\exp(-ikx)$ (eigenvalue k); they not do vanish at $x = 0$.

One can show that the necessary and sufficient condition for the convergence of the expansion of the right-hand side is that the series $\sum_n |c_n|^2$ and the integral $\int |\gamma(\nu)|^2 \, d\nu$ converge.

In all generality

$$F(A)\Psi = \sum_n c_n \, F(a_n) \, \varphi_n + \int \gamma(\nu) \, F(a_\nu) \, \varphi(\nu) \, d\nu$$

and this definition has a meaning if

$$\sum_n |c_n|^2 \, |F(a_n)|^2 \quad \text{and} \quad \int |\gamma(\nu)|^2 \, |F(a_\nu)|^2 \, d\nu$$

converge. In particular, the action of the operator $\exp(i\xi A)$ is always well defined since the expression

$$e^{i\xi A} \, \Psi = \sum_n c_n \, e^{i\xi a_n} \, \varphi_n + \int \gamma(\nu) \, e^{i\xi a_\nu} \, \varphi(\nu) \, d\nu \tag{V.41}$$

always converges.

In order to arrive at the characteristic function of the distribution of A, we make use of the relation

$$\langle \Psi, e^{i\xi A} \Psi \rangle = \sum_n |c_n|^2 \, e^{i\xi a_n} + \int |\gamma(\nu)|^2 \, e^{i\xi a_\nu} \, d\nu$$

obtained by using the expansions (V.40) and (V.41) and the orthonormality relations. The characteristic function $f(\xi)$ can be put in the form:

$$f(\xi) = \frac{\langle \Psi, e^{i\xi A} \Psi \rangle}{\langle \Psi, \Psi \rangle} = \sum_n w_n \, e^{i\xi a_n} + \int \varpi(\nu) \, e^{i\xi a_\nu} \, d\nu$$

with the notation

$$w_n = \frac{|c_n|^2}{\langle \Psi, \Psi \rangle}, \qquad \varpi(\nu) = \frac{|\gamma(\nu)|^2}{\langle \Psi, \Psi \rangle}. \tag{V.42}$$

Inspection of this characteristic function (cf. footnote p. 177) leads to the following conclusions:

1) *the only values the quantity $\mathscr{A}$ can take are the eigenvalues of the operator associated with it;*

2) *the probability that $\mathscr{A}$ takes one of the values a_n of the discrete spectrum is equal to w_n;*

3) *the probability that $\mathscr{A}$ takes one of the values of the continuous spectrum contained in the interval* $[a(\nu), a(\nu+\mathrm{d}\nu)]$ *is equal to* $\varpi(\nu)\,\mathrm{d}\nu$.

The sum of all these probabilities $\sum w_n + \int \varpi(\nu)\,\mathrm{d}\nu$ is actually equal to 1 (Parseval relation). One can also verify that the average value of A, if it exists, is actually equal to $\langle\Psi, A\Psi\rangle/\langle\Psi, \Psi\rangle$, in accordance with the fundamental postulate from which we started out.

If the spectrum is degenerate, one obtains the same result with a suitable modification of the quantities w_n and $\varpi(\nu)$. To be definite, let us assume that expansion (V.40) is replaced by expansion (V.34); the new definitions of w_n and $\varpi(\nu)$ are:

$$w_n = \frac{\sum_r |c_n^{(r)}|^2}{\langle\Psi, \Psi\rangle} \tag{V.43}$$

$$\varpi(\nu) = \frac{\sum_r \int \mathrm{d}\varrho\, |\gamma^{(r)}(\nu, \varrho)|^2}{\langle\Psi, \Psi\rangle}. \tag{V.44}$$

As it stands here, the desired probability law involves explicitly a particular set of eigenfunctions of A. There is a high degree of arbitrariness in the choice of that set. It is clear, however, that the probability law, just as its characteristic function, must be independent of that choice. This property is easily verified directly in expressions (V.43) and (V.44) (Problem V.5).

11. Other Ways of Treating the Continuous Spectrum

The main advantage of the foregoing treatment lies in its great formal simplicity. This advantage compensates to a large extent for the lack of mathematical rigor when using the δ-"function". In fact, all operations carried out with the δ-"function" can be made rigorous with the help of Distribution Theory (cf. Appendix A).

However, it is possible to overcome the difficulties of the continuous spectrum by having recourse to more classical mathematical procedures. Rather than base this entire analysis upon the eigenvalue problem and to make use of eigensolutions which may not belong to Hilbert space, one can, following a method due to von Neumann, treat the problem rigorously without ever leaving Hilbert space. The method consists in seeking what mathematicians call the decompositions of unity in Hilbert space, and to show that with every observable of

Wave Mechanics there is associated a particular decomposition of unity. This treatment is strictly equivalent to the treatment given above. We mention it here only for completeness [1]).

Another way of treating problems connected with the continuous spectrum consists in replacing the eigenvalue problem (V.9) by another problem where the sequence of eigenvalues is entirely discrete, and of which the initial problem may be considered as a limiting case when the conditions are suitably modified. Although such a procedure cannot claim to be rigorous, it has the advantage of being intuitively very simple to grasp. We shall treat the operators q and $(\hbar/\mathrm{i})\mathrm{d}/\mathrm{d}q$ by this method. The treatment should be compared to that of § 8.

To treat the problem of the position measurement, we divide the interval $(-\infty, +\infty)$ into equal segments of length η, and replace the wave functions $\psi(q)$ by approximate wave functions $\psi^{\times}(q)$ which are constant in each segment and are defined by the condition

$$\psi^{\times}(q) = \psi(n_q \eta),$$

where n_q denotes the largest integer contained in q/η; in other words: $q - \eta < n_q \eta \leqslant q$. Similarly, we replace the operator q by the operator

$$q^{\times} = \text{multiplication by } n_q \eta.$$

In the limit where $\eta \to 0$, $q^{\times} \to q$ and $\psi^{\times}(q) \to \psi(q)$.

The set of functions $\psi^{\times}$ forms a Hilbert space in which the operator $q^{\times}$ is well defined and possesses a discrete eigenvalue spectrum

$$n\eta \qquad (n = 0, \pm 1, \pm 2, \pm 3, \ldots, \pm\infty).$$

With each eigenvalue $n\eta$ is associated the eigenfunction u_n normalized to unity

$$u_n = \begin{cases} \eta^{-\frac{1}{2}} & \text{if} \quad n\eta \leqslant q < (n+1)\eta \\ 0 & \text{in the opposite case.} \end{cases}$$

The eigenfunctions u_n satisfy the orthonormality relations

$$\langle u_n, u_{n'} \rangle = \delta_{nn'}.$$

---

[1]) The reader who is interested in the mathematical aspects of Quantum Theory, will find a complete presentation thereof in J. von Neumann, *loc. cit.* (footnote p. 166).

Furthermore, they form a complete set since any function can be represented by an expansion in a series of the u_n:

$$\psi^{\times} = \sum_{n=-\infty}^{+\infty} \eta^{\frac{1}{2}}\, \psi(n\eta)\, u_n. \tag{V.45}$$

One can therefore apply the theory of §§ 5 and 6 with the result that the probability that $q^{\times}=n\eta$ is equal to $\eta|\psi(n\eta)|^2$.

In the limit $\eta \to 0$, the distance between neighboring eigenvalues goes to zero; the spectrum becomes continuous. The value n corresponding to a given non-zero value $q'=n\eta$ goes to infinity; however, the probability of finding this precise value of the position is proportional to η and therefore tends to zero. Actually, this probability is of no concern since the position spectrum is continuous. One actually looks for the probability $P(q')\,\delta q'$ of finding the particle in the interval $(q', q'+\delta q')$, namely

$$P(q')\,\delta q' = \sum\nolimits^{(q', q'+\delta q')} \eta|\psi(n\eta)|^2,$$

the summation being extended over all n such that $n\eta$ be contained in the interval $(q', q'+\delta q')$. As $\delta q'$ remains constant but sufficiently small, the terms of that sum, $\delta q'/\eta$ in number, are practically all equal to $\eta|\psi(q')|^2$. Hence, in the limit $\eta \to 0$,

$$P(q')\,\delta q' = |\psi(q')|^2\,\delta q'.$$

Note that expansion (V.45) can also be written

$$\begin{aligned}\psi^{\times} &= \sum_{n=-\infty}^{+\infty} \psi(n\eta)\,\frac{u_n}{\sqrt{\eta}} \times \eta \\ &= \sum_{q'} \psi(q')\, v_\eta(q') \times \eta,\end{aligned}$$

where $\sum_{q'}$ indicates a summation over the discrete sequence of values $q'=n\eta$ and where

$$v_n(q') = \eta^{-\frac{1}{2}}\, u_n = \begin{cases} \eta^{-1} & \text{if} \quad q-\eta < q' \leqslant q \\ 0 & \text{otherwise.} \end{cases}$$

When $\eta \to 0$, the series tends to the integral of the product of $\psi(q')$ by the limit of the function $v_\eta(q')$; this limit is precisely $\delta(q'-q)$. One recognizes the expansion (V.32).

In an anologous way the momentum measurement may be treated by restricting the interval of variation of q to the domain $(-L/2,$

$+L/2$), where L is a quantity which is finally allowed to tend to infinity. In order that the operator $p=(\hbar/\mathrm{i})\mathrm{d}/\mathrm{d}q$ be Hermitean in this finite space, one must impose suitable conditions upon the functions $\psi(q)$ at the boundaries of the interval. The hermiticity condition reads

$$\int_{-L/2}^{+L/2} \varphi^* \left(\frac{\hbar}{\mathrm{i}}\frac{\mathrm{d}\psi}{\mathrm{d}q}\right) \mathrm{d}q - \int_{-L/2}^{+L/2} \left(\frac{\hbar}{\mathrm{i}}\frac{\mathrm{d}\varphi}{\mathrm{d}q}\right)^* \psi \,\mathrm{d}q \equiv \frac{\hbar}{\mathrm{i}} \varphi^*\psi \Big|_{-L/2}^{+L/2} = 0,$$

for any functions $\varphi(q)$ and $\psi(q)$, or still

$$\frac{\psi(L/2)}{\psi(-L/2)} = \frac{\varphi^*(-L/2)}{\varphi^*(L/2)} = \text{constant, independent of } \psi \text{ and } \varphi.$$

In other words, it is necessary that for any function $\psi(q)$:

$$\psi(L/2) = \mathrm{e}^{\mathrm{i}\alpha}\psi(-L/2),$$

where $\exp(\mathrm{i}\alpha)$ is a phase factor chosen once and for all. Let us make it equal to 1, hence the *condition of periodicity*

$$\psi(L/2) = \psi(-L/2).$$

The eigenvalue problem of the operator $(\hbar/\mathrm{i})\mathrm{d}/\mathrm{d}q$ is then easily solved. The eigenvalue spectrum is discrete:

$$p_n = \frac{2\pi\hbar}{L} n \qquad (n=0, \pm 1, \pm 2, ..., \pm\infty).$$

To the eigenvalue p_n corresponds the eigenfunction normalized to unity

$$u_n = L^{-\frac{1}{2}} \mathrm{e}^{\mathrm{i}p_n q/\hbar}.$$

The functions u_n are mutually orthogonal; furthermore, they form a complete set since, according to the theory of Fourier series, any square-integrable function $\psi(q)$ of the interval $(-L/2, +L/2)$ can be represented by the series

$$\psi(q) = \sum_{n=-\infty}^{+\infty} c_n\, u_n, \tag{V.46}$$

with

$$c_n = \langle u_n, \psi(q)\rangle = L^{-\frac{1}{2}} \int_{-L/2}^{+L/2} \mathrm{e}^{-\mathrm{i}p_n q/\hbar}\, \psi(q)\, \mathrm{d}q.$$

Applying the theory of §§ 5 and 6 it is found that the probability of finding $p = p_n$ is equal to $|c_n|^2$.

In the limit $L \to \infty$, the distance

$$\varepsilon = \frac{2\pi\hbar}{L}$$

between neighboring eigenvalues goes to zero and the momentum spectrum $p_n = n\varepsilon$ becomes continuous. The study of the limiting process is entirely analogous to that carried out for q. When $\varepsilon \to 0$, $p' = n\varepsilon$ remaining constant, $\varepsilon^{-\frac{1}{2}} c_n$ tends toward the Fourier transform

$$\varphi(p') = \frac{1}{\sqrt{2\pi\hbar}} \int_{-\infty}^{+\infty} \mathrm{e}^{-\mathrm{i}p'q/\hbar}\, \psi(q)\, \mathrm{d}q.$$

We leave it to the reader to find, by this limiting process, the statistical distribution of the momentum measurements, and to show how the representation of $\psi(q)$ by the Fourier series (V.46) tends toward the Fourier integral

$$\psi(q) = \int_{-\infty}^{+\infty} \varphi(p')\, u(p'; q)\, \mathrm{d}p'$$

where $u(p'; q)$ is the limit of $\varepsilon^{-\frac{1}{2}} u_n$, that is to say

$$u(p'; q) = \frac{\mathrm{e}^{\mathrm{i}p'q/\hbar}}{\sqrt{2\pi\hbar}}.$$

12. Comments and Examples

To summarize, we have shown how the statistical distribution of the results of measurement of a dynamical variable can be deduced from the knowledge of its wave function. It is particularly satisfying that this result can be stated in an entirely general manner. With each dynamical variable, one associates a Hermitean operator possessing a complete orthonormal set of eigenfunctions. Starting from a very natural and formally very simple postulate about average values, we have shown that the only possible results of measurement are the eigenvalues of the observable A, and that the desired probability law is directly related to the squares of the moduli of the expansion coefficients of the wave function in a series of eigenfunctions of A.

Among the dynamical variables most commonly considered one should mention, besides the position and momentum coordinates, the energy represented by the Schrödinger Hamiltonian, and the angular momentum.

The *energy* spectrum can be, depending upon the case, either purely discrete (cf. Ch. III, § 5), purely continuous (cf. Ch. III, § 3), or mixed (cf. Ch. III, § 6). The eigenvalue problem of the Hamiltonian H is important in Quantum Theory not only in relation to the definition of the energy, but also because of the role it plays in the solution of the Schrödinger equation. *When H is time-independent*, the only case where the concept of energy has a real meaning, the wave function $\Psi(t)$ at time t is deduced from the wave function $\Psi(t_0)$ at the initial time t_0 by the operation

$$\Psi(t) = \mathrm{e}^{-\mathrm{i}H(t-t_0)/\hbar}\, \Psi(t_0).$$

One knows how to calculate the right-hand side of this equation by using the expansion of $\Psi(t_0)$ in a series of eigenfunctions of H [this is a special case of eq. (V.41)]; it is easily shown, making use of this expansion, that

$$\frac{\partial}{\partial t}\Psi(t) = -\frac{\mathrm{i}}{\hbar} H\, \mathrm{e}^{-\mathrm{i}H(t-t_0)/\hbar}\, \Psi(t_0) = -\frac{\mathrm{i}}{\hbar} H\Psi(t),$$

hence that $\Psi(t)$ satisfies the Schrödinger equation, and furthermore that the initial conditions are actually fulfilled when $t = t_0$.

Suppose, for simplicity, that the spectrum of H is entirely discrete and non-degenerate; denote by E_n ($n = 1, 2, \ldots$) the eigenvalues of H, and by ψ_n the corresponding eigenfunctions. The expansion of $\Psi(t_0)$ is written

$$\Psi(t_0) = \sum_n c_n\, \psi_n, \qquad c_n = \langle \psi_n, \Psi(t_0) \rangle.$$

The function $\Psi(t)$ is then given by

$$\Psi(t) = \sum_n c_n\, \mathrm{e}^{-\mathrm{i}E(t-t_0)/\hbar}\, \psi_n. \tag{V.47}$$

Note that the absolute value of the coefficient of ψ_n in this expansion is independent of t. Hence the interesting property: *the statistical distribution of the energy of a system* (whose Hamiltonian is time-independent) *is constant in time.*

The *angular momentum* of a particle in Classical Mechanics is the vector $\boldsymbol{r} \times \boldsymbol{p}$. In Quantum Theory, there corresponds to it the vector operator

$$\boldsymbol{l} \equiv \frac{\hbar}{\mathrm{i}} (\boldsymbol{r} \times \nabla). \tag{V.48}$$

Let us write one of its components explicitly:

$$l_z = \frac{\hbar}{\mathrm{i}} \left(x \frac{\partial}{\partial y} - y \frac{\partial}{\partial x} \right).$$

Take Oz as polar axis and denote by (r, θ, φ) the polar coordinates of the particle; one easily verifies that $\partial/\partial\varphi = x\partial/\partial y - y\partial/\partial x$. Consequently,

$$l_z = \frac{\hbar}{\mathrm{i}} \frac{\partial}{\partial \varphi}. \tag{V.49}$$

The eigenvalue problem of l_z appears particularly simple if one expresses the eigenfunction in polar coordinates

$$l_z \psi(r, \theta, \varphi) \equiv \frac{\hbar}{\mathrm{i}} \frac{\partial}{\partial \varphi} \psi(r, \theta, \varphi) = l_z' \psi(r, \theta, \varphi).$$

We have

$$\psi(r, \theta, \varphi) = F(r, \theta)\, \mathrm{e}^{\mathrm{i} l_z' \varphi / \hbar},$$

where $F(r, \theta)$ is an arbitrary function of r and θ. Since the eigenfunction is a single-valued function of $\boldsymbol{r}$, $\psi(r, \theta, \varphi)$ must take on the same value when one changes φ to $\varphi + 2\pi$, hence

$$l_z' = m\hbar \qquad (m \text{ integer}). \tag{V.50}$$

Consequently, the eigenvalue spectrum of any component of the angular momentum of a particle is an entirely discrete spectrum. This result is easily extended to the components of the total angular momentum of a system of particles, in good agreement with the experimental fact of space quantization.

IV. DETERMINATION OF THE WAVE FUNCTION

13. Measuring Process and "Filtering" of the Wave Packet. Ideal Measurements

The statistical distributions defined in the preceding sections are directly amenable to experiment. Thus the distribution associated with a given dynamical variable $\mathscr{A}$ is the distribution of the results obtained when one carries out the measurement of $\mathscr{A}$ on a very large number $\mathscr{N}$ of identical systems, independent of each other and in the same dynamical state at the instant of measurement. Each system is represented at that instant [1]) by the same wave function Ψ (defined to within a constant) to which there corresponds a well-defined theoretical distribution. The latter may be compared with the distribution observed experimentally.

For the physical interpretation of the theory to be complete, one must further specify:

(*i*) how previous observations, made on a system, allow us to know its dynamical state, and, in particular, how one can be sure that the $\mathscr{N}$ systems considered above are actually all in the dynamical state represented by Ψ;

(*ii*) what becomes of each of these systems once the measurement is completed.

These two questions are closely related. We shall deal with the second one first.

In contrast to what happens during the process of measurement [2]), the system, once the measurement is completed, can again be treated as an entity completely separated from the measuring apparatus. It becomes possible again to describe it by means of a wave function involving only its own dynamical variables. This wave function of the system after measurement is certainly different from the wave function immediately before the measurement, except perhaps if the latter is an eigenfunction of the observable A associated with the measured quantity. We shall call this (*non-causal*) change of the wave function by the measurement process the *filtering of the wave packet*.

We know that this non-causal change appears as an uncontrollable perturbation in the motion of the system by its interaction with the measuring device; its effect is roughly to make the complementary

---

[1]) The instant in question here is the *instant at which the measurement begins*. Actually, as soon as the measurement has started, the system is in interaction with the measuring device and any description of its evolution by means of its wave function alone becomes impossible.

[2]) The detailed study of the mechanism of measurement will not be made in this book. On that subject, see the references cited in the footnote of p. 157.

variable to the one measured the less well-defined the more precise the measurement that is carried out. This uncontrollable perturbation should not be confused with any modifications – in principle exactly calculable – which the system might undergo during the measuring process. In particular, it often happens that the measured variable is modified during the measuring process. The two momentum measurements of Chapter IV are examples of such a situation. The modification of the measured quantity evidently depends upon the type of measuring device used. Thus in the momentum measurement by Compton collision contemplated in Chapter IV, p. 147, the difference $p' - p$ is smaller the lower the frequency ν of the photon; it vanishes if ν is chosen infinitely small. Clearly, no general statement can be made concerning the modification of the dynamical state of the system during the measuring operation, since this modification depends in each case upon the particular conditions under which the measurement was carried out. One can, however, imagine idealized conditions of measurement, in which all exactly calculable modifications, which we have just mentioned, are strictly compensated, and where only the specific, uncontrollable perturbation of the quantum phenomena enters. We shall assume that such *ideal measurements* can actually be realized, or, at the very least, that they are limiting cases of actually realizable measurements (for instance: position measurements discussed in Chapter IV in the limit where the duration of the measurement is zero; momentum measurement by Compton collision in the limit where $\nu = 0$).

Consider therefore an ideal measurement of the quantity $\mathscr{A}$, and suppose at the start that the value found a_i is a non-degenerate eigenvalue. According to our hypothesis, we know with certainty that $\mathscr{A} = a_i$ once the measurement is completed, hence that the wave function of the system is the eigenfunction ψ_i (to within a constant) corresponding to the eigenvalue a_i. The arbitrary constant has no physical significance since the statistical distribution of the results of any subsequent measurement is independent of the choice of that constant. The wave function of the system after measurement is thus known without ambiguity. The measuring device works in some sense like a "perfect filter". The wave function before measurement is a function $\Psi = \sum_n c_n \psi_n$. There is a probability $|c_i|^2$ that the result of measurement is a_i. Assuming that the measurement gives a_i, the net effect of the measuring process is to "pass" (without distortion) only the term $c_i \psi_i$ of the expansion of Ψ in a series of eigenfunctions of A.

In the more general case where the eigenvalues of the observable A, and especially the value a_i, are degenerate, the wave function before measurement (assumed normalized to unity) can be put in the form [cf. eq. (V.20)]

$$\Psi = \sum_p \Psi_p, \qquad \Psi_p = \sum_r c_p{}^{(r)} \psi_p{}^{(r)}. \tag{V.51}$$

There is a probability $\langle \Psi_i, \Psi_i \rangle = \sum_r |c_p{}^{(r)}|^2$ of obtaining the result a_i. In the case of an ideal measurement, the wave function after measurement is an eigenfunction of A corresponding to the eigenvalue a_i: it is a linear combination of the functions $\psi_i{}^{(r)}$ (r is variable). However, this information is not sufficient to specify it completely; it is less so the higher the degree of degeneracy. The ideal measurement is one where the measuring device acts as a "perfect filter" and "passes" without distortion and to the exclusion of all the rest, the portion of the expansion (V.51) of Ψ pertaining to eigenvalue a_i, i.e. the function

$$\Psi_i = \sum_r c_i{}^{(r)} \psi_i{}^{(r)}.$$

When the measurement is not ideal, the "passing" of these terms is accompanied by some distortion. That distortion is in principle exactly calculable and depends upon the measuring device used.

In the remainder of this chapter, all contemplated measurements will be assumed ideal. All discussions are thereby greatly simplified. Actually, such a restriction is not essential to the argument, and one could relax it without bringing about fundamental changes in the physical interpretation of the theory.

14. Commuting Observables and Compatible Variables

Consider two observables A and B. For simplicity the eigenvalue spectra are assumed to be entirely discrete, although the properties stated below are quite general. Let us suppose that they possess a common eigenfunction Ψ_0

$$A\Psi_0 = a\Psi_0$$
$$B\Psi_0 = b\Psi_0.$$

The physical meaning of these two equations is the following: if the physical system is in state Ψ_0 at a given time t, a precise measurement of the quantities $\mathscr{A}$ and $\mathscr{B}$ will yield with certainty the results a and b, respectively. A necessary condition for these two equations to hold simultaneously is that

$$(AB - BA)\Psi_0 = [A, B]\Psi_0 = 0; \tag{V.52}$$

that is to say, that the commutator of A and B have Ψ_0 as eigenfunction corresponding to the eigenvalue 0.

An example in which this condition never holds is given by the observables x and p_x, since their commutator is a non-zero constant. More precisely [cf. eq. (II.10)]

$$[x, p_x] = \frac{\hbar}{\mathrm{i}}\left[x, \frac{\partial}{\partial x}\right] = \mathrm{i}\hbar \neq 0, \tag{V.53}$$

and it is indeed well known that these two quantities can never be defined simultaneously with infinite precision.

On the other hand, equation (V.52) is automatically satisfied when the observables A and B commute. In that case one has the important theorem:

If two observables commute, they possess a complete orthonormal set of common eigenfunctions, and conversely.

Physically, this means that the dynamical variables represented by these two observables may be defined simultaneously in a precise way: they are *compatible variables.* In particular, it is possible to carry out simultaneously an ideal measurement of the variables $\mathscr{A}$ and $\mathscr{B}$; in that case, the wave function after measurement is an eigenfunction common to A and B.

The proof of the direct theorem is as follows. Assume that the observables A and B commute:

$$[A, B] = 0.$$

Let ψ_a be an eigenfunction of A, and a its eigenvalue. ψ_a can be expanded into a complete orthonormal set of eigenfunctions of B. It can therefore be written in the form

$$\psi_a = \sum_m \varphi(a; b_m),$$

where $\varphi(a; b_m)$ is an eigenfunction of B corresponding to the eigenvalue b_m. One can always arrange matters in such a way that every function occurring in that sum corresponds to a different eigenvalue [cf. eq. (V.20)]. Let us show that

$$\hat{\varphi}_m \equiv (A-a)\,\varphi(a\,;b_m)=0.$$

Since A and B commute, we have

$$B\hat{\varphi}_m=(A-a)\,B\varphi(a\,;b_m)=(A-a)\,b_m\varphi(a\,;b_m)=b_m\hat{\varphi}_m.$$

The functions $\hat{\varphi}_m$ are therefore eigenfunctions of B; since the corresponding eigenvalues are all different, these functions are linearly independent. However, one has

$$\sum_m \hat{\varphi}_m = (A-a)\psi_a = 0.$$

This is possible only if each of the functions $\hat{\varphi}_m$ vanishes. In other words, the functions $\varphi(a\,;b_m)$ are simultaneously eigenfunctions of A and of B.

Consider now a *complete* orthonormal set $\{\psi_n{}^{(r)}\}$ of eigenfunctions of A:

$$A\psi_n{}^{(r)}=a_n\psi_n{}^{(r)}.$$

According to the above argument, these functions can be put in the form

$$\psi_n{}^{(r)}=\sum_m \varphi^{(r)}(a_n\,;b_m). \tag{V.54}$$

The functions $\varphi^{(r)}(a_n\,;b_m)$ are eigenfunctions common to A and B. The ensemble of functions $\varphi^{(r)}(a_n\,;b_m)$ corresponding to the same pair of eigenvalues a_n, b_m may not be linearly independent. But it is always possible to choose (by means of the orthogonalization process of Schmidt, for instance) a set of orthonormal functions $\chi^{(s)}(a_n\,;b_m)$ corresponding to the same pair of eigenvalues and such that the functions $\varphi^{(r)}(a_n\,;b_m)$ are linear combinations of these functions

$$\varphi^{(r)}(a_n\,;b_m)=\sum_s c_{rs}\,\chi^{(s)}(a_n\,;b_m). \tag{V.55}$$

The ensemble $\{\chi\}$ of all these functions constitutes an orthonormal set of eigenfunctions common to A and B. Moreover, it is a complete set, since any wave function Ψ can be expanded in a series of χ; to form this expansion, it suffices to expand Ψ in a series of functions of the complete set $\{\psi_n{}^{(r)}\}$, and then to substitute for the functions

$\psi_n^{(r)}$ in this expansion, their expressions as function of the $\chi^{(s)}(a_n; b_m)$ obtained with the aid of eqs. (V.54) and (V.55). Q.E.D.

Conversely, if A and B possess a complete orthonormal set of common eigenfunctions $\chi^{(s)}(a_n; b_m)$, one has

$$AB\chi^{(s)}(a_n; b_m) = a_n b_m \chi^{(s)}(a_n; b_m) = BA\chi^{(s)}(a_n; b_m),$$

hence

$$[A, B]\, \chi^{(s)}(a_n; b_m) = 0.$$

The action of the commutator $[A, B]$ on all functions of the set $\{\chi\}$ yields zero. Since by hypothesis all wave functions Ψ can be developed in a series of the χ, one has $[A, B]\, \Psi = 0$, for any Ψ. Therefore

$$[A, B] = 0.$$

Using the commuting observables A, B one can build up new observables of the form $f(A, B)$, $f(x, y)$ being an arbitrarily chosen real function. By definition, the action of $f(A, B)$ on an eigenfunction $\chi(a; b)$ common to A and B yields

$$f(A, B)\, \chi(a; b) = f(a, b)\, \chi(a; b).$$

Its action on an arbitrary function Ψ is obtained by expanding Ψ in a series of the χ's and applying the operator $f(A, B)$ to each of the terms of the expansion. This procedure is valid provided the series converges; otherwise the function $f(A, B)\, \Psi$ does not exist. It is evident from the way in which it was defined, that $f(A, B)$ possesses, along with A and B, a complete orthonormal set of common eigenfunctions, namely the set $\{\chi\}$; hence $f(A, B)$ commutes with A and B.

All these results are easily extended to an arbitrary number R of pairwise commuting observables. If R observables commute in pairs, they possess (at least) a complete orthonormal set of common eigenfunctions, and conversely. Furthermore, any (real) function of these observables is an observable which commutes with each of them and possesses the same set of eigenfunctions.

15. Complete Sets of Commuting Observables

Consider an observable A. With its eigenfunctions one can build up a complete orthonormal set of eigenfunctions; we shall henceforth call it the *basis of A*. In general, this basis is not unique. The arbitrary part in the choice of this basis was discussed in § 9. We shall by convention consider two bases as identical if their functions differ only in the phase and (in the case of the continuous spectrum) in the norm. With this convention, the basis of A is unique if none of its eigenvalues are degenerate. On the other hand, suppose, to be specific, that the eigenvalue a is doubly degenerate, and let ψ_1, ψ_2 be two orthonormal eigenfunctions corresponding to that eigenvalue. As can be easily verified, the functions

$$\varphi_1 = \psi_1 \cos \alpha + \psi_2 \sin \alpha$$
$$\varphi_2 = -\psi_1 \sin \alpha + \psi_2 \cos \alpha$$

also have that property. Thus one can form the basis of A just as well with the pair (φ_1, φ_2) as with the pair (ψ_1, ψ_2), Now, assume that the basis of A is not unique. Let B be an observable that commutes with A. It may happen that the basis common to A and B, whose existence we have demonstrated in § 14, is unique. One then says that the observables A and B form a complete set of commuting observables.

If A and B do not share a unique common basis, one is led to associate with them a third observable C which commutes with both of them, and so forth.

More generally, one says that *the observables $A, B, ..., L$ form a complete set of commuting observables if they possess one and only one common basis*. In that case, any observable which commutes with each of the observables of the set, necessarily has this set as its basis. Its eigenvalues are therefore well-defined functions of the eigenvalues $a, b, ..., l$ of the observables $A, B, ..., L$. In other words, this observable can be considered as a function of the observables of the set.

The dynamical variables represented by the observables of a complete set of commuting observables, can always be all simultaneously defined with precision; moreover, they form a *complete set of compatible variables* (Ch. IV, § 17). If one carries out simultaneously a precise measurement of the values assumed by these variables, one can be sure that the wave function of the system is an eigenfunction of the observables $A, B, ..., L$ corresponding to the eigenvalues $a, b, ..., l$ found in the operation of measurement. *Since there exists but one eigenfunction possessing this property, the specification of these measurements completely defines the wave function of the physical system.* One says that *the dynamical state of the system is completely specified by giving the quantum numbers a, b, ..., l.* In fact, this function is only determined to within a constant. Since the only physically measurable quantities, that is to say the statistical distributions of the results of various possible measurements, are independent of the choice of this constant, one can fix the latter as one wishes without changing the physical meaning of the wave function. If, as is often done, the function is normalized to unity, there still remains an arbitrary phase without physical significance.

The time sequence of an experiment in physics can be viewed in the following manner. At the initial time t_0 one *prepares* the system by performing on it the simultaneous measurement of a complete set of compatible variables. Its dynamical state is thus completely determined at time t_0.

Once this preparation is completed, the wave function of the system evolves in time in a manner exactly determined by the Schrödinger equation. At all later instants, the dynamical state of the system is thus perfectly known, at least as long as it is not perturbed by the intervention of a measuring device. Eventually, at a later time t, one carries out a given measurement. Since one knows the wave function $\Psi(t)$ at the instant where the measurement is performed, one can exactly predict the statistical distribution of the results of the measurement. Repetition of this experiment on a very large number $\mathcal{N}$ of identical systems, yields an experimental distribution which can be compared with the theoretical distribution.

16. Pure States and Mixtures

In practice, a complete "preparation" of the system like the one contemplated above, is rarely achieved. Most frequently, the dynamical variables measured during the preparation do not constitute a complete set. As a consequence, the dynamical state of the system is known incompletely and one must have recourse to the methods of statistics. Instead of a given dynamical state, one is dealing with a statistical mixture of states; instead of assigning to the system a well-defined wave function, one assigns to it a statistical mixture of wave functions each having a suitable statistical weight. There exists a *Quantum Statistical Mechanics*, just as there is a Classical Statistical Mechanics.

When the preparation is complete, and consequently the dynamical state of the system is exactly known, one says that one is dealing with a *pure state*, in contrast to the statistical mixtures which characterize incomplete preparations.

In the prediction of the results of measurements carried out on a mixture, statistics enters in two ways: first for specific quantum reasons connected with the uncontrollable perturbation of the system in the measuring operation, and secondly because the dynamical state of the system is incompletely known.

Suppose that at the time of preparation t_0, the system could be represented by the set of wave functions $\Psi^{(1)}(t_0), \ldots, \Psi^{(k)}(t_0), \ldots$, with statistical weights $p_1, \ldots, p_k, \ldots$ ($\sum_k p_k = 1$), respectively. Let $\Psi^{(1)}(t), \ldots, \Psi^{(k)}(t), \ldots$, be the solutions of the Schrödinger equation corresponding to the initial conditions $\Psi^{(1)}(t_0), \ldots, \Psi^{(k)}(t_0), \ldots$ respectively. At time t, the system is represented by the set of these functions $\Psi^{(1)}(t), \ldots, \Psi^{(k)}(t), \ldots$, with the same statistical weights $p_1, \ldots, p_k, \ldots$. Let $\langle A \rangle_k$ be the average value of the results of measurement of a given quantity $\mathscr{A}$, performed on the system when it is in the dynamical state $\Psi^{(k)}(t)$:

$$\langle A \rangle_k = \frac{\langle \Psi^{(k)}(t), A\Psi^{(k)}(t) \rangle}{\langle \Psi^{(k)}(t), \Psi^{(k)}(t) \rangle}.$$

The average value of the results of measurement of $\mathscr{A}$ carried out on the statistical mixture at time t is given by

$$\langle A \rangle = \sum_k p_k \langle A \rangle_k.$$

Similarly, if $w_i^{(k)}$ is the probability of finding the result a_i when the dynamical state of the system is represented by $\Psi^{(k)}(t)$, the probability of finding this result when carrying out the same measurement on the mixture at time t is

$$w_i = \sum_k p_k w_i^{(k)}. \tag{V.56}$$

A very important case of quantum-statistical mixture is that of a system in thermodynamic equilibrium with a heat reservoir at temperature T. The various possible dynamical states are the eigenstates of the Hamiltonian H of the system. The statistical weight of a given eigenstate depends only upon the corresponding eigenvalue of H; it is proportional to the Boltzmann factor $\exp(-E/kT)$, in which E is the eigenvalue of H and k is the Boltzmann constant.

V. COMMUTATOR ALGEBRA AND ITS APPLICATIONS

17. Commutator Algebra and Properties of Basic Commutators

As long as one deals only with commuting observables the rules of ordinary algebra may be used without restriction. However, the observables of a given quantum system do not all commute. More precisely, the observables of a quantum system in R dimensions are functions of the position observables q_i $(i=1, 2, \ldots, R)$ and the momentum observables p_i $(i=1, 2, \ldots, R)$ [1]), all pairs of which do not commute. The commutators of the q's and the p's play a fundamental role in the theory. One has:

$$[q_i, q_j]=0, \qquad [p_i, p_j]=0 \tag{V.57}$$

$$[q_i, p_j]=\mathrm{i}\hbar\, \delta_{ij}. \tag{V.58}$$

Relations (V.57) are obvious; in particular the second merely states that operations of differentiation commute with each other. Relation (V.58) is a generalization of eq. (V.53); it is readily obtained by using the explicit form of the operators p:

$$p_i = \frac{\hbar}{\mathrm{i}} \frac{\partial}{\partial q_i}.$$

From the fact that the q's and the p's do not commute in pairs, the precise definition of a dynamical variable $\mathscr{A} \equiv A(q_1, \ldots, q_R; p_1, \ldots, p_R)$ requires that one properly specifies the order of the q's and the p's in the explicit expression of the function $A(q_1, \ldots, q_R; p_1, \ldots, p_R)$. In practice, A is put in the form of a polynomial in p – or possibly in the form of a power series in p – whose coefficients are functions of q. Each term is a product of components p_i and functions of the q arranged in a certain order. The function A, considered as an operator, is well defined only when the order in each of its terms is specified.

It is interesting to know the commutators of the q's and the p's with a given function A. For functions of the q's alone, or of the p's

---

[1]) Actually, this is true only to the extent that the quantum system has a classical analogue. In what follows we shall come to introduce supplementary variables, namely the spin variables which have no classical analogue.

alone, one obtains the relations

$$[q_i, F(q_1, \ldots, q_R)] = 0 \tag{V.59}$$

$$[p_i, G(p_1, \ldots, p_R)] = 0 \tag{V.60}$$

$$[p_i, F(q_1, \ldots, q_R)] = \frac{\hbar}{\mathrm{i}} \frac{\partial F}{\partial q_i} \tag{V.61}$$

$$[q_i, G(p_1, \ldots, p_R)] = \mathrm{i}\hbar \frac{\partial G}{\partial p_i}. \tag{V.62}$$

The relations (V.59) and (V.60) are particular cases of the property stated at the end of § 14. To prove equation (V.61), it suffices to write down the operator p_i explicitly and to verify that the action of each side of the equation on an arbitrary wave function gives the same result [cf. eq. (II.9)]. Equation (V.62) is proved by making an analogous verification in momentum space; let us recall that if $\Phi(p_1, \ldots, p_R)$ is the wave function of momentum space corresponding to $\Psi(q_1, \ldots, q_R)$, the function of momentum space corresponding to $q_i\Psi(q_1, \ldots, q_R)$ is

$$\mathrm{i}\hbar \frac{\partial}{\partial p_i} \Phi(p_1, \ldots, p_R).$$

One arrives at the same result using the rules of *commutator algebra*. Let us give here the four principal rules. These rules are direct consequences of the definition of commutators. Their proofs are left to the reader. If A, B, and C denote three arbitrary linear operators, one has

$$[A, B] = -[B, A] \tag{V.63}$$

$$[A, B+C] = [A, B] + [A, C] \tag{V.64}$$

$$[A, BC] = [A, B]C + B[A, C] \tag{V.65}$$

$$[A, [B, C]] + [B, [C, A]] + [C, [A, B]] = 0. \tag{V.66}$$

By repeated application of rule (V.65), one has

$$[A, B^n] = \sum_{s=0}^{n-1} B^s[A, B]\, B^{n-s-1}.$$

In particular, for a one-dimensional system one has

$$[q, p^n] = n\, \mathrm{i}\hbar\, p^{n-1}.$$

Equation (V.62) is thus verified when F is an arbitrary power of the p; it is thus also verified [rule (V.64)] when F is a polynomial, or else a convergent power series in p.

For general functions of the q's and p's, one can also write

$$[p_i, A] = \frac{\hbar}{\mathrm{i}} \frac{\partial A}{\partial q_i} \tag{V.67}$$

$$[q_i, A] = \mathrm{i}\hbar \frac{\partial A}{\partial p_i}, \tag{V.68}$$

$\partial A/\partial q_i$, $\partial A/\partial p_i$ being defined by partial differentiation of A, it being understood that the order of the p's and q's in their explicit expression has been suitably chosen.

Let us illustrate this by an example dealing with one-dimensional quantum systems. Let $f(q)$ be a function of q. The three commutators of q and of each of the functions $p^2 f(q)$, $pf(q)p$, and $f(q)p^2$ may all be identified (to within the factor $\mathrm{i}\hbar$) with the derivative with respect to p of these functions, but they are not the same operators. Indeed, by repeated application of rule (V.62),

$$[q, p^2 f(q)] = 2\mathrm{i}\hbar\, pf(q)$$
$$[q, pfp] = \mathrm{i}\hbar(fp + pf)$$
$$[q, fp^2] = 2\mathrm{i}\hbar\, fp.$$

In the same way

$$[p, p^2 f] = \frac{\hbar}{\mathrm{i}}\, p^2 f'$$

$$[p, pfp] = \frac{\hbar}{\mathrm{i}}\, pf'p$$

$$[p, fp^2] = \frac{\hbar}{\mathrm{i}}\, f'p^2.$$

18. Commutation Relations of Angular Momentum

As an application of rules (V.63) to (V.65) of commutator algebra, let us calculate the commutators of the components of the angular momentum of a particle:

$$\boldsymbol{l} \equiv \boldsymbol{r} \times \boldsymbol{p}.$$

One has

$$\begin{aligned}[l_x, l_y] &= [yp_z - zp_y, zp_x - xp_z] \\ &= [yp_z, zp_x] + [zp_y, xp_z] && \text{[rule (V.64)]} \\ &= y[p_z, z]p_x + p_y[z, p_z]x && \text{[rule (V.65)]} \\ &= \mathrm{i}\hbar(xp_y - yp_x) \\ &= \mathrm{i}\hbar\, l_z.\end{aligned}$$

The other two commutators are calculated by cyclic permutation. Thus

$$[l_x, l_y] = \mathrm{i}\hbar\, l_z, \qquad [l_y, l_z] = \mathrm{i}\hbar\, l_x, \qquad [l_z, l_x] = \mathrm{i}\hbar\, l_y. \tag{V.69}$$

The three components of the angular momentum do not commute in pairs. There is no complete orthonormal set common to any two of them. In other words, two components of angular momentum cannot, in general [1]), be defined simultaneously with infinite precision.

Note that [rule (V.65)]

$$\begin{aligned}&[l_z, l_x^2] = \mathrm{i}\hbar(l_y l_x + l_x l_y) \\ &[l_z, l_y^2] = -\mathrm{i}\hbar(l_y l_x + l_x l_y) \\ &[l_z, l_z^2] = 0.\end{aligned}$$

Adding term by term [rule (V.64)], we obtain

$$[l_z, \boldsymbol{l}^2] = 0, \tag{V.70}$$

where the operator

$$\boldsymbol{l}^2 = l_x^2 + l_y^2 + l_z^2 \tag{V.71}$$

is the square of the length of the vector $\boldsymbol{l}$.

The operators $\boldsymbol{l}^2$ and l_z commute: they can therefore be simultaneously defined with infinite precision. The pairs $(\boldsymbol{l}^2, l_x)$ and $(\boldsymbol{l}^2, l_y)$ obviously possess the same property.

---

[1]) The words "in general" are important. The three components have no common basis, but have common eigenfunctions, those for which $l_x = l_y = l_z = 0$. These are the functions which depend only upon the length $r = \sqrt{x^2 + y^2 + z^2}$ and not upon the direction of the vector $\boldsymbol{r}$.

19. Time Dependence of the Statistical Distribution. Constants of the Motion

Consider the Schrödinger equation and the complex conjugate equation:

$$\mathrm{i}\hbar \frac{\partial \Psi}{\partial t} = H\Psi, \qquad \mathrm{i}\hbar \frac{\partial \Psi^*}{\partial t} = -(H\Psi)^*.$$

If Ψ is normalized to unity at the initial instant, it remains normalized at any later time. The mean value of a given observable A is equal at every instant to the scalar product

$$\langle A\rangle = \langle \Psi, A\Psi\rangle = \int \Psi^* A \Psi \, \mathrm{d}\tau,$$

and one has

$$\frac{\mathrm{d}}{\mathrm{d}t}\langle A\rangle = \left\langle \frac{\partial \Psi}{\partial t}, A\Psi \right\rangle + \left\langle \Psi, A\frac{\partial \Psi}{\partial t} \right\rangle + \left\langle \Psi, \frac{\partial A}{\partial t}\Psi \right\rangle.$$

The last term of the right-hand side, $\langle \partial A/\partial t\rangle$, is zero if A does not depend upon the time explicitly.

Taking into account the Schrödinger equation and the hermiticity of the Hamiltonian, one has

$$\begin{aligned}\frac{\mathrm{d}}{\mathrm{d}t}\langle A\rangle &= -\frac{1}{\mathrm{i}\hbar}\langle H\Psi, A\Psi\rangle + \frac{1}{\mathrm{i}\hbar}\langle \Psi, AH\Psi\rangle + \left\langle \frac{\partial A}{\partial t}\right\rangle \\ &= \frac{1}{\mathrm{i}\hbar}\langle \Psi, [A, H]\Psi\rangle + \left\langle \frac{\partial A}{\partial t}\right\rangle.\end{aligned}$$

Hence we obtain the general equation giving the time-dependence of the mean value of A:

$$\mathrm{i}\hbar \frac{\mathrm{d}}{\mathrm{d}t}\langle A\rangle = \langle [A, H]\rangle + \mathrm{i}\hbar \left\langle \frac{\partial A}{\partial t}\right\rangle. \tag{V.72}$$

When we replace A by the operator $\exp(\mathrm{i}\xi A)$, we obtain an analogous equation for the time-dependence of the characteristic function of the statistical distribution of A.

In particular, for any variable C which commutes with the Hamiltonian

$$[C, H] = 0$$

and which does not depend explicitly upon the time, one has the result

$$\frac{\mathrm{d}}{\mathrm{d}t} \langle C \rangle = 0.$$

The mean value of C remains constant in time. More generally, if C commutes with H, the function $\exp(\mathrm{i}\xi C)$ also commutes with H, and, consequently

$$\frac{\mathrm{d}}{\mathrm{d}t} \langle \mathrm{e}^{\mathrm{i}\xi C} \rangle = 0.$$

The characteristic function, and hence the statistical distribution of the observable C, remain constant in time.

By analogy with Classical Analytical Mechanics, C is called a *constant of the motion*. In particular, if at the initial instant the wave function is an eigenfunction of C corresponding to a given eigenvalue c, this property continues to hold in the course of time. One says that c is a "good quantum number". If, in particular, H does not explicitly depend upon the time, and if the dynamical state of the system is represented at time t_0 by an eigenfunction common to H and C, the wave function remains unchanged in the course of time, to within a phase factor. The energy and the variable C remain well defined and constant in time.

20. Examples of Constants of the Motion. Energy. Parity

There exists an observable which always commutes with the Hamiltonian: the Hamiltonian itself. The energy is therefore a constant of the motion of all systems whose Hamiltonian does not depend explicitly upon the time. This result was already shown in § 12.

As another possible constant of the motion, let us mention *parity* (cf. Ch. III, § 14). We denote under the name of parity the observable P defined by

$$P\psi(q) = \psi(-q).$$

It is easily verified that P is Hermitean. Moreover, $P^2 = 1$ and, consequently, the only possible eigenvalues of P are $+1$ and -1; even functions are associated with $+1$, and odd functions with -1.

When the Hamiltonian is invariant under the substitution of $-q$ *for* q, we obviously have

$$[P, H] = 0.$$

Indeed, if

$$H\left(\frac{\hbar}{\mathrm{i}}\frac{\mathrm{d}}{\mathrm{d}q}, q\right) = H\left(-\frac{\hbar}{\mathrm{i}}\frac{\mathrm{d}}{\mathrm{d}q}, -q\right),$$

one has, for any $\psi(q)$,

$$PH\psi = H\left(-\frac{\hbar}{\mathrm{i}}\frac{\mathrm{d}}{\mathrm{d}q}, -q\right)\psi(-q) = H\left(\frac{\hbar}{\mathrm{i}}\frac{\mathrm{d}}{\mathrm{d}q}, q\right)\psi(-q) = HP\psi.$$

Under these conditions, if the wave function has a definite parity at a given initial instant of time, it conserves the same parity in the course of time.

This property is easily extended to a system having an arbitrary number of dimensions; in particular, it applies to systems of particles for which the parity operation amounts to a reflection in space ($\boldsymbol{r}_i \to -\boldsymbol{r}_i$) and for which the observable parity is defined by

$$P\Psi(\boldsymbol{r}_1, \boldsymbol{r}_2, \ldots) = \Psi(-\boldsymbol{r}_1, -\boldsymbol{r}_2, \ldots).$$

EXERCISES AND PROBLEMS

1. Prove the Schwarz inequality $|\langle\varphi, \psi\rangle| \leqslant \sqrt{\langle\varphi, \varphi\rangle\langle\psi, \psi\rangle}$ from the properties (*a*), (*b*) and (*c*) of the scalar product (§ 2). [Write down that the norm of an arbitrary linear combination of φ and ψ is necessarily positive or zero]. Show that the equality between the two sides holds if, and only if the functions φ and ψ are multiples of each other.

2. Consider the eigenvalue problem of the operator $p = (\hbar/\mathrm{i})\mathrm{d}/\mathrm{d}q$ acting on the functions $\psi(q)$ defined in the interval $(-\infty, +\infty)$. Verify that the spectrum is continuous and that the eigenfunctions have an infinite norm. Show that by superposition of eigenfunctions corresponding to neighboring eigenvalues, one can build up functions of finite norm (eigendifferentials) for which the root-mean-square deviation Δp can be made as small as desired. Consider the same question for the operator

$$\frac{p^2}{2m} = -\frac{\hbar^2}{2m}\frac{\mathrm{d}^2}{\mathrm{d}q^2}.$$

3. Consider the operator q acting on the functions $\psi(q)$ of finite norm in the interval $(-\infty, +\infty)$. (*a*) Form a sequence of continuous functions depending

on a parameter η such that their norm and the mean value $\langle q \rangle$ remain independent of η, and that the root-mean-square deviation Δq vanishes in the limit $\eta \to 0$ (there exists a large number of sequences of this type). One notes that this sequence does not converge toward a function of Hilbert space. (*b*) In the same way, build the sequence of eigendifferentials (they are functions of finite norm, but are not continuous) depending on the parameter δq and obtained by superposing the eigen-"functions" $\delta(q - q_0)$ corresponding to eigenvalues q_0 located in the interval δq; verify that the root-mean-square deviation Δq corresponding to these eigendifferentials also vanishes in the limit $\delta q \to 0$.

4. Verify that the statistical distribution of the measurements of a given quantity is uniquely defined, despite the large degree of arbitrariness in the choice of the complete orthonormal set of eigenfunctions associated with the operator which represents this physical quantity (arbitrariness in the choice of the phases of the eigenfunctions; in the choice of the eigenfunctions of the same degenerate eigenvalue; in the normalization of the functions of the continuous spectrum).

5. How can one extend equations (V.20) and (V.21) of this chapter to the case of the continuous spectrum?

6. In polar coordinates (r, θ, φ) the component l_z of the angular momentum of a particle takes the form $(\hbar/i)\partial/\partial\varphi$. This suggests that l_z and φ form a pair of complementary variables. However, the uncertainty relation $\Delta l_z \Delta\varphi \gtrsim \hbar$ has no meaning. Explain why. Show that when $\Delta l_z = 0$, the angle φ is completely indeterminate. Expand the wave packet

$$\psi(\varphi) = \sum_{n=\infty}^{+\infty} \exp\left[-\sigma(\varphi - \varphi_0 + 2n\pi)^2\right]$$

in a series of eigenfunctions of l_z. Compare the uncertainty in φ and the uncertainty in l_z in this example and discuss the complementary character of these two variables.

CHAPTER VI

CLASSICAL APPROXIMATION AND THE WKB METHOD

I. THE CLASSICAL LIMIT OF WAVE MECHANICS

1. General Remarks

In the limit where $\hbar \to 0$, the laws of Quantum Mechanics must reduce to those of Classical Mechanics. The correspondence principle, which played such an important role in the establishment of the theory, had as its exact goal the fulfilment of this fundamental requirement.

Classical Mechanics must furnish a good description of phenomena in all those circumstances where the quantum of action may be considered infinitely small. The purpose of the present chapter is to indicate under what circumstances and to what extent this classical approximation is justified.

One of the manifestations of the finite character of $\hbar$ is the existence of a discrete spectrum of eigenvalues for certain observables; the spacing between neighboring eigenvalues is of the order of $\hbar$ [1]). In order that the approximation be justified, it is necessary that this spacing could be considered negligible; that is the case if *large quantum numbers* are involved. This condition, based on the correspondence principle, has already been used in the Old Quantum Theory to calculate the Rydberg constant and to establish the Bohr-Sommerfeld quantization rules (Ch. I, §§ 13 and 15). The condition is certainly not sufficient; thus, some purely quantum-mechanical effects such as the uncertainty relations are not related to the discreteness of certain spectra. More generally, the conditions of validity of the classical approximation are those of *geometrical optics.*

[1]) Thus, in the hydrogen atom,

$$E_n = -\tfrac{1}{2}\left(\frac{e^2}{\hbar c}\right)^2 \frac{mc^2}{n^2}, \qquad \Delta E = E_{n+1} - E_n = -\frac{2n+1}{(n+1)^2} E_n.$$

In the limit where $\hbar \to 0$, the distance from a given energy level $E_n = E$ to its nearest neighbor is $\dfrac{2n+1}{(n+1)^2}E$; in the limiting process, $n^2\hbar^2 =$ constant, hence $n \to \infty$ as $1/\hbar$ and $\Delta E \approx \mathrm{O}(1/n) = \mathrm{O}(\hbar)$.

The classical approximation can be formulated in two different ways. The first, which is most readily grasped intuitively, consists in defining the dynamical state of each particle at a given instant by its position and its velocity. If $\hbar$ were zero, this would be rigorously justified, since the components of the position and the momentum of each particle would be represented by commuting observables. In fact, the existence of non-zero commutators

$$[q_i, p_j] = \mathrm{i}\hbar\,\delta_{ij} \tag{VI.1}$$

sets a limit to the precision to which position and momentum can be defined simultaneously, a limit which is given by the uncertainty relations. The dynamical state of the system is rigorously represented by its wave function, and the best one can do is to form the minimum wave packet for which $\Delta q_i \cdot \Delta p_i = \frac{1}{2}\hbar$. The classical picture consists in attributing to each particle a position and a momentum equal to the respective mean values of these quantities in the corresponding quantum state. It systematically ignores any fluctuation about these mean values. In order that this picture be satisfactory, it is necessary that:

(*a*) the mean values follow the classical laws of motion to a good approximation;

(*b*) the dimensions of the wave packet be small with respect to the characteristic dimensions of the problem, and that they remain small in the course of time.

These points will be considered in the two succeeding paragraphs. In particular, we shall see that, except under very special circumstances, any wave packet "spreads" indefinitely in the course of time, and may occupy as large a portion of space as one wishes provided one waits a sufficiently long time. The above classical picture therefore only holds during a finite interval of time.

The second way to formulate the classical approximation consists in likening the system to a statistical mixture of classical systems. More precisely, one defines by means of the wave function a classical statistical mixture whose density at each point of configuration space is equal to the probability density of the presence of the quantum system at that point; one can show that in the limit $\hbar \to 0$, the evolution of this mixture is just the one predicted by Classical

Mechanics. This will be done in § 4. From the mathematical point of view, this formulation [1]) is more satisfactory than the first, because it views the equations of Classical Mechanics as limits of the Schrödinger equation. The conditions of validity of this classical approximation are exactly those of geometrical optics.

This approximation is closely related to an approximate method of solution of the Schrödinger equation, known under the name of WKB method [2]), which is applicable when the Schrödinger equation can be replaced by its classical limit except for limited regions of space surrounding singular points. The WKB method is outlined in the second section of this chapter.

2. Ehrenfest's Theorem

The Ehrenfest Theorem gives the law of motion of the mean values of the coordinates q and the conjugate momenta p of a quantum system. It stipulates that the equations of motion of these mean values are formally identical to the Hamilton equations of Classical Mechanics, except that the quantities which occur on both sides of the classical equations must be replaced by their average values.

This theorem results directly from the application of the general equation (V.72) to the position and momentum coordinates. Let $q_1, \ldots, q_R$ be the (cartesian) coordinates of position, $p_1, \ldots, p_R$ their conjugate momenta, and $H(q_1, \ldots, q_R; p_1, \ldots, p_R)$ the Hamiltonian of the system.

In accordance with eq. (V.72),

$$i\hbar \frac{\mathrm{d}}{\mathrm{d}t} \langle q_i \rangle = \langle [q_i, H] \rangle$$

$$i\hbar \frac{\mathrm{d}}{\mathrm{d}t} \langle p_j \rangle = \langle [p_j, H] \rangle .$$

The calculation of the commutators of the right-hand sides was

---

[1]) Actually, there exist several variants, since the definition of the classical statistical mixture from the wave function is not unique (cf. Problem VI.4). The simplest one is given in § 4 below.

[2]) The WKB technique was introduced by Lord Rayleigh (1912) in the solution of problems on wave propagation. Its first application to quantum mechanics is due to H. Jeffreys (1923); it was then developed simultaneously by G. Wentzel, H. A. Kramers, and L. Brillouin (1926).

carried out in Chapter V [eqs. (V.67) and (V.68)]. Whence the announced results [1])

$$\begin{aligned} \frac{\mathrm{d}}{\mathrm{d}t}\langle q_i\rangle &= \left\langle\frac{\partial H}{\partial p_i}\right\rangle, \qquad (i=1, 2, \ldots, R) \\ \frac{\mathrm{d}}{\mathrm{d}t}\langle p_j\rangle &= -\left\langle\frac{\partial H}{\partial q_j}\right\rangle, \qquad (i=1, 2, \ldots, R). \end{aligned} \tag{I}$$

One must clearly understand the connection between the system of equations (I) and the canonical equations of Hamilton. It is generally not correct to state that the mean values $\langle q_i\rangle$ and $\langle p_i\rangle$ follow the laws of Classical Mechanics. The derivatives with respect to time of the *classical* quantities q_i and p_j are well-defined functions $\partial H/\partial p_i$, $-\partial H/\partial q_j$ of these quantities. The values taken by these quantities in the course of time can be exactly derived from their initial values. According to equations (I) on the other hand, the derivatives $\mathrm{d}\langle q_i\rangle/\mathrm{d}t$, $\mathrm{d}\langle p_j\rangle/\mathrm{d}t$ are equal to certain average values whose calculation generally necessitates the knowledge of the wave function $\Psi(t)$. The mean values $\langle q_i\rangle$, $\langle p_j\rangle$ do not follow the classical laws of motion unless one can replace the mean values in (I) of the functions of the right-hand sides by the functions of the mean values, namely

$$\left\langle\frac{\partial}{\partial p_i}H(q_1, \ldots, q_R; p_1, \ldots, p_R)\right\rangle \text{ by } \frac{\partial}{\partial p_i}H(\langle q_1\rangle \ldots \langle q_R\rangle; \langle p_1\rangle \ldots \langle p_R\rangle) \tag{VI.2a}$$

$$\left\langle\frac{\partial}{\partial q_j}H(q_1, \ldots, q_R; p_1, \ldots, p_R)\right\rangle \text{ by } \frac{\partial}{\partial q_j}H(\langle q_1\rangle \ldots \langle q_R\rangle; \langle p_1\rangle \ldots \langle p_R\rangle). \tag{VI.2b}$$

This replacement is rigorously justified only if the Hamiltonian is a polynomial of the second degree with respect to the q's and the p's (free particle, harmonic oscillator, charged particle in a constant electric or magnetic field, cf. Problems VI.1 and VI.2). Apart from these particular cases, this can only happen if the fluctuations of the q's and the p's about their average values are negligible.

By way of an example, let us consider the case of a particle in a potential

$$H = \frac{p^2}{2m} + V(\boldsymbol{r}).$$

---

[1]) Provided one exercises caution in the definition of the operators $\partial H/\partial p_i$, $\partial H/\partial q_i$ (cf. Ch. V, § 17).

Let us introduce the force

$$\mathbf{F} = -\operatorname{grad} V(\mathbf{r}).$$

The Ehrenfest equations are in that case

$$\frac{\mathrm{d}}{\mathrm{d}t}\langle \mathbf{r} \rangle = \frac{\langle \mathbf{p} \rangle}{m}, \qquad \frac{\mathrm{d}}{\mathrm{d}t}\langle \mathbf{p} \rangle = \langle \mathbf{F} \rangle,$$

or else

$$\langle \mathbf{F} \rangle = m \frac{\mathrm{d}^2}{\mathrm{d}t^2} \langle \mathbf{r} \rangle, \qquad \text{(VI.3)}$$

the quantum analogue of Newton's law.

In order that the average position

$$\langle \mathbf{r} \rangle \equiv \int \Psi^*(\mathbf{r}, t)\, \mathbf{r} \Psi(\mathbf{r}, t)\, \mathrm{d}\mathbf{r}$$

actually follow Newton's classical equation, one must be able to replace in eq. (VI.3) the mean value of the force

$$\langle \mathbf{F} \rangle = \int \Psi^*(\mathbf{r}, t)\, F(\mathbf{r})\, \Psi(\mathbf{r}, t)\, \mathrm{d}\mathbf{r}$$

by its value $\mathbf{F}(\langle \mathbf{r} \rangle)$ at the point $\mathbf{r}$. When the force vanishes (free particle) or depends linearly upon the coordinates of $\mathbf{r}$ (harmonic oscillator), one has rigorously $\mathbf{F} = \mathbf{F}(\langle \mathbf{r} \rangle)$. In other cases the substitution is justified only if the wave function remains localized in a sufficiently small region of space so that the force has a practically constant value over that entire region.

3. Motion and Spreading of Wave Packets

In order that the motion of a wave packet may be likened to the motion of a classical particle, it is first of all necessary that its position and momentum follow the laws of classical mechanics; but the dimensions of this packet must also remain sufficiently small at all times. In fact, as Ehrenfest's Theorem suggests, the first requirement is rarely satisfied without the second. We shall especially focus our attention on the latter.

In order to see the essential features, it suffices to study the motion of a one-dimensional wave packet $\psi(q, t)$. Let H be the Hamiltonian:

$$H = \frac{p^2}{2m} + V(q).$$

The quantities whose evolution in time we wish to study are the mean values $\langle q \rangle$ and $\langle p \rangle$ and the mean-square deviations

$$\chi \equiv (\Delta q)^2 = \langle q^2 \rangle - \langle q \rangle^2, \qquad \varpi \equiv (\Delta p)^2 = \langle p^2 \rangle - \langle p \rangle^2.$$

In the classical approximation, the packet represents a particle of position and momentum

$$q_{\text{cl.}} = \langle q \rangle, \qquad p_{\text{cl.}} = \langle p \rangle$$

respectively [1]. Note that the energy of this classical particle

$$E_{\text{cl.}} = \frac{\langle p \rangle^2}{2m} + V(\langle q \rangle)$$

is not equal to the average value $\langle H \rangle$. Actually, if the classical approximation is justified, $E_{\text{cl.}}$ is constant in time, as well as the difference

$$\varepsilon = \langle H \rangle - E_{\text{cl.}}. \tag{VI.4}$$

Since the extension $\sqrt{\chi}$ of the wave packet is to remain small, it is natural to replace functions such as $V(q)$, $V'(q)$ by their Taylor expansion about $\langle q \rangle$, namely

$$V(q) = V_{\text{cl.}} + (q - \langle q \rangle)\, V'_{\text{cl.}} + \tfrac{1}{2}(q - \langle q \rangle)^2\, V''_{\text{cl.}} + \dots \tag{VI.5}$$

$$V'(q) = V'_{\text{cl.}} + (q - \langle q \rangle)\, V''_{\text{cl.}} + \tfrac{1}{2}(q - \langle q \rangle)^2\, V'''_{\text{cl.}} + \dots \tag{VI.6}$$

In these expressions, $V_{\text{cl.}}$, $V'_{\text{cl.}}$, ... denote the values assumed by the functions V, V', ... at the point $q = \langle q \rangle$. Taking the average values of these quantities, we obtain, so to speak, their expansions in series of powers of χ, namely

$$\langle V \rangle = V_{\text{cl.}} + \tfrac{1}{2}\chi V''_{\text{cl.}} + \dots \tag{VI.7}$$

$$\langle V' \rangle = V'_{\text{cl.}} + \tfrac{1}{2}\chi V'''_{\text{cl.}} + \dots \tag{VI.8}$$

With the help of these expansions, we are in a position to obtain entirely general results, valid for any $V(q)$.

---

[1] In the study of the one-dimensional wave packet of Chapter II, the center of the wave packet was defined by a condition of stationary phase, while the definition adopted here is the mean value $\langle q \rangle$. In the classical approximation, the difference between these two definitions is negligible.

The quantities $\langle q \rangle$ and $\langle p \rangle$ follow Ehrenfest's equations

$$\frac{\mathrm{d}}{\mathrm{d}t}\langle q \rangle = \frac{\langle p \rangle}{m} \qquad \text{(VI.9 } a\text{)}$$

$$\frac{\mathrm{d}}{\mathrm{d}t}\langle p \rangle = -\langle V' \rangle. \qquad \text{(VI.9 } b\text{)}$$

They reduce to the classical equations if the right-hand side of eq. (VI.9*b*) is replaced by the first term of its expansion (VI.8). This replacement is justified if the term $\frac{1}{2}\chi V'''_{\mathrm{cl.}}$ and terms of higher order are negligible. This holds rigorously true when $V'''(q)$ is zero everywhere, i.e. when $V(q)$ is a polynomial of at most the second degree in q, and especially when $V(q) = cq^2$ (harmonic oscillator) and when $V(q) = 0$ (free particle). Otherwise, $V(q)$ must vary sufficiently slowly over a distance of the order of the extension $\chi^{\frac{1}{2}}$ of the wave packet, so that the effect of V''' and higher order derivatives in the expansion (VI.8) is sufficiently small.

Assuming these conditions to be fulfilled [which amounts to assuming the expansions (VI.7) and (VI.8) to be rapidly converging] we obtain for the constant ε [cf. eq. (VI.4)] the expression

$$\varepsilon \equiv \frac{\varpi}{2m} + \langle V \rangle - V_{\mathrm{cl.}} \simeq \frac{1}{2m}(\varpi + m V''_{\mathrm{cl.}}\, \chi) = \text{constant} \qquad \text{(VI.10)}$$

connecting the mean-square deviations ϖ and χ.

We shall restrict ourselves therefore to a study of χ as a function of time. χ is the mean value of the operator $q^2 - \langle q \rangle^2$ (depending explicitly upon the time since $\langle q \rangle$ is a function of time). Applying relation (V.72) to this operator, one obtains after some calculation

$$\frac{\mathrm{d}}{\mathrm{d}t}\chi = \frac{1}{m}(\langle pq + qp \rangle - 2\langle p \rangle \langle q \rangle).$$

By an analogous procedure for $\mathrm{d}\chi/\mathrm{d}t$, one has

$$\frac{\mathrm{d}^2\chi}{\mathrm{d}t^2} = \frac{2\varpi}{m^2} - \frac{1}{m}(\langle V'q + qV' \rangle - 2\langle q \rangle \langle V' \rangle).$$

By replacing the operator V' in the bracket of the right-hand side by the first two terms of expansion (VI.6), we obtain the approximate equation

$$\frac{\mathrm{d}^2\chi}{\mathrm{d}t^2} \simeq \frac{2}{m^2}(\varpi - m V''_{\mathrm{cl.}}\, \chi) \qquad \text{(VI.11)}$$

which we can write, taking into account equation (VI.10)

$$\frac{d^2\chi}{dt^2} \approx \frac{4}{m}(\varepsilon - V''_{cl.}\chi). \tag{VI.12}$$

Knowing the deviations χ_0, ϖ_0, and $\dot{\chi}_0 \equiv d\chi_0/dt$ at time t_0, one obtains χ at any later time by solving the equation (VI.12) – noting carefully that $V''_{cl.}$ may be time-dependent – and ϖ by means of equation (VI.10). In view of this fact, the error introduced by replacing $\langle V' \rangle$ by $V'_{cl.}$ in equation (VI.9*b*) can be evaluated. All the elements are therefore at hand for deciding if the assimilation of the wave packet to a classical particle is justified.

The two most interesting cases are those of the harmonic oscillator and the free particle, cases for which the motion of the center of the packet is rigorously identical to that of a classical particle. In the case of the harmonic oscillator ($V = \frac{1}{2}m\omega^2 q^2$), $\langle q \rangle$ oscillates about zero with the frequency $\omega/2\pi$ and χ oscillates about $\varepsilon/m\omega^2$ with twice that frequency. (Cf. Problem VI.1.)

If one is dealing with a free particle ($V = 0$), $\langle q \rangle$ undergoes uniform rectilinear motion of velocity $\langle p \rangle/m$, the deviation ϖ remains strictly constant ($\varpi = \varpi_0 = 2m\varepsilon$) and since one has rigorously $d^2\chi/dt^2 = 2\varpi_0/m^2$ [the equations (VI.11) and (VI.12) are exact in that case]

$$\chi = \chi_0 + \dot{\chi}_0 t + \frac{\varpi_0}{m^2} t^2. \tag{VI.13}$$

Consequently, after a sufficiently long time, χ becomes as large as desired: the free wave packet "spreads" indefinitely.

The appearance of an indefinite spreading is an important fact because it sets a limit for the time interval during which the wave packet may be likened to a classical particle. Indeed, apart from very special cases such as that of harmonic oscillation, the wave packet always spreads, especially in collision problems where the laws of motion for the packet approach those of a free wave packet as soon as the colliding particle is sufficiently far from the scattering center.

The spreading law for a free wave packet turns out to be quite simple if the wave packet is taken to be minimum at time t_0 (cf. Problem IV.4), i.e. $\varpi_0\chi_0 = \frac{1}{2}\hbar^2$ (in which case $\dot{\chi}_0 = 0$); then one has

$$\chi = \chi_0 + \frac{\varpi_0}{m^2} t^2$$

or else

$$\Delta q = \left[(\Delta q_0)^2 + \left(\frac{\Delta p_0}{m} t\right)^2\right]^{\frac{1}{2}} . \tag{VI.14}$$

The very form of the "spreading term" $\Delta p_0 t/m$ suggests a simple classical picture for this free wave packet, that of a swarm of projectiles initially bunched within an interval Δq_0 about the average value $\langle q_0 \rangle$, the velocities of these projectiles being distributed over an interval

$$\Delta v = \Delta p_0 / m$$

about the group velocity of the packet $v = \langle p_0 \rangle / m$. Owing to their dispersion in velocity, projectiles initially located at the same point, come to be uniformly distributed over a band Δvt at time t; consequently, the projectiles do not retain their original clustering, and the width of the swarm increases appreciably according to the law (VI.14).

This same law can be put into other forms. In particular, it may be written

$$\Delta q = \Delta q_0 \left[1 + \frac{1}{4}\left(\frac{D\lambda\!\!\!^{-}}{(\Delta q_0)^2}\right)^2\right]^{\frac{1}{2}}$$

where $D \equiv vt$ is the distance covered by the wave packet during the time t and $\lambda\!\!\!^{-} \equiv \hbar/mv$ is the average wave length. The spread of the free wave packet is thus negligible provided

$$\sqrt{D\lambda\!\!\!^{-}} \ll \Delta q_0. \tag{VI.15}$$

In fact, it is easy to show that the extension Δq always exceeds the length $\sqrt{D\lambda\!\!\!^{-}}$. (Note that: $D\lambda\!\!\!^{-} = \hbar t/m$.)

4. Classical Limit of the Schrödinger Equation

We shall now study the second formulation of the classical approximation mentioned in the introduction of this chapter.

To be definite, we treat the case of a particle in a potential $V(\boldsymbol{r})$. Let us separate modulus and phase in its wave function:

$$\Psi(\boldsymbol{r}) = A(\boldsymbol{r}) \exp\left(\frac{i}{\hbar} S(\boldsymbol{r})\right). \tag{VI.16}$$

Substituting expression (VI.16) in the Schrödinger equation and separating real and imaginary parts, we obtain the two equations

$$\frac{\partial S}{\partial t} + \frac{(\nabla S)^2}{2m} + V = \frac{\hbar^2}{2m}\frac{\triangle A}{A} \tag{VI.17}$$

$$m\frac{\partial A}{\partial t} + (\nabla A \cdot \nabla S) + \frac{A}{2}\triangle S = 0. \tag{VI.18}$$

These two equations are strictly equivalent to the Schrödinger equation. In fact, equation (VI.18) is just the continuity equation (IV.11). Indeed, the probability density of presence $P(\mathbf{r})$ and the current density $\mathbf{J}(\mathbf{r})$ (cf. Ch. IV, §§ 2 and 4) are respectively given by

$$P = A^2, \qquad \mathbf{J} = A^2 \frac{\nabla S}{m}.$$

After multiplying term by term by $2A$, eq. (VI.18) can be put in the form

$$m\frac{\partial}{\partial t}A^2 + \operatorname{div}(A^2 \nabla S) = 0. \tag{VI.19}$$

The classical approximation consists in setting $\hbar$ equal to zero in eq. (VI.17), that is

$$\frac{\partial S}{\partial t} + \frac{(\nabla S)^2}{2m} + V = 0. \tag{VI.20}$$

We deduce the following result:

In the classical approximation, Ψ describes a fluid of non-interacting classical particles of mass m, (statistical mixture), *and subject to the potential $V(\mathbf{r})$: the density and current density of this fluid at each point of space are at all times respectively equal to the probability density P and the probability current density $\mathbf{J}$ of the quantum particle at that point* [1]).

Indeed, since the continuity equation of this fluid is satisfied [eq. (VI.19)], it suffices to show that the velocity field

$$\mathbf{v} = \frac{\mathbf{J}}{P} = \frac{\nabla S}{m} \tag{VI.21}$$

[1]) The density in phase space of this classical statistical mixture is somewhat peculiar since to each point of configuration space corresponds a well-defined momentum ΔS. The solution S of equation (VI.20) is the "principal function" of Hamilton used in the Hamilton-Jacobi formulation of Classical Mechanics.

of this fluid actually follows the law of motion of the classical fluid in question. Now, with the definition (VI.21), eq. (VI.20) is written

$$\frac{\partial S}{\partial t} + \frac{mv^2}{2} + V = 0.$$

Writing down that the gradient of the left-hand side is zero, one has, taking into account eq. (VI.21),

$$\left(\frac{\partial}{\partial t} + (\mathbf{v} \cdot \text{grad})\right) m\mathbf{v} + \text{grad } V = 0,$$

from which one concludes that the particles of the fluid obey the equation of motion

$$m \frac{\mathrm{d}\mathbf{v}}{\mathrm{d}t} = - \text{grad } V. \qquad \text{Q.E.D.}$$

It is important to stress the very great generality of this result which remains valid for systems with any number of dimensions. The density $P = |\Psi|^2$ is a well-defined function of configuration space; similarly, the current $\boldsymbol{J}$ is a well-defined vector field of that space. The proof for the general case will be left to the reader.

When Ψ represents a stationary state of energy E,

$$\frac{\partial A}{\partial t} = 0, \qquad \frac{\partial S}{\partial t} = - E.$$

Equations (VI.17) and (VI.18) reduce to

$$(\nabla S)^2 - 2m(E - V) = \hbar^2 \frac{\triangle A}{A} \tag{VI.22}$$

$$\text{div } (A^2 \nabla S) = 0. \tag{VI.23}$$

In the classical approximation, the right-hand side of eq. (VI.22) is neglected and the result proved above still holds. Ψ then represents a stationary flow of the fluid of classical particles.

Optical analogy is even more suggestive than this hydrodynamical analogy, especially for stationary solutions. Since the velocities of the particles are proportional to the gradient of S, the trajectories of these particles are orthogonal to the surfaces of equal phase $S = const.$ In the language of optics, the latter are the wave fronts and the trajectories of the particles are the rays [1]). Hence, the classical approximation is equivalent to the geometrical optics approximation: we find once again, as a consequence of the Schrödinger equation, the basic postulate of the theory of matter waves.

The optical analogy is extremely useful whenever one wishes to determine if the conditions of validity of the classical approximation are actually fulfilled. Let $\lambda\!\!\!^{-}$ be the reduced wavelength:

$$\lambda\!\!\!^{-} = \frac{\hbar}{\sqrt{2m\,[E - V(\mathbf{r})]}}.$$

It is a well-defined function of $\mathbf{r}$. Eq. (VI.22) can be written in the form

$$(\nabla S)^2 = \frac{\hbar^2}{\lambda\!\!\!^{-2}}\left(1 + \lambda\!\!\!^{-2}\,\frac{\triangle A}{A}\right). \qquad \text{(VI.22')}$$

In the classical approximation

$$\lambda\!\!\!^{-2}\,\frac{\triangle A}{A} \ll 1 \qquad \text{(VI.24)}$$

over all space; more precisely, the regions in which this condition is not fulfilled must be sufficiently small so that eq. (VI.22′) can be replaced by the approximate equation

$$(\nabla S)^2 = \frac{\hbar^2}{\lambda\!\!\!^{-2}}. \qquad \text{(VI.25)}$$

This is the equation of the wave fronts of geometrical optics.

When $V=0$ over all space, $\lambda\!\!\!^{-}$ is constant and the function $S= \mathbf{p}\cdot\mathbf{r}+$*constant*, where $\mathbf{p}$ is a given vector of length $\hbar/\lambda\!\!\!^{-}$, is a particular solution of eq. (VI.25) [the ensemble of functions of this type forms a complete integral of this first-order partial differential equation]; the wave fronts corresponding to this solution are planes perpen-

---

[1]) In the presence of a vector potential A, the definition (VI.21) of the classical velocity must be replaced by

$$\mathbf{v} = \frac{1}{m}\left(\nabla S - \frac{e}{c}\,A\right).$$

With this new definition, the above theorem remains true. The notions of wave front and ray remain valid, but the rays are no longer orthogonal to the wave fronts; this situation is analogous to that of anisotropic media in geometrical optics.

dicular to $\boldsymbol{p}$, and the rays are straight lines parallel to $\boldsymbol{p}$. In the general case, the wave fronts and the rays are curved.

Once the energy E is fixed, the specification of a wave front $S(x, y, z) = S_0$ determines one and only one solution of eq. (VI.25). This solution can be obtained in the following manner. With the surface (S_0) is associated a (two parameter) family of trajectories of the classical particle; these trajectories correspond to the energy E and are orthogonal to (S_0); the other wave fronts are surfaces orthogonal to these trajectories (Fig. VI.1). To find the value taken

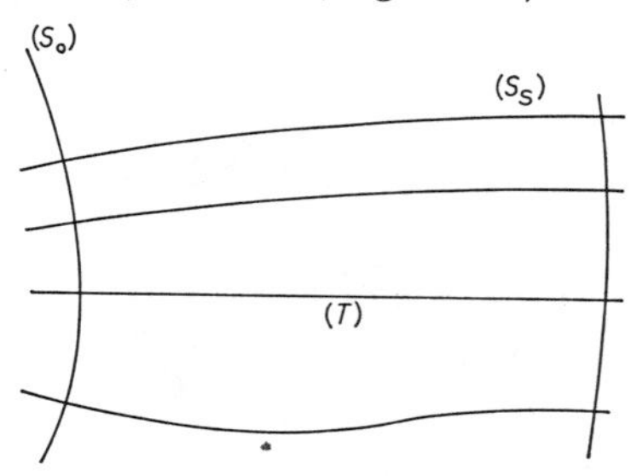

Fig. VI.1. Family of trajectories (of given energy E) associated with a wave front (S_0) in the approximation of geometrical optics: $(\nabla S)^2 = \hbar^2/\lambda\!\!\!^{-2}$.

by S on each of these wave fronts, one considers one of the trajectories (T). Each point of (T) is labelled by its curvilinear coordinate s; we locate the origin $s=0$ at the intersection of (T) with (S_0). One has, according to eq. (VI.25),

$$S(s) = S_0 + \int_0^s \frac{\hbar}{\lambda\!\!\!^{-}} \, \mathrm{d}s\,;$$

S being thus determined over all space, specifying A on the surface (S_0) uniquely determines the function A over all space. Actually eq. (VI.23) relates only the values $A(s)$ taken by A along the same trajectory; indeed it may be written

$$\frac{\hbar}{\lambda\!\!\!^{-}} \frac{\mathrm{d}}{\mathrm{d}s} (A^2) + A^2 \, \triangle S = 0.$$

Since the functions $\lambda\!\!\!^{-}$ and $\triangle S$ are well-defined functions of s on that trajectory, this equation defines $A(s)$ in a unique manner when its value $A(0)$ at the intersection of (T) and (S_0) is known.

Knowing the solutions A and S in the classical approximation, it is possible to evaluate the perturbation due to the term $\lambda\!\!\!^{-2} \triangle A/A$ in the rigorous equation (VI.22′). The effect depends not only upon the "optical properties" of the medium through which the wave propagates, but also upon the particular solution of the wave equation As a general rule, the classical approximation is justified if the "transverse" dimensions of the wave are everywhere large compared to $\lambda\!\!\!^{-}$ and if one has

$$|\text{grad}\ \lambda\!\!\!^{-}| \ll 1 \tag{VI.26}$$

if not over all space, then at least in that portion of space where the density A^2 takes on non-negligible values.

These conditions of validity are deduced from the following semi-quantitative arguments. As suggested by the optical analogy, the curvature of the light rays, i.e. of the particle trajectories, must be small compared to the wavelength. Now, the radius of curvature R is related to the velocity v of the particles and to the transverse component of the force $-(\text{grad}\ V)_\perp$ by the relation

$$\frac{mv^2}{R} = |(\text{grad}\ V)_\perp|.$$

It is therefore necessary that

$$\frac{\lambda\!\!\!^{-}}{R} = \frac{\lambda\!\!\!^{-}|(\text{grad}\ V)_\perp|}{mv^2} = \frac{m\lambda\!\!\!^{-3}}{\hbar^2}|(\text{grad}\ V)_\perp| \ll 1.$$

Taking into account the expression for $\lambda\!\!\!^{-}$ as a function of the potential V, this condition is written

$$|(\text{grad}\ \lambda\!\!\!^{-})_\perp| \ll 1.$$

Now, the curvature of the wave fronts must be small with respect to $1/\lambda\!\!\!^{-}$ (except possibly in certain restricted regions of space, in particular near the focal surfaces) and this may in general be realized if the trajectories also have small curvatures, provided we make an appropriate choice of the surface (S_0). Similarly the relative variation of A on each wave front must be negligible in a region of order $\lambda\!\!\!^{-}$; in other words, the "transverse" dimensions of the wave must be large compared to $\lambda\!\!\!^{-}$. If we assume that these conditions on the solution are fulfilled, the function $A(s)$ along a trajectory (T) is approximately given by

$$A(s) \approx A(0)\sqrt{\frac{\lambda\!\!\!^{-}(s)}{\lambda\!\!\!^{-}(0)}},$$

which yields, after a short calculation,

$$\lambda\!\!\!^{-2}\frac{\triangle A}{A} \approx \frac{\lambda\!\!\!^{-2}}{A}\frac{\mathrm{d}^2A}{\mathrm{d}s^2} \approx \frac{1}{4}\left[2\lambda\!\!\!^{-}\frac{\mathrm{d}^2\lambda\!\!\!^{-}}{\mathrm{d}s^2} - \left(\frac{\mathrm{d}\lambda\!\!\!^{-}}{\mathrm{d}s}\right)^2\right].$$

In order that this term be small it is necessary that $d\lambda\!\!\!^{-}/ds$ and $\lambda\!\!\!^{-} d^2\lambda\!\!\!^{-}/ds^2$ be much less than 1. In practice, the second condition is always fulfilled if the first one holds; that is to say if the component of grad $\lambda\!\!\!^{-}$ along the trajectory is very much less than 1, that is

$$|(\text{grad } \lambda\!\!\!^{-})_{\|}| \ll 1.$$

5. Application to Coulomb Scattering. The Rutherford Formula

By way of an application, we shall present a brief outline of the classical theory of Coulomb scattering and its conditions of validity.

Consider the scattering of a particle of mass m by a Coulomb potential

$$V = \frac{Ze^2}{r},$$

where r is the distance of the particle from the center of force C. This particle is, for instance, a proton of charge e in the Coulomb field of an atomic nucleus of charge Ze (repulsive potential). The theory applies equally well if the charges are of opposite sign (attractive potential). We shall thus treat the constant Ze^2 as an algebraic quantity capable of taking on either sign.

Since the energy E of the particle is fixed once and for all by the relation

$$E = \frac{p_0^2}{2m} = \tfrac{1}{2} m v_0^2,$$

and the direction of propagation of the incident particle is also given, specification of the impact parameter b completes the determination of the initial conditions of the motion. Let

$$a = \frac{1}{2} \frac{Ze^2}{E}. \qquad \text{(VI.27)}$$

As is well known, the trajectory is a branch of an hyperbola (Fig. VI.2) with focus C, semi-axis $OA = |a|$, and focal distance $OC = \sqrt{a^2 + b^2}$. The deflection angle is given by the relation

$$b = |a| \cot \frac{\theta}{2} \qquad \text{(VI.28)}$$

(the direction of the deflection depends on the sign of the potential, but the absolute value of the angle is the same for two potentials equal in absolute value).

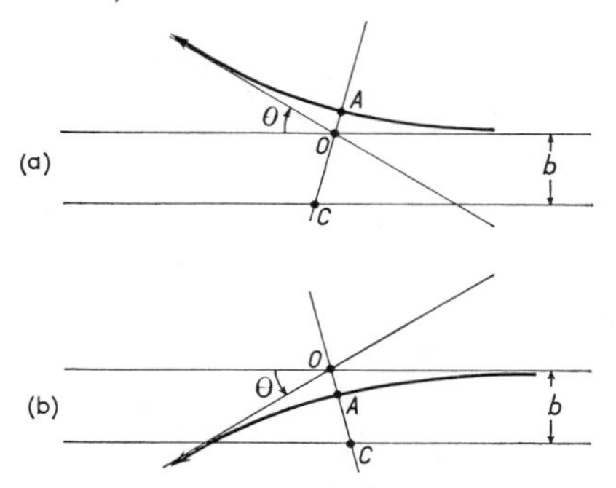

Fig. VI.2. Trajectory followed (heavy line) by a particle in a (*a*) repulsive; (*b*) attractive Coulomb field.

The important quantity one must know in practice is the differential cross section $\mathrm{d}\sigma/\mathrm{d}\Omega$. Consider a parallel beam of incident monoërgic particles directed toward the center of force; one observes the number of particles scattered into a given solid angle $(\Omega, \Omega+\mathrm{d}\Omega)$. $\mathrm{d}\sigma/\mathrm{d}\Omega$ is by definition the number of particles scattered in that direction, per unit solid angle and per unit time, when the particles are uniformly distributed over the incident beam with a flux that is constant in time and equal to unity; that is to say, when any surface located very far from C in the incident direction of propagation, and perpendicular to that direction, is crossed by the particles at the rate of one particle per unit time and per unit area.

Take the incident direction as polar axis, and call θ and φ the polar angles of the direction of propagation of the scattered particle. θ is the deflection angle introduced above; it is connected with the impact parameter b by relation (VI.28). The solid angle $\mathrm{d}\Omega$ in the direction (θ, φ) is equal to

$$\mathrm{d}\Omega = \sin\theta \; \mathrm{d}\theta \; \mathrm{d}\varphi.$$

The number $\mathrm{d}\sigma$ of particles scattered per unit time into this solid angle is equal to the number of incident particles crossing the surface $b \; \mathrm{d}b \; \mathrm{d}\varphi$ per unit time; that is, since the incident flux is equal to 1,

$$\mathrm{d}\sigma = b \; \mathrm{d}b \; \mathrm{d}\varphi = b \, \frac{\mathrm{d}b}{\mathrm{d}\theta} \, \mathrm{d}\theta \; \mathrm{d}\varphi = \frac{b}{\sin\theta} \frac{\mathrm{d}b}{\mathrm{d}\theta} \, \mathrm{d}\Omega.$$

Replacing in this equation the quantities b and $\mathrm{d}b/\mathrm{d}\theta$ by their expressions in terms of θ calculated from relation (VI.28), one obtains the *Rutherford formula* [1])

$$\frac{\mathrm{d}\sigma}{\mathrm{d}\Omega} = \frac{a^2}{4\sin^4\frac{\theta}{2}} = \frac{(Ze^2)^2}{16E^2\sin^4\frac{\theta}{2}}. \tag{VI.29}$$

We shall now discuss the validity of the classical approximation. Note first that $|a|$ is a characteristic length of this collision problem. Moreover, the wavelength of the particle

$$\lambda\!\!\!^{-} = \frac{\hbar}{p} = \frac{\hbar}{p_0}\left(1 - \frac{2a}{r}\right)^{-\frac{1}{2}}$$

is of the order of magnitude of its initial value $\lambda\!\!\!^{-}_0 = \hbar/mv_0$. The ratio of the two quantities $|a|$ and $\lambda\!\!\!^{-}$ is therefore of order

$$\gamma = \frac{|a|}{\lambda\!\!\!^{-}_0} = \frac{|Ze^2|}{\hbar v_0}. \tag{VI.30}$$

One expects the classical approximation to be justified in the limit where

$$\gamma \gg 1. \tag{VI.31}$$

Let us examine to what extent condition (VI.26) is fulfilled. One has

$$|\operatorname{grad} \lambda\!\!\!^{-}| = \left|\frac{\mathrm{d}\lambda\!\!\!^{-}}{\mathrm{d}r}\right| = \frac{1}{\gamma}\frac{a^2}{\sqrt{r(r-2a)^3}} \ll 1\,;$$

$|\operatorname{grad} \lambda\!\!\!^{-}|$ is larger the smaller r. Since the classical trajectory approaches the scattering center the closer the larger the angle of deflection, we conclude that the classical approximation holds for small deflection angles (that is to say for large impact parameters) and ceases to be valid for large angles. In order to reach a more quantitative conclusion, consider the expression for the maximum of $|\operatorname{grad} \lambda\!\!\!^{-}|$ (i.e. its value at the vertex of the hyperbolic trajectory) as a function of θ; after some calculation

$$|\operatorname{grad} \lambda\!\!\!^{-}|_{\max.} = \frac{F(\theta)}{\gamma}, \qquad F(\theta) = \tan\tfrac{1}{2}\theta\,\frac{\sin\frac{1}{2}\theta}{1-(\operatorname{sgn} a)\sin\frac{1}{2}\theta}.$$

---

[1]) This formula, due to Rutherford, is of great historical importance; upon it rested the entire interpretation of Rutherford's famous experiments on α-particle scattering.

[By definition, $\operatorname{sgn} a \equiv |a|/a$.] When θ increases from 0 to π, $F(\theta)$ increases from 0 to $+\infty$. Let θ_c be an angle such that

$$F(\theta_c) = \gamma.$$

The classical approximation is justified as long as $\theta < \theta_c$ and ceases to be valid when $\theta > \theta_c$. Note that θ_c lies the closer to π the larger γ, in agreement with the rough predictions made earlier.

II. THE WKB METHOD [1]

6. Principle of the Method

The method consists in introducing an expansion in powers of $\hbar$ and in neglecting terms of higher order than $\hbar^2$. One thus replaces (at least in some regions of space) the Schrödinger equation by its classical limit. However, the method has a wider range of applicability than the classical approximation proper because this procedure can be carried out even in regions of space where the classical interpretation is meaningless (regions $E < V$ which are inaccessible to classical particles). In order to include these regions in our treatment, we must slightly modify the definitions of A and S of § 4 and write

$$\Psi(\boldsymbol{r}) = \exp\left(\frac{\mathrm{i}}{\hbar} W(\boldsymbol{r})\right) \tag{VI.32}$$

$$W(\boldsymbol{r}) = S(\boldsymbol{r}) + \frac{\hbar}{\mathrm{i}} T(\boldsymbol{r}) \tag{VI.33}$$

$$A(\boldsymbol{r}) = \exp\,[T(\boldsymbol{r}).] \tag{VI.34}$$

Requiring S and T to be even functions of $\hbar$ defines A and S in unique fashion. With these new definitions, equations (VI.17) and (VI.18) as well as (VI.22) and (VI.23) remain valid although A and S are no longer necessarily real. The WKB approximation consists in expanding $W(r)$ in powers of $\hbar$ and in neglecting in the Schrödinger equation terms of order equal to or greater than $\hbar^2$.

---

[1] Cf. § 1, footnote 2, p. 216. For more details, see R. E. Langer, Phys. Rev. **51** (1937) 669; W. H. Furry, Phys. Rev. **71** (1947) 360. See also P. M. Morse and H. Feshbach, *Methods of Theoretical Physics* (New York, McGraw-Hill, 1953) pp. 1092 ff.

7. One-Dimensional WKB Solutions

The WKB method can be easily applied only to *one-dimensional problems.* We shall therefore restrict ourselves to treating one-dimensional problems and we shall look for the stationary solutions of the time-independent Schrödinger equation [eqs. (VI.22) and (VI.23)]. The method developed here may eventually serve to solve the three-dimensional Schrödinger equation, since the latter can be reduced in many practical cases to the solution of a one-dimensional equation by separating angular and radial variables (cf. Ch. IX).

Let $y(x)$ be the wave function satisfying the Schrödinger equation

$$y'' + \frac{2m}{\hbar^2}[E - V(x)]y = 0.$$

Setting

$$y = e^{iw/\hbar}, \qquad w = S + \frac{\hbar}{i} \ln A$$

(S and $\ln A$ are even functions of $\hbar$) one obtains the system of equivalent equations

$$S'^2 - 2m(E - V) = \hbar^2 \frac{A''}{A} \tag{VI.35}$$

$$2A'S' + AS'' = 0. \tag{VI.36}$$

The equation of continuity (VI.36) can be integrated to yield

$$A = \text{const.}\,(S')^{-\frac{1}{2}}. \tag{VI.37}$$

Substituting this expression for A into eq. (VI.35), one obtains the equation

$$S'^2 = 2m(E - V) + \hbar^2 \left[\frac{3}{4}\left(\frac{S''}{S'}\right)^2 - \frac{1}{2}\frac{S'''}{S'}\right]. \tag{VI.38}$$

This differential equation of the third order in S is *rigorously* equivalent to the initial Schrödinger equation.

In the WKB *approximation* one expands S in a power series in $\hbar^2$:

$$S = S_0 + \hbar^2 S_1 + \ldots, \tag{VI.39}$$

then one substitutes this expansion into eq. (VI.38) and keeps only zero-order terms:

$$S'^2 \approx S_0'^2 = 2m[E - V(x)]. \tag{VI.40}$$

This approximate equation is easily integrated.

One must distinguish two cases according to whether $E \gtrless V(x)$.

First Case: $E > V(x)$.

We define the wavelength

$$\lambda\!\!\!^{-}(x) = \frac{\hbar}{\sqrt{2m[E - V(x)]}}. \tag{VI.41}$$

Equation (VI.40) is satisfied if $S' \approx \pm\hbar/\lambda\!\!\!^{-}$. The WKB solution is a linear combination of oscillating functions

$$y(x) = \alpha\sqrt{\lambda\!\!\!^{-}} \cos\left(\int^{x} \frac{\mathrm{d}x}{\lambda\!\!\!^{-}} + \varphi\right) \tag{VI.42}$$

(α and φ are arbitrary constants).

Second Case: $E < V(x)$ (forbidden region for classical particles).

$$l(x) = \frac{\hbar}{\sqrt{2m[V(x) - E]}}. \tag{VI.43}$$

Eq. (VI.40) is satisfied if $S' \approx \pm \mathrm{i}\hbar/l$. The WKB solution is a linear combination of real exponentials

$$y(x) = \sqrt{l}\left[\gamma \exp\left(+\int^{x} \frac{\mathrm{d}x}{l}\right) + \delta \exp\left(-\int^{x} \frac{\mathrm{d}x}{l}\right)\right]. \tag{VI.44}$$

8. Conditions for the Validity of the WKB Approximation

The theory of the WKB approximation is rather involved. Let us merely indicate here without proof that the expansion (VI.39) in powers of $\hbar^2$ does not in general converge, but that it is an asymptotic expansion which, if broken off after a finite number of terms, gives S to a good approximation if $\hbar$ is sufficiently small.

In order to find a criterion for the validity of the WKB approximation, one can calculate the second term $\hbar^2 S_1$ of the expansion (VI.39). The correction of order $\hbar^2$ consists in multiplying the WKB solution by the factor $\exp(\mathrm{i}\hbar S_1)$. The effect is negligible if $\hbar S_1 \ll 1$.

By substituting expansion (VI.39) into equation (VI.38) and equating the coefficients of $\hbar^2$ on the two sides, one obtains the differential equation for S_1

$$2S_0'S_1' = \frac{[(S_0')^{-\frac{1}{2}}]''}{(S_0')^{-\frac{1}{2}}} = \frac{3}{4}\left(\frac{S_0''}{S_0'}\right)^2 - \frac{1}{2}\frac{S_0'''}{S_0'}.$$

When $E > V$, $S_0' = \pm\hbar/\lambda\!\!\!^{-}$. After some calculation this yields

$$\hbar S_1' = \pm\tfrac{1}{2}\sqrt{\lambda\!\!\!^{-}}(\sqrt{\lambda\!\!\!^{-}})'' = \pm\left(\tfrac{1}{4}\lambda\!\!\!^{-}{}'' - \frac{1}{8}\frac{\lambda\!\!\!^{-}{}'^2}{\lambda\!\!\!^{-}}\right),$$

from which one obtains

$$\hbar S_1 = \pm\left(\tfrac{1}{4}\lambda\!\!\!^{-}{}' - \frac{1}{8}\int^x \frac{\lambda\!\!\!^{-}{}'^2}{\lambda\!\!\!^{-}}\,\mathrm{d}x\right). \tag{VI.45}$$

When $E > V$, one obtains an identical expression with $l(x)$ instead of $\lambda\!\!\!^{-}(x)$. The condition $\hbar S_1 \ll 1$ is fulfilled if

$$\begin{aligned} \lambda\!\!\!^{-}{}'(x) &\ll 1 \quad \text{when} \quad E > V(x),\\ l'(x) &\ll 1 \quad \text{when} \quad E < V(x). \end{aligned} \tag{VI.46}$$

One can compare the criterion (VI.46) with condition (VI.26) for the validity of the classical approximation in general.

This criterion may just as well be expressed by the following inequality which involves the potential $V(x)$ and its first derivative

$$\frac{|\,m\hbar V'\,|}{|\,2m\,(E-V)\,|^{3/2}} \ll 1. \tag{VI.47}$$

9. Turning Points and Connection Formulae

In most applications of the WKB approximation condition (VI.47) is fulfilled everywhere except in the vicinity of the points for which

$$E = V(x).$$

These are the *turning points* of the classical motion, points where the velocity of the particle vanishes and changes sign.

From the mathematical point of view, the WKB approximation consists in replacing the Schrödinger equation

$$y'' + \frac{y}{\lambda\!\!\!^{-}{}^2} = 0$$

by the equation

$$y'' + \left(\frac{1}{\lambda\!\!\!^{-}{}^2} - \frac{(\sqrt{\lambda\!\!\!^{-}})''}{\sqrt{\lambda\!\!\!^{-}}}\right) y = 0 \tag{VI.48}$$

[both in the region $E > V$ as well as in the region $E < V$, where $\lambda\!\!\!^{-} = il$]. In fact one can easily verify that expressions (VI.42) and (VI.44) are actually the general solutions of equation (VI.48). This equation has a singular point – a singularity in $(x-a)^{-2}$ – at any point a where the wavelength becomes infinite, i.e. at each of the turning points. In the vicinity of these points, the replacement of the Schrödinger equation by eq. (VI.48) is certainly not justified. To obtain the complete solution, one must therefore solve the Schrödinger equation in a region of suitable extension surrounding the turning point, and then smoothly join this solution with the solutions (VI.42) or (VI.44) which are the wave functions in neighboring regions where the WKB approximation is valid.

In practice, it is of little importance to know the particular form of the solution in the region of the turning point as long as one knows how to join the WKB solutions together on either side of that point. This joining problem is a difficult mathematical problem, a thorough and lucid account of which may be found in Langer's article (see footnote p. 231). The solution proposed by Langer consists in replacing the Schrödinger equation, not by equation (VI.48), but by another equation which is regular at the turning point and which asymptotically approaches (VI.48) on either side. The general solution of this equation asymptotically approaches one or the other of solutions (VI.42) and (VI.44) on either side of this point. We shall merely give the *connection formulae* between the exponential WKB solution and the oscillatory WKB solution on either side of a turning point.

Suppose, to be definite, that $E \gtrless V$ according to whether $x \gtrless a$ (barrier to the left). The general solution is a linear combination of two solutions y_1 and y_2 whose asymptotic forms are

$$x \ll a \qquad\qquad\qquad x \gg a$$

$$y_1 \sim \sqrt{l}\exp\left(+\int_x^a \frac{\mathrm{d}x}{l}\right) \qquad y_1 \sim -\lambda\!\!\!^{-\frac{1}{2}} \sin\left(\int_a^x \frac{\mathrm{d}x}{\lambda\!\!\!^{-}} - \frac{\pi}{4}\right) \tag{VI.49}$$

$$y_2 \sim \frac{\sqrt{l}}{2}\exp\left(-\int_x^a \frac{\mathrm{d}x}{l}\right) \qquad y_2 \sim \lambda\!\!\!^{-\frac{1}{2}} \cos\left(\int_a^x \frac{\mathrm{d}x}{\lambda\!\!\!^{-}} - \frac{\pi}{4}\right). \tag{VI.50}$$

Let us define the "number of wavelengths" contained in a given interval (x_1, x_2) by the integral $(1/2\pi)\int_{x_1}^{x_2}(\mathrm{d}x/\lambda\!\!\!^{-})$ or $(1/2\pi)\int_{x_1}^{x_2}(\mathrm{d}x/l)$ according to whether we are to the right or to the left of the turning point. The conditions for the validity of these connection formulae are as follows:

(*i*) At the turning point the kinetic energy $E-V$ tends to zero as $(x-a)$ and remains to a good approximation proportional to $(x-a)$ in a region extending over at least one, but preferably several "wavelengths" on either side.

(*ii*) Each of these turning regions joins on, on either side of its turning point, to an asymptotic region extending over several "wavelengths", in which the WKB approximation proper is justified.

Care must be taken in using formulae (VI.49) and (VI.50). The difficulty arises from the fact that the solution Ay_1+By_2 has, except in the very special case where $A=0$, the same asymptotic form as Ay_1 in the region $x \ll a$; the exponentially increasing term Ay_1 always predominates over the exponentially decreasing term By_2, no matter how small A compared to B, provided A does not vanish. Consequently, knowledge of the asymptotic form is not sufficient to specify the solution unless this form is of the exponentially decreasing type (type y_2); conversely, if the coefficients A and B are only known approximately and if $|A| \ll |B|$, any determination, even though approximate, of the asymptotic form is impossible.

Now suppose that we know the WKB solution in this asymptotic region $(x \ll a)$ and that we ask on to what particular oscillatory WKB solution it joins. The question cannot be answered unless this solution is of the exponentially decreasing type $\frac{1}{2}Bl^{\frac{1}{2}} \exp\left[-\int_x^a (\mathrm{d}x/l)\right]$; in that case, the solution is certainly of the form By_2 in the region of the turning point and its behavior in the region $x \gg a$ is given by formula (VI.50). The result may be written

$$\tfrac{1}{2}\sqrt{l}\, \exp\left(-\int_x^a \frac{\mathrm{d}x}{l}\right) \to \sqrt{\lambda}\, \cos\left(\int_a^x \frac{\mathrm{d}x}{\lambda} - \frac{\pi}{4}\right), \qquad \text{(VI.51)}$$

the arrow indicating the direction in which the connection is made.

Suppose, on the other hand, that we know the WKB solution in the oscillatory region $(x \gg a)$. This solution is of the form (VI.42), namely

$$C\sqrt{\lambda}\, \cos\left(\int_a^x \frac{\mathrm{d}x}{\lambda} - \frac{\pi}{4} + \varphi\right)$$

(C and φ are complex constants). According to formulae (VI.49) and (VI.50), this is the asymptotic form of the solution defined by

$$A \approx C \sin \varphi \qquad B \approx C \cos \varphi.$$

We must insist here that the constants A and B can be calculated only approximately from the asymptotic form. Because of this fact, if $|\tan \varphi| \ll 1$, any determination of the asymptotic form of this solution in the region $x \ll a$ is impossible. If $|\tan \varphi|$ is not small, it is given by formula (VI.49). The result may be written

$$\sqrt{\lambda} \cos \left(\int_a^x \frac{\mathrm{d}x}{\lambda} - \frac{\pi}{4} + \varphi \right) \rightarrow \sin \varphi \sqrt{l} \exp \left(\int_x^a \frac{\mathrm{d}x}{l} \right) \qquad \text{(VI.52)}$$

the arrow indicating the direction in which the connection is made.

In the case where the barrier is to the right, i.e. when $E \gtrless V$ according to whether $a \gtrless x$, the connection formulae (VI.51) and (VI.52) remain valid provided that one interchanges x and a in the inequalities and in the limits of integration; in particular, the direction of the arrows must be preserved.

10. Penetration of a Potential Barrier

By way of an illustration we shall apply the WKB method to the calculation of the transmission coefficient through the potential barrier shown in Fig. VI.3. In the region $x < a$ (region I), $V(x) = V_0 = \text{const.}$; when $x > a$, $V(x)$ is a positive function decreasing monotonically from the positive value $V_a = V(a)$ to $V(\infty) = 0$.

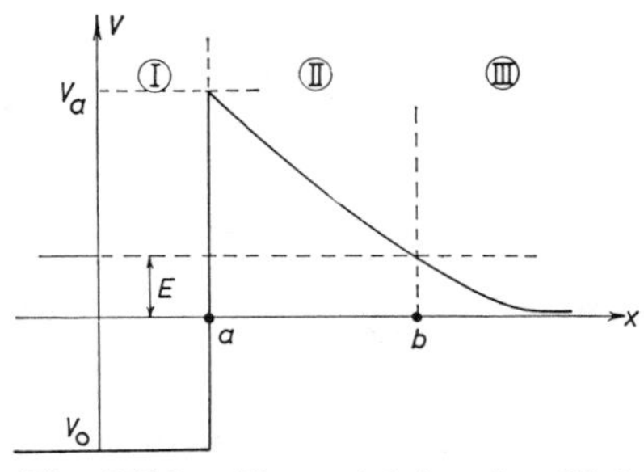

Fig. VI.3. Potential barrier $V(x)$.

Let E be the energy of the particle, b ($>a$) the point of the x axis such that $E = V(b)$. The point of discontinuity a and the turning point

b divide the x axis into three regions, I, II, and III. We assume the WKB method to be applicable in regions II and III.

To find the transmission coefficient, one must construct the solution of the Schrödinger equation whose asymptotic form in region III represents a purely transmitted wave (propagating in the direction of increasing x). In that region, the WKB solution is of the form (VI.42). The condition we impose upon its asymptotic form determines that solution (to within a constant), namely

$$y_{\mathrm{III}} = \sqrt{\lambda}\,\exp\left(\mathrm{i}\int_b^x \frac{\mathrm{d}x}{\lambda} - \mathrm{i}\,\frac{\pi}{4}\right) \qquad (x \gg b)$$

(the phase $\pi/4$ has been added to simplify subsequent calculations), or else

$$y_{\mathrm{III}} = \sqrt{\lambda}\left[\cos\left(\int_b^x \frac{\mathrm{d}x}{\lambda} - \frac{\pi}{4}\right) + \mathrm{i}\sin\left(\int_b^x \frac{\mathrm{d}x}{\lambda} - \frac{\pi}{4}\right)\right].$$

According to connection formulae (VI.49) and (VI.50), this particular solution extends into region II in the form

$$y_{\mathrm{II}} = -\,\mathrm{i}\sqrt{l}\,\exp\left(\int_x^b \frac{\mathrm{d}x}{l}\right) \qquad (a < x \ll b)$$

$$= -\,\mathrm{i}\sqrt{l}\,\mathrm{e}^{\tau}\exp\left(-\int_a^x \frac{\mathrm{d}x}{l}\right),$$

an expression in which

$$\tau = \int_a^b \frac{\mathrm{d}x}{l} = \int_a^b \frac{\sqrt{2m[V(x)-E]}}{\hbar}\,\mathrm{d}x.$$

Let us put

$$l_a = l(a) = \frac{\hbar}{\sqrt{2m(V_a - E)}}, \qquad k = \frac{\sqrt{2m(E - V_0)}}{\hbar}.$$

In region I, the solution of the Schrödinger equation is (rigorously) of the form

$$y_{\mathrm{I}} = A\sin[k(x-a)+\delta].$$

The constants A and δ are obtained by applying the continuity conditions to the wave function and its logarithmic derivative at the point a, namely

$$k \cot \delta = -\frac{1}{l_a}, \qquad A \sin \delta = -\mathrm{i}\sqrt{l_a}\, \mathrm{e}^{\tau}. \tag{VI.53}$$

y_{I} is the sum of an incident and a reflected wave. Since δ is real, the incident wave $-\frac{1}{2}\mathrm{i}A \exp[\mathrm{i}(k(x-a)+\delta)]$ has a flux equal to $\frac{1}{4}|A|^2\hbar k/m$. According to equations (VI.53)

$$\frac{k|A|^2}{4} = \frac{kl_a}{4}\frac{\mathrm{e}^{2\tau}}{\sin^2\delta} = \frac{kl_a}{4}\mathrm{e}^{2\tau}(1+\cot^2\delta) = \mathrm{e}^{2\tau}\frac{1+k^2l_a^{\,2}}{4kl_a}.$$

Since the flux of the transmitted wave y_{III} is equal to $\hbar/m$, the transmission coefficient is

$$\begin{aligned} T &= \frac{4kl_a}{1+k^2l_a^{\,2}}\mathrm{e}^{-2\tau} \\ &= 4\frac{\sqrt{(V_a-E)(E-V_0)}}{V_a-V_0}\mathrm{e}^{-2\tau}. \end{aligned}$$

This method of calculation is justified if $V(x)$ varies sufficiently slowly in regions II and III where the WKB approximation was made [condition (VI.47)]. Furthermore, in a region of several "wave-lengths" about the turning point, $V(x)$ must be correctly approximated by a linear function of x. This requires in particular that the barrier have a "thickness" of at least several "wavelengths", hence that $\tau \gg 2\pi$ and consequently that T be extremely small ($\lesssim 10^{-5}$).

11. Energy Levels of a Potential Well

As a second application, we consider the potential well of Fig. VI.4 and we calculate the energy levels of the discrete spectrum.

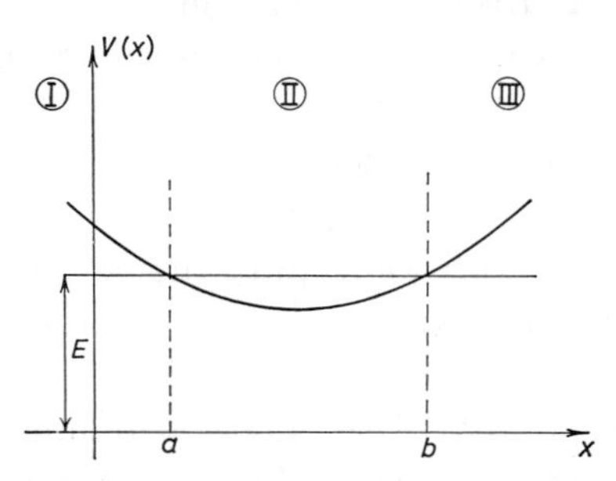

Fig. VI.4. Potential well $V(x)$.

For a given energy E there are two turning points a and b of the classical motion. They divide the x axis into three regions, I, II, and III. We look for the WKB solution which decreases exponentially in regions I and III, namely

$$y_{\mathrm{I}} = \tfrac{1}{2}c\sqrt{l}\,\exp\left(-\int_x^a \frac{\mathrm{d}x}{l}\right) \qquad (x \ll a)$$

$$y_{\mathrm{III}} = \tfrac{1}{2}c'\sqrt{l}\,\exp\left(-\int_b^x \frac{\mathrm{d}x}{l}\right) \qquad (x \gg b)$$

(c and c' are adjustable constants). In accordance with connection formula (VI.50) and the analogous formula corresponding to the barrier to the right, these functions are respectively continued into region II by the functions

$$\left.\begin{aligned} y_a(x) &= c\sqrt{\lambda}\cos\left(\int_a^x \frac{\mathrm{d}x}{\lambda} - \frac{\pi}{4}\right) \\ y_b(x) &= c'\sqrt{\lambda}\cos\left(\int_x^b \frac{\mathrm{d}x}{\lambda} - \frac{\pi}{4}\right). \end{aligned}\right\} \qquad (a \ll x \ll b).$$

These functions are equal, $y_a(x) = y_b(x) = y_{\mathrm{II}}$, if

$$I \equiv \int_a^b \frac{\mathrm{d}x}{\lambda} \equiv \int_a^b \frac{\sqrt{2m[E - V(x)]}}{\hbar}\,\mathrm{d}x = (N + \tfrac{1}{2})\pi, \tag{VI.54}$$

where N is an integer ($\geqslant 0$). Rule (VI.54) fixes the energy levels of the discrete spectrum. In that case, one has $c' = (-1)^N c$. Since y_{I}, y_{II} and y_{III} are approximate solutions of the Schrödinger equation, they are, apart from a constant which can be easily calculated, good approximations of the eigensolutions in regions I, II and III, respectively. Thus, the method gives at the same time the energy levels and approximate expressions of the corresponding eigensolutions valid everywhere except in small regions surrounding the two turning points a and b.

The conditions of validity of the method require that both turning regions in which $V(x)$ is a reasonably linear function of x, extend over several wavelengths; hence the method is applicable only for very large quantum numbers

$$N \gg 1.$$

In certain privileged cases, such as that of the harmonic oscillator, the method yields exact values of all the energy levels up to and including the ground state (cf. Problem VI.6). This must be regarded as an accident.

Quantization rule (VI.54) may be viewed as a condition for a standing wave: the interval (a, b) must contain a "half-integral" (i.e. $\frac{1}{2}$+integer) number of half-wavelengths. In fact, it differs from the Bohr-Sommerfeld quantization rule only by the presence of "half-integral" quantum numbers instead of integral quantum numbers. The integral I is just equal to $2\hbar$ times the action integral $\oint p\, \mathrm{d}q$ of the corresponding classical phase space. On the other hand, the action integral is equal to the extension in phase $\omega(E) \equiv \iint_{H \leqslant E} \mathrm{d}p\, \mathrm{d}q$ of the points whose energy is less than E. Rule (VI.54) can therefore be stated in the form

$$\omega(E) \equiv \oint_{H=E} p\, \mathrm{d}q = (N + \tfrac{1}{2})\, h. \tag{VI.55}$$

Following formula (VI.55), the extension in phase $\omega(E)$ increases by h when one goes from one energy level to the one immediately above. We deduce therefrom the important result concerning the distribution of energy levels:

The extension in phase of the energy band $(E, E+\delta E)$ measured in units of h *is equal to the number of bound states of the corresponding quantum system, located within that energy band.*

The conditions of validity of this result are actually less restrictive than those of the WKB method; in practice, it is always valid "in the limit of large quantum numbers". One commonly assumes – and it has actually been shown in the simplest cases – that this result also applies to systems whose number of dimensions R is greater than unity provided that one takes h^r as unit of extension in phase.

EXERCISES AND PROBLEMS

1. The Hamiltonian of the harmonic oscillator is $H = (p^2 + m^2\omega^2 q^2)/2m$. Show that the mean values $\langle q \rangle$, $\langle p \rangle$ carry out sinusoidal oscillations of frequency

$\omega/2\pi$ about the origin; show that the squares of the deviations ϖ, χ (notation of § 3) oscillate sinusoidally with half the period about some (positive) average values and calculate these values. Under what conditions do ϖ and χ remain constant?

2. Show that the motion of the center of a wave packet representing a charged particle in an electromagnetic field is rigorously the same as that of a classical particle in the two following cases: (*a*) constant electric field; (*b*) constant magnetic field.

3. Show that the "spreading" of the wave packet representing a charged particle in a constant electric field obeys the same laws as that of a free wave packet (ϖ = const., χ = quadratic function of time).

4. With the wave function $\psi(\boldsymbol{r})$ of a particle, one forms the function

$$D(\boldsymbol{R}, \boldsymbol{P}) = \frac{1}{(2\pi\hbar)^3} \int \exp\left(-\frac{\mathrm{i}}{\hbar}\boldsymbol{P}\cdot\boldsymbol{r}\right) \psi^*\left(\boldsymbol{R}-\frac{\boldsymbol{r}}{2}\right) \psi\left(\boldsymbol{R}+\frac{\boldsymbol{r}}{2}\right) \mathrm{d}\boldsymbol{r}$$

which one interprets (after Wigner) as the density in phase of a classical statistical mixture associated with this wave function [1]). Show that:

(*i*) the distributions of position and momentum of this mixture are the same as those of the quantum particle in the state $\psi(\boldsymbol{r})$:

$$\int D(\boldsymbol{R}, \boldsymbol{P})\, \mathrm{d}\boldsymbol{P} = |\psi(\boldsymbol{R})|^2, \qquad \int D(\boldsymbol{R}, \boldsymbol{P})\, \mathrm{d}\boldsymbol{R} = |\varphi(\boldsymbol{P})|^2$$

[$\varphi(\boldsymbol{P})$ is wave function in momentum space];

(*ii*) if the particle is free, the evolution in time of the mixture is strictly that of a statistical mixture of free classical particles of the same mass;

(*iii*) find the "spreading" law of a free wave packet.

5. Calculate by the WKB method the transmission coefficient T of a particle of mass m and energy E through a slowly varying potential barrier $V(x)$ approaching limits less than E at the two boundaries of the interval $(-\infty, +\infty)$; assume that there exist only two turning points a and b. Show that

$$T = \exp\left(-2\int_a^b \frac{\sqrt{2m\,[V(x)-E]}}{\hbar}\,\mathrm{d}x\right).$$

6. Calculate the energy levels of the harmonic oscillator by the WKB method (Hamiltonian of Problem VI.1). Discuss the conditions of validity of the method.

---

[1]) In contrast to the density in phase of a true classical statistical mixture, $D(\boldsymbol{R}, \boldsymbol{P})$ may take on negative values.

CHAPTER VII

GENERAL FORMALISM OF THE QUANTUM THEORY (A) MATHEMATICAL FRAMEWORK

1. Superposition Principle and Representation of Dynamical States by Vectors

The entire interpretation of Wave Mechanics developed in Chapters IV and V had as its starting point the definition of the probability densities of position and momentum by means of the wave functions Ψ and Φ referring to configuration space and momentum space, respectively. We have already indicated the symmetrical role played in the theory by these two functions. The parallelism can be pushed all the way. Indeed, the fundamental postulates concerning the mean values (Ch. V, § 3) are just as well expressed by means of operations in momentum space. Instead of generalizing, as we have done, expressions (IV.13) and (IV.20) of the mean values of functions of the form $F(\mathbf{r})$ and $G(\mathbf{p})$, respectively, one can equally well carry out the same generalization in the corresponding expressions (IV.21) and (IV.14) constructed with the functions Φ. For the postulates (*a*) and (*b*) of Ch. V, § 3, one substitutes the equivalent postulates:

(*a*′) With any dynamical variable $\mathscr{A} = A(q_1, \ldots, q_R;\, p_1, \ldots, p_R)$ there is associated the linear operator

$$A\left(\mathrm{i}\hbar\frac{\partial}{\partial p_1}, \ldots, \mathrm{i}\hbar\frac{\partial}{\partial p_R};\, p_1, \ldots, p_R\right).$$

(*b*′) The average value taken by this dynamical variable when the system is in the dynamical state defined by the function $\Phi(p_1, \ldots, p_R)$ is

$$\langle A\rangle = \frac{\langle\Phi, A\Phi\rangle}{\langle\Phi, \Phi\rangle},$$

an expression in which the brackets on the right-hand side are scalar products of momentum space

$$\langle\Phi, A\Phi\rangle = \int\ldots\int\Phi^*(A\Phi)\,\mathrm{d}p_1\ldots\mathrm{d}p_R, \qquad \langle\Phi, \Phi\rangle = \int\ldots\int\Phi^*\Phi\,\mathrm{d}p_1\ldots\mathrm{d}p_R.$$

The equivalence of postulates (*a*), (*b*), and postulates (*a*′), (*b*′) is based upon the properties of the Fourier transformation (cf. Appendix A). The proof of this equivalence is not difficult; we shall not undertake it here.

Just as the functions Φ and Ψ are equivalent *representations* of one and the same dynamical state, the observables

$$A\left(\mathrm{i}\hbar\frac{\partial}{\partial p_1}, \ldots, \mathrm{i}\hbar\frac{\partial}{\partial p_R}; p_1, \ldots, p_R\right) \text{ and } A\left(q_1, \ldots, q_R; \frac{\hbar}{\mathrm{i}}\frac{\partial}{\partial q_1}, \ldots, \frac{\hbar}{\mathrm{i}}\frac{\partial}{\partial q_R}\right)$$

are equivalent *representations* of the same dynamical variable (cf. Ch. IV, remark at the end of § 5). The entire theory of observables can be developed indifferently in one or the other of these representations. We thus obtain two equivalent formulations of the Quantum Theory.

These considerations become clearer if one imagines the wave functions Φ and Ψ to represent the same vector of a space with infinitely many dimensions. All the concepts introduced in Chapter V, such as function space, scalar product, norm, orthogonality, etc., then have a very simple geometrical interpretation. According to this picture, the values taken by the function $\Psi(q_1, \ldots, q_R)$ at each point of configuration space are the components of this vector along a certain system of orthogonal axes. Similarly, the values taken by the function $\Phi(p_1, \ldots, p_R)$ at each point of momentum space are the components of this same vector along another system of axes. Moreover, the coefficients $c_p^{(r)}$ of the expansion (V.14) of Ψ in a series of eigenfunctions of a given complete orthonormal system are the components of this vector along a third system of axes.

One is thus led to build up the entire Quantum Theory by starting directly from the vector concept, without reference to the particular representation which can be made thereof. In the spirit of such a formulation, one states as a fundamental principle of the theory the *superposition principle of dynamical states*, according to which the possible dynamical states of a given quantized system possess the following property characteristic of any wave in general: they may be superposed linearly and may consequently be represented by vectors of a certain linear space yet to be defined. Therefore, with each dynamical state is associated a vector of this abstract space. Likewise, with each dynamical variable is associated a linear operator of this space. In this presentation, the theory is formally simpler and more elegant than Wave Mechanics; it is also more general, because it applies equally well to quantum systems having no classical analogues.

We shall now present this general formulation of the Quantum Theory. We essentially follow the presentation of Dirac, making use of the particularly convenient notation he has introduced for this purpose [1]). The present chapter is devoted to the concepts of linear algebra [2]) which constitute the mathematical framework of the theory. The description of physical phenomena by means of this formalism constitutes the subject of Chapter VIII.

I. VECTORS AND OPERATORS

2. Vector Space. "Ket" Vectors

Following the hypotheses of the foregoing paragraph, we associate with every dynamical state a certain type of vector which we call, following Dirac, *ket vector* or *ket* and which we represent by the symbol $|\rangle$. In order to distinguish the kets from each other, we complete each symbol by inserting either a particular letter, or one or even several indices capable of taking, depending upon the case, discrete or continuous values. Thus the ket u is represented by the symbol $|u\rangle$.

The kets form a linear vector space: any linear combination of several ket vectors is also a ket vector. For instance, let us take two kets $|1\rangle$ and $|2\rangle$ and two arbitrary complex numbers λ_1 and λ_2. The linear combination

$$|v\rangle = \lambda_1|1\rangle + \lambda_2|2\rangle \tag{VII.1}$$

is a vector of ket space.

In analogous fashion, if $|\xi\rangle$ depends on a continuous index, ξ, and if $\lambda(\xi)$ is an arbitrary complex function of ξ, the integral

$$|w\rangle = \int_{\xi_1}^{\xi_2} \lambda(\xi)\,|\xi\rangle\,\mathrm{d}\xi \tag{VII.2}$$

is a vector of ket space. We shall say, by an obvious extension of the terminology, that $|w\rangle$ is a linear combination (or linear superposition) of the kets $|\xi\rangle$.

---

[1]) P. A. M. Dirac, *loc. cit.*, footnote 3, p. 47, Ch. II.

[2]) The reader will find a rigorous and complete treatment of these questions in M. H. Stone, *loc. cit.*, footnote 1, p. 166, Ch. V; also, Paul R. Halmos, *Finite Dimensional Vector Spaces* (Princeton, Princeton University Press, 1948).

By definition, the kets of a given ensemble are *linearly independent* if none of them can be expressed in the form of a linear combination of the others [this linear combination can be of type (VII.1), of type (VII.2), or a mixture].

If the vector space contains at most n linearly independent vectors, it is a space with a finite number of dimensions, and the number of dimensions is by definition equal to n. Take n particular, linearly independent vectors in that space; all other vectors are equal to linear combinations of these n particular vectors.

If there is no limit to the number of linearly independent vectors of the vector space under consideration, the latter has an infinite number of dimensions. This is the case of Hilbert space, and as we have already seen, it is also the case of the wave functions of Wave Mechanics. One can nevertheless always choose a (denumerably or non-denumerably infinite) set of linearly independent vectors such that any vector of that space is equal to a linear combination (possibly an infinite series or an integral) of these "basis vectors".

Consider a set of ket vectors belonging to a given space $\mathscr{E}$. The ensemble of the kets of this set and of all linear combinations of these kets form a vector space $\mathscr{E}'$. $\mathscr{E}'$ is by definition the space subtended by the kets of this set. Any ket of $\mathscr{E}'$ belongs to $\mathscr{E}$. One says that $\mathscr{E}'$ is a *subspace* of $\mathscr{E}$. If the space has a finite number of dimensions n, the number of dimensions of $\mathscr{E}'$ is certainly finite and smaller than n. If $\mathscr{E}$ has an infinite number of dimensions, there exists no restriction whatsoever on the number of dimensions of $\mathscr{E}'$.

3. Dual Space. "Bra" Vectors

It is well known in linear algebra that with every vector space can be associated a dual vector space. Indeed, any linear function $\chi(|u\rangle)$ of the kets $|u\rangle$ possesses the superposition property characteristic of vectors [1]) and therefore defines a new type of vector which, following Dirac, we shall call *bra vector* or *bra* and which we shall represent by

---

[1]) The property of χ of being a linear function of the kets $|u\rangle$ implies that

$$\chi(\lambda_1|1\rangle + \lambda_2|2\rangle) = \lambda_1\chi(|1\rangle) + \lambda_2\chi(|2\rangle).$$

It is evident that if two functions χ_1, χ_2 have this property, any linear combination $\mu_1\chi_1 + \mu_2\chi_2$ of these two functions also has this property.

the symbol $\langle|$. Thus the function $\chi(|u\rangle)$ defines the bra $\langle\chi|$; the value taken by this function for a particular ket $|u\rangle$ is a number (generally complex) which we denote by the symbol $\langle\chi|u\rangle$.

By definition, the bra $\langle\Phi|$ vanishes if the function $\Phi(|u\rangle)$ vanishes for any $|u\rangle$:

$$\langle\Phi| = 0, \quad \text{if} \quad \langle\Phi|u\rangle = 0 \quad \text{for any} \quad |u\rangle. \tag{VII.3}$$

Similarly, two bras are equal,

$$\langle\Phi_1| = \langle\Phi_2|, \quad \text{if} \quad \langle\Phi_1|u\rangle = \langle\Phi_2|u\rangle \quad \text{for any} \quad |u\rangle.$$

If the ket space has a finite number of dimensions, the dual space has the same number of dimensions. If the number of dimensions of ket space is infinite, the dual space has the same property.

In order to introduce a metric in the vector space we have just defined, we make the hypothesis that there exists a one-to-one correspondence between the vectors of this space and those of the dual space. Bra and ket thus associated by this one-to-one correspondence are said to be conjugates of each other and are labelled by the same letter (or the same indices). Thus the bra conjugate to the ket $|u\rangle$ is represented by the symbol $\langle u|$.

We assume also that this correspondence is *antilinear*. In other words, the bra conjugate to the ket

$$|v\rangle = \lambda_1|1\rangle + \lambda_2|2\rangle \tag{VII.4}$$

is

$$\langle v| = \lambda_1{}^*\langle 1| + \lambda_2{}^*\langle 2|. \tag{VII.5}$$

Likewise, the bra conjugate to the ket

$$|w\rangle = \int_{\xi_1}^{\xi_2} \lambda(\xi)\, |\xi\rangle\, \mathrm{d}\xi \tag{VII.6}$$

is

$$\langle w| = \int_{\xi_1}^{\xi_2} \lambda^*(\xi)\, \langle\xi|\, \mathrm{d}\xi. \tag{VII.7}$$

Thus the correspondence between kets and bras is analogous to the correspondence between the wave functions of wave mechanics and their complex conjugates. Let us note in passing that if a ket vanishes, its conjugate bra is also zero, and conversely.

The ensemble of the bras conjugate to the kets of a subspace $\mathscr{E}'$ of the space $\mathscr{E}$ forms the dual subspace of $\mathscr{E}'$.

4. Scalar Product

By definition, the *scalar product* of the ket $|u\rangle$ by the ket $|v\rangle$ is the number (generally complex) $\langle v|u\rangle$, that is to say the value $v(|u\rangle)$ taken by the linear function associated with the bra conjugate to $|v\rangle$.

As an immediate consequence of the definition, the scalar product $\langle v|u\rangle$ is linear with respect to $|u\rangle$ and antilinear with respect to $|v\rangle$.

We adopt the hypothesis that the scalar product possesses all the other characteristic properties of the scalar product of the wave functions of Wave Mechanics (Ch. V, § 2), namely:

1) The scalar product of $|v\rangle$ by $|u\rangle$ is the complex conjugate of the scalar product of $|u\rangle$ by $|v\rangle$

$$\langle u|v\rangle = \langle v|u\rangle^*. \tag{VII.8}$$

2) Any vector u has a real non-negative norm $N_u \equiv \langle u|u\rangle$

$$\langle u|u\rangle \geqslant 0. \tag{VII.9}$$

It vanishes if and only if the vector u vanishes.

From these properties comes the *Schwarz inequality*: for any $|u\rangle$ and $|v\rangle$

$$|\langle u|v\rangle|^2 \leqslant \langle u|u\rangle\, \langle v|v\rangle. \tag{VII.10}$$

The equality holds if and only if the vectors $|u\rangle$ and $|v\rangle$ are collinear (i.e. proportional).

The preceding axioms must be complemented by the postulate that the space of the ket vectors $\mathscr{E}$ (as well as its dual space, bra space) is complete and separable (cf. Ch. V, § 2): *it is a Hilbert space.*

By definition two vectors are *orthogonal* if their scalar product vanishes. Two subspaces $\mathscr{E}_1$, $\mathscr{E}_2$ of $\mathscr{E}$ are orthogonal if each of the vectors in one is orthogonal to each of the vectors in the other. It is evident that $\mathscr{E}_1$ and $\mathscr{E}_2$ have no vector in common; indeed, any vector common to $\mathscr{E}_1$ and $\mathscr{E}_2$ can only be zero since, being orthogonal to itself, it has a vanishing norm.

The ensemble of vectors orthogonal to $\mathscr{E}_1$ form a subspace $\mathscr{E}_1^\times$ orthogonal to $\mathscr{E}_1$, the *complementary subspace* to $\mathscr{E}_1$. $\mathscr{E}_1^\times$ reduces to zero if the subspace $\mathscr{E}_1$ coincides with the space $\mathscr{E}$ itself. One can show (cf. footnote 2, p. 245) that *any vector of* $\mathscr{E}$ *may be written*, and *in a unique manner, as the sum of a vector located in* $\mathscr{E}_1$ *and a vector located in its complementary space*:

$$|u\rangle = |u_1\rangle + |\,u_1^{\times}\rangle$$

$|u_1\rangle$ is by definition the *projection* of $|u\rangle$ in the subspace $\mathscr{E}_1$. We shall have occasion to return at length to this concept of projection in Section II.

In all our considerations on the scalar product, it is implicitly assumed that the vectors (ket or bra) have finite norm, lest the axiom concerning the norm of vectors lose all meaning. As long as this is understood, the space $\mathscr{E}$ of the ket vectors under consideration is a Hilbert space. We have seen in Chapter V that the vectors capable of representing dynamical states must actually have finite norm, but that the treatment of the continuous spectra in the eigenvalue problems requires the introduction of eigenvectors of infinite length. We must therefore also include in our vector space vectors $|\xi\rangle$ of infinite norm, depending on (at least) one continuous index, and extend the concept of scalar product to this category of vectors.

We assume that $|\xi\rangle$ has a finite scalar product:

$$\langle u|\xi\rangle$$

with any vector $|u\rangle$ of finite norm; moreover, this scalar product is linear in $|\xi\rangle$ and antilinear in $|u\rangle$. One defines the scalar product $\langle\xi|u\rangle$ in analogous fashion and we state as an axiom that

$$\langle\xi|u\rangle = \langle u|\xi\rangle^*.$$

On the other hand, the scalar product of two vectors of the type $|\xi\rangle$ may not converge. In particular, the norm of $|\xi\rangle$ is infinite. But we assume that the eigendifferential

$$|\xi, \delta\xi\rangle \equiv (\delta\xi)^{-\frac{1}{2}} \int_{\xi}^{\xi+\delta\xi} |\xi'\rangle\, \mathrm{d}\xi' \qquad \text{(VII.11)}$$

has a positive definite norm which tends toward a finite limit as $\delta\xi \to 0$. *Rigorously speaking*, the vector $|\xi\rangle$ does not belong to space $\mathscr{E}$; but its eigendifferentials, and more generally, the linear combinations of the type (VII.2) do belong to this space and satisfy all the axioms relating to the vectors of a Hilbert space.

5. Linear Operators

Once we have defined the space of the ket vectors, we can define the linear operators of that space (cf. Ch. II, § 11).

Let us suppose that to each ket $|u\rangle$ of vector space there corresponds a certain ket $|v\rangle$: one says that $|v\rangle$ results from the action of a certain operator on the ket $|u\rangle$. If in addition this correspondence is linear, the operator thus defined is a certain *linear* operator A. One writes

$$|v\rangle = A|u\rangle.$$

Such an operator vanishes if the ket $|v\rangle$ is zero for any $|u\rangle$.

For an operator A to be zero, it is necessary and sufficient that for any $|u\rangle$,

$$\langle u|A|u\rangle = 0.$$

The proof of this property presents no serious difficulty and wil not be given here. One deduces immediately that:

For two operators A and B to be equal, it is necessary and sufficient that for any $|u\rangle$

$$\langle u|A|u\rangle = \langle u|B|u\rangle. \tag{VII.12}$$

The main algebraic operations on the operators have already been defined (Ch. II, § 11): multiplication by a constant, sum, and product. The *sum* of linear operators is associative and commutative. The *product* is associative, distributive with respect to the sum, *but it is not commutative* – therein lies the fundamental difference between the algebra of linear operators and ordinary algebra. Let us recall that the commutator of two linear operators A and B is represented by the symbol

$$[A, B] \equiv AB - BA.$$

The main properties of commutator algebra were studied in Ch. V, § 17 [eqs. (V.63) to (V.66)]; they all remain valid and will not be re-examined here.

Note that the multiplication of a ket by a given constant c also defines a linear operator. This operator c commutes with all linear operators: $[A, c] = 0$.

In particular, the multiplication by unity is the "identity" operator.

If the correspondence defined above between $|u\rangle$ and $|v\rangle$ is reciprocal, it defines two linear operators A and B:

$$|v\rangle = A|u\rangle, \qquad |u\rangle = B|v\rangle. \tag{VII.13}$$

These operators are by definition the *inverse* of each other. One may also say that the operators A and B are each other's inverse if they simultaneously satisfy the equations

$$AB = 1, \qquad BA = 1. \tag{VII.14}$$

These two definitions are equivalent.

The inverse of a given operator A does not always exist. When it exists, one commonly denotes it by the symbol A^{-1}. Making use of the defining equations (VIII.14), one can easily prove the following property:

If two operators P and Q each possess an inverse, the product PQ has an inverse and one has

$$(PQ)^{-1} = Q^{-1}\, P^{-1} \tag{VII.15}$$

[note the reversal of the order of the factors on the right-hand side of (VII.15)].

Once we know the action of the linear operator A in the space of the ket vectors, its action in the dual vector space is defined unambiguously as follows. A bra $\langle\chi|$ having been given, the scalar product $\langle\chi|(A|u\rangle)$ is obviously a linear function of $|u\rangle$, since the operator A is linear. Let $\langle\eta|$ be the bra defined by this function; to each bra $\langle\chi|$ corresponds a bra $\langle\eta|$. It is clear that this correspondence is linear (property of the scalar product). One says that $\langle\eta|$ results from the operation of A on $\langle\chi|$ and one writes

$$\langle\eta| = \langle\chi|A. \tag{VII.16}$$

Following this definition, one has the important identity between scalar products

$$(\langle\chi|A)|u\rangle \equiv \langle\chi|(A|u\rangle). \tag{VII.17}$$

The parentheses are therefore unnecessary in these two expressions and we shall henceforth write $\langle\chi|A|u\rangle$ to denote indifferently one or the other of these two equal scalar products.

By means of the identity (VII.17) one can define the various operations of the algebra of linear operators acting on the bras. In particular, one has for the three fundamental operations:

(*a*) *multiplication of A by a complex constant c*:

$$(cA)|u\rangle = c(A|u\rangle) \quad \text{hence} \quad \langle\chi|(cA) = c(\langle\chi|A);$$

(*b*) *sum* of operators $S = A + B$:

$$S|u\rangle = A|u\rangle + B|u\rangle \quad \text{hence} \quad \langle\chi|S = \langle\chi|A + \langle\chi|B;$$

(*c*) *product* of operators $P = AB$:

$$P|u\rangle = A(B|u\rangle), \quad \text{hence} \quad \langle\chi|P = (\langle\chi|A)B.$$

We shall always adopt the convention of placing the bras to the left and the kets to the right of the operators; the algebraic manipulations on the linear operators are the same in the two cases.

Certain operators are particularly easy to handle when using the foregoing notations: they are the operators of the type $|u\rangle\langle v|$, whose action on any ket $|w\rangle$ yields a ket proportional to $|u\rangle$, namely the ket $|u\rangle\langle v|w\rangle$ (constant of proportionality $\langle v|w\rangle$), and whose action on any bra $\langle w|$ yields a bra proportional to $\langle v|$, namely the bra $\langle w|u\rangle\langle v|$. The operator $|u\rangle\langle v|$ has no inverse.

6. Tensor Product [1] of Two Vector Spaces

To complete this introduction to vector algebra, we must still define a frequently used operation, namely the tensor product of two vector spaces.

The purpose and interest of this operation are illustrated by the following example.

Let us consider a quantum system of two particles. The product $\Psi_1(\boldsymbol{r}_1)\,\Psi_2(\boldsymbol{r}_2)$ of a wave function $\Psi_1(\boldsymbol{r}_1)$ relative to particle 1 by a wave function $\Psi_2(\boldsymbol{r}_2)$ relative to particle 2 represents a very special dynamical state of this system (Ch. IV, § 6). The most general wave function $\Psi(\boldsymbol{r}_1, \boldsymbol{r}_2)$ is not of this form, but can always be written as a linear combination of wave functions of this form. One of the many ways of doing this consists in expanding Ψ in a series of functions

[1] This type of product is often called Kronecker product.

of a complete orthonormal system of functions of $\mathbf{r}_1$; since the expansion coefficients are functions of $\mathbf{r}_2$, each term of the series actually has the required form. The space of the wave functions of the total system is therefore formed by linear combination of products of wave functions relating to each of the partial systems $\Psi_1(\mathbf{r}_1)$ and $\Psi_2(\mathbf{r}_2)$. One says that the space of functions $\Psi(\mathbf{r}_1, \mathbf{r}_2)$ is the tensor product of the space of functions $\Psi_1(\mathbf{r}_1)$ and the space of functions $\Psi_2(\mathbf{r}_2)$.

In fact, the products $\Psi_1(\mathbf{r}_1)\, \Psi_2(\mathbf{r}_2)$ play a privileged role in the study of the total system. Indeed, the dynamical variables of particle 1 are represented by observables A_1 acting on the wave function $\Psi(\mathbf{r}_1, \mathbf{r}_2)$ considered as a function or $\mathbf{r}_1$; those of particle 2, by observables A_2 acting on this same function considered as a function of $\mathbf{r}_2$. Clearly, each of the observables A_1 commutes with each of the observables A_2. When Ψ is of the form $\Psi_1(\mathbf{r}_1)\, \Psi_2(\mathbf{r}_2)$ the action of observables of this type is especially simple; thus $A_1(\Psi_1 \Psi_2)$ is equal to the product of $(A_1 \Psi_1)$ by Ψ_2.

The foregoing remarks apply most generally to any quantized system capable of being analyzed into two simpler systems.

In the abstract language which we are developing in this chapter, the tensor product is defined in the following manner. Let $\mathscr{E}_1$, $\mathscr{E}_2$ be two vector spaces. Taking a ket $|u\rangle^{(1)}$ of the one and a ket $|u\rangle^{(2)}$ of the other, one can form the product ket $|u\rangle^{(1)}\, |u\rangle^{(2)}$. This multiplication is commutative and we use the notation

$$|u^{(1)}\, u^{(2)}\rangle \equiv |u\rangle^{(1)}\, |u\rangle^{(2)}. \tag{VII.18}$$

Moreover, we assume it to be distributive with respect to the sum. If

$$|u\rangle^{(1)} = \lambda |v\rangle^{(1)} + \mu |w\rangle^{(1)},$$
$$|u^{(1)}\, u^{(2)}\rangle = \lambda |v^{(1)}\, u^{(2)}\rangle + \mu |w^{(1)}\, u^{(2)}\rangle.$$

Similarly, if

$$|u\rangle^{(2)} = \lambda |v\rangle^{(2)} + \mu |w\rangle^{(2)},$$
$$|u^{(1)}\, u^{(2)}\rangle = \lambda |u^{(1)}\, v^{(2)}\rangle + \mu |u^{(1)}\, w^{(2)}\rangle.$$

The kets $|u^{(1)}\, u^{(2)}\rangle$ span a new vector space, the space $\mathscr{E}^{(1)} \otimes \mathscr{E}^{(2)}$, which one calls the *tensor product of the vector spaces $\mathscr{E}^{(1)}$ and $\mathscr{E}^{(2)}$.* If the latter have N_1 and N_2 dimensions, respectively, the number of dimensions of the product space is $N_1 N_2$. However, the tensor product may also be carried out on spaces possessing an infinite number of dimensions, as shown by the example given at the beginning of this paragraph.

To each linear operator $A^{(1)}$ of the space $\mathscr{E}^{(1)}$ there corresponds a linear operator of the product space which we designate by the same symbol. The action of $A^{(1)}$ on any ket $|u\rangle^{(1)}$ being known, namely

$$A^{(1)}|u\rangle^{(1)} = |v\rangle^{(1)},$$

its action on the particular kets $|u^{(1)}\, u^{(2)}\rangle$ of the product space is defined by

$$A^{(1)}|u^{(1)}\, u^{(2)}\rangle = |v^{(1)}\, u^{(2)}\rangle \tag{VII.19}$$

and its action on the general ket of the product space is deduced from it by linear superposition. Likewise, every linear operator $A^{(2)}$ of the space $\mathscr{E}^{(2)}$ permits the definition of a linear operator in the product space.

Each operator $A^{(1)}$ commutes with each of the operators $A^{(2)}$:

$$[A^{(1)}, A^{(2)}] = 0.$$

Indeed one verifies, using the definition of the operators $A^{(1)}$ and $A^{(2)}$ itself, that the action of $[A^{(1)}, A^{(2)}]$ on each vector $|u^{(1)}\, u^{(2)}\rangle$ spanning the product space yields a vanishing result:

$$A^{(1)}\, A^{(2)}|u^{(1)}\, u^{(2)}\rangle = |v^{(1)}\, v^{(2)}\rangle = A^{(2)}\, A^{(1)}|u^{(1)}\, u^{(2)}\rangle.$$

In the product space one can define the correspondence between kets and bras, the action of linear operators on the bras, and so on. The algebraic rules stated above remain valid for all algebraic operations in the product space. The proof of these results is not difficult and will not be given here.

II. HERMITEAN OPERATORS, PROJECTORS AND OBSERVABLES

7. Adjoint Operators and Conjugation Relations

From the one-to-one correspondence between bras and conjugate kets one can derive an analogous conjugation relation between linear operators.

Let A be a linear operator. Let $|v\rangle$ be the ket conjugate to the bra

$\langle u|A.\ |v\rangle$ depends antilinearly upon the bra $\langle u|$; it is therefore a linear function of $|u\rangle$. This linear correspondence defines a linear operator which goes by the name of *Hermitean conjugate* operator of A or *adjoint* operator of A, and which one denotes by the symbol $A^\dagger$:

$$|v\rangle = A^\dagger|u\rangle.$$

It is clear that $A^\dagger = 0$ if $A = 0$, and conversely.

Since $A^\dagger|u\rangle$ is the ket conjugate to the bra $\langle u|A$, the scalar product of this ket by a given arbitrary bra $\langle t|$ is the complex conjugate of the scalar product of $|t\rangle$ by $\langle u|A$ (property (VII.8)). We therefore have the very important conjugation relation

$$\langle t|A^\dagger|u\rangle = \langle u|A|t\rangle^*. \qquad \text{(VII.20)}$$

Since this relation holds for any $|u\rangle$ and $|t\rangle$, the ket conjugate to $\langle t|A^\dagger$ is necessarily equal to $A|t\rangle$. Consequently, the Hermitean conjugate of $A^\dagger$ is the operator A itself. Hermitean conjugation is a *reciprocal* operation:

$$(A^\dagger)^\dagger = A. \qquad \text{(VII.21}$$

In an analogous manner one obtains the following fundamental relations:

$$(cA)^\dagger = c^* A^\dagger \qquad \text{(VII.22)}$$

$$(A+B)^\dagger = A^\dagger + B^\dagger \qquad \text{(VII.23)}$$

$$(AB)^\dagger = B^\dagger A^\dagger. \qquad \text{(VII.24)}$$

Note the reversal in the order of the operators in the expression (VII.24) for the adjoint of the product AB. Moreover, the adjoint of the operator $|u\rangle\langle v|$ is

$$(|u\rangle\langle v|)^\dagger = |v\rangle\langle u|. \qquad \text{(VII.25)}$$

Hermitean conjugation is for operators what the conjugation between bras and kets is for vectors, and complex conjugation for ordinary numbers. All these conjugation operations play an important role in the formalism. The notation of Dirac makes it possible to carry them out without difficulty for any algebraic expression. One merely has to obey the following very simple rule: *replace everywhere the numbers by their complex conjugate, bras by the conjugate kets and* vice versa, *the operators by their Hermitean conjugates, and reverse in each term the order of the various symbols occurring there, that is to say the order of the bras, kets, and operators.*

This rule is the obvious generalization of the properties (VII.20), (VII.24) and (VII.25). Let us give some examples of its application. The Hermitean conjugate of the operator $AB|u\rangle\,\langle v|C$ is $C^\dagger|v\rangle\,\langle u|B^\dagger A^\dagger$; the ket $AB|u\rangle\,\langle v|C|w\rangle$ has as its conjugate bra $\langle w|C^\dagger|v\rangle\,\langle u|B^\dagger A^\dagger$; the complex conjugate of $\langle t|AB|u\rangle\,\langle v|C|w\rangle$ is $\langle w|C^\dagger|v\rangle\,\langle u|B^\dagger A^\dagger|t\rangle$, and so forth.

8. Hermitean (or Self-Adjoint) Operators, Positive Definite Hermitean Operators, Unitary Operators

By definition a linear operator H is *Hermitean* if it is its own adjoint:

$$H = H^\dagger.$$

An operator I is *anti-Hermitean* if it has the opposite sign of its adjoint:

$$I = -I^\dagger.$$

From these definitions, one easily derives the following properties.

Any linear operator can be put uniquely in the form of a sum of two operators, one Hermitean, the other anti-Hermitean

$$A = H_A + I_A \tag{VII.26}$$

and one has

$$H_A = \frac{A+A^\dagger}{2}, \qquad I_A = \frac{A-A^\dagger}{2}. \tag{VII.27}$$

Any linear combination with *real* coefficients of Hermitean operators is Hermitean. *The product HK of two Hermitean operators H and K is not necessarily Hermitean* since, according to equation (VII.24)

$$(HK)^\dagger = KH \tag{VII.28}$$

HK is Hermitean if and only if H and K commute. Moreover, the commutator $[H, K]$ is anti-Hermitean and the decomposition (VII.26) of the product HK is written

$$HK = \frac{HK+KH}{2} + \frac{1}{2}[H, K]. \tag{VII.29}$$

The operator $|a\rangle\,\langle a|$ is a particular Hermitean operator. With two distinct kets $|a\rangle$ and $|b\rangle$, one can form the two Hermitean operators

$|a\rangle\langle a|$ and $|b\rangle\langle b|$, but the product of these two operators $|a\rangle\langle a|b\rangle\langle b|$ is proportional to $|a\rangle\langle b|$ which is not Hermitean; the product is therefore not Hermitean (except if $|a\rangle$ and $|b\rangle$ are orthogonal, in which case this product vanishes).

By definition the *Hermitean* operator H is *positive definite* if

$$\langle u|H|u\rangle \geqslant 0 \quad \text{for any} \quad |u\rangle.$$

The operator $|a\rangle\langle a|$ is a particular positive definite Hermitean operator.

The operators of this type have remarkable properties (cf. Problems VII.7 and VII.8). In particular, if H is positive definite Hermitean, one has the generalized Schwarz inequality:

$$|\langle u|H|v\rangle|^2 \leqslant \langle u|H|u\rangle\langle v|H|v\rangle,$$

which holds for any $|u\rangle$ and $|v\rangle$; the equality obtains if and only if $H|u\rangle$ and $H|v\rangle$ are proportional. Furthermore, the equality

$$\langle u|H|u\rangle = 0$$

necessarily implies: $H|u\rangle = 0$.

An operator U is *unitary* if it is the inverse of its own adjoint:

$$UU^\dagger = U^\dagger U = 1.$$

The product $W = UV$ of two unitary operators U, V is a unitary operator. In fact [properties (VII.15) and (VII.24)],

$$W^{-1} = V^{-1}U^{-1} = V^\dagger U^\dagger = W^\dagger.$$

9. Eigenvalue Problem and Observables

Let A be a linear operator. By definition the complex number a is an eigenvalue of A and the ket $|u\rangle$ an eigenket associated with a if

$$A|u\rangle = a|u\rangle.$$

Similarly $\langle u'|$ is an eigenbra associated with the eigenvalue a' if

$$\langle u'|A = a'\langle u'|.$$

If $|u\rangle$ is an eigenket of A, any multiple $c|u\rangle$ of this ket vector is an eigenket of A belonging to the same eigenvalue; if there exist several linearly independent eigenkets belonging to the same eigenvalue a, any linear combination of these kets is an eigenket of A belonging to this eigenvalue. In other words, the ensemble of eigenkets of A belonging to a given eigenvalue a forms a vector space; we shall call it the *subspace of the eigenvalue a*. If this subspace has but one dimension, the eigenvalue is said to be single or non-degenerate. If not, we have *degeneracy*, the order of the degeneracy being by definition the number of dimensions of this subspace; the degeneracy may be of infinite order.

Similar remarks apply to the eigenbras of A. If A is arbitrary, there is no simple relation between the eigenvalue problem for the kets, and the eigenvalue problem for the bras. However, these two problems are closely related if A is Hermitean, the only case of practical interest.

If A is Hermitean, one has in fact the following properties:

(*i*) the two eigenvalue spectra are identical;

(*ii*) all the eigenvalues are *real*;

(*iii*) any bra conjugate to an eigenket of A is an eigenbra belonging to the same eigenvalue, and *vice versa*; in other words, the subspace of the eigenbras of a given eigenvalue is the dual space of the subspace of the eigenkets of the same eigenvalue.

The proof of property (*ii*) is the same as the one given in Ch. V, § 5, except for notation. If $A=A^\dagger$ and if $A|u\rangle=a|u\rangle$,

$$\langle u|A|u\rangle = a\langle u|u\rangle,$$

and since

$$\langle u|A|u\rangle^* = \langle u|A^\dagger|u\rangle = \langle u|A|u\rangle,$$

$\langle u|A|u\rangle$ is real, as well as $\langle u|u\rangle$; hence a is real. An analogous proof can be given for an eigenvalue belonging to a bra.

Moreover, since any eigenvalue is real, $A|u\rangle=a|u\rangle$ necessarily implies $\langle u|A=a\langle u|$, and conversely; from this one easily derives the properties (*i*) and (*iii*).

Another important property of the eigenvectors of a Hermitean operator is the *orthogonality* property of the eigenvectors belonging to distinct eigenvalues. The proof here is also patterned after the one of Ch. V, § 5. If $|u\rangle$ and $|v\rangle$ are eigenkets belonging to the eigenvalues a and b, respectively,

$$A|u\rangle = a|u\rangle, \qquad \langle v|A = b\langle v|.$$

Multiplying the first equation from the left by $\langle v|$ (scalar product), the second equation by $|u\rangle$ from the right, and subtracting term by term, one obtains

$$0 = (a-b)\,\langle v|u\rangle.$$

Consequently, if $a \neq b$,

$$\langle v|u\rangle = 0.$$

In all these arguments it is implicitly assumed that the eigenvectors belong to Hilbert space. Stated in this way, the eigenvalue problem is too restrictive to satisfy all the requirements of the Theory. One must also consider as acceptable eigensolution, the vectors of infinite norm subject to the conditions stated at the end of Ch. VII, § 4. These vectors are associated with the eigenvalues of the *continuous spectrum.*

The difficulties of the continuous spectrum were discussed at length in Chapter V, and we shall not return to them. The principal results of this discussion are easily translated into our new language. The properties (*i*), (*ii*), and (*iii*) remain true in the case of the continuous spectrum. As to the orthogonality property, it is conveniently expressed by means of the Dirac δ-"function".

By way of an example, let us return to the Hermitean operator of Ch. V, § 9 whose eigenvalue spectrum is made up of a series of discrete eigenvalues a_n and a continuous part $a(\nu)$. The eigenfunctions $\varphi_n{}^{(r)}$ belonging to the eigenvalue a_n represent orthonormal kets which we denote by the symbols $|nr\rangle$, respectively. Likewise, the eigenfunctions $\varphi^{(r)}(\nu, \varrho)$ represent the kets $|\nu\varrho r\rangle$. The orthonormality relations between these various kets are written [eq. (V.38)]:

$$\langle nr|n'r'\rangle = \delta_{nn'}\,\delta_{rr'} \qquad \text{(VII.30}a\text{)}$$

$$\langle nr|\nu'\varrho'r'\rangle = 0 \qquad \text{(VII.30}b\text{)}$$

$$\langle \nu\varrho r|\nu'\varrho'r'\rangle = \delta(\nu-\nu')\,\delta(\varrho-\varrho')\,\delta_{rr'}. \qquad \text{(VII.30}c\text{)}$$

If the ensemble of these vectors spans the entire space, in other words if any vector of finite norm can be expanded in a series (or integral) of these vectors, they are said to form a *complete* set and the Hermitean operator is called an *observable*. Observables are the only Hermitean operators of $\mathscr{E}$-space capable of having a physical interpretation.

The important problem of knowing if a given Hermitean operator is an observable is in most cases a difficult mathematical problem. However, for a very important class of operators this problem is easily solved; namely the projection operators, or projectors for short.

10. Projectors (Projection Operators)

Let $\mathscr{S}$ be a subspace of Hilbert space $\mathscr{E}$, and $\mathscr{S}^{\times}$ its complementary subspace. Any ket $|u\rangle$ possesses a projection in $\mathscr{S}$ and a projection in $\mathscr{S}^{\times}$; these two vectors, $|u_S\rangle$ and $|u_S^{\times}\rangle$, are uniquely defined and one has

$$|u\rangle = |u_S\rangle + |u_S^{\times}\rangle. \tag{VII.31}$$

To each ket $|u\rangle$ there thus corresponds one and only one ket $|u_S\rangle$. This correspondence is easily seen to be linear. The linear operator P_S thus defined is called the *projector* (or *projection operator) on the subspace* $\mathscr{S}$

$$P_S|u\rangle = |u_S\rangle.$$

It is a Hermitean operator. Indeed, we have for any $|v\rangle$

$$\langle u|P_S|v\rangle = \langle u|v_S\rangle = \langle u_S|v_S\rangle = \langle u_S|v\rangle$$

hence

$$\langle u|P_S = \langle u_S|.$$

P_S is evidently an observable possessing a total of two eigenvalues 0 and 1, whose subspaces are respectively $\mathscr{S}^{\times}$ and $\mathscr{S}$.

Moreover, since, for any $|u\rangle$

$$P_S^2|u\rangle = P_S(P_S|u\rangle) = P_S|u_S\rangle = |u_S\rangle = P_S|u\rangle,$$

P_S satisfies the operator equation

$$P_S^2 = P_S.$$

Conversely, *any Hermitean operator P satisfying the equation*

$$P^2 = P \tag{VII.32}$$

is a projector. The subspace $\mathscr{S}$ on which it projects is the subspace of its eigenvalue 1.

Indeed, if p is an eigenvalue of this operator, and $|p\rangle$ one of the corresponding eigenvectors,

$$P|p\rangle = p|p\rangle.$$

According to eq. (VII.32)

$$0 = (P^2 - P)|p\rangle = (p^2 - p)|p\rangle$$

and since the ket $|p\rangle$ does not vanish, $p^2 - p = 0$. In other words, the only possible eigenvalues of P are equal to 0 or 1.

Furthermore, P is an observable since any vector $|u\rangle$ can be put in the form of a sum of eigenvectors of P. Indeed one has

$$|u\rangle = P|u\rangle + (1 - P)|u\rangle. \qquad \text{(VII.33)}$$

The vector $P|u\rangle$ is an eigenket of P belonging to the eigenvalue 1, since by virtue of eq. (VII.32)

$$P^2|u\rangle \equiv P(P|u\rangle) = P|u\rangle.$$

The vector $(1-P)|u\rangle$ is an eigenket of P belonging to the eigenvalue 0, since likewise

$$P(1 - P)|u\rangle = (P - P^2)|u\rangle = 0.$$

One can easily verify that the vectors $P|u\rangle$ and $(1-P)|u\rangle$ are orthogonal and, therefore, that the sum of their norms is equal to the norm of $|u\rangle$. These two vectors thus certainly have a finite norm: they belong to Hilbert space.

Let $\mathscr{S}$ be the subspace of the eigenvectors of P belonging to the eigenvalue 1. The subspace $\mathscr{S}^\times$, which is complementary to $\mathscr{S}$, i.e. the subspace of the vectors which are orthogonal to the vectors of the subspace $\mathscr{S}$, is composed of the sum total of the eigenvectors of P belonging to the eigenvalue 0. Following the decomposition (VII.33), the action of P on an arbitrary vector $|u\rangle$ yields the projection of this vector on $\mathscr{S}$. P is therefore just the projector on $\mathscr{S}$. It is clear that $(1-P)$ is the projector on $\mathscr{S}^\times$.

The property relating to the norm of $P|u\rangle$ stated earlier can be rewritten

$$0 \leqslant \langle u|P|u\rangle \leqslant \langle u|u\rangle. \qquad \text{(VII.34)}$$

If $\langle u|P|u\rangle = 0$, $|u\rangle$ lies entirely in $\mathscr{S}^\times$.

If $\langle u|P|u\rangle = \langle u|u\rangle$, $|u\rangle$ lies entirely in $\mathscr{S}$.

Two extreme cases deserve mention. When the projection space $\mathscr{S}$

is the space $\mathscr{E}$ itself, any ket $|u\rangle$ is its own projection: one has $\langle u|P|u\rangle = \langle u|u\rangle$ for any $|u\rangle$; the complementary space $\mathscr{S}^\times$ is zero. This is the case $P=1$.

The other extreme case is the one where the space $\mathscr{S}$ vanishes (the complementary space $\mathscr{S}^\times$ then being the space $\mathscr{E}$ itself): $\langle u|P|u\rangle = 0$ for any $|u\rangle$. This is the case $P=0$.

As an illustration, we give some typical examples of projectors.

Let $|a\rangle$ be a ket normalized to unity. It spans a one-dimensional subspace. Let us denote by $|u_a\rangle$ the projection of an arbitrary vector $|u\rangle$ in this subspace:

$$|u\rangle = |u_a\rangle + |u_a^\times\rangle. \tag{VII.35}$$

By hypothesis

$$\langle a|u_a^\times\rangle = 0 \qquad |u_a\rangle = c|a\rangle.$$

Multiplying both sides of eq. (VII.35) from the left by $\langle a|$ (scalar product), we have $c = \langle a|u\rangle$. Hence

$$|u_a\rangle = |a\rangle\,\langle a|u\rangle.$$

Consequently, the projector on $|a\rangle$ is the operator

$$P_a \equiv |a\rangle\,\langle a| \qquad (\langle a|a\rangle = 1). \tag{VII.36}$$

Projection operators of this type will be called *elementary*.

Consider now a set of orthonormal vectors $|1\rangle, |2\rangle, \ldots, |N\rangle$

$$\langle m|n\rangle = \delta_{mn} \qquad (m, n = 1, 2, \ldots, N).$$

These vectors span a certain subspace $\mathscr{E}_1$ (with N dimensions) of the ket vector space to which they belong. One easily verifies that the operator

$$P_1 \equiv \sum_{m=1}^{N} |m\rangle\,\langle m| \tag{VII.37}$$

is the projector on $\mathscr{E}_1$.

All the vectors considered thus far have finite norm. But one might also contemplate a set of ket vectors $|\xi\rangle$ depending upon the continuous index ξ varying in a certain domain (ξ_1, ξ_2). We assume that the eigendifferentials formed with these kets have finite norm and belong to the Hilbert space under study. Therefore, any linear combination of these vectors also belongs to Hilbert space. The ensemble of these linear combinations forms a certain subspace $\mathscr{E}_2$ of the total Hilbert space: $\mathscr{E}_2$ is the subspace spanned by the vectors $|\xi\rangle$. Let us further suppose that these vectors satisfy the "orthonormality" condition

$$\langle\xi'|\xi\rangle = \delta(\xi'-\xi). \tag{VII.38}$$

Clearly

$$P_2 \equiv \int_{\xi_1}^{\xi_2} |\xi\rangle \, \mathrm{d}\xi \, \langle\xi| \tag{VII.39}$$

is the projector upon $\mathscr{E}_2$. Indeed, the vector

$$P_2|u\rangle \equiv \int_{\xi_1}^{\xi_2} |\xi\rangle \, \mathrm{d}\xi \, \langle\xi|u\rangle$$

obtained by letting P_2 act on an arbitrarily chosen vector $|u\rangle$ is certainly located in $\mathscr{E}_2$ since it is a linear combination of vectors $|\xi\rangle$; the difference $(1-P_2)|u\rangle$ is orthogonal to all vectors of the set $|\xi\rangle$:

$$\begin{aligned}\langle\xi'|(1-P_2)|u\rangle &= \langle\xi'|u\rangle - \int_{\xi_1}^{\xi_2} \langle\xi'|\xi\rangle \, \mathrm{d}\xi \, \langle\xi|u\rangle \\ &= \langle\xi'|u\rangle - \int_{\xi_1}^{\xi_2} \delta(\xi'-\xi) \, \mathrm{d}\xi \, \langle\xi|u\rangle = 0.\end{aligned}$$

and hence orthogonal to $\mathscr{E}_2$.

11. Projector Algebra

The projectors of Hilbert space are of great interest in view of their very simple geometrical interpretation [1]). We shall give here the main properties of the algebra of these operators. Since the proofs are for the most part elementary, we shall merely state the principle of these proofs, leaving the task of completing them to the reader.

---

[1]) The treatment of the continuous spectrum by the method of von Neumann is based on the systematic study of the properties of the projectors of Hilbert space; in this way one succeeds in surmounting all the difficulties of the continuous spectrum without ever leaving Hilbert space; cf. von Neumann, *loc. cit.*, p. 157, Ch. V.

Let P_i, P_j be the projectors upon the subspaces $\mathscr{E}_i$, $\mathscr{E}_j$ of the Hilbert space $\mathscr{E}$. In order that the *product*

$$P_{[ij]} \equiv P_i P_j$$

also be a projector, it is necessary and sufficient that P_i and P_j commute.

The condition is necessary, otherwise $P_{[ij]}$ would not be Hermitean. It is sufficient since, in that case, $P_{[ij]}$ is Hermitean and

$$P_{[ij]}^2 = P_i P_j P_i P_j = P_i^2 P_j^2 = P_i P_j = P_{[ij]}.$$

The subspace $\mathscr{E}_{[ij]}$ which corresponds to this is the *intersection* of the subspaces $\mathscr{E}_i$ and $\mathscr{E}_j$, that is to say the subspace of the vectors common to $\mathscr{E}_i$ and $\mathscr{E}_j$. Two extreme cases can arise, the one where $\mathscr{E}_{[ij]}$ is identical with one of the two subspaces from which we started, $\mathscr{E}_j$, say, and the one where $\mathscr{E}_{[ij]}$ vanishes. In the first case, $\mathscr{E}_j$ is a subspace of $\mathscr{E}_i$; in the second case $\mathscr{E}_i$ and $\mathscr{E}_j$ are orthogonal.

One easily derives the two following properties.

In order that $\mathscr{E}_j$ be a subspace of $\mathscr{E}_i$ (i.e. in order that any vector of the subspace $\mathscr{E}_j$ be a vector of the subspace $\mathscr{E}_i$), the necessary and sufficient condition is that

$$P_i P_j = P_j.$$

For $\mathscr{E}_i$ and $\mathscr{E}_j$ to be orthogonal, it is necessary and sufficient that

$$P_i P_j = 0. \tag{VII.40}$$

In that case, by an extension of the meaning of the term, we say that the *projectors are orthogonal.*

·Concerning the *sum of projectors*, one has the important theorem:

Let P_i, P_j, P_k, ... be the projectors upon the subspaces $\mathscr{E}_i$, $\mathscr{E}_j$, $\mathscr{E}_k$, ... respectively. In order that their sum $P_i + P_j + P_k + \ldots$ likewise be a projector, it is necessary and sufficient that these operators be mutually orthogonal. The subspace upon which the projection is carried out is the *direct sum* of the subspaces $\mathscr{E}_i$, $\mathscr{E}_j$, $\mathscr{E}_k$, ... (i.e. the ensemble of vectors formed by linear superposition of vectors belonging to any one of these partial subspaces).

The condition of orthogonality is obviously sufficient.

To show that it is necessary, it suffices to show that, if the sum $S = P_i + P_j + P_k + \dots$ is a projector, the action of $P_j P_i$ on every ket vector $|u\rangle$ of ξ_i gives 0. To this effect, we apply property (VII.34) to the projectors S and P_i, P_j, P_k, ... The upper limit of the sum

$$\langle u|S|u\rangle = \langle u|P_i|u\rangle + \langle u|P_j|u\rangle + \langle u|P_k|u\rangle + \dots$$

is $\langle u|u\rangle$ and all its terms are positive. If $P_i|u\rangle = |u\rangle$, the first term is equal to this upper limit and all the other terms vanish. Further, since $\langle u|P_j|u\rangle = 0$ implies $P_j|u\rangle = 0$, we have:

$$P_j P_i|u\rangle = P_j|u\rangle = 0.$$

The operator P_1 defined by eq. (VII.37) is a sum of orthogonal projectors. Specifically, the projectors $|m\rangle\langle m|$ which occur in this sum are elementary projectors. It is quite clear that the space $\mathscr{E}_1$ upon which the projection is carried out is the direct sum of the spaces upon which the projections relating to each term of the sum are carried out. Being the direct sum of N one-dimensional spaces, $\mathscr{E}_1$ is a space with N dimensions, while the operator P_1 is the sum of N elementary orthogonal projectors.

If $N \neq 1$, P_1 can be put into this form in infinitely many ways. Indeed, designate by $\{n\}$ a set $|1\rangle, |2\rangle, \dots, |N\rangle$ of N orthonormal vectors of $\mathscr{E}_1$. The set $\{n\}$ forms a basis in $\mathscr{E}_1$, in the sense that any vector of $\mathscr{E}_1$ can be expressed as a linear function of these N vectors. We adopt the convention of not considering two bases distinct whose vectors differ only by a phase factor or by the relative order in which they occur. Now it is clear that

$$P_1 = \sum_{n=1}^{N} |n\rangle\langle n|$$

and that there are as many expressions of P_1 of this type as there are distinct bases.

These considerations can be extended with only minor changes to the case where the subspace $\mathscr{S}$ upon which the projection is performed possesses an infinite number of dimensions. One shows in the theory of Hilbert space that it is always possible to choose in $\mathscr{S}$, in an infinite number of ways, a basis $\{n\} \equiv \{|1\rangle, |2\rangle, \dots, |n\rangle, \dots\}$ containing a denumerably infinite set of orthonormal vectors. The projector P on $\mathscr{S}$ can then be put *in the form of a series of elementary, orthogonal projectors*:

$$P = \sum_{n=1}^{\infty} |n\rangle\langle n|.$$

However, one may also form a basis in $\mathscr{S}$ containing kets depending upon a continuous index. Let us suppose, for instance, that there exists a (non-denumerably infinite) set of vectors $|\xi\rangle$ of infinite norm, depending on the continuous index ξ and satisfying the "orthonormality" relations (VII.38); let us further suppose that the subspace $\mathscr{S}$ is made up of the sum total of the vectors of finite norm formed by linear superposition of the kets $|\xi\rangle$ of a certain domain (ξ_1, ξ_2). In that case, one can also write P in the form of the type (VII.39),

$$P = \int_{\xi_1}^{\xi_2} |\xi\rangle \, \mathrm{d}\xi \, \langle\xi|.$$

In that form P can still be regarded as a sum of orthogonal projectors. Let us subdivide the domain of integration (ξ_1, ξ_2) into a number of partial domains. P is the sum of the projectors obtained by integrating $|\xi\rangle \, \mathrm{d}\xi \, \langle\xi|$ over each of these partial domains. The latter can actually be arbitrarily small. Denote by δP the operator obtained by integration over the infinitesimal domain $(\xi, \xi+\delta\xi)$:

$$\delta P = \int_{\xi}^{\xi+\delta\xi} |\xi'\rangle \, \mathrm{d}\xi' \, \langle\xi'|.$$

P is the sum of an infinitely large number of operators of the type δP. We shall call the operators of the type δP *differential projectors*; the projection space corresponding to this type of operator has an infinity of dimensions.

12. Observables Possessing an Entirely Discrete Spectrum

Let A be a Hermitean operator. In this section, we shall study the eigenvalue problem, limiting ourselves to the eigenvectors located in Hilbert space. The eigenvalues then form a discrete set $a_1, a_2, \ldots, a_n, \ldots$. We designate by $\mathscr{E}_n$ the subspace belonging to the eigenvalue a_n, by P_n the projector upon this subspace. If the eigenvalue a_n is not degenerate, $\mathscr{E}_n$ has only one dimension and P_n is an elementary projector. If it is degenerate, one can always choose in an infinity of ways a basis in $\mathscr{E}_n$, $|n1\rangle, |n2\rangle, \ldots, |nr\rangle, \ldots$, and one has

$$P_n = \sum_r |nr\rangle \langle nr|. \qquad \text{(VII.41)}$$

The subspaces $\mathscr{E}_n, \mathscr{E}_{n'}$, belonging to two distinct eigenvalues $a_n, a_{n'}$, are orthogonal, consequently

$$P_n P_{n'} = 0 \qquad (n \neq n'). \qquad \text{(VII.42)}$$

Carrying out the summation of the projectors P_n corresponding to all the eigenvalues of the discrete spectrum, one obtains the projector

$$P_A \equiv \sum_n P_n \tag{VII.43}$$

whose subspace of projection $\mathscr{E}_A$ is formed of the direct sum of all the $\mathscr{E}_n$; $\mathscr{E}_A$ is the vector space formed by linear superposition of the eigenkets of A located in Hilbert space.

If A is an observable and if its spectrum is entirely discrete, $\mathscr{E}_A$ by definition coincides with the total space $\mathscr{E}$, in other words

$$P_A \equiv \sum_n P_n = 1. \tag{VII.44}$$

One sometimes calls the left-hand side of (VII.44) the *decomposition of unity* with respect to the eigenvalues of A. It is clear that this decomposition is unique, hence that any ket $|u\rangle$ can be written in a unique way as a sum of eigenkets of A each of which belongs to a different eigenvalue. To write down this sum, one merely operates on $|u\rangle$ with each side of eq. (VII.44) and equates the two kets thus obtained, namely:

$$|u\rangle = \sum_n P_n|u\rangle. \tag{VII.45}$$

According to the definition of P_n, $P_n|u\rangle$ is either zero, or an eigenket of A belonging to the eigenvalue a_n, for any $|u\rangle$. Therefore

$$(A - a_n)P_n = 0. \tag{VII.46}$$

Multiplying eq. (VII.44) term by term by A, one obtains, taking into account eqs. (VII.46)

$$A = \sum_n a_n P_n. \tag{VII.47}$$

According to this equation the observable A is completely defined once its eigenvalues and their respective subspaces are given. Furthermore, from expression (VII.47) it is obvious that the operator A commutes with all the projectors P_n.

The relations (VII.44), (VII.45), and (VII.47) are characteristic of observables possessing an entirely discrete spectrum, whether the number of their eigenvalues be finite or infinite. We shall not elaborate here on the question of convergence of the series; in practice that convergence always obtains.

Especially convenient expressions result by substituting everywhere for every P_n its expression (VII.41). Thus the left-hand side of eq. (VII.44) is written in the form of a sum of elementary projectors, and one obtains the *closure relation*

$$P_A \equiv \sum_{n,r} |nr\rangle \langle nr| = 1. \tag{VII.48}$$

Together with the orthogonality relations

$$\langle nr|n'r'\rangle = \delta_{nn'}\,\delta_{rr'} \tag{VII.49}$$

this expresses the fact that the assembly of vectors $|nr\rangle$ forms a complete orthonormal set.

Upon applying the closure relation to any vector $|u\rangle$, one obtains the expansion of $|u\rangle$

$$|u\rangle = \sum_{n,r} |nr\rangle \langle nr|u\rangle \tag{VII.50}$$

in a series of eigenvectors $|nr\rangle$. The coefficients of the expansion are equal to the scalar products $\langle nr|u\rangle$ [cf. eqs. (V.14) and (V.15)]. Moreover

$$\langle u|u\rangle = \langle u|P_A|u\rangle = \sum_{n,r} \langle u|nr\rangle \langle nr|u\rangle = \sum_{n,r} |\langle nr|u\rangle|^2. \tag{VII.51}$$

The norm of $|u\rangle$ is equal to the sum of the squares of the absolute values of the expansion coefficients: this is the Parseval relation [cf. eq. (V.16)].

The observable A can be written in the form of a series of elementary, orthogonal projectors. By a procedure analogous to the one which yielded eq. (VII.47), one actually obtains:

$$A = AP_A = \sum_{n,r} |nr\rangle a_n \langle nr|. \tag{VII.52}$$

13. Observables in the General Case. Generalized Closure Relation

The Hermitean operator A is an observable if the space $\mathscr{E}_A$ of all the vectors of finite norm, which can be obtained by superposing the eigenvectors of A, coincides with the total Hilbert space $\mathscr{E}$, that is if the operator P_A of projection upon $\mathscr{E}_A$ is equal to unity.

When the spectrum is entirely discrete, P_A can be written in the form of an expansion in a series of elementary, orthogonal projectors formed with the eigenvectors of A. The condition that A be an observable is conveniently expressed by the closure relation (VII.48).

The extension of this relation to the general case necessitates the introduction of differential projectors; they play, in the treatment of the continuous portion of the spectrum, the role which the elementary projectors play in the treatment of the discrete spectrum.

Let us deal first with the case where the spectrum of A possesses no degeneracy. We assume that this spectrum contains a continuous portion labelled by the continuous index ν and a discrete portion labelled by the integral index n. Thus a_n is an eigenvalue of the discrete spectrum, $a(\nu)$ an eigenvalue of the continuous spectrum; $a(\nu)$ is a monotonic function of ν taking on all values of some interval $[a(\nu_1), a(\nu_2)]$. We denote by $|n\rangle$ and $|\nu\rangle$ the eigenkets belonging to the eigenvalues a_n and $a(\nu)$, respectively. These kets are orthonormal; in particular

$$\langle\nu|\nu'\rangle = \delta(\nu-\nu').$$

The operator

$$\delta P = \int_{\nu}^{\nu+\delta\nu} |\nu'\rangle\, \mathrm{d}\nu'\, \langle\nu'|$$

is the projector upon the subspace spanned by the kets $|\nu'\rangle$ of the interval $(\nu, \nu+\delta\nu)$. By adding projectors of this type, one forms the projector

$$P_{\mathrm{c}} = \int_{\nu_1}^{\nu_2} |\nu\rangle\, \mathrm{d}\nu\, \langle\nu|$$

on the subspace $\mathscr{E}_{\mathrm{c}}$ spanned by the eigenkets of the continuous spectrum. This subspace is orthogonal to the subspace $\mathscr{E}_{\mathrm{d}}$ spanned by the eigenkets of the discrete spectrum, a subspace whose projector is

$$P_{\mathrm{d}} = \sum_n |n\rangle\, \langle n|.$$

The condition that A be an observable can be written

$$P_A \equiv P_{\mathrm{c}} + P_{\mathrm{d}} = 1,$$

or else

$$P_A \equiv \sum_n |n\rangle\, \langle n| + \int_{\nu_1}^{\nu_2} |\nu\rangle\, \mathrm{d}\nu\, \langle\nu| = 1. \qquad \text{(VII.53)}$$

The closure relation (VII.53) is the necessary and sufficient condition that the ensemble of the orthonormal vectors $|n\rangle$, $|\nu\rangle$ form a complete set.

The extension to the most general case where all or part of the spectrum of A is degenerate is quite straightforward. As an illustration let us take up once again the example mentioned at the end of Ch. VII, § 9. The eigenkets of A, $|nr\rangle$, $|\nu\varrho r\rangle$ satisfy the orthonormality relations (VII.30). Furthermore, if A is an observable, in other words, if these kets form a complete set, they satisfy the *closure relation*

$$\boxed{P_A \equiv \sum_{n,r} |nr\rangle \langle nr| + \sum_r \iint |\nu\varrho r\rangle \, \mathrm{d}\nu \, \mathrm{d}\varrho \, \langle \nu\varrho r| = 1} \qquad \text{(VII.54)}$$

As in the case where the spectrum is entirely discrete, it is very convenient to use the closure relation to form the expansion of any vector $|u\rangle$ of Hilbert space in a series of the basic kets of the observable A. For ease of writing, we suppose that the spectrum of A is non-degenerate [relation (VII.53)]. One obtains

$$|u\rangle = P_A|u\rangle = \sum_n |n\rangle \langle n|u\rangle + \int_{\nu_1}^{\nu_2} |\nu\rangle \, \mathrm{d}\nu \, \langle \nu|u\rangle. \qquad \text{(VII.55)}$$

Similarly one finds the generalized Parseval relation

$$\langle u|u\rangle = \langle u|P_A|u\rangle = \sum_n |\langle n|u\rangle|^2 + \int_{\nu_1}^{\nu_2} |\langle \nu|u\rangle|^2 \, \mathrm{d}\nu \qquad \text{(VII.56)}$$

and the expansion of A in a series of projectors

$$A = AP_A = \sum_n |n\rangle \, a_n \, \langle n| + \int_{\nu_1}^{\nu_2} |\nu\rangle \, a(\nu) \, \mathrm{d}\nu \, \langle \nu|. \qquad \text{(VII.57)}$$

In conclusion, note that it is sometimes convenient to replace the normalization condition of the eigenvectors of the continuous spectrum by the more general condition

$$\langle \nu|\nu'\rangle = f(\nu) \, \delta(\nu - \nu'),$$

where $f(\nu)$ is a real, positive function of ν. This amounts to multiplying each vector $|\nu\rangle$ by a constant of modulus $\sqrt{f}$. In that case, all the above remains valid provided that one replaces in all formulae the differential element

$$|\nu\rangle \, \mathrm{d}\nu \, \langle \nu| \quad \text{by} \quad |\nu\rangle \frac{\mathrm{d}\nu}{f(\nu)} \langle \nu|.$$

Likewise, if one replaces the normalization condition (VII.30*c*) by

$$\langle \nu\varrho r|\nu'\varrho' r'\rangle = F_r(\nu, \varrho) \, \delta(\nu - \nu') \, \delta(\varrho - \varrho') \, \delta_{rr'},$$

the expression for P_A in the closure relation (VII.54) remains valid provided that one divides the integrand there by the function $F_r(\nu, \varrho)$.

14. Functions of an Observable

A linear operator is completely defined by its action upon the vectors of a complete set of orthonormal eigenvectors. Its action on any arbitrary linear superposition of these vectors can be derived immediately therefrom, provided that it is convergent, if one is dealing with an infinite series (the conditions of convergence have already been set forth in Chapter V). In particular, any function $f(a)$ of the eigenvalues of an observable A enables us to define a linear operator $f(A)$, a function of this observable. The action of $f(A)$ on an eigenvector $|a\rangle$ of A belonging to the eigenvalue a is by definition

$$f(A)|a\rangle = f(a)|a\rangle. \tag{VII.58}$$

When the function f is a polynomial, this definition coincides with the one obtained by simply applying the rules of operator algebra; but it remains meaningful in more general cases.

From its very definition it follows that any eigenvector of A is an eigenvector of $f(A)$.

Conversely, if every eigenvector of an observable A is an eigenvector of a linear operator F, the latter is a function of A.

This is rather obvious if none of the eigenvalues of A are degenerate. Let us therefore consider a degenerate eigenvalue, and let $|a1\rangle$, $|a2\rangle$ be two linearly independent eigenvectors belonging to this eigenvalue. By hypothesis, they are eigenvectors of F

$$F|a1\rangle = f_a^{(1)}|a1\rangle \qquad F|a2\rangle = f_a^{(2)}|a2\rangle.$$

Moreover, any linear combination of these two vectors is an eigenvector of F

$$F(\lambda_1|a1\rangle + \lambda_2|a2\rangle) = f_a^{(\lambda)}(\lambda_1|a1\rangle + \lambda_2|a2\rangle);$$

consequently

$$\lambda_1(f_a^{(\lambda)} - f_a^{(1)})|a1\rangle + \lambda_2(f_a^{(\lambda)} - f_a^{(2)})|a2\rangle = 0,$$

and since $|a1\rangle$ and $|a2\rangle$ are linearly independent,

$$f_a^{(1)} = f_a^{(\lambda)} = f_a^{(2)}.$$

Hence all the eigenfunctions of A belonging to the same eigenvalue a are eigenfunctions of F belonging to the same eigenvalue f; the latter is a certain function of a, namely $f(a)$, and one actually has $F=f(A)$.

Any function $f(A)$ of the observable A may be expressed like A itself, in the form of a linear combination of elementary or differential projectors. Suppose, to be definite, that A satisfies equation (VII.57). One likewise has

$$f(A) = f(A)P_A = \sum_n |n\rangle\, f(a_n)\, \langle n| + \int_{\nu_1}^{\nu_2} |\nu\rangle\, f[a(\nu)]\, \mathrm{d}\nu\, \langle \nu|. \qquad \text{(VII.59)}$$

As examples of functions of the observable A, let us mention the projector upon the subspace of a particular eigenvalue, the projector upon the space spanned by the eigenvectors whose eigenvalues are located in a given region. Let us also mention the exponential $\exp(\mathrm{i}\xi A)$ (ξ = given constant), and the inverse A^{-1}. The function $\exp(\mathrm{i}\xi A)$ is always defined; the inverse A^{-1} is well defined only if none of the eigenvalues of A vanish.

15. Operators which Commute with an Observable. Commuting Observables

The functions of the observable A belong to a more general class of operators, *the operators which commute with* A. Such operators have a particularly simple effect upon the eigenvectors of A. Indeed, if $|a\rangle$ is an eigenvector,

$$A|a\rangle = a|a\rangle,$$

and if

$$[A, X] = 0,$$

it follows that

$$0 = (AX - XA)|a\rangle = (A - a)\, X|a\rangle.$$

$X|a\rangle$ *is an eigenvector of* A *belonging to the same eigenvalue* (unless it vanishes).

Conversely, in order that an operator X commute with the observable A, it suffices that its action on each of the vectors of a complete orthonormal set of eigenvectors of A yield an eigenvector of A belonging to the same eigenvalue.

Indeed, according to our earlier hypothesis, the action of the commutator $[A, X]$ on each vector of this complete set gives zero. Therefore, we have $[A, X]=0$.

All this applies especially to *commuting observables*. Moreover, all the considerations developed in Ch. V, §§ 14 and 15 on the subject of commuting observables may be taken up here once again, except for some details of terminology. We shall merely state the results.

We call *basis* of a given observable any complete orthonormal set of eigenvectors of this observable, having adopted the convention to consider two bases to be identical if their eigenvectors differ only by phase factors, by the order in which the vectors of the discrete spectrum are arranged, or by the choice of continuous indices relating to those of the continuous spectrum.

We have the following important theorem:

If two observables A and B commute, they have at least one basis in common, and, conversely, if two observables A and B possess a common basis, they commute.

Any function $f(a, b)$ of the eigenvalues of two commuting observables, A and B, makes it possible to define a linear operator $f(A, B)$, function of these two observables, by an obvious generalization of the concept of function of an observable. One easily shows that if every eigenvector common to the observables A and B is an eigenvector of a linear operator F, the latter is some function of A and B.

All this can be easily generalized to an arbitrary number of pairwise commuting observables.

Finally, one says that a set $A, B, C, \ldots$ of observables form a *complete set of commuting observables*, if these observables all commute in pairs, and if their common basis is uniquely defined. To each set of eigenvalues $a, b, c, \ldots$ corresponds one and only one common eigenvector (defined to within a constant). This vector may be regarded as a function of the eigenvalues $a, b, c, \ldots$. One commonly denotes it by the symbol $|abc\ldots\rangle$.

III. REPRESENTATION THEORY

16. General Remarks on Finite Matrices

By definition a matrix A of the type $M \times N$ is a set of MN quantities A_{mn} $(m=1, 2, \ldots, M;\ n=1, 2, \ldots, N)$ which one usually arranges in a rectangular array with M rows and N columns

$$(A) \equiv \begin{pmatrix} A_{11} & A_{12} & \cdots & A_{1N} \\ A_{21} & \cdots & & \vdots \\ \vdots & & & \vdots \\ A_{M1} & \cdots & & A_{MN} \end{pmatrix}$$

A_{mn} is the matrix element of the mth row and the nth column.

If $M=N$, it is called a *square matrix*; the number of its rows and of its columns is the number of dimensions or *order* of that matrix. If one of the two integers M or N is equal to 1, the matrix elements can be likened to the components of a vector. We shall call *column vector* a matrix with one column (M = number of dimensions of the vector; $N=1$), and *row vector* a matrix with one row ($M=1$; N = number of dimensions of the vector). A *scalar* is a special matrix for which $M=N=1$.

From a matrix A of the type $M \times N$ one can derive new matrices by carrying out certain *conjugation operations*, namely:

(*i*) the *complex conjugate* matrix A^*, matrix of the type $M \times N$ whose elements are the complex conjugates of the elements of A: $(A^*)_{kl} = A^*_{kl}$;

(*ii*) the *transposed* matrix $\tilde{A}$, matrix of the type $N \times M$, derived from A by interchanging rows and columns:

$$(\tilde{A})_{kl} = A_{lk};$$

(*iii*) the *Hermitean conjugate* matrix $A^\dagger$, matrix of the type $N \times M$ obtained by performing upon A both of the foregoing operations: $(A^\dagger)_{kl} = A^*_{lk}$.

The complex conjugate of a column vector is a column vector. The transposed and Hermitean conjugate matrix of a column vector are row vectors, and *vice versa*. The complex conjugate, transpose, and Hermitean conjugate of a square matrix of order N are square matrices of order N.

One defines the following *algebraic operations* with matrices:

(*a*) *multiplication* of a matrix A *by a constant* c; the product cA is a matrix of the same type as A:

$$(cA)_{mn} = cA_{mn};$$

(*b*) *sum* $S=A+B$ of two matrices of the same type; S is a matrix of the same type as A and B:

$$S_{mn} = A_{mn} + B_{mn};$$

(*c*) *product* (from the left) $P=AB$ of a matrix B of type $M_B \times N_B$ by a matrix A of type $M_A \times N_A$ whose number of columns is equal to the number of rows of B: $N_A = M_B = K$. It is a matrix of type $M_A \times N_B$ whose elements are:

$$P_{mn} = \sum_{k=1}^{K} A_{mk} B_{kn}.$$

The product from the right of B by A (if it exists) is equal to the product from the left of A by B and is written BA.

Clearly

$$(A+B)^* = A^* + B^* \qquad \widetilde{(A+B)} = \tilde{A} + \tilde{B} \qquad (A+B)^\dagger = A^\dagger + B^\dagger$$

$$(AB)^* = A^* B^* \qquad \widetilde{(AB)} = \tilde{B}\tilde{A} \qquad (AB)^\dagger = B^\dagger A^\dagger.$$

Note the reversal of the order of factors in the last two equations.

The product from the left of an N-dimensional column vector by an N-dimensional row vector is a scalar. The product from the left of an N-dimensional row vector by an N-dimensional column vector is a square matrix of order N.

Another important operation is the *tensor product* of two matrices. With a matrix $A^{(1)}$ of type $M_1 \times N_1$ and a matrix $A^{(2)}$ of type $M_2 \times N_2$ one can form, by tensor product, a matrix $A^{(12)} \equiv A^{(1)} \otimes A^{(2)}$ of the type $M_1 M_2 \times N_1 N_2$. The $M_1 M_2$ rows of this matrix are labelled by two indices m_1 and m_2 ($m_1 = 1, 2, \ldots, M_1$; $m_2 = 1, 2, \ldots, M_2$), its $N_1 N_2$ columns by two indices n_1 and n_2 ($n_1 = 1, 2, \ldots, N_1$; $n_2 = 1, 2, \ldots, N_2$):

$$A^{(12)}_{m_1 m_2;\, n_1 n_2} = A^{(1)}_{m_1 n_1} A^{(2)}_{m_2 n_2}.$$

17. Square Matrices

In this paragraph, we shall state a number of definitions and properties peculiar to square matrices.

In a square matrix A of order N, one distinguishes the diagonal elements A_{nn} ($n = 1, 2, \ldots, N$) from the off-diagonal elements A_{kl} ($k \neq l$). The *trace* of A is the sum of its diagonal elements:

$$\mathrm{Tr}\, A \equiv \sum_n A_{nn}. \tag{VII.60}$$

The *determinant* of A, $\det A$, is the determinant of the array of its elements.

The *unit matrix* I is a matrix whose diagonal elements are all equal to 1 and whose off-diagonal elements all vanish,

$$I_{mn} = \delta_{mn}.$$

The product of the unit matrix by a constant is, by definition, a *constant matrix*. A *diagonal matrix* is a matrix whose off-diagonal elements all vanish.

A square matrix is real, symmetrical or Hermitean according to whether it is equal to its complex conjugate, its transpose or its Hermitean conjugate.

The sum, and the product of two matrices of order N are always defined; they are matrices of order N. The sum is associative and commutative. The product is associative, distributive with respect to the sum, but it is not necessarily commutative. The algebra of matrices of order N is a *non-commutative algebra*.

In order that a matrix of order N commute with all matrices of order N, it is necessary and sufficient that it be constant (Problem VII.4). In particular, the unit matrix I is such that for any A

$$IA = AI = A. \tag{VII.61}$$

Two diagonal matrices necessarily commute. In order that a matrix of order N commute with all diagonal matrices of order N, it is necessary and sufficient that it be diagonal (Problem VII.4).

The *trace* of a product of matrices is *invariant under cyclic permutation* of these matrices

$$\mathrm{Tr}\, ABC = \mathrm{Tr}\, CAB. \tag{VII.62}$$

The determinant of a matrix product is equal to the product of their determinants:

$$\det ABC = \det A \cdot \det B \cdot \det C. \tag{VII.63}$$

A matrix B is by definition the *inverse* of a matrix A if one has

$$AB = 1 \quad \text{and} \quad BA = 1. \tag{VII.64}$$

Indeed, if one of these inequalities is realised, so is the other. One writes:

$$B = A^{-1}.$$

A necessary and sufficient condition for matrix A to possess an inverse is that its determinant be different from zero: $\det A \neq 0$ If the determinant vanishes, the matrix is said to be *singular*.

One easily verifies that

$$(\tilde{A})^{-1} = (\widetilde{A^{-1}}) \qquad (A^*)^{-1} = (A^{-1})^* \qquad (A^\dagger)^{-1} = (A^{-1})^\dagger$$

and that

$$(PQ)^{-1} = Q^{-1}P^{-1}.$$

A matrix O is *orthogonal* if its transpose $\tilde{O}$ is equal to its inverse:

$$O\tilde{O} = \tilde{O}O = I.$$

A matrix U is *unitary* if its Hermitean conjugate equals its inverse:

$$UU^\dagger = U^\dagger U = I.$$

The product from the left of an N-dimensional column vector by a matrix of order N is an N-dimensional column vector. The product from the right of an N-dimensional row vector by a matrix of order N is an N-dimensional row vector.

The action of a diagonal matrix upon such vectors is particularly simple. Let

$$D_{mn} = d_m\, \delta_{mn}$$

be the elements of such a matrix, and u_n the components of a column vector u

$$(Du)_n = d_n\, u_n.$$

Likewise, if v_n are the components of a row vector v

$$(vD)_n = v_n\, d_n.$$

If a matrix is singular, there exists at least one column vector u such that one has $Au = 0$, and conversely.

From this result we obtain the important theorem:

A and B being two matrices of order N, in order that there exist a column vector u (with N dimensions) such that

$$Au = \lambda Bu$$

it is necessary and sufficient that the constant λ be a solution of the equation

$$\det(A - \lambda B) = 0.$$

In particular:

A being a matrix of order N, in order that there exist a column vector u such that

$$Au = \lambda u$$

it is necessary and sufficient that the constant λ be a solution of the equation

$$\det(A - \lambda I) = 0.$$

This algebraic equation, of the Nth degree at most, is called the *secular equation* of A.

Analogous properties hold for the row vectors.

The tensor product of two matrices of order N_1 and N_2, respectively, is a matrix of order N_1N_2. In particular, the tensor product of the unit matrices $I^{(1)}$, $I^{(2)}$, is a unit matrix $I^{(12)}$ of order N_1N_2.

By way of an example, let us mention the fourth order matrices obtained by the tensor product of second-order matrices by second-order matrices. It is common practice to introduce the following two-dimensional matrices (Pauli matrices):

$$1^{(\sigma)} = \begin{pmatrix} 1 & 0 \\ 0 & 1 \end{pmatrix}, \quad \sigma_1 = \begin{pmatrix} 0 & 1 \\ 1 & 0 \end{pmatrix}, \quad \sigma_2 = \begin{pmatrix} 0 & -i \\ i & 0 \end{pmatrix}, \quad \sigma_3 = \begin{pmatrix} 1 & 0 \\ 0 & -1 \end{pmatrix}. \tag{VII.65}$$

All matrices in two dimensions can be put in the form of a linear combination of these four Hermitean matrices. On the other hand, let us consider the matrices of another two-dimensional space

$$1^{(\varrho)} = \begin{pmatrix} 1 & 0 \\ 0 & 1 \end{pmatrix}, \quad \varrho_1 = \begin{pmatrix} 0 & 1 \\ 1 & 0 \end{pmatrix}, \quad \varrho_2 = \begin{pmatrix} 0 & -i \\ i & 0 \end{pmatrix}, \quad \varrho_3 = \begin{pmatrix} 1 & 0 \\ 0 & -1 \end{pmatrix}. \tag{VII.66}$$

Forming the tensor product of a matrix of the type (σ) by a matrix of the type (ϱ) one obtains a matrix with four rows and four columns. Let us give some explicit examples of matrices of the type $(\varrho\sigma)$ thus formed:

$$\varrho_1 \otimes \sigma_1 = \begin{pmatrix} 0 & \sigma_1 \\ \sigma_1 & 0 \end{pmatrix} = \begin{pmatrix} 0 & 0 & 0 & 1 \\ 0 & 0 & 1 & 0 \\ 0 & 1 & 0 & 0 \\ 1 & 0 & 0 & 0 \end{pmatrix}$$

$$\varrho_1 \otimes \sigma_2 = \begin{pmatrix} 0 & \sigma_2 \\ \sigma_2 & 0 \end{pmatrix} = \begin{pmatrix} 0 & 0 & 0 & -\mathrm{i} \\ 0 & 0 & \mathrm{i} & 0 \\ 0 & -\mathrm{i} & 0 & 0 \\ \mathrm{i} & 0 & 0 & 0 \end{pmatrix}$$

$$\varrho_3 \otimes 1^{(\sigma)} = \begin{pmatrix} 1 & 0 \\ 0 & -1 \end{pmatrix} = \begin{pmatrix} +1 & 0 & 0 & 0 \\ 0 & +1 & 0 & 0 \\ 0 & 0 & -1 & 0 \\ 0 & 0 & 0 & -1 \end{pmatrix}$$

One can regard these tensor product matrices as matrices of one of the spaces, the (ϱ)-space for instance, each matrix element of which is a matrix of the other space. This was done in the center of each of the equations above. The right-hand sides give the explicit expressions of the matrices; if one agrees to label the rows (and the columns) by two indices $m_\varrho m_\sigma$, the first of which refers to the components of (ϱ)-space, and the second to those of (σ)-space, the rows (and columns) will be arranged in the order: 11, 12, 21, 22.

By linear combination of tensor product matrices, one forms square matrices with two indices

$$A_{m_1 m_2; n_1 n_2} \qquad (m_1, n_1 = 1, 2, ..., N_1;\ m_2, n_2 = 1, 2, ..., N_2)$$

and with $N_1 N_2$ dimensions. As is shown by the foregoing example, they may be regarded as matrices of type (1) whose elements are matrices of type (2). Forming the sum of the diagonal elements of such a matrix, one obtains a matrix of type (2) in the ordinary sense of the term. It is by definition the partial trace in space (1) of the matrix from which we started out:

$$(\mathrm{Tr}_1\, A)_{m_2 n_2} \equiv \sum_{n=1}^{N_1} A_{n\, m_2;\, n\, n_2}. \tag{VII.67}$$

Likewise one defines the partial trace in space (2). It is obvious that

$$\mathrm{Tr}\, A = \mathrm{Tr}_2\, (\mathrm{Tr}_1\, A) = \mathrm{Tr}_1\, (\mathrm{Tr}_2\, A) \tag{VII.68}$$

and that, if A is equal to the tensor product $A^{(1)} \otimes A^{(2)}$,

$$\mathrm{Tr}\, (A^{(1)} \otimes A^{(2)}) = (\mathrm{Tr}_1\, A^{(1)})\, (\mathrm{Tr}_2\, A^{(2)}). \tag{VII.69}$$

18. Extension to Infinite Matrices

Most of the remarks on finite matrices may be extended to infinite matrices, whose rows and columns are labelled by one or several discrete indices or, still more generally, by a number of indices which can take on a finite or denumerably infinite number of discrete values, and a number of continuous indices which can assume all values located in a given interval. An infinite matrix is square if its rows and columns are labelled by the same system of indices. It is a column vector if it has but one column, and a row vector it if has but one row.

The operations of complex conjugation, transposition, and Hermitean conjugation extend to infinite matrices without change. The same is true for multiplication by a constant, and for the sum. As far as the product (from the left) of B by A is concerned, it is understood that the rows of B and the columns of A must be labelled by the same system of indices. Moreover, if some indices are continuous, the summation must be replaced by an integration. Let us suppose, to be definite, that B and A are square matrices depending upon a continuous index q which can take any value in the interval (q_1, q_2). The matrix element $P(q; q')$ of the product $P = AB$ is

$$P(q; q') = \int_{q_1}^{q_2} A(q; q'')\, B(q''; q')\, \mathrm{d}q''.$$

The product is well defined only if the summations or integrals which occur in the definition of its elements converge.

Aside from any questions of convergence, the considerations of § 17 on square matrices may be extended to matrices of infinite order, with the exception of the notion of determinant. The only points to be made concern the definition of diagonal matrices in the case where the indices are continuous, and the conditions concerning the existence of an inverse matrix.

By definition, a continuous matrix $D(q; q')$ is *diagonal* if it is of the form

$$D(q; q') = d(q)\, \delta(q - q') \tag{VII.70}$$

where $d(q)$ is an arbitrary function of the index q. In this way the two characteristic properties of diagonal matrices are preserved, namely that two diagonal matrices commute, and that the action of a diagonal matrix upon a vector consists in multiplying each of its components by the corresponding diagonal element. Thus, the action of the diagonal matrix defined by expression (VII.70) upon the column vector g with components $g(q)$ yields the vector $h = Dg$ with components

$$h(q) = \int D(q, q')\, g(q')\, \mathrm{d}q' = d(q)\, g(q).$$

Note that the continuous matrix $\delta'(q-q')$ *is not diagonal.*

As far as *the existence of the inverse* of a given matrix is concerned, in contrast to the case of finite matrices, the fact that

$$AB - I \qquad \text{(VII.71}a\text{)}$$

does not necessarily imply that

$$BA = I. \qquad \text{(VII.71}b\text{)}$$

The two equations (VII.71a) and (VII.71b) must be satisfied simultaneously in order that one may assert that A and B are the inverse of each other.

Actually, it is not necessary that A be a square matrix for it to possess an inverse; for instance, it may happen that a matrix A whose rows are labelled by a discrete index and whose columns are labelled by a continuous index, possesses an inverse. In that case, the inverse A^{-1} has a discrete column index and a continuous row index. In particular, in a *unitary matrix* U, a matrix satisfying by definition the two equations

$$UU^{\dagger} = 1, \qquad U^{\dagger}U = 1 \qquad \text{(VII.72}$$

the row indices and the column indices are not necessarily of the same kind. However, the unit matrices of the right-hand sides in the two defining equations (VII.72) are necessarily square matrices. If the matrix U is not a square matrix, the systems of indices of each of these unit matrices are different.

19. Representation of Vectors and Operators by Matrices

Consider a vector space $\mathscr{E}$ and choose a basis in this space. The latter may possibly be the eigenvectors of a complete set of commuting observables. To simplify the writing, we shall make our arguments on a basis whose vectors are labelled by a discrete index n. We assume, for instance, that they are the eigenvectors of some observable Q

$$Q|n\rangle = q_n|n\rangle.$$

We shall say that they are the basis vectors of the representation $\{Q\}$. These vectors form a complete orthonormal set

$$\langle m|n\rangle = \delta_{mn} \tag{VII.73}$$

$$P_Q \equiv \sum_n |n\rangle \langle n| = 1. \tag{VII.74}$$

Equations (VII.73) and (VII.74) are the fundamental equations of representation $\{Q\}$.

For any ket vector $|u\rangle$,

$$|u\rangle = P_Q|u\rangle = \sum_n |n\rangle \langle n|u\rangle.$$

The quantities $u_n = \langle n|u\rangle$ can be regarded as the elements of a matrix with one column whose row index is n. This column vector completely defines $|u\rangle$: it is the matrix representing $|u\rangle$ in representation $\{Q\}$.

For any bra vector $\langle v|$,

$$\langle v| = \langle v|P_Q = \sum_n \langle v|n\rangle \langle n|.$$

The quantities $\langle v|n\rangle$ are the complex conjugates of the components v_n of the column vector representing the ket $|v\rangle$ in the representation $\{Q\}$. They may be regarded as the elements of a row vector; this row vector defines $\langle v|$ completely; it is the vector representing $\langle v|$ in representation $\{Q\}$. With this convention, the bra conjugate to a given ket is represented by the Hermitean conjugate of the vector representing that ket.

Any linear operator A can be expanded in a unique manner in a double series of operators of basis $|m\rangle \langle n|$

$$A = P_Q\, AP_Q = \sum_{mn} |m\rangle \langle m|A|n\rangle \langle n|.$$

The coefficients of the expansion $A_{mn} = \langle m|A|n\rangle$ completely define A and can be regarded as the elements of a square matrix whose row index is m and whose column index is n: it is the matrix representing A in representation $\{Q\}$.

Having established a one-to-one correspondence between vectors and operators on the one hand, and matrices on the other hand, we shall examine how each operation concerning the operators and vectors of the space $\mathscr{E}$ is translated into the language of the matrices that represent them.

To the conjugation relations between vectors or between operators correspond relations of Hermitean conjugation between matrices. We have already noted this fact in connection with the conjugation between bras and kets. Likewise, the matrices representing two Hermitean-conjugate operators A and $A^\dagger$ are Hermitean conjugate; their elements actually satisfy the relations characteristic of Hermitean conjugation:

$$A^\dagger_{mn} \equiv \langle m|A^\dagger|n\rangle = \langle n|A|m\rangle^* = A^*_{nm}.$$

As for the different algebraic operations between vectors and operators, they are translated into matrix language *by the same algebraic operations upon their representative matrices.* One can convince oneself by merely examining each of the elementary operations defined in the two preceding sections.

This is evident for the multiplication by a constant and the sum; thus to any linear combination $\lambda_1 A_1 + \lambda_2 A_2$ of two operators corresponds the same linear combination of their representative matrices

$$\langle m|\ (\lambda_1 A_1 + \lambda_2 A_2)\ |n\rangle = \lambda_1 \langle m|A_1|n\rangle + \lambda_2 \langle m|A_2|n\rangle.$$

Likewise the different products defined for vectors and operators are represented by the corresponding matrix products, namely

(*i*) scalar product of $|u\rangle$ by $|v\rangle$:

$$\langle v|u\rangle = \langle v|P_Q|u\rangle = \sum_n \langle v|n\rangle \langle n|u\rangle = \sum_n v^*_n u_n;$$

$\langle v|u\rangle$ is equal to the product (from the left) of the matrix (column vector) representing $|u\rangle$ by the Hermitean conjugate of the matrix representing $|v\rangle$;

(*ii*) action of an operator A on a ket $|u\rangle$ or on a bra $\langle v|$

$$\langle n|A|u\rangle = \langle n|AP_Q|u\rangle = \sum_k \langle n|A|k\rangle \langle k|u\rangle$$

$$\langle v|A|n\rangle = \langle v|P_Q A|n\rangle = \sum_l \langle v|l\rangle \langle l|A|n\rangle.$$

The matrix (column vector) representing $A|u\rangle$ is the product (from the left) of the matrix representing $|u\rangle$ by the matrix representing A.

The matrix (row vector) representing $\langle v|A|$ is the product (from the right) of the matrix representing $\langle v|$ by the matrix representing A;

(*iii*) the product AB of the operator B by the operator A:

$$\langle m|AB|n\rangle = \langle m|AP_Q B|n\rangle = \sum_k \langle m|A|k\rangle \langle k|B|n\rangle$$

the matrix representing AB is the product (from the left) of the matrix representing B by the matrix representing A;

(*iv*) operator $|u\rangle \langle v|$; the element (m, n) of its representative matrix is $\langle m|u\rangle \langle v|n\rangle$; the latter therefore results from the product (from the left) of the matrix (row vector) representing $\langle v|$ by the matrix (column vector) representing $|u\rangle$ (which actually yields a square matrix).

To sum up, we have succeeded in defining a representation of vectors and operators of $\mathscr{E}$ space by matrices, in such a way that there exist very simple correspondence rules between the various operations on the vectors and operators, and the operations on matrices. Any geometrical problem in $\mathscr{E}$ space may be treated either by the methods of pure geometry by directly manipulating the vectors and operators which occur there, or by manipulations of algebra or analysis on the matrices representing them in a suitable representation. In this last case, a more or less judicious choice of representation will lead to a more or less simple solution of the problem, in the same way as a more or less happy choice of coordinates makes the solution of a problem of analytical geometry more or less simple. In practice one will choose the representation in which the vectors and operators under study are represented by the simplest possible matrices.

Note in this connection that the observable Q is represented by a particularly simple matrix in representation $\{Q\}$: it is a diagonal matrix. More generally, any function $f(Q)$ is represented by a diagonal matrix,

$$\langle m|f(Q)|n\rangle = f(q_n)\, \delta_{mn}.$$

Operators which commute with Q are also represented by very simple matrices. In fact, if $[X, Q] = 0$,

$$(q_n - q_m) \langle m|X|n\rangle = 0$$

and consequently $\langle m|X|n\rangle = 0$ for any pair (m, n) such that $q_m \neq q_n$. In other words, all the matrix elements whose row and column indices refer to distinct eigenvalues of Q, are necessarily zero (cf. Ch. VII, § 15).

All the foregoing considerations can be extended without difficulty to the entire space $\mathscr{E}_1 \otimes \mathscr{E}_2$ resulting from the tensor product of the two spaces $\mathscr{E}_1$ and $\mathscr{E}_2$. The vectors and operators formed by tensor product can be represented by the tensor product matrices of the matrices representing the vectors and operators of the spaces $\mathscr{E}_1$ and $\mathscr{E}_2$.

20. Matrix Transformations

Consider once again matrices of finite order. In what follows, we shall designate by a capital letter square matrices of order N, and by a small letter N-dimensional column or row vectors. Let T be a non-singular matrix (T^{-1} exists). This matrix enables us to define a transformation in which the matrix A', the transform of A, is defined by

$$A' = TAT^{-1}. \tag{VII.75}$$

The correspondence between A and A' is one-to-one, A being deduced from A' by the inverse transformation

$$A = T^{-1}A'T. \tag{VII.76}$$

Such a transformation conserves the trace and the determinant

$$\mathrm{Tr}\, A = \mathrm{Tr}\, A', \qquad \det A = \det A' \tag{VII.77}$$

(properties of the trace and the determinant of a matrix product). It is likewise clear that it conserves any algebraic equation between matrices. If, for instance,

$$A = \lambda BC + \mu DEF,$$

then, multiplying term by term by T from the left and by T^{-1} from the right, and inserting into each monomial the expression $T^{-1}T$ as many times as necessary, we have

$$TAT^{-1} = \lambda TBT^{-1}\, TCT^{-1} + \mu TDT^{-1}\, TET^{-1}\, TFT^{-1},$$

that is

$$A' = \lambda B'C' + \mu D'E'F'.$$

One likewise defines the transform u' of a column vector u:

$$u' = Tu, \qquad u = T^{-1}u' \tag{VII.78}$$

and the transform v' of a row vector v:

$$v' = vT^{-1}, \qquad v = v'T. \tag{VII.79}$$

As may be easily verified, *the transformation conserves in all generality the algebraic equations* involving square matrices and vectors of one or the other type. On the other hand, if c is an arbitrary constant, the transforms of square matrices do not change upon replacing T by cT; however, the column vectors are multiplied by c, and the row vectors by $1/c$.

A transformation of the kind just described does not in general conserve the various conjugation relations between matrices (Problem VII.5). In particular, let us look for the condition which T must satisfy in order that the transformation conserve the Hermitean conjugation. In order that $A' = TAT^{-1}$ imply

$$A'^{\dagger} = TA^{\dagger}T^{-1}$$

for any A, one must have

$$TAT^{-1} = (TA^{\dagger}T^{-1})^{\dagger} = (T^{-1})^{\dagger}AT^{\dagger},$$

or else, multiplying term by term from the left by $T^{\dagger}$ and from the right by T,

$$T^{\dagger}TA = AT^{\dagger}T.$$

It is therefore necessary that $T^{\dagger}T$ commute with all matrices A, and thus that $T^{\dagger}T$ be a multiple of unity:

$$T^{\dagger}T = cI.$$

Moreover, in order that $u' = Tu$ imply $u'^{\dagger} = u^{\dagger}T^{-1}$ for any u, one must have $u = T^{\dagger}Tu$ for any u, hence that $c = 1$. In conclusion, the matrix T must be *unitary*. It is clear that this condition which is necessary for the Hermitean conjugation relations to be conserved, is also sufficient.

One calls *unitary transformation* a transformation whose matrix U

is unitary. Since, in that case, $U^{-1} = U^\dagger$, the definitions of the transforms of a matrix A, of a column vector u, and of a row vector v are, respectively,

$$\begin{aligned} A' &= UAU^\dagger & A &= U^\dagger A' U \\ u' &= Uu & u &= U^\dagger u' \\ v' &= vU^\dagger & v &= v'U. \end{aligned} \qquad \text{(VII.80)}$$

As all transformations, *a unitary transformation conserves the trace and the determinant of the matrices* and *the algebraic equations between matrices and vectors. In addition, it conserves the Hermitean conjugation relations.*

Furthermore, one has the two following fundamental theorems which we state here without proof.

A) *Any Hermitean matrix H may be diagonalized by a unitary transformation*

$$H' = UHU^\dagger, \qquad H' \text{ diagonal.}$$

The diagonal elements of H' are the "eigenvalues" of H. They are all real (H' is Hermitean) and are solutions of the secular equation

$$\det (H - xI) = 0.$$

B) *In order that two Hermitean matrices H, K might be diagonalized by one and the same unitary transformation, it is necessary and sufficient that they commute.*

All definitions and properties relating to matrices of finite order are extended without difficulty to infinite matrices. Any infinite matrix T possessing an inverse defines a transformation of square matrices or (column or row) vectors, provided that the sums or integrals which enter in the definitions converge. In contrast to the transformations of finite matrices it is not necessary that T be a square matrix. Of course the rows and columns of the (square) matrices which one transforms are labelled by the same system of indices as the columns of T; the same holds for the components of the column and row vectors. As for the rows and columns of the transformed matrices and the components of the transformed vectors, their system of indices is that of the rows of T.

The properties of conservation of the trace (with the proviso that it converges), of the algebraic equations and, in the case of unitary transformations, of Hermitean conjugation persist in the transformations of infinite matrices. However, the two fundamental theorems concerning the diagonalization of Hermitean matrices by a unitary transformation do not hold for all Hermitean matrices; we shall assume that they apply to all those we shall encounter.

21. Change of Representation

Let us return to the problem of the representation of vectors and operators of a vector space $\mathscr{E}$ by matrices. To each basis in this space corresponds a given representation. It is convenient to know how to relate the matrices representing one and the same operator or vector in each of the representations one can thus form. We shall see that one passes from one to the other by *unitary transformation.*

Consider two bases, one formed with the eigenvectors $|n\rangle$ $(n = 1, 2, ..., \infty)$ of the observable Q of § 19, the other with the eigenvectors $|\xi\rangle$ of another observable Ξ whose spectrum we suppose to be continuous. These two bases define the representations $\{Q\}$ and $\{\Xi\}$. The fundamental equations of the representation $\{Q\}$ have already been written down [eqs. (VII.73) and (VII.74)]. Those of the representation $\{\Xi\}$ are

$$\langle\xi|\xi'\rangle = \delta(\xi - \xi') \tag{VII.81}$$

$$P_{\Xi} \equiv \int |\xi\rangle\, \mathrm{d}\xi\, \langle\xi| = 1. \tag{VII.82}$$

The basis vectors of each of the representations can be expanded in a series of basis vectors of the other:

$$|n\rangle = \int |\xi\rangle\, \mathrm{d}\xi\, \langle\xi|n\rangle, \qquad |\xi\rangle = \sum_n |n\rangle\, \langle n|\xi\rangle. \tag{VII.83}$$

The scalar product $\langle\xi|n\rangle$ occurring as coefficient in the expansion of $|n\rangle$ may be regarded as the element $S(\xi; n)$ of a matrix S whose row index is ξ and column index is n. The scalar product $\langle n|\xi\rangle$ occurring as coefficient in the expansion of $|\xi\rangle$ may be regarded as the element $T(n; \xi)$ of a matrix T whose row index is n and column index is ξ. Moreover, since $\langle\xi|n\rangle = \langle n|\xi\rangle^*$,

$$T = S^{\dagger}.$$

Furthermore,

$$\sum_n \langle\xi|n\rangle\, \langle n|\xi'\rangle = \langle\xi|\xi'\rangle = \delta(\xi - \xi'),$$

$$\int \langle n|\xi\rangle\, \mathrm{d}\xi\, \langle\xi|n'\rangle = \langle n|n'\rangle = \delta_{nn'}.$$

In other words

$$SS^\dagger = I \tag{VII.84a}$$

$$TT^\dagger \equiv S^\dagger S = I. \tag{VII.84b}$$

The matrix S is unitary.

Now designate by $(u)_Q$ the column vector with components $\langle 1|u\rangle, \langle 2|u\rangle, \ldots$ representing the ket $|u\rangle$ in representation $\{Q\}$, and by $(u)_\Xi$ the column vector with components $\langle \xi|u\rangle$ representing this same ket in representation $\{\Xi\}$. By application of relation (VII.74),

$$\langle \xi|u\rangle = \sum_k \langle \xi|k\rangle \langle k|u\rangle,$$

in other words,

$$(u)_\Xi = S(u)_Q. \tag{VII.85}$$

Likewise designate by $(A)_Q$ and by $(A)_\Xi$ the matrices representing a given operator A in the representations $\{Q\}$ and $\{\Xi\}$, respectively. One has

$$\langle \xi|A|\xi'\rangle = \sum_{kl} \langle \xi|k\rangle \langle k|A|l\rangle \langle l|\xi'\rangle$$

or else

$$(A)_\Xi = S(A)_Q S^\dagger. \tag{VII.86}$$

One would likewise obtain between the row vectors $(v)_Q$ and $(v)_\Xi$ representing the same bra $\langle v|$:

$$(v)_\Xi = (v)_Q S^\dagger. \tag{VII.87}$$

Eqs. (VII.85), (VII.86), and (VII.87) are the characteristic transformation equations of the unitary transformation S [eq. (VII.80)].

The elements of this matrix have the following remarkable properties:

– considered as functions of the column index n, the elements $\langle \xi|n\rangle$ of the ξth row are the components of the row vector $(\xi)_Q$ representing the eigenbra $\langle \xi|$ of Ξ in representation $\{Q\}$;

– considered as functions of the row index ξ, the elements $\langle \xi|n\rangle$ of the nth column are the components of the column vector $(n)_\Xi$ representing the eigenket $|n\rangle$ of Q in representation $\{\Xi\}$.

In particular, the solution in representation $\{Q\}$ of the eigenvalue problem of the operator Ξ is a problem mathematically equivalent to the determination of the transformation S which diagonalizes the matrix $(\Xi)_Q$. Likewise, the solution in the representation $\{\Xi\}$ of the eigenvalue problem of the operator Q is equivalent to the problem of determining the transformation S which diagonalizes the matrix $(Q)_\Xi$.

It is important to recognize the quantities and relations which can be defined independently of any representation. All quantities and all relations defined directly by means of vectors and operators obviously possess this property. Thus the *scalar product* of two vectors is invariant under a change of representation. The *Hermitean conjugation relations,* and the *algebraic equations* between vectors and between operators likewise possess this invariance property.

Let us also mention the *conservation of the trace*: the trace (if it converges) of the matrix representing an operator retains the same value no matter what representation is used; it is a quantity characteristic of the operator itself. In particular it is easy to show that (Problem VII.6)

$$\mathrm{Tr}\, |u\rangle \langle u| = \langle u|u\rangle \tag{VII.88}$$

$$\mathrm{Tr}\, |u\rangle \langle v| = \langle v|u\rangle. \tag{VII.89}$$

22. Unitary Transformations of Operators and Vectors

The matrix S of the preceding paragraph does not represent an operator. The matrix representative of an operator is defined in a given representation whereas the transformation matrix straddles, so to speak, two representations. In the particular example examined earlier this is revealed especially since the matrix S is not a square matrix.

However, and here we have an important special case, it may happen that a one-to-one correspondence exists between the basis vectors of the first representation and those of the second. In that case the vectors of the two bases are labelled by the same set of indices. Let us consider, to be definite, a representation $\{Q\}$ whose basis vectors $|n\rangle$ are labelled by a discrete index n, and a representation $\{\bar{Q}\}$ whose basis vectors $|\bar{n}\rangle$ are labelled by the same index. Two kets $|n\rangle$, $|\bar{n}\rangle$ labelled by the same index correspond to each other. Let U be the linear operator defined by this correspondence:

$$|n\rangle = U|\bar{n}\rangle.$$

One has

$$U = U(\sum_n |\bar{n}\rangle \langle \bar{n}|) = \sum_n |n\rangle \langle \bar{n}| \tag{VII.90}$$

and

$$U^\dagger = \sum_n |\bar{n}\rangle \langle n|$$

taking into account the orthonormality relations of the $|n\rangle$ and the $|\bar{n}\rangle$,

$$UU^\dagger = U^\dagger U = 1. \tag{VII.91}$$

Hence, U is a *unitary operator*. In fact, the unitary matrix $\langle \bar{m}|n\rangle$ defining the change of representation from $\{Q\}$ to $\{\bar{Q}\}$ is the matrix representing U in representation $\{\bar{Q}\}$.

In the case where one can form a unitary operator U, one can define a manipulation which is to some extent complementary to the change of representation. Instead of transforming the basis $\{Q\}$ into a new basis $\{\bar{Q}\}$ whose vectors are given by the equation

$$|\bar{n}\rangle = U^\dagger|n\rangle \tag{VII.92}$$

one can carry out the transformation on the vectors and the operators of the space $\mathscr{E}$ themselves, and associate with each vector $|u\rangle$, the vector $|\hat{u}\rangle = U|u\rangle$, and with each operator A, the operator $\hat{A} = UAU^\dagger$.

Taking into account the fact that U is unitary, it is clear that the transformation U conserves the conjugation relations and the equations between vectors and operators. In particular,

(*i*) the scalar product is conserved: $\langle \hat{u}|\hat{A}|\hat{v}\rangle = \langle u|A|v\rangle$;

(*ii*) the hermiticity is conserved.

In fact, if A is an observable, its transform $\hat{A}$ is an observable possessing the same eigenvalue spectrum, since the eigenvalue equation of A

$$A|a\rangle = a|a\rangle$$

transforms into the equation

$$\hat{A}|\hat{a}\rangle = a|\hat{a}\rangle. \tag{VII.93}$$

The eigenkets of $\hat{A}$ corresponding to a given eigenvalue a are the transforms of the eigenkets of A corresponding to the same eigenvalue. Note that the matrix representing $\hat{A}$ in representation $\{Q\}$ is the same as that which represents A in representation $\{\bar{Q}\}$. Likewise the vector $|\hat{a}\rangle$ has the same components in $\{Q\}$ as $|a\rangle$ has in $\{\bar{Q}\}$.

To carry out successively two transformations defined, respectively, by the operators U and V is equivalent to the transformation defined by the operator $W = VU$. Since W is unitary, the resulting transformation is unitary. In other words, *the product of two unitary transformations is a unitary transformation.*

If the operator U defining a unitary transformation is infinitely close to 1, the transformation is said to be *infinitesimal.* U takes the form

$$U \equiv 1 + \mathrm{i}\varepsilon F \qquad \text{(VII.94)}$$

where ε is a real, infinitesimal quantity. The unitarity condition (VII.91) reads

$$(1 - \mathrm{i}\varepsilon F^\dagger)(1 + \mathrm{i}\varepsilon F) = (1 + \mathrm{i}\varepsilon F)(1 - \mathrm{i}\varepsilon F^\dagger) = 1,$$

or else, retaining only terms of the first order in ε,

$$F = F^\dagger.$$

The operator F is *Hermitean.*

In such an infinitesimal transformation, the transformed vectors and operators are given by the expressions

$$|\hat{u}\rangle \equiv |u\rangle + \delta|u\rangle = (1 + \mathrm{i}\varepsilon F)\,|u\rangle$$

$$\hat{A} \equiv A + \delta A = (1 + \mathrm{i}\varepsilon F)\,A(1 - \mathrm{i}\varepsilon F) = A + \mathrm{i}\varepsilon[F, A]$$

or

$$\delta|u\rangle = \mathrm{i}\varepsilon F|u\rangle \qquad \text{(VII.95)}$$

$$\delta A = \mathrm{i}\varepsilon[F, A]. \qquad \text{(VII.96)}$$

EXERCISES AND PROBLEMS

1. By definition, a projector P_i is less than or equal to another projector P_j, if $P_iP_j = P_i$; one then uses the notation $P_i \leqslant P_j$. Show that if $P_i \leqslant P_j$, one necessarily has $\langle u|P_i|u\rangle \leqslant \langle u|P_j|u\rangle$ for any $|u\rangle$, and conversely. Show either directly or by making use of this last property, that the inequality thus defined actually satisfies the characteristic axioms of an inequality, namely that (*i*) $P_i \leqslant P_j$ and $P_j \leqslant P_i$ imply $P_i = P_j$; (*ii*) $P_i \leqslant P_j$ and $P_j \leqslant P_k$ imply $P_i \leqslant P_k$.

2. $P_1, P_2, \ldots, P_K$ being projectors, show that their sum is likewise a projector if, and only if

$$\sum_{i=1}^{K} \langle u|P_i|u\rangle \leqslant \langle u|u\rangle$$

for any vector $|u\rangle$ of Hilbert space.

3. (*i*) An observable A possesses a finite number N of eigenvalues. One denotes them by $a_1, a_2, \ldots, a_N$ and sets

$$f(A) \equiv (A - a_1)(A - a_2) \ldots (A - a_N)$$
$$\equiv (A - a_n)\, g_n(A).$$

Show that $f(A) = 0$ and that the projector P_n upon the subspace of the nth eigenvalue is given by the expression

$$P_n = g_n(A)/g_n(a_n).$$

(*ii*) Derive the converse property, namely:

If A is a Hermitean operator satisfying the algebraic equation of the Nth degree

$$f(A) \equiv (A - a_1)(A - a_2) \ldots (A - a_N) = 0$$

and if it satisfies no other algebraic equation of degree less than N, it is an observable possessing N eigenvalues, and these are the N necessarily real and distinct roots of the equation $f(x) = 0$.

4. Show that a matrix of order N

(*i*) is necessarily a constant if it commutes with all matrices of order N;

(*ii*) is necessarily diagonal if it commutes with all diagonal matrices of order N.

5. Show that: (*a*) in order that a transformation conserve complex conjugation between matrices, it is necessary and sufficient that the transformation matrix be real; (*b*) in order that a transformation conserve the transposition relation between matrices, it is necessary and sufficient that the transformation matrix be orthogonal.

6. Let $|u\rangle$ and $|v\rangle$ be two vectors of finite norm. Show that

$$\mathrm{Tr}\ |u\rangle\langle u| = \langle u|u\rangle$$
$$\mathrm{Tr}\ |u\rangle\langle v| = \langle v|u\rangle.$$

7. Let H be a positive definite, Hermitean operator. Show that for any $|u\rangle$ and $|v\rangle$,

$$|\langle u|H|v\rangle|^2 \leqslant \langle u|H|u\rangle\langle v|H|v\rangle$$

and that the equality $\langle u|H|u\rangle = 0$ necessarily implies $H|u\rangle = 0$. Show also that $\mathrm{Tr}\, H \geqslant 0$ and that the equality implies $H = 0$.

8. Show that if H and K are two positive definite observables, $\mathrm{Tr}\, HK \geqslant 0$ and that the equality implies $HK = 0$.

9. A being a linear operator, show that $A^\dagger A$ is a positive definite Hermitean operator and that its trace is equal to the sum of the squares of the moduli of the matrix elements representing A in an arbitrarily chosen representation. Deduce that $\mathrm{Tr}\, A^\dagger A \geqslant 0$, and that the equality $\mathrm{Tr}\, A^\dagger A = 0$ implies $A = 0$.

CHAPTER VIII

GENERAL FORMALISM
(B) DESCRIPTION OF PHYSICAL PHENOMENA

1. Introduction

In Classical Physics, the dynamical state of a given system is defined at every instant, once the values assumed by the assembly of quantities or dynamical variables associated with the system are known at that instant. The latter may, in principle, all be determined simultaneously with infinite precision. The object of Classical Theory is to enumerate these dynamical variables, and then to discover and to study their equations of motion.

In the Quantum Theory, the relationship between dynamical states and dynamical variables is much less direct. In the measurement process of a given dynamical variable, the dynamical state of the system on which the measurement is performed, is in general modified by the intervention of the measuring device. This modification, which is usually neglected in Classical Physics, ceases to be negligible on the microscopic scale; it appears as an unpredictable and uncontrollable disturbance of the system and sets a limit to the precision with which the dynamical variables can all be measured simultaneously. One therefore abandons the fundamental postulate of Classical Physics, according to which all the various quantities belonging to a system take on well-defined values at each instant of time. One can only determine for each of these variables a *statistical distribution* of values, which is the probability law of the results of measurement in the eventuality that such a measurement is performed.

According to the usual terminology (Ch. VI, § 17) all the dynamical variables of a quantum system are not compatible with each other. One assumes, however, that one can always add to every dynamical variable of the system a certain number of others and thus form a *complete set of compatible variables*; by definition, all the variables of such a set are compatible with each other and there exist no other variables compatible with each of them, aside from functions of these variables themselves. The precise determination of the variables of a complete set constitutes the largest possible amount of information one can obtain on the dynamical state of a quantum system. Consequently, the dynamical state of a quantum system is no longer defined, as in Classical Theory, by the precise specification of all the dynamical variables associated with the system, but rather by the specification of those which occur in one of the various complete sets of compatible variables one can construct.

One establishes the principle that the dynamical states of a quantum system are *linearly superposable*. In accordance with this principle (Ch. VII, § 1), one can associate with such a system a certain vector space $\mathscr{E}$ such that each dynamical state of the system is represented by a vector of this space. One further supposes that $\mathscr{E}$ is a Hilbert space. Hereafter, we shall use the notation and properties of Hilbert spaces as they were stated in Chapter VII. Hence, *to each dynamical state corresponds a certain ket* $|u\rangle$ *of space* $\mathscr{E}$ [1]). Likewise, *to each dynamical variable is attached an observable of space* $\mathscr{E}$. According to whether wo variables are or are not compatible, the observables corresponding to them do or do not commute.

The general formalism of the Quantum Theory is based upon this correspondence between dynamical states and physical quantities on the one hand, and between vectors and operators on the other. In Section I we shall define this correspondence in precise fashion, and we shall indicate a practical way to construct the Hilbert space $\mathscr{E}$, and the physical significance to be attached to its vectors and operators. In Section II, this general theoretical scheme is complemented by giving the *equations of motion*. In Section III, we show that there exist as many particular formulations of the theory as there are particular matrix representations of vectors and operators in $\mathscr{E}$ space. Wave Mechanics is one of these particular formulations. When the dynamical state of a quantum system is incompletely known, one can represent it by a statistical mixture of vectors, following the usual methods of statistics. An equivalent procedure consists in representing this state by an operator of a special type, the *density operator*; the formalism of the density operator is outlined in the fourth and last section of this chapter.

---

[1]) The space $\mathscr{E}$ plays a similar role in Quantum Theory to that of phase space in Classical Theory. Each point of phase space represents a classical dynamical state; likewise each vector of $\mathscr{E}$ space represents a dynamical quantum state. In this latter case, however, it is not a one-to-one correspondence since two vectors of $\mathscr{E}$ space which are multiples of each other represent the same state; cf. § 2 below.

I. DYNAMICAL STATES AND PHYSICAL QUANTITIES

2. Definition of Probabilities. Postulates Concerning Measurement

In a given dynamical state, a definite statistical distribution of values is associated with each dynamical variable of the system. In order to define these probabilities, one starts out from the following fundamental postulate:

The mean value of any function $F(A)$ of a given physical quantity A is

$$\boxed{\langle F(A)\rangle = \langle u|\,F(A)\,|u\rangle,} \tag{VIII.1}$$

an expression in which the ket $|u\rangle$ represents the dynamical state of the system, and the observable A represents the physical quantity in question.

In particular, the characteristic function $f(\xi)$ of the statistical distribution of A is the mean value of the function $\exp(\mathrm{i}\xi A)$:

$$f(\xi) = \langle u|\, \mathrm{e}^{\mathrm{i}\xi A}\, |u\rangle. \tag{VIII.2}$$

Since a statistical distribution is completely determined upon specifying its characteristic function, the fundamental postulate we have just stated completely defines the statistical distributions of all the dynamical variables of the system.

Let us see how this postulate affects the correspondence between dynamical states and ket vectors. Whatever the operator $F(A)$, expression (VIII.1) remains unchanged when multiplying vector $|u\rangle$ by an arbitrary phase factor $\exp(\mathrm{i}\alpha)$ (α real and arbitrary). Hence the statistical distributions calculated for two vectors which differ only by a phase factor, are strictly identical: two such vectors represent the same dynamical state. In other words, *to each dynamical state corresponds a vector defined to within a phase factor*. On the other hand, one necessarily has $f(0) = 1$ (the average value of 1 is equal to 1); the vector $|u\rangle$ must therefore be *normalized to unity*

$$\langle u|u\rangle = 1. \tag{VIII.3}$$

It is sometimes convenient to relax this last condition. To this end, one replaces the definition (VIII.1) of the mean values by the more general definition

$$\langle F(A)\rangle = \frac{\langle u|F(A)|u\rangle}{\langle u|u\rangle}. \tag{VIII.4}$$

With this definition, two vectors which are multiples of each other represent the same dynamical state (it being understood that the vectors with which we are dealing have finite norm).

To obtain the statistical distribution of A explicitly, one calculates the expression (VIII.2) of the characteristic function $f(\xi)$ [or the expression (VIII.4) if $|u\rangle$ is not normalized to unity] in a representation where A is diagonal. Except for slight differences in terminology, this method has already been outlined in Chapter V. It will not be restated here. We shall merely state the results:

1) *The only precise values which the quantity A may assume are those of the eigenvalue spectrum of the observable associated with A.*

2) *Let $\mathscr{E}_D$ be the subspace spanned by the eigenvectors of A corresponding to the eigenvalues located in a certain domain D of the spectrum of A; denote by $|u_D\rangle \equiv P_D|u\rangle$ the projection of ket $|u\rangle$ on $\mathscr{E}_D$. The probability w_D that the result of a measurement of A belongs to the domain D is* [1])

$$\boxed{w_D = \langle P_D\rangle = \frac{\langle u_D|u_D\rangle}{\langle u|u\rangle}} \qquad \text{(VIII.5)}$$

Relation (VIII.5) summarizes the results obtained in each particular case studied in Chapter V (Problem VIII.1). Indeed, D may simply be an eigenvalue of the discrete spectrum, in which case relation (VIII.5) is identical to relation (V.21). But D can equally well be made up of an assembly of several discrete eigenvalues, or of a portion of the continuous spectrum, or even of a combination of both. In particular, when D is an infinitesimal portion $[a(\nu), a(\nu+\mathrm{d}\nu)]$ of the continuous spectrum, as in the example at the end of Ch. V, § 10, $w_D = \omega(\nu)\,\mathrm{d}\nu$ and the probability density $\omega(\nu)$, calculated by means of relation (VIII.5) is precisely the one given by equation (V.44).

The next point to consider is how to define the dynamical state of the system, once the measurement is completed. The answer to this question depends upon the particular conditions under which the measurement was carried out; it is simple only in the case of an

[1]) w_D is the average value of the projector P_D, that is to say of the function of A equal to 1 for all eigenvectors of A located in $\mathscr{E}_D$, and equal to 0 for all eigenvectors of A orthogonal to $\mathscr{E}_D$.

ideal measurement (cf. Ch. V, § 13). If, with the hypothesis of an ideal measurement, the observation carried out on the system indicates that it is in an eigenstate of A belonging to the domain D defined above, *its dynamical state after measurement is represented by the projection of vector $|u\rangle$ upon space $\mathscr{E}_D$*. In other words, the (non-causal) evolution of the state vector in the course of the measurement corresponds to the scheme

$$|u\rangle \rightarrow \text{ideal measurement yielding result } D \rightarrow P_D|u\rangle$$

This postulate of the filtering of the wave packet may be regarded as a genuine definition of the ideal measurement.

With the convention that dynamical states are always represented by vectors normalized to unity – $|u\rangle$ is then assumed normalized to unity – the state vector of the system after measurement is the product $P_D|u\rangle$ by a normalization factor defined to within a phase, the square of whose modulus is equal to $1/w_D$, that is $1/\langle u|P_D|u\rangle$.

3. Observables of a Quantized System and Their Commutation Relations

The first step in the study of a quantized system consists in enumerating the dynamical variables of the system and in defining the algebra of the observables associated with the variables. In fact, the various observables can be expressed as function of some set of "fundamental observables"; one completes the definition of their algebra by giving the commutation relations of these fundamental observables.

When the considered quantum system possesses a classical analogue, as is the case for all those we have mentioned so far, one follows a general procedure based upon the correspondence principle.

In a classical system with N dimensions, the most general dynamical variable is a function of $2N$ independent variables, the N coordinates $q_1, q_2, \ldots, q_N$ and the N momenta $p_1, p_2, \ldots, p_N$. One assigns the same dynamical variables to the corresponding quantum system. One thus introduces N position variables and N momentum variables. To these variables correspond observables which we shall denote by the same symbols $q_1, q_2, \ldots, q_N, p_1, p_2, \ldots, p_N$ as the dynamical variables themselves. One postulates further that the only observables which do not commute are the N pairs formed by associating each coordinate with its conjugate momentum; for such pairs one has $[q_r, p_r] = i\hbar$. In other words,

$$[q_r, q_s] = 0, \qquad [p_r, p_s] = 0 \tag{VIII.6}$$

$$[q_r, p_s] = i\hbar\delta_{rs} \tag{VIII.7}$$

$$(r, s = 1, 2, \ldots, N).$$

Since the most general observable is a function of the q's and the p's, the commutator of any two observables is defined unambiguously by specifying the fundamental commutator relations (VIII.6) and (VIII.7); one can actually calculate it explicitly by making use of the rules of commutator algebra (Chapter V, § 17). This correspondence between observables of a quantum system and quantities of the analogous classical system has already been discussed several times (Ch. II, § 15 and Ch. V, § 3). In order to remove all ambiguities, one always starts from the cartesian coordinates in configuration space and conforms to the empirical rules of Ch. II, § 15. In particular, the "symmetrization" rule given in that paragraph insures that with every real quantity connected with the system, there is associated a Hermitean operator.

All quantum systems cannot be treated by this correspondence method. It often happens that the dynamical variables introduced through correspondence with an appropriate classical analogue are not sufficient to exhaust all physical properties of the quantum system one wishes to study. It is then necessary to introduce additional variables. The choice of these new variables and of the commutation relations associated with them is purely a matter of intuition.

Among the physical quantities attached to a quantum system, special mention must be made of its energy. The observable H representing it is called the Hamiltonian of the system. When the system has a classical analogue, H is derived by correspondence from the Hamiltonian function of Classical Mechanics.

4. Heisenberg's Uncertainty Relations

The position-momentum uncertainty relations of Heisenberg result directly from the commutation relations (VIII.7).

Indeed, let us show in all generality that if two observables A and B satisfy the equation

$$[A, B] = i\hbar \tag{VIII.8}$$

the product of their root-mean-square deviations will always remain greater or equal to $\hbar/2$:

$$\Delta A \cdot \Delta B \geqslant \tfrac{1}{2}\hbar. \tag{VIII.9}$$

The proof we shall give is essentially the same as that of Ch. IV, § 8.

By definition,

$$\Delta A = (\langle A^2 \rangle - \langle A \rangle^2)^{\frac{1}{2}}, \qquad \Delta B = (\langle B^2 \rangle - \langle B \rangle^2)^{\frac{1}{2}}.$$

Let us introduce the observables

$$\hat{A} = A - \langle A \rangle, \qquad \hat{B} = B - \langle B \rangle.$$

Clearly

$$[\hat{A}, \hat{B}] = i\hbar$$

and

$$\Delta A = \Delta \hat{A} = \langle \hat{A}^2 \rangle^{\frac{1}{2}}, \qquad \Delta B = \Delta \hat{B} = \langle B^2 \rangle^{\frac{1}{2}}.$$

Assume that the dynamical state of the system is represented by the ket $|u\rangle$ normalized to unity, and apply the Schwarz inequality to the vectors $\hat{A}|u\rangle$ and $\hat{B}|u\rangle$:

$$(\Delta A)^2 (\Delta B)^2 \equiv \langle u|\hat{A}^2|u\rangle \langle u|\hat{B}^2|u\rangle \geqslant |\langle u|\hat{A}\hat{B}|u\rangle|^2.$$

Separating in $\hat{A}\hat{B}$ the Hermitean from the anti-Hermitean part [cf. eq. (VII.29)]:

$$\hat{A}\hat{B} = \frac{\hat{A}\hat{B}+\hat{B}\hat{A}}{2} + \frac{\hat{A}\hat{B}-\hat{B}\hat{A}}{2} = \frac{\hat{A}\hat{B}+\hat{B}\hat{A}}{2} + \frac{i\hbar}{2},$$

one can separate in $\langle u|\hat{A}\hat{B}|u\rangle$ the real from the imaginary part:

$$\langle u|\hat{A}\hat{B}|u\rangle = \left\langle \frac{\hat{A}\hat{B}+\hat{B}\hat{A}}{2} \right\rangle + \frac{i\hbar}{2},$$

and rewrite the Schwarz inequality

$$(\Delta A)^2 \cdot (\Delta B)^2 \geqslant \left\langle \frac{\hat{A}\hat{B}+\hat{B}\hat{A}}{2} \right\rangle^2 + \frac{\hbar^2}{4}.$$

A fortiori

$$\Delta A \cdot \Delta B \geqslant \tfrac{1}{2}\hbar. \qquad \text{Q.E.D.}$$

In order that the product $\Delta A \cdot \Delta B$ be equal to its minimum value $\frac{1}{2}\hbar$, it is necessary on the one hand that the Schwarz inequality reduce to an equality, hence that $\hat{A}|u\rangle = c\hat{B}|u\rangle$ (c an arbitrary constant), and on the other hand that the mean value of $\hat{A}\hat{B}+\hat{B}\hat{A}$ be zero, that is

$$\langle u|\hat{A}\hat{B}|u\rangle + \langle u|\hat{B}\hat{A}|u\rangle = (c^* + c)\,\langle u|\hat{B}^2|u\rangle = 0,$$

or Re $c=0$. To sum up, for inequality (VIII.9) to reduce to an equality, it is necessary and sufficient that $|u\rangle$ satisfy the equation

$$(A - \alpha)|u\rangle = \mathrm{i}\gamma(B - \beta)|u\rangle \qquad \text{(VIII.10)}$$

in which α, β, and γ are arbitrary real constants.

The application of this general result to the position-momentum pairs (q_r, p_r) of the foregoing paragraph yields the uncertainty relations

$$\Delta q_r \cdot \Delta p_r \geqslant \tfrac{1}{2}\hbar \qquad (r = 1, 2, \ldots, N), \qquad \text{(VIII.11)}$$

the equality being actually realized if $|u\rangle$ is a solution of the equation

$$(p_r - \mathrm{i}\gamma q_r)|u\rangle = (\alpha - \mathrm{i}\gamma\beta)|u\rangle$$

(α, β, γ are arbitrary real constants).

5. Definition of the Dynamical States and Construction of Space $\mathscr{E}$

Once the observables of our quantum system have been enumerated and their commutation relations established, one must precisely define the various possible quantum states. One must construct the Hilbert space $\mathscr{E}$ in which these observables act. For this it is sufficient to select a basis of this space, and then to define each observable by its action upon the vectors of this basis. This action must be defined in such a way that all operators representing physical quantities actually turn out to be observables, and that the rules of algebra of these observables are actually obeyed.

To define the basis, one selects from the entire collection of observables a complete set of commuting observables $A, B, C, \ldots$. The simultaneous measurement of the dynamical variables which they represent constitutes the maximum of information one can obtain on the state of the system; it therefore completely defines a particular dynamical state of the system. Consequently, each set of eigenvalues a, b, c, ... of these observables defines a vector of $\mathscr{E}$ to within a constant: fixing this constant arbitrarily, one obtains a certain representative vector $|a\,b\,c\ldots\rangle$. The ensemble of vectors $|a\,b\,c\ldots\rangle$

obtained by letting each of the eigenvalues a, b, c, ... vary over the entire range of the respective spectra of A, B, C, ... forms a complete orthogonal set in $\mathscr{E}$-space. In fact, if one fixes the normalizations of the vectors $|a\, b\, c\, ...\rangle$ in an appropriate manner – normalization to unity for all vectors of finite norm, normalization by means of the Dirac δ-function for vectors of infinite norm – it is a complete *orthonormal* set in $\mathscr{E}$. One thus obtains a basis of $\mathscr{E}$ by specifying the respective spectra of the observables A, B, C, ...

The action of the basic observables A, B, C, ... on each of these vectors is obvious. The action of the various other observables susceptible of representing physical quantities is yet to be determined.

The mere knowledge of the commutation relation is generally sufficient –

(*i*) to show that the set A, B, C, ... forms a complete set of commuting observables;

(*ii*) to deduce their respective spectra;

(*iii*) to deduce the action of the other observables upon the vectors of their basis.

In other words, the knowledge of the algebra of the observables of the system in general suffices to define unambiguously the space $\mathscr{E}$ in which they act [1]).

It remains then to check the internal consistency of the construction thus defined; in other words, one must verify that the operators representing physical quantities are actually observables.

Note that at this stage the theory already lends itself to experimental check. The physical quantities are defined in principle by well-defined operations of measurement, and their spectrum is directly accessible to experiment. It is necessary that the theoretical spectrum, that is to say the eigenvalue spectrum of the observable associated with each physical quantity, coincide with this experimental spectrum.

6. One-Dimensional Quantum System Having a Classical Analogue

Let us apply the method of construction of § 5 to a one-dimensional quantum system having a classical analogue. Its observables are

[1]) This is correct only if $\mathscr{E}$ is irreducible with respect to the said observables; this condition, which we merely point out here for the record, is implicitly postulated in all the arguments below. We shall discuss more fully the notion of irreducibility and its physical significance in Chapter XV (§ 6).

functions of two of them, q and p, obeying the commutation relation

$$[q, p] = \mathrm{i}\hbar. \tag{VIII.12}$$

The position q constitutes a complete set of commuting observables by itself. Indeed, the commutator of q and any given function $A(q, p)$ of p and q is [eq. (V.68)]

$$[q, A] = \mathrm{i}\hbar \frac{\partial A}{\partial p}; \tag{VIII.13}$$

q commutes with A if, and only if A is independent of p; in other words, the only observables which commute with q are functions of q.

Simple considerations of internal consistency impose very restrictive conditions upon the eigenfunctions and the eigenvalue spectrum of q. Indeed, let $|q_0\rangle$ be an eigenket of q

$$q|q_0\rangle = q_0|q_0\rangle.$$

Let us write that the identical operators $[q, p]$ and $\mathrm{i}\hbar$ have the same diagonal element corresponding to $|q_0\rangle$:

$$\mathrm{i}\hbar \langle q_0|q_0\rangle = \langle q_0|qp|q_0\rangle - \langle q_0|pq|q_0\rangle.$$

$|q_0\rangle$ certainly does not have a finite norm; otherwise the right-hand side would vanish identically, while the left-hand side would be finite and non-zero.

On the other hand, consider the operator

$$S(\xi) = \mathrm{e}^{-\mathrm{i}p\xi/\hbar}. \tag{VIII.14}$$

It is a function of the observable p depending upon the parameter ξ. Clearly, it is a unitary operator:

$$S^\dagger S = SS^\dagger = 1,$$

since its Hermitean conjugate is

$$S^\dagger(\xi) = S(-\xi) = \mathrm{e}^{-\mathrm{i}p\xi/\hbar}.$$

Applying equation (VIII.12) yields

$$[q, S] = \mathrm{i}\hbar \frac{\partial S}{\partial p} = \xi S.$$

In other words

$$qS = S(q+\xi) \tag{VIII.15}$$

and consequently,

$$qS|q_0\rangle = S(q+\xi)|q_0\rangle = (q_0+\xi)S|q_0\rangle. \tag{VIII.16}$$

Hence $S|q_0\rangle$ is an eigenvector of q belonging to the eigenvalue $(q_0+\xi)$. This vector is certainly not zero (otherwise S would not have an nverse); in fact its (infinite) norm is the same as that of $|q_0\rangle$ since S is unitary:

$$\langle q_0|S^\dagger S|q_0\rangle = \langle q_0|q_0\rangle.$$

This operation can be performed no matter what the value taken by ξ in the entire interval $(-\infty, +\infty)$. Thus, by a suitable unitary transformation on $|q_0\rangle$, one can form an eigenket of q belonging to any given eigenvalue in the interval $(-\infty, +\infty)$.

In conclusion, the spectrum of q is necessarily continuous, non-degenerate, and extends from $-\infty$ to $+\infty$; its eigenvectors necessarily have infinite norm.

Denote by $|q'\rangle$ one of the eigenkets of q belonging to the eigenvalue q':

$$q|q'\rangle = q'|q'\rangle;$$

$|q'\rangle$ is defined to within a constant whose modulus we fix by the normalization condition

$$\langle q'|q''\rangle = \delta(q'-q''). \tag{VIII.17}$$

The space $\mathscr{E}$ is by definition the space formed by linear superposition of the vectors $|q'\rangle$.

q is obviously an observable of that space. Actually, the vectors $|q'\rangle$ form the basis of a certain representation of vectors and operators of $\mathscr{E}$, the representation $\{q\}$, in which q is diagonal:

$$\langle q'|q|q''\rangle = q'\,\delta(q'-q''). \tag{VIII.18}$$

Let us show that p is a well-defined Hermitean operator in space $\mathscr{E}$; to this effect one merely has to determine its matrix in the representation $\{q\}$.

Consider first the unitary operator $S(\xi)$ defined by equation (VIII.14).

Since this operator satisfies eq. (VIII.15), $S(\xi)|q'\rangle$ is an eigenvector of q belonging to the eigenvalue $(q'+\xi)$:

$$S(\xi)|q'\rangle = c|q'+\xi\rangle,$$

c is a phase factor [1]) which may depend on ξ and on q'. We choose the phases of the vectors of the basis in such a way that

$$|q'\rangle = S(q')\,|0\rangle.$$

In this way the phase factor c is equal to 1, no matter what ξ and q'. Indeed

$$\begin{aligned} S(\xi)|q'\rangle &= S(\xi)\,S(q')|0\rangle = \mathrm{e}^{-\mathrm{i}p\xi/\hbar}\,\mathrm{e}^{-\mathrm{i}pq'/\hbar}\,|0\rangle \\ &= \mathrm{e}^{-\mathrm{i}p(\xi+q')/\hbar}|0\rangle = S(q'+\xi)|0\rangle \\ &= |q'+\xi\rangle, \end{aligned} \qquad \text{(VIII.19)}$$

or else

$$\langle q'|S(\xi)|q''\rangle = \langle q'|q''+\xi\rangle = \delta(q'-q''-\xi).$$

From the matrix elements of $S(\xi)$ calculated above, we deduce the representative matrix of p by noting that in the limit where ξ is equal to an infinitesimal quantity ε,

$$S(\varepsilon) \sim 1 - \frac{\mathrm{i}}{\hbar}\,p\varepsilon$$

and, consequently

$$\delta(q'-q''-\varepsilon) = \langle q'|S(\varepsilon)|q''\rangle \sim \delta(q'-q'') - \frac{\mathrm{i}}{\hbar}\,\varepsilon\,\langle q'|\,p\,|q''\rangle,$$

whence

$$\langle q'|\,p\,|q''\rangle = \frac{\hbar}{\mathrm{i}}\lim_{\varepsilon\to 0}\frac{\delta(q'-q'')-\delta(q'-q''-\varepsilon)}{\varepsilon} = \frac{\hbar}{\mathrm{i}}\,\delta'(q'-q''). \qquad \text{(VIII.20)}$$

As the "function" $\delta'(x)$ is odd, it is clear that $\langle q''|p|q'\rangle = \langle q'|p|q''\rangle^*$, hence that the operator p is Hermitean.

---

[1]) Since S is unitary, one has

$$\langle q''|S^\dagger(\xi)\,S(\xi)|q'\rangle = c^*(\xi, q'')\,c(\xi, q')\,\delta(q''-q') = \delta(q''-q'),$$

from which $|c(\xi, q')| = 1$.

Let us likewise verify that q and p actually satisfy the commutation relation (VIII.12)

$$\begin{aligned}\langle q'|(qp-pq)|q''\rangle &= (q'-q'')\langle q'|p|q''\rangle \\ &= \frac{\hbar}{\mathrm{i}}(q'-q'')\,\delta'(q'-q'') \\ &= \mathrm{i}\hbar\,\delta(q'-q'')\end{aligned}$$

[we have used the identity (A.30) of Appendix A].

It remains to be shown that p is an observable. To see this, we solve the eigenvalue problem of p in the representation $\{q\}$. Let $|p'\rangle$ be the eigenket belonging to the eigenvalue p'. The equation

$$p|p'\rangle = p'|p'\rangle$$

is written in the representation $\{q\}$, taking into account eq. (VIII.20),

$$\begin{aligned}p'\langle q'|p'\rangle = \langle q'|p|p'\rangle &= \int \langle q'|p|q''\rangle\,\mathrm{d}q''\,\langle q''|p'\rangle \\ &= \frac{\hbar}{\mathrm{i}}\int \delta'(q'-q'')\,\langle q''|p'\rangle\,\mathrm{d}q'' \\ &= \frac{\hbar}{\mathrm{i}}\frac{\mathrm{d}}{\mathrm{d}q'}\,(\langle q'|p'\rangle).\end{aligned}$$

This is a differential equation of the function $\langle q'|p\rangle$ of the variable q', whose general solution is

$$\langle q'|p'\rangle = a\,\mathrm{e}^{\mathrm{i}p'q'/\hbar},$$

where a is an arbitrary constant. We thus verify that p has a continuous spectrum of eigenvalues p' extending from $-\infty$ to $+\infty$. The eigenvectors have infinite norm. They satisfy the conditions of orthonormality

$$\langle p'|p''\rangle = \delta(p'-p'')$$

if one takes for the value of the constant: $a=(2\pi\hbar)^{-\frac{1}{2}}$. Now p is obviously an observable, since the vectors $|p'\rangle$ satisfy the closure relation. Indeed, the projection operator

$$P_p \equiv \int_{-\infty}^{+\infty} |p'\rangle\,\mathrm{d}p'\,\langle p'|$$

has for its matrix elements, in the representation $\{q'\}$,

$$\langle q'|P_p|q''\rangle = \int_{-\infty}^{+\infty} \langle q'|p'\rangle \, \mathrm{d}p' \, \langle p'|q''\rangle$$

$$= \frac{1}{2\pi\hbar} \int_{-\infty}^{+\infty} \mathrm{e}^{ip'(q'-q'')/\hbar} \, \mathrm{d}p' = \delta(q' - q'');$$

whence the closure relation:

$$P_p = 1.$$

With the fundamental observables p, q one can build each of the operators $F(p, q)$ representing the various dynamical variables of the system. One can easily ensure that these operators are Hermitean. To be complete, one must show further that they actually are observables. It is traditional in Quantum Theory to pass over these points of rigor and to admit without proof that all Hermitean operators representing physical quantities are observables.

7. Construction of the $\mathscr{E}$ Space of a System by Tensor Product of Simpler Spaces

Knowing how to build up the $\mathscr{E}$ space for a system having a one-dimensional classical analogue, it is simple to solve the same problem for a system having a classical analogue in N dimensions.

In the latter case, the dynamical variables are functions of the $2N$ fundamental variables of position and momentum. The observables representing the latter obey the commutation relations (VIII.6) and (VIII.7). They may be grouped in N pairs (q_1, p_1), (q_2, p_2), ..., (q_N, p_N), each made up of one of the position components and its conjugate momentum. Each observable of a pair commutes with all the observables of the other pairs.

The observables of a given pair, (q_i, p_i) for instance, may be regarded as the fundamental observables of a one-dimensional system of the type studied in the preceding paragraph. One knows how to build up the ket space $\mathscr{E}_i$ of such a system. According to the results of § 6, $\mathscr{E}_i$ is spanned by the orthonormal kets $|q_i'\rangle$ whose index q_i' is continuous and may take on all values in the interval $(-\infty, +\infty)$.

The space $\mathscr{E}$ of the dynamical states of the N-dimensional system is the tensor product (cf. Ch. VII, § 6) of the spaces $\mathscr{E}_1$, $\mathscr{E}_2$, ..., $\mathscr{E}_N$

thus constructed

$$\mathscr{E} = \mathscr{E}_1 \otimes \mathscr{E}_2 \otimes \dots \otimes \mathscr{E}_N,$$

that is to say the space spanned by the kets:

$$|q_1'q_2' \dots q_N'\rangle \equiv |q_1'\rangle\, q_2'\rangle \dots |q_N'\rangle. \tag{VIII.21}$$

To each operator q_i, p_i of the space $\mathscr{E}_i$ corresponds a well-defined operator q_i, p_i of the product space $\mathscr{E}$. To represent the $2N$ fundamental variables, we thus obtain $2N$ well-defined operators in $\mathscr{E}$. In fact, in accordance with the rules of tensor multiplication, to every observable of a partial space corresponds an observable of the product space; two observables coming from different partial spaces commute; two observables coming from the same partial space $\mathscr{E}_i$ obey the same commutation relations in $\mathscr{E}$ as in $\mathscr{E}_i$. Therefore, the $2N$ operators $q_1, \dots, q_N, p_1, \dots, p_N$ which we established in $\mathscr{E}$ are observables and actually obey the commutation relations (VIII.6) and (VIII.7).

The ensemble of vectors $|q_1', \dots, q_N'\rangle$ obtained by varying each of the eigenvalues $q_1', \dots, q_N'$ from $-\infty$ to $+\infty$ forms a basis in $\mathscr{E}$-space and defines a certain representation, the representation $\{q\}$. It is instructive to write down explicitly the matrix elements representing the q's and the p's in this representation. To this effect, we make use of the shorthand notation

$$|q'\rangle \equiv |q_1'\, q_2' \dots q_N'\rangle \tag{VIII.22}$$

$$\left.\begin{aligned} \delta(q'-q'') &\equiv \prod_{i=1}^{N} \delta(q_i'-q_i'') \\ &\equiv \delta(q_1'-q_1'')\delta(q_2'-q_2'') \dots \delta(q_N'-q_N'') \end{aligned}\right\} \tag{VIII.23}$$

$$\frac{\partial}{\partial q_n'}\,[\delta(q'-q'')] \equiv \delta'(q_n'-q_n'') \prod_{i\neq n} \delta(q_i'-q_i''). \tag{VIII.24}$$

In the last expression $\prod_{i\neq n}$ designates a product of the $(N-1)$ terms obtained by letting the index i assume all integral values from 1 to N with the exception of the value n.

Upon applying the relations (VIII.17), (VIII.18), and (VIII.20) of § 6, we successively obtain the orthonormality relations,

$$\langle q'|q''\rangle = \prod_{i=1}^{N} \langle q_i'|q_i''\rangle = \delta(q'-q''), \tag{VIII.25}$$

the elements of the (diagonal) matrices representing the coordinates

$$\begin{aligned} \langle q'|q_n|q''\rangle &= \langle q_n'|q_n|q_n''\rangle \prod_{i\neq n} \langle q_i'|q_i\rangle \\ &= q_n' \, \delta(q' - q''), \end{aligned} \tag{VIII.26}$$

and the elements of the (non-diagonal) matrices representing the momenta

$$\begin{aligned} \langle q'|p_n|q''\rangle &= \langle q_n'|p_n|q_n''\rangle \prod_{i\neq n} \langle q_i'|q_i''\rangle \\ &= \frac{\hbar}{i} \delta'(q_n' - q_n'') \prod_{i\neq n} \delta(q_i' - q_i'') \\ &= \frac{\hbar}{i} \frac{\partial}{\partial q_n'} [\delta(q' - q'')]. \end{aligned} \tag{VIII.27}$$

One easily verifies the commutation relations (VIII.6) and (VIII.7) using these explicit expressions for the matrices representing the p's and the q's.

Any dynamical variable of the system is a function of the p's and the q's; therefore a well-defined operator of $\mathscr{E}$-space corresponds to it. One must make sure that this operator is an observable. In keeping with the tradition mentioned above, this point is in most cases admitted without proof.

The construction of the ket space of a system from the tensor product of simpler spaces is a very general procedure. In practice, one can always express the dynamical variables of a system as a function of a certain number of "basis" variables; now, the latter can usually be classified into a number of sets such that any variable belonging to a given set is compatible with all the variables of the other sets. To be specific, suppose that these "basis" variables have been grouped into two sets (A_1, B_1, ...) and (A_2, B_2, ...) and that each variable of the type (1) is compatible with every variable of type (2). Each set taken separately defines a partial system, whose ket space one knows how to construct. Let $\mathscr{E}_1$, $\mathscr{E}_2$ be the spaces of the partial systems (1) and (2), respectively. Clearly the ket space $\mathscr{E}$ of the total system is the tensor product of the two partial spaces

$$\mathscr{E} = \mathscr{E}_1 \otimes \mathscr{E}_2.$$

II. THE EQUATIONS OF MOTION

8. Evolution Operator and the Schrödinger Equation

From the fact that on the scale of precision of quantum phenomena, there exists no net separation between the system itself and the observing instrument, the evolution of a quantum system ceases to be strictly causal as soon as it is subjected to an observation. However, a quantum system isolated from any external influence evolves in an exactly predictable manner. Let $|\psi(t_0)\rangle$ be the ket vector representing its dynamical state at time t_0; the ket vector $|\psi(t)\rangle$ representing its state at the later time t is exactly determined by specifying $|\psi(t_0)\rangle$ if, as we shall assume henceforth, the system was not subjected to any observation during the time interval (t_0, t). This fundamental law of evolution will be discussed in this paragraph.

In the first place, we postulate that *the linear superposition of states is preserved in the course of time*. Consequently, the correspondence between $|\psi(t_0)\rangle$ and $|\psi(t)\rangle$ is linear and defines a certain linear operator $U(t, t_0)$ which is called the *evolution operator*:

$$|\psi(t)\rangle = U(t, t_0)\, |\psi(t_0)\rangle. \tag{VIII.28}$$

If the system is conservative, that is to say if its energy, as represented by the Hamiltonian H, does not depend explicitly upon the time, $U(t, t_0)$ can be deduced from the requirement that the motion of a system of energy E be periodic, and that its (angular) frequency ω be given by Einstein's law

$$E = \hbar\omega. \tag{VIII.29}$$

Indeed, since the eigenvectors of H span the space $\mathscr{E}$, it is sufficient, for a determination of U, to know its action upon each of these vectors. Let $|u_E(t_0)\rangle$ be an eigenvector of H corresponding to the energy E

$$H|u_E(t_0)\rangle = E|u_E(t_0)\rangle. \tag{VIII.30}$$

According to Einstein's law, we postulate that this vector evolves in the course of time according to the law

$$|u_E(t)\rangle = \mathrm{e}^{-\mathrm{i}\omega(t-t_0)}\, |u_E(t_0)\rangle = \mathrm{e}^{-\mathrm{i}E(t-t_0)/\hbar}\, |u_E(t_0)\rangle$$

or else, taking into account eq. (VIII.30),

$$|u_E(t)\rangle = \mathrm{e}^{-\mathrm{i}H(t-t_0)/\hbar}\, |u_E(t_0)\rangle.$$

Consequently

$$U(t, t_0) = \mathrm{e}^{-\mathrm{i}H(t-t_0)/\hbar}. \tag{VIII.31}$$

Upon differentiating [1]) the two sides of this equation with respect to t, we obtain the differential equation

$$\boxed{\mathrm{i}\hbar \frac{\mathrm{d}}{\mathrm{d}t} U(t, t_0) = HU(t, t_0)} \tag{VIII.32}$$

$U(t, t_0)$ is the solution of this equation satisfying the initial condition

$$\boxed{U(t_0, t_0) = 1.} \tag{VIII.33}$$

By a very natural extension, we postulate that the operator $U(t_0, t_0)$ is the solution of the differential equation (VIII.32) satisfying the initial condition (VIII.33), *even when* the quantum system is not conservative. In the latter case, H depends upon the time explicitly, relation (VIII.29) loses all meaning, and the operator U is no longer given by eq. (VIII.31).

Note that U is likewise defined by the integral equation

$$\boxed{U(t, t_0) = 1 - \frac{\mathrm{i}}{\hbar} \int_{t_0}^{t} HU(t', t_0)\, \mathrm{d}t'} \tag{VIII.34}$$

The equations (VIII.32) and (VIII.33), or the integral equation (VIII.34) express the fundamental law of evolution of the quantum system. An equivalent expression of this law is the *Schrödinger equation*, that is, the differential equation of motion of the dynamical states of the system. One obtains this equation by differentiating equation (VIII.28) term by term,

$$\frac{\mathrm{d}}{\mathrm{d}t} |\psi(t)\rangle = \left(\frac{\mathrm{d}}{\mathrm{d}t} U(t, t_0)\right) |\psi(t_0)\rangle$$

---

[1]) The derivative with respect to t of an operator $X(t)$ depending upon a continuous parameter t is defined just as the derivative of a function:

$$\frac{\mathrm{d}X}{\mathrm{d}t} = \lim_{\varepsilon \to 0} \frac{X(t+\varepsilon) - X(t)}{\varepsilon}.$$

(Cf. Problem VIII.3.)

and substituting for $\mathrm{d}/\mathrm{d}t\, U(t, t_0)$ its expression (VIII.32). We obtain

$$\boxed{i\hbar \frac{\mathrm{d}}{\mathrm{d}t} |\psi(t)\rangle = H|\psi(t)\rangle.} \tag{VIII.35}$$

In order that the norm of the vector $|\psi(t)\rangle$ remain constant in the course of time, it is necessary and sufficient that H be Hermitean; this is easily shown starting from the Schrödinger equation. The Hermitean property of the Hamiltonian will always be assumed to hold.

In fact, if H is Hermitean, $U(t, t_0)$ *is a unitary operator*. When H does not depend upon time, this can immediately be verified in the explicit expression for U, (VIII.31). However, even if H depends upon the time, we have according to the Schrödinger equation

$$|\psi(t + \mathrm{d}t)\rangle = \left(1 - \frac{i}{\hbar} H\, \mathrm{d}t\right) |\psi(t)\rangle.$$

Since H is Hermitean, the operator

$$U(t + \mathrm{d}t, t) \equiv 1 - \frac{i}{\hbar} H\, \mathrm{d}t$$

is an infinitesimal unitary operator (cf. Ch. VII, § 22): one passes from the ket vector at the time t, to the ket vector at the time $t + \mathrm{d}t$ by an infinitesimal unitary transformation. The transformation $U(t, t_0)$ which enables us to pass from $|\psi; t_0\rangle$ to $|\psi; t\rangle$ is thus a succession of infinitesimal unitary transformations; $U(t, t_0)$, a product of infinitesimal unitary operators, is unitary.

9. Schrödinger "Representation"

The Schrödinger equation completes the general scheme of description of quantum phenomena which we intended to present in this chapter. If we limit ourselves to the essentials, this scheme may be summarized in the following manner.

1) DEFINITION OF DYNAMICAL STATES.

The dynamical state of a quantum system is defined by a collection of precisely defined quantities, namely the particular values taken by the dynamical variables of a complete set of compatible variables. By carrying out the simultaneous measurement of the variables of such a set, one defines unambiguously the state of the system at time t when the measurement was performed.

2) DEFINITION OF THE KET SPACE OF THE SYSTEM.

Each state can be represented (principle of superposition) by a ket vector $|\chi\rangle$ (normalized to unity and defined to within a phase factor) of a certain vector space $\mathscr{E}$. Each dynamical variable is represented by an observable of this space; the only states for which this variable has a well-defined value are the states represented by the eigenvectors of this observable, the value of the variable then being the eigenvalue belonging to the eigenvector in question. The observables satisfy certain rules of algebra which one can specify completely by giving the commutation relations. Compatible variables are represented by commuting observables.

3) DEFINITION OF PROBABILITIES.

If one performs on the quantum system a simultaneous measurement of a complete set of compatible dynamical variables, the probability of finding the system in the state $|\chi\rangle$ (i.e. of finding the particular values of these variables defining the dynamical state represented by $|\chi\rangle$) is equal to the square of the modulus of the scalar product of the vector $|\psi\rangle$ (normalized to unity) representing the dynamical state of the system at the instant the measurement is carried out, by $|\chi\rangle$, namely

$$|\langle\chi|\psi\rangle|^2.$$

More generally, the probability of finding the system in the subspace $\mathscr{E}_D$ (i.e. of finding the system in any one of the states of that subspace) is equal to the average of the projector P_D on that subspace, that is

$$\langle P_D\rangle = \langle\psi|P_D|\psi\rangle.$$

4) EQUATION OF MOTION.

In the absence of any external interference, the dynamical state of a system evolves in a strictly causal manner in the course of time. The vector $|\psi(t)\rangle$ which represents it in the space $\mathscr{E}$ moves continuously according to the Schrödinger equation (VIII.35). In other words, one goes over from $|\psi(t_0)\rangle$ to $|\psi(t)\rangle$ by the unitary transformation (VIII.28) in which $U(t, t_0)$ is a unitary operator defined by eqs. (VIII.32) and (VIII.33).

Knowing the dynamical state $|\psi\rangle$ of the system at a given initial time t_0, one is thus able to predict the statistical distribution of the results of any measurement performed upon the system at a given later time t_1. Indeed, the dynamical state of the system at the instant when the measurement begins is

$$|\psi(t_1)\rangle = U(t_1, t_0)|\psi\rangle,$$

and consequently, the probability of finding the system in a given state $|\chi\rangle$ is

$$|\langle\chi|\psi(t_1)\rangle|^2 = |\langle\chi|U(t_1, t_0)|\psi\rangle|^2. \tag{VIII.36}$$

In the above description of phenomena, the state of the system is represented by a moving ket vector $|\psi(t)\rangle$. On the other hand, the physical quantities, or at least those which do not depend upon the time explicitly, are represented by stationary observables of $\mathscr{E}$ space; likewise, the eigenvectors of these observables are stationary vectors of $\mathscr{E}$ space, as is the case for the vectors $|\chi\rangle$ and $|\psi\rangle$ of expression (VIII.36). This mode of description of quantum phenomena is called the *Schrödinger "representation"* [1]).

10. Heisenberg "Representation"

One obtains a mode of description of phenomena that is strictly equivalent to the foregoing one by performing a unitary transformation on the kets and observables of the Schrödinger "representation", and assigning to the transformed quantities the same physical significance as to those from which they originated. In such a transformation the observables transform into observables possessing the same eigenvalue spectrum; the eigenvectors transform into eigenvectors; the algebraic relations, the conjugation relations, and the scalar products are conserved. Since the only measurable quantities are moduli of scalar products [cf. eq. (VIII.36)], it is clear that the

---

[1]) One must not confuse this concept of "representation" with the notion of the representation of vectors and operators of vector space by matrices. The "representation" with which we are dealing here is that of the motion of the quantum system. In order to avoid confusion, it would be preferable here to speak of the Schrödinger "mode of description". The term "representation" is unfortunately entrenched by usage. In order to distinguish the two concepts, we agree to place the word "representation" between quotation marks whenever it is used in its present sense. The distinction to be made here is analogous to the distinction between unitary transformation of matrices and unitary transformation of vectors and operators (cf. Ch. VII, Sec. III).

predictions made by means of the new quantities are identical to the predictions made with the old quantities.

In particular, one defines the *Heisenberg "representation"* by performing upon the kets and observables of the Schrödinger "representation" the unitary, time-dependent transformation $U^\dagger(t, t_0)$. Let us attach the subscript S to the old quantities, and the subscript H to the new quantities. The ket

$$|\psi_S(t)\rangle = U(t, t_0)\,|\psi_S(t_0)\rangle$$

which represents the dynamical state of the system at time t is transformed into a stationary ket

$$|\psi_H\rangle = U^\dagger(t, t_0)|\psi_S(t)\rangle = |\psi_S(t_0)\rangle. \qquad \text{(VIII.37)}$$

Conversely, an observable A_S of the Schrödinger "representation" transforms into

$$A_H(t) = U^\dagger(t, t_0)\, A_S U(t, t_0). \qquad \text{(VIII.38)}$$

In general, A_H is not stationary, even when A_S does not depend upon the time explicitly. Indeed, if one takes into account differential equation (VIII.32) and its Hermitean conjugate equation, one obtains, on differentiating the last equation term by term

$$\begin{aligned} i\hbar \frac{\mathrm{d}A_H}{\mathrm{d}t} &= -\, U^\dagger H A_S\, U + i\hbar U^\dagger \frac{\partial A_S}{\partial t} U + U^\dagger A_S H U \\ &= U^\dagger [A_S, H] U + i\hbar U^\dagger \frac{\partial A_S}{\partial t} U. \end{aligned} \qquad \text{(VIII.39)}$$

In this equation, H is the Hamiltonian of the Schrödinger "representation". Introducing the Hamiltonian of the Heisenberg "representation":

$$H_H = U^\dagger H U,$$

one has

$$U^\dagger[A_S, H]\, U = [A_H, H_H].$$

On the other hand, A_S, a function of the fundamental observables of the Schrödinger "representation", may depend upon the time explicitly; the right-hand side of (VIII.39) takes account of this fact. $\partial A_S/\partial t$ is some function of the observables of the Schrödinger "represetantion". If $\partial A_H/\partial t$ is the function obtained by replacing the latter observables by the corresponding observables of the Heisenberg "representation", it is clear that

$$\frac{\partial A_{\mathrm{H}}}{\partial t} = U^{\dagger} \frac{\partial A_{\mathrm{S}}}{\partial t} U.$$

Equation (VIII.39) is therefore written

$$\boxed{\mathrm{i}\hbar \frac{\mathrm{d}A_{\mathrm{H}}}{\mathrm{d}t} = [A_{\mathrm{H}}, H_{\mathrm{H}}] + \mathrm{i}\hbar \frac{\partial A_{\mathrm{H}}}{\partial t}.} \qquad \text{(VIII.40)}$$

This equation is known as the *Heisenberg equation.*

In conclusion, the Heisenberg "representation" is obtained by imposing upon the vector space of the Schrödinger "representation" an overall motion chosen in such a way that the dynamical state of the quantum system is represented by a stationary ket $|\psi_{\mathrm{H}}\rangle$. In other words, *any stationary ket of the Heisenberg "representation" represents a possible motion of the quantum system.* Conversely, the various physical quantities are represented by observables evolving in the course of time according to the law (VIII.38) or, what amounts to the same thing, according to the Heisenberg equation (VIII.40) with the initial condition $A_{\mathrm{H}}(t_0) = A_{\mathrm{S}}(t_0)$.

Equations (VIII.38) and (VIII.40) apply equally well to any function of the observables of the Heisenberg representation, and especially to the expression $\exp(\mathrm{i}\xi A_{\mathrm{H}})$ or to the projector $P_D^{(\mathrm{H})}$ upon the subspace of the eigenvectors belonging to the eigenvalues of a given domain D of the spectrum of A_{H}.

Likewise, a ket $|\chi_{\mathrm{H}}\rangle$ representing an ensemble of compatible variables generally depends upon the time and is deduced from its homologue $|\chi_{\mathrm{S}}\rangle$ of the Schrödinger "representation" by the formula

$$|\chi_{\mathrm{H}}(t)\rangle = U^{\dagger}(t, t_0)|\chi_{\mathrm{S}}\rangle. \qquad \text{(VIII.41)}$$

Let us suppose that the motion of the quantum system is represented after time t_0 by the (stationary) ket $|\psi_{\mathrm{H}}\rangle$. The probability of finding it in the state $|\chi_{\mathrm{H}}\rangle$ upon performing a measurement at the later time t_1 is

$$|\langle\chi_{\mathrm{H}}(t_1)|\psi_{\mathrm{H}}\rangle|^2,$$

a quantity which is obviously equal to the one obtained with the corresponding kets of the Schrödinger "representation" [eq. (VIII.36)], since the scalar product is invariant under the unitary transformation $U^{\dagger}(t_1, t_0)$.

11. Heisenberg "Representation" and Correspondence Principle

As shown above, the "representations" of Schrödinger and of Heisenberg are strictly equivalent. In practice, the Schrödinger "representation" is more often used, because it lends itself better to calculations. In fact, the Schrödinger equation, an equation between vectors, is *a priori* easier to solve than the Heisenberg equation which is an equation between operators. However, certain general properties of quantum systems are more immediately apparent in the Heisenberg "representation".

The formal analogy between the Classical Theory and the Quantum Theory is particularly striking in the Heisenberg "representation". The motion of a quantum system in fact, just like the motion of a classical system, appears in the Heisenberg "representation" as a motion of the dynamical variables which are associated with it.

Let us consider a quantum system possessing a classical analogue, and let us compare the motions of the two systems. To every physical quantity of the classical system corresponds a physical quantity of the quantum system. The only difference lies in the fact that the physical quantities of the classical system are quantities obeying the rules of ordinary algebra, whereas their quantum analogues are operators obeying the rules of a non-commutative algebra. But, to the extent that one can identify the expressions of a non-commutative algebra with expressions of ordinary algebra, the equations of motion of the quantized quantities are identical to those of their classical analogues. Indeed, the Heisenberg equations for the variables $q_1, \ldots, q_N$ and $p_1, \ldots, p_N$ are written

$$\begin{aligned} \frac{\mathrm{d}q_i}{\mathrm{d}t} &= \frac{1}{\mathrm{i}\hbar}[q_i, H] = \frac{\partial H}{\partial p_i} \qquad (i = 1, 2, \ldots, N) \\ \frac{\mathrm{d}p_i}{\mathrm{d}t} &= \frac{1}{\mathrm{i}\hbar}[p_i, H] = -\frac{\partial H}{\partial q_i} \qquad (i = 1, 2, \ldots, N). \end{aligned} \tag{I}$$

To obtain these expressions, the fundamental commutation relations between the q's and the p's as well as the properties (V.67) and (V.68) to which they give rise, have been taken into account. The system of equations (I) is formally identical to Hamilton's canonical equations of Classical Mechanics.

More generally, a classical dynamical variable $A_{\mathrm{cl.}} = A(q_1, ..., q_N; p_1, ..., p_N; t)$ obeys the equation of motion

$$\frac{\mathrm{d}A_{\mathrm{cl.}}}{\mathrm{d}t} = \{A_{\mathrm{cl.}}, H_{\mathrm{cl.}}\} + \frac{\partial A_{\mathrm{cl.}}}{\partial t}, \tag{VIII.42}$$

where $\{A_{\mathrm{cl.}}, H_{\mathrm{cl.}}\}$ designates the Poisson bracket of $A_{\mathrm{cl.}}$ and $H_{\mathrm{cl.}}$ according to the definition

$$\{A, H\} \equiv \sum_i \left(\frac{\partial A}{\partial q_i}\frac{\partial H}{\partial p_i} - \frac{\partial A}{\partial p_i}\frac{\partial H}{\partial q_i}\right).$$

We see that the classical equation (VIII.42) is identical to the corresponding Heisenberg equation to the extent that one can identify the Poisson bracket $\{A, H\}$ with the commutator $[A_{\mathrm{H}}, H_{\mathrm{H}}]/i\hbar$. Making use of the fundamental commutation relations and of the similarity between the rules of commutator algebra and the rules of Poisson bracket algebra, one can actually prove the identity of these two expressions provided one makes a suitable choice of the order of the q's and the p's in the explicit expression of the Poisson bracket.

12. Constants of the Motion

The notion of constant of the motion is particularly simple to grasp in the Heisenberg "representation". A dynamical variable *which does not depend upon the time explicitly* is a constant of the motion if the observable C_{H} representing it in the Heisenberg "representation" remains constant in time. Consequently, its system of eigenvectors remains stationary and the statistical distribution of the results of a possible measurement of this quantity is always independent of the time at which this measurement is undertaken.

According to the above definition of the constant of the motion

$$i\hbar \frac{\mathrm{d}}{\mathrm{d}t} C_{\mathrm{H}} = [C_{\mathrm{H}}, H_{\mathrm{H}}] = 0.$$

The constants of the motion are thus represented by *observables which commute with the Hamiltonian*. This result holds true equally well in the Schrödinger "representation" and in the Heisenberg "representation" since the commutation relations are conserved in passing from one to the other.

Moreover, since C_{H} is time-independent, it is equal to its initial value C_{S}

$$C_{\mathrm{H}}(t) = C_{\mathrm{H}}(t_0) = C_{\mathrm{S}} = C.$$

If, in particular, the dynamical state of the system is represented by an eigenvector of C in the Heisenberg "representation",

$$C|\psi_{\mathrm{H}}\rangle = c|\psi_{\mathrm{H}}\rangle,$$

the variable C keeps the same well-defined value c in the course of time; the eigenvalue c is then said to be a *good quantum number*. As can easily be shown, C commutes with the evolution operator $U(t, t_0)$; hence, the ket $|\psi_{\mathrm{S}}(t)\rangle$ of the Schrödinger "representation" remains forever in the subspace of the eigenvalue c,

$$C|\psi_{\mathrm{S}}(t)\rangle = c|\psi_{\mathrm{S}}(t)\rangle.$$

13. Equations of Motion for the Mean Values. Time-Energy Uncertainty Relation

Starting from the Heisenberg "representation", it is particularly simple to write down a differential equation for the mean value of a given observable A_{H}. Indeed, since $|\psi_{\mathrm{H}}\rangle$ is time-independent,

$$\frac{\mathrm{d}\langle A\rangle}{\mathrm{d}t} = \frac{\mathrm{d}}{\mathrm{d}t}\langle\psi_{\mathrm{H}}|A_{\mathrm{H}}|\psi_{\mathrm{H}}\rangle = \langle\psi_{\mathrm{H}}|\frac{\mathrm{d}A_{\mathrm{H}}}{\mathrm{d}t}|\psi_{\mathrm{H}}\rangle.$$

Using the Heisenberg equation, we arrive at eq. (V.72) once again:

$$\frac{\mathrm{d}}{\mathrm{d}t}\langle A\rangle = \frac{1}{\mathrm{i}\hbar}\langle[A, H]\rangle + \left\langle\frac{\partial A}{\partial t}\right\rangle. \qquad \text{(VIII.43)}$$

In particular, one obtains the Ehrenfest equations (Ch. VI, § 2) by carrying out this manipulation on system (I).

As an application of eq. (VIII.43), we shall give a precise statement of the time-energy uncertainty relation (cf. Ch. IV, § 10). Consider a system whose Hamiltonian H does not explicitly depend upon the time, and let A be another observable of this system which does not depend upon the time explicitly. We consider the dynamical state of the system at a given time t. Let $|\psi\rangle$ be the vector representing that state. Call ΔA, ΔE the root-mean-square deviations of A and of H, respectively. Applying the Schwarz inequality to the vectors $(A-\langle A\rangle)|\psi\rangle$ and $(H-\langle H\rangle)|\psi\rangle$ and carrying out the same manipulations as in § 4, we find after some calculations

$$\Delta A \cdot \Delta E \geqslant \tfrac{1}{2}|\langle [A, H]\rangle|, \tag{VIII.44}$$

the equality being realized when $|\psi\rangle$ satisfies the equation

$$(A-\alpha)|\psi\rangle = \mathrm{i}\gamma(H-\varepsilon)|\psi\rangle$$

where α, γ, and ε are arbitrarily real constants [cf. eq. (VIII.10)]. However, according to eq. (VIII.43), $\langle [A, H]\rangle = \mathrm{i}\hbar\, \mathrm{d}\langle A\rangle/\mathrm{d}t$; the inequality (VIII.44) may equally well be written

$$\frac{\Delta A}{|\mathrm{d}\langle A\rangle/\mathrm{d}t|} \cdot \Delta E \geqslant \tfrac{1}{2}\hbar,$$

or else

$$\tau_A \cdot \Delta E \geqslant \tfrac{1}{2}\hbar \tag{VIII.45}$$

if one puts

$$\tau_A = \frac{\Delta A}{|\mathrm{d}\langle A\rangle/\mathrm{d}t|}\,; \tag{VIII.46}$$

τ_A appears as a time characteristic of the evolution of the statistical distribution of A. It is the time required for the center $\langle A\rangle$ of this distribution to be displaced by an amount equal to its width ΔA; in other words, it is the time necessary for this statistical distribution to be appreciably modified. In this manner we can define a characteristic evolution time for each dynamical variable of the system.

Let τ be the shortest of the times thus defined. τ may be considered as a characteristic time of evolution of the system itself: whatever the measurement carried out on the system at an instant of time t', the statistical distribution of the results is essentially the same as would be obtained at the instant t, as long as the difference $|t-t'|$ is less than τ.

According to the inequality (VIII.45), this time τ and the energy spread ΔE satisfy the time-energy uncertainty relation

$$\tau \cdot \Delta E \geqslant \tfrac{1}{2}\hbar. \tag{VIII.47}$$

If, in particular, the system is in a stationary state, $\mathrm{d}\langle A\rangle/\mathrm{d}t = 0$ no matter what A, and consequently τ is infinite; however, $\Delta E = 0$, in conformity with relation (VIII.47).

14. Intermediate "Representations"

The Schrödinger and Heisenberg "representations" are not the only possible ones. Any unitary transformation of the vectors and the observables of the Schrödinger (or Heisenberg) "representations" defines a new "representation". All these "representations" furnish strictly equivalent descriptions of quantum phenomena. In practice, one therefore adopts the "representation" which lends itself best to the solution of each particular problem.

Any problem of Quantum Mechanics essentially consists of a more or less complete and more or less precise determination of the properties of the unitary operator $U(t, t_0)$; indeed, all the predictions of the theory are given by matrix elements of $U(t, t_0)$ such as the one occurring in eq. (VIII.36). The solution of equation (VIII.32) is thus the central problem of the theory. When one knows an approximate solution $U^{(0)}(t, t_0)$ of this equation, it is often convenient to set

$$U = U^{(0)} U'. \tag{VIII.48}$$

Substituting this expression in eq. (VIII.32) and multiplying both sides from the left by the unitary operator $U^{(0)\dagger}$, we obtain the differential equation

$$i\hbar \frac{\mathrm{d}}{\mathrm{d}t} U' = U^{(0)\dagger} \left(H U^{(0)} - i\hbar \frac{\mathrm{d} U^{(0)}}{\mathrm{d}t} \right) U'; \tag{VIII.49}$$

U' is the solution of this equation satisfying the initial condition

$$U'(t_0, t_0) = 1.$$

If the approximation is justified, U' is an operator changing slowly as a function of time; this is actually quite evident from eq. (VIII.49), since in that case, the operator $HU^{(0)} - i\hbar(\mathrm{d}U^{(0)}/\mathrm{d}t)$ almost vanishes. Equation (VIII.49), therefore, lends itself better than equation (VIII.32) to an approximate solution [1].

Since $U^{(0)}$ is unitary, the operator

$$H^{(0)}(t) \equiv i\hbar \left[\frac{\mathrm{d}}{\mathrm{d}t} U^{(0)}(t, t_0) \right] U^{(0)\dagger}(t, t_0)$$

[1] The procedure discussed here is the generalization to the differential equations between operators, of the method of variation of constants of the elementary theory of differential equations.

is Hermitean (Problem VIII.6). $U^{(0)}(t, t_0)$ is therefore the rigorous solution of the Schrödinger equation:

$$i\hbar \frac{\mathrm{d}}{\mathrm{d}t} U^{(0)} = H^{(0)} U^{(0)}, \qquad U^{(0)}(t_0, t_0) = 1. \tag{VIII.50}$$

The Hamiltonian H is thus the sum of two Hermitean operators

$$H = H^{(0)} + H'$$

of which one, H', may be considered as a small perturbation, under the hypothesis considered here, while the other, $H^{(0)}$, is the Hamiltonian of a Schrödinger equation whose solution is known. With this notation, eq. (VIII.49) can then simply be written

$$i\hbar \frac{\mathrm{d}}{\mathrm{d}t} U' = H_{\mathrm{I}}' U', \tag{VIII.51}$$

an expression in which H_{I}' is deduced from H' by the time-dependent unitary transformation

$$H_{\mathrm{I}}' = U^{(0)\dagger} H' U^{(0)}. \tag{VIII.52}$$

It is then convenient to adopt, throughout the discussion, a "representation" intermediate between that of Schrödinger and of Heisenberg; namely the one obtained by applying to the vectors and observables of the Schrödinger "representation" the unitary transformation $U^{(0)\dagger}(t, t_0)$. Let us label the vectors and observables of this new representation by a subscript I:

$$|\psi_{\mathrm{I}}(t)\rangle = U^{(0)\dagger} |\psi_{\mathrm{S}}(t)\rangle. \tag{VIII.53}$$

$$A_{\mathrm{I}}(t) = U^{(0)\dagger} A_{\mathrm{S}} U^{(0)}. \tag{VIII.54}$$

In this intermediate "representation", the vector $|\psi_{\mathrm{I}}(t)\rangle$ representing a possible motion of the quantum system is equal to $U'|\psi_{\mathrm{S}}(t_0)\rangle$. According to eq. (VIII.51), this vector moves (slowly) in time satisfying a Schrödinger equation whose Hamiltonian is the perturbation energy H_{I}':

$$i\hbar \frac{\mathrm{d}}{\mathrm{d}t} |\psi_{\mathrm{I}}(t)\rangle = H_{\mathrm{I}}' |\psi_{\mathrm{I}}(t)\rangle. \tag{VIII.55}$$

On the other hand, the physical quantities are represented by moving observables; these observables are in fact subject to the Heisenberg equations of motion written with the "unperturbed" Hamiltonian $H_{\mathrm{I}}^{(0)}$

$$i\hbar \frac{\mathrm{d}}{\mathrm{d}t} A_{\mathrm{I}} = [A_{\mathrm{I}}, H_{\mathrm{I}}^{(0)}] + i\hbar \frac{\partial A_{\mathrm{I}}}{\partial t}, \qquad \text{(VIII.56)}$$

as one can easily show by carrying out on eq. (VIII.54) a procedure analogous to the one applied to eq. (VIII.38) in order to establish the Heisenberg equation.

III. VARIOUS REPRESENTATIONS OF THE THEORY

15. Definition of a Representation

According to the theory developed in the first two sections, all the elements for the description of a quantum system are present when one has defined its fundamental dynamical variables, the commutation relations obeyed by the representative observables, and the explicit expression, as a function of these fundamental observables, of the Hamiltonian which governs the motion of the system. One can then build up the $\mathscr{E}$ space of the vectors representing the various possible dynamical states of the system; define the physical meaning of the vectors of $\mathscr{E}$ by solving the eigenvalue problems for the various observables associated with the system, establish and solve the fundamental equations of motion, and finally, perform the calculation of the statistical distribution of the results of measurement which the theory is in a position to predict exactly.

In order to solve all these problems of analysis or of algebra in $\mathscr{E}$ space, one can always, and in an infinite number of ways at that, choose in that space a complete, orthonormal set of vectors and represent the operators and vectors of $\mathscr{E}$ by their matrices in the representation having this set of vectors as a basis.

In this way, any dynamical variable of the system is represented by a square Hermitean matrix, and any dynamical state by a column vector (or equally well by the Hermitean conjugate row vector) defined to within a constant.

There are as many possible representations of the Theory as there are distinct bases. One passes from one to the other by unitary transformation. These unitary transformations of matrices must not be confused with the unitary transformations of operators and vectors which permit, according to the discussion of Sec. II, to change the "representation" of the motion of the quantum system itself.

Most often a representation is defined by giving a complete set of commuting observables; their common eigenvectors are the basis of the representation. The basis observables of the representation and all functions of these observables are represented by diagonal matrices.

16. Wave Mechanics

Wave Mechanics is obtained by formulating the Quantum Theory in the Schrödinger "representation" and in a representation where the position variables are diagonal.

Let us return to the quantum system having a classical analogue in N dimensions, studied in § 7. The position coordinates $q_1, q_2, \ldots, q_N$ form a complete set of commuting observables and define a certain representation, the $\{q\}$ representation. This representation has already been used to construct the $\mathscr{E}$ space itself. With a suitable choice of phases and normalization for the basis vectors, we obtained very simple expressions for the matrix elements of the q's and the p's [eqs. (VIII.26) and (VIII.27)].

The fundamental equations of the $\{q\}$ representation are the orthonormality relations (VIII.25) and the closure relation which, following the abbreviated notation of § 7 ($\mathrm{d}q \equiv \mathrm{d}q_1\, \mathrm{d}q_2 \ldots \mathrm{d}q_N$), is written

$$P_q \equiv \int |q'\rangle\, \mathrm{d}q'\, \langle q'| = 1. \qquad \text{(VIII.57)}$$

Any ket $|\psi\rangle$ is represented by the single-column matrix with components $\langle q'|\psi\rangle$. This function of the coordinates $q_1', q_2', \ldots, q_N'$ of configuration space, which can equally well be written $\psi(q_1', q_2', \ldots, q_N')$, is the wave function representing the dynamical state of the system in the language of Wave Mechanics:

$$\langle q'|\psi\rangle \equiv \langle q_1'\, q_2' \ldots q_N'|\psi\rangle \equiv \psi(q_1', q_2', \ldots, q_N'). \qquad \text{(VIII.58)}$$

The scalar product of $|\psi\rangle$ by $|\varphi\rangle$ is equal to the scalar product of the corresponding wave functions as defined in Wave Mechanics:

$$\langle\varphi|\psi\rangle = \langle\varphi|P_q|\psi\rangle = \int \langle\varphi|q'\rangle\, \mathrm{d}q'\, \langle q'|\psi\rangle = \int \varphi^*(q')\, \psi(q')\, \mathrm{d}q'. \qquad \text{(VIII.59)}$$

Let us verify the identity between the operators of Wave Mechanics and the matrices representing the observables in the $\{q\}$ representation.

Using expression (VIII.26) of the matrix representing the observable q_n, we verify that $q_n|\psi\rangle$ is represented by the wave function

$$\langle q'|q_n|\psi\rangle = q_n'\langle q'|\psi\rangle = q_n'\psi(q').$$

More generally, the action of an arbitrary function $V(q) \equiv V(q_1, q_2, \ldots, q_N)$ of the coordinates of configuration space, on a ket $|\psi\rangle$ is represented by the function obtained by multiplying $\psi(q')$ by $V(q')$,

$$\langle q'|\, V(q)\, |\psi\rangle = V(q')\, \psi(q'). \tag{VIII.60}$$

Using the explicit expression (VIII.27) of the matrix representing the observable p_n, we verify that $p_n|\psi\rangle$ is represented by the wave function

$$\begin{aligned} \langle q'|p_n|\psi\rangle &= \int \langle q'|p_n|q''\rangle\, \mathrm{d}q''\, \langle q''|\psi\rangle \\ &= \frac{\hbar}{\mathrm{i}} \int \frac{\partial}{\partial q_n'} [\delta(q'-q'')]\, \psi(q'')\, \mathrm{d}q'' \\ &= \frac{\hbar}{\mathrm{i}} \frac{\partial}{\partial q_n'} \psi(q'). \end{aligned} \tag{VIII.61}$$

p_n is therefore represented by the operation of partial differentiation $-\mathrm{i}\hbar\partial/\partial q_n$ applied to the wave function to the right.

The identification which we intended to make is therefore actually verified for the functions of position coordinates [eq. (VIII.60)] and for the components of the momentum [eq. (VIII.61)]. Now, since any observable is an algebraic function of the p's, and of functions of the q's, this identification holds true for any observable; hence, any physical quantity $A(q; p)$ is represented in Wave Mechanics by he operator

$$A\left(q; \frac{\hbar}{\mathrm{i}} \frac{\partial}{\partial q}\right).$$

As an illustration, let us consider the energy H of the system, assuming a purely static potential energy. H reads

$$H(q; p) = \sum_i \frac{p^2}{2m_i} + V(q_1, \ldots, q_N).$$

The matrix which represents the energy in the $\{q\}$ representation is of the form

$$\begin{aligned} \langle q'|H|q''\rangle &= H\left(q'; \frac{\hbar}{\mathrm{i}} \frac{\partial}{\partial q'}\right) \delta(q'-q'') \\ &= \left[\sum_i \left(-\frac{\hbar^2}{2m_i}\right) \frac{\partial^2}{\partial q_i'^2} + V(q')\right] \delta(q'-q''). \end{aligned}$$

In the last line, the expression in brackets must be regarded as an operator acting on $\delta(q'-q'')$ considered as a function of the q'. Therefore, the vector $H|\psi\rangle$ is represented by the wave function

$$\langle q'|H|\psi\rangle \equiv H\psi(q') = \Big[\sum_i -\frac{\hbar^2}{2m_i}\frac{\partial}{\partial q_i'^2} + V(q')\Big]\psi(q').$$

Finally, if we write, in the framework of the Schrödinger "representation", the fundamental equation of motion (VIII.35) in the $\{q\}$ representation, we again arrive at the Schrödinger equation in its usual form:

$$i\hbar\frac{\partial}{\partial t}\psi(q;t) = H\psi(q;t).$$

This completes the proof of the identity between Wave Mechanics and the formulation of Quantum Theory in the $\{q\}$ representation and in the Schrödinger "representation".

17. Momentum Representation. ($\{p\}$ Representation)

As another example of representation, we mention the $\{p\}$ representation, in which the components of the momentum are diagonal. Let us denote by $|p'\rangle \equiv |p_1'\rangle\,|p_2'\rangle \dots |p_N'\rangle$ the basis vectors of this representation. They are the eigenvectors common to $p_1, p_2, \dots, p_N$ belonging to the eigenvalues $p_1', p_2', \dots, p_N'$. These vectors are assumed to be orthonormal:

$$\langle p'|p''\rangle = \delta(p'-p'').$$

They satisfy the closure relation

$$P_p \equiv \int |p'\rangle\,\mathrm{d}p'\,\langle p'| = 1,$$

(we make use in this whole section of abbreviated notations analogous to those introduced in §§ 7 and 16).

In accordance with the results of § 6, the wave function of the vector $|p'\rangle$ in the $\{q\}$ representation is

$$\langle q'|p'\rangle \equiv \prod_{i=1}^{N}\langle q_i'|p_i'\rangle = (2\pi\hbar)^{-\frac{1}{2}N}\,\mathrm{e}^{\mathrm{i}(p_1'q_1'+\dots+p_N'q_N')/\hbar}.$$

Considered as a function of the q' and the p', $\langle p'|q'\rangle$ $(=\langle q'|p'\rangle^*)$ is the unitary matrix S which transforms the matrices of the $\{q\}$

representation into the matrices of the $\{p\}$ representation. A ket vector $|\psi\rangle$ is represented in the latter by its "wave function in momentum space"

$$\Phi(p') \equiv \langle p'|\psi\rangle.$$

Clearly $\Phi(p')$ is the Fourier transform (suitably normalized) of the wave function $\Psi(q') \equiv \langle q'|\psi\rangle$ of configuration space:

$$\begin{aligned}\Phi(p') = \langle p'|\psi\rangle &= \int \langle p'|q'\rangle \,\mathrm{d}q' \,\langle q'|\psi\rangle \\ &= (2\pi\hbar)^{-\frac{1}{2}N} \int \Psi(q')\, \mathrm{e}^{-\mathrm{i}(p_1'q_1'+\ldots+p_N'q_N')/\hbar}\, \mathrm{d}q'.\end{aligned}$$

It is easy to show (one could in fact prove it directly) that the action of the operator p_n on the function $\Phi(p')$ reduces to a multiplication by p_n' and that the action of q_n is represented by the operation of partial differentiation $\mathrm{i}\hbar\partial/\partial p_n'$.

As an example, let us write the Schrödinger equation of a particle of mass m in a static potential $V(\mathbf{r})$ in the $\{p\}$ representation. The energy of the system is represented by the observable

$$H(\mathbf{r}, \mathbf{p}) \equiv \frac{p^2}{2m} + V(\mathbf{r}).$$

The basis vectors $|\mathbf{p}'\rangle$ depend upon the three momentum coordinates p_x', p_y', p_z' and satisfy the orthonormality and closure relations:

$$\delta(\mathbf{p}' - \mathbf{p}'') \equiv \delta(p_x' - p_x'')\, \delta(p_y' - p_y'')\, \delta(p_z' - p_z'') = \langle \mathbf{p}'|\mathbf{p}''\rangle$$

$$P_p \equiv \int |\mathbf{p}'\rangle \,\mathrm{d}\mathbf{p}'\, \langle \mathbf{p}'| = 1.$$

The unitary transformation matrix S which transforms the matrices of the $\{\mathbf{r}\}$ representation into matrices of the $\{\mathbf{p}\}$ representation is

$$\langle \mathbf{p}'|\mathbf{r}'\rangle = (2\pi\hbar)^{-3/2}\, \mathrm{e}^{-\mathrm{i}(\mathbf{p}'\cdot\mathbf{r}')/\hbar}.$$

Therefore, the matrix elements of $V(\mathbf{r})$ in the $\{\mathbf{p}\}$ representation read explicitly

$$\begin{aligned}\langle \mathbf{p}'|V|\mathbf{p}''\rangle &= \iint \langle \mathbf{p}'|\mathbf{r}'\rangle \,\mathrm{d}\mathbf{r}'\, \langle \mathbf{r}'|V(\mathbf{r})|\mathbf{r}''\rangle \,\mathrm{d}\mathbf{r}''\, \langle \mathbf{r}''|\mathbf{p}''\rangle \\ &= (2\pi\hbar)^{-3} \iint \mathrm{e}^{-\mathrm{i}(\mathbf{p}'\cdot\mathbf{r}')/\hbar}\, \mathrm{d}\mathbf{r}'\, V(\mathbf{r}')\, \delta(\mathbf{r}' - \mathbf{r}'')\, \mathrm{d}\mathbf{r}''\, \mathrm{e}^{\mathrm{i}(\mathbf{p}''\cdot\mathbf{r}'')/\hbar} \\ &= (2\pi\hbar)^{-3} \int V(\mathbf{r}')\, \mathrm{e}^{-\mathrm{i}(\mathbf{p}'-\mathbf{p}'')\cdot\mathbf{r}'/\hbar}\, \mathrm{d}\mathbf{r}'.\end{aligned}$$

Let us put

$$\mathscr{V}(\mathbf{k}) = (2\pi\hbar)^{-3} \int V(\mathbf{r})\, \mathrm{e}^{-\mathrm{i}(\mathbf{k}\cdot\mathbf{r})/\hbar}\, \mathrm{d}\mathbf{r}.$$

We then have

$$\langle \mathbf{p}' | V | \mathbf{p}'' \rangle = \mathscr{V}(\mathbf{p}' - \mathbf{p}'')$$

and the matrix elements in the $\{\mathbf{p}\}$ representation of the operator H are:

$$\langle \mathbf{p}' | H | \mathbf{p}'' \rangle = \frac{p'^2}{2m} \delta(\mathbf{p}' - \mathbf{p}'') + \mathscr{V}(\mathbf{p}' - \mathbf{p}'').$$

Let us denote by $\Phi(\mathbf{p}')$ the wave function in momentum space of a dynamical state $|\psi\rangle$ whose wave function in the usual sense of the term is $\Psi(\mathbf{r}')$:

$$\begin{aligned} \Phi(\mathbf{p}') &= \langle \mathbf{p}' | \psi \rangle = \int \langle \mathbf{p}' | \mathbf{r}' \rangle \, \mathrm{d}\mathbf{r}' \, \langle \mathbf{r}' | \psi \rangle \\ &= (2\pi\hbar)^{-3/2} \int \mathrm{e}^{-\mathrm{i}(\mathbf{p}' \cdot \mathbf{r}')/\hbar} \, \Psi(\mathbf{r}') \, \mathrm{d}\mathbf{r}'. \end{aligned}$$

The Schrödinger equation which is written

$$\mathrm{i}\hbar \frac{\partial}{\partial t} \Psi(\mathbf{r}; t) = \left[-\frac{\hbar^2}{2m} \triangle + V(\mathbf{r}) \right] \Psi(\mathbf{r}; t)$$

in Wave Mechanics, takes the form of an integro-differential equation in the $\{\mathbf{p}\}$ representation.

$$\mathrm{i}\hbar \frac{\partial}{\partial t} \Phi(\mathbf{p}; t) = \frac{p^2}{2m} \Phi(\mathbf{p}; t) + \int \mathscr{V}(\mathbf{p} - \mathbf{p}') \, \Phi(\mathbf{p}'; t) \, \mathrm{d}\mathbf{p}'.$$

18. An Example: Motion of a Free Wave Packet

As an application of the foregoing considerations, consider the motion of the free wave packet ($V = 0$).

Denote by $|\psi\rangle$ the state vector at the time $t = 0$, by $\psi(\mathbf{r})$ and $\varphi(\mathbf{p})$ the wave functions representing this vector in the $\{\mathbf{r}\}$ and $\{\mathbf{p}\}$ representations, respectively. At some later time t, the dynamical state of the system is given by the vector

$$|\Psi\rangle = \mathrm{e}^{-\mathrm{i}Ht/\hbar} |\psi\rangle,$$

where $H \equiv \mathbf{p}^2/2m$ is the Hamiltonian of the free particle. The momentum being a constant of the motion, its average value remains constant in time; the same holds true for the group velocity

$$\mathbf{v} = \frac{\langle \mathbf{p} \rangle}{m}.$$

One knows (Ch. VI, § 3) that the spreading of the wave packet remains negligible if t is sufficiently small. We intend to make this statement more precise and to show that when these conditions of negligible spreading are fulfilled, the wave packet propagates practically without distortion, and that it may, to a very good approximation, be written

$$\psi(\mathbf{r} - \mathbf{v}t).$$

This approximate wave function represents the vector

$$|\overline{\Psi}\rangle \equiv \mathrm{e}^{-\mathrm{i}(\mathbf{p}\cdot\mathbf{v})t/\hbar}\,|\psi\rangle$$

as can be easily shown either by making use of and generalizing the property (VIII.16), or by mere inspection of the corresponding wave function in the $\{\mathbf{p}\}$ representation. The approximation is better the closer to unity the probability for the system to be in the state $|\overline{\Psi}\rangle$; in other words, one must have

$$1-|\langle\Psi|\overline{\Psi}\rangle|^2 \ll 1.$$

Replacing $|\Psi\rangle$ and $|\overline{\Psi}\rangle$ by the expressions given above, one finds

$$|\langle\Psi|\overline{\Psi}\rangle| = \left|\langle\psi|\exp\left(\frac{\mathrm{i}}{\hbar}\frac{(\mathbf{p}-m\mathbf{v})^2}{2m}t\right)|\psi\rangle\right|.$$

The matrix element of the right-hand side is particularly simple to evaluate in the $\{\mathbf{p}\}$ representation; we have

$$\langle\psi|\exp\left(\frac{\mathrm{i}}{\hbar}\frac{(\mathbf{p}-m\mathbf{v})^2}{2m}t\right)|\psi\rangle = \int |\varphi(\mathbf{p})|^2\,\mathrm{e}^{\mathrm{i}(\mathbf{p}-m\mathbf{v})^2t/2m\hbar}\,\mathrm{d}\mathbf{p}.$$

If one assumes that the wave packet $\varphi(\mathbf{p})$ is of the type described in Fig. IV.1, it is a function exhibiting a very pronounced peak of linear dimensions Δp around the mean value $\mathbf{p}=m\mathbf{v}$, Δp being the magnitude of the vector $\Delta\mathbf{p}$ representing the root-mean-square deviation of the momentum of the particle. Under this hypothesis, the exponential of the right-hand side stays very close to unity over the entire region of this peak as long as

$$(\Delta p)^2 t/2m\hbar \ll 1$$

or else

$$\frac{\Delta p}{m}t \ll \frac{2\hbar}{\Delta p}.$$

The approximation is therefore quite justified when this condition is fulfilled. Now this condition of validity is just the condition

$$\text{spreading} \ll \text{width}$$

which we have determined in the study of the spreading of the wave packet of Chapter VI, § 3 [condition (VI.15)].

19. Other Representations. Representations in which the Energy is Diagonal

The preceding examples show us that the various equations of the Theory take on very different forms according to which representation is adopted. Hence the procedures of calculation can vary greatly from one representation to another.

Among the various representations of the Quantum Theory, some turn out to be particularly useful in the treatment of conservative systems, by reason of the very simple form taken by the Schrödinger equation: they are the representations in which the energy H is diagonal [1]). The basis vectors $|E\alpha\rangle$ of such a representation can be labelled by the eigenvalue E of the energy and by the assembly α of the eigenvalues of other constants of the motion forming with H the basis of observables. The vector $|\psi(t)\rangle$ representing the dynamical state of the system in the Schrödinger "representation" is represented by the "wave function"

$$\psi(E, \alpha; t) \equiv \langle E\alpha|\psi(t)\rangle.$$

The latter satisfies the Schrödinger equation

$$i\hbar \frac{\partial}{\partial t} \psi(E, \alpha; t) = \langle E\alpha|H|\psi(t)\rangle = E\psi(E, \alpha; t).$$

We therefore have simply

$$\psi(E\alpha; t) = \psi(E\alpha; t_0)\, e^{-iE(t-t_0)/\hbar}.$$

It is thus very easy to find the motion of the vector representing the dynamical state of a quantum system if one knows its representation $\{E\alpha\}$ at the initial instant. In practice, the initial position of this vector is often given by its components in some other representation, $\{q\}$ for instance. The solution of its equation of motion is then readily obtained if one knows how to go over from the $\{q\}$ representation to a representation in which H is diagonal. Mathematically, the construction of the unitary matrix of this change of representation is a problem equivalent to the eigenvalue problem of H in the $\{q\}$ representation, in other words, to the solution in this representation of the time-independent Schrödinger equation.

---

[1]) In its original form, the Matrix Mechanics of Born, Heisenberg, and Jordan was a particular formulation of the Quantum Theory in the Heisenberg "representation", expressed in a representation where the energy is diagonal.

IV. QUANTUM STATISTICS

20. Incompletely Known Systems and Statistical Mixtures

When the dynamical state of a quantum system is incompletely known, certain predictions can still be made concerning its behavior if one has recourse to the usual methods of statistics. The discussion on this subject, begun in Chapter V (§ 16), may be easily translated into our new language.

The dynamical state of a quantum system is completely known if one has succeeded in determining precisely the variables of one of the complete sets of compatible variables associated with the system. It can then be represented by a certain vector $|\rangle$. When the information about the system is incomplete, one merely states that the system has certain probabilities $p_1, p_2, \ldots, p_m, \ldots$, of being in the dynamical states represented by the ket vectors $|1\rangle, |2\rangle, \ldots, |m\rangle, \ldots$, respectively. In other words, the dynamical state of the system must no longer be represented by a unique vector, but by a statistical mixture of vectors.

Suppose that we perform the measurement of a given quantity A on the system. The mean value $\langle A \rangle$ of the results of measurement has a probability p_m of being equal to $\langle A \rangle_m = \langle m|A|m\rangle / \langle m|m\rangle$. One can therefore write, assuming the vectors $|1\rangle, |2\rangle, \ldots, |m\rangle, \ldots$ to be normalized to unity,

$$\langle A \rangle = \sum_m p_m \langle m|A|m\rangle. \qquad \text{(VIII.62)}$$

The mean value of an arbitrary function $F(A)$ is similarly obtained by replacing A by $F(A)$ in this formula; one easily derives therefrom the statistical distribution of the results of measurement of A.

21. The Density Operator [1]

It is especially convenient to describe the statistical mixture which was just described by means of the operator

$$\varrho = \sum_m |m\rangle\, p_m \,\langle m|. \qquad \text{(VIII.63)}$$

In this expression, the vectors $|m\rangle$ are normalized to unity (but not

[1] For a more complete presentation of the properties of the density operator, see U. Fano, *Description of States in Quantum Mechanics by Density Matrix and Operator Techniques*, Rev. Mod. Phys. 29 (1957) 74.

necessarily orthogonal) and the quantities p_m have the characteristic properties of statistical weights, namely

$$p_m \geqslant 0, \qquad \sum_m p_m = 1. \tag{VIII.64}$$

The operator ϱ is called *density operator* or *statistical operator.*

The average value of the observable A is the trace of ϱA:

$$\langle A \rangle = \operatorname{Tr} \varrho A. \tag{VIII.65}$$

Indeed

$$\operatorname{Tr} \varrho A = \sum_m p_m \operatorname{Tr} (|m\rangle \langle m|A).$$

To prove the equivalence of equations (VIII.62) and (VIII.65) it suffices to show that

$$\operatorname{Tr} (|m\rangle \langle m|A) = \langle m|A|m\rangle.$$

Now, since the operator $P_m \equiv |m\rangle \langle m|$ is a projector of trace unity [eq. (VII.88)],

$$\begin{aligned} \operatorname{Tr} P_m A &= \operatorname{Tr} P_m{}^2 A = \operatorname{Tr} P_m A P_m = \operatorname{Tr} |m\rangle \langle m|A|m\rangle \langle m| \\ &= \langle m|A|m\rangle \operatorname{Tr} P_m = \langle m|A|m\rangle. \end{aligned}$$

The same proof applied to the special case $A=1$ yields the normalization condition

$$\operatorname{Tr} \varrho = 1.$$

This argument applies equally well to any function of the observable A and gives

$$\boxed{\langle F(A) \rangle = \operatorname{Tr} \varrho\, F(A).}$$

Knowing ϱ, it is thus possible to derive the statistical distribution of the results of measurement of A.

Very generally, if P_D is the projector upon the subspace spanned by the eigenvectors of A belonging to the eigenvalues located in a certain domain D of the spectrum of A, the probability w_D that the result of measurement belongs to the domain D is $\sum_m p_m \langle m|P_D|m\rangle$ [cf. eq. (VIII.5)], that is

$$\boxed{w_D = \operatorname{Tr} \varrho\, P_D.} \tag{VIII.66}$$

In particular, the probability of finding the system in the quantum state represented by the vector $|\chi\rangle$ (of norm unity) is

$$w_\chi = \mathrm{Tr}\,(\varrho|\chi\rangle\,\langle\chi|) = \langle\chi|\varrho|\chi\rangle. \qquad \text{(VIII.67)}$$

Since it is sufficient to give ϱ to be able to calculate all measurable physical quantities, average values, and statistical distributions of measurements, *we shall henceforth consider identical two statistical mixtures possessing the same density operator*: every quantized statistical mixture is exactly and completely defined by its density operator.

To complete the generalization of the postulates of § 2 to statistical mixtures, one still has to define which density operator represents the dynamical state of the system once the measurement is completed. We restrict ourselves, as in § 2, to the case of an ideal measurement. If the observation carried out on the system indicates that it is in an eigenstate of A belonging to the domain D defined above, the density operator after measurement is, to within a normalization constant, the projection $P_D\varrho P_D$ of the operator ϱ representing the statistical mixture before measurement. The constant is determined by the condition that this operator have a trace equal to unity. It is therefore equal to the inverse of the quantity $\mathrm{Tr}\,P_D\,\varrho P_D = \mathrm{Tr}\,\varrho P_D = w_D$. The (non-causal) evolution of the density operator in the course of the measurement, therefore, ultimately corresponds to the scheme [1]).

$$\boxed{\varrho \to \begin{array}{l}\text{ideal measurement}\\ \text{yielding result } D\end{array} \to \frac{P_D\,\varrho P_D}{\mathrm{Tr}\,\varrho P_D}}$$

22. Evolution in Time of a Statistical Mixture

We start out in the Schrödinger "representation". Suppose that at time t_0 the dynamical state of the system is represented by the mixture of vectors (of norm 1) $|1\rangle_0$, $|2\rangle_0$, ..., $|m\rangle_0$, ... with respective statistical weights $p_1, p_2, ..., p_m, ...$. Every component of the mixture evolves in time according to the law

$$|m\rangle_t = U(t, t_0)|m\rangle_0$$

---

[1]) To justify this extension of the postulate of filtering of a wave packet, one must have recourse to the detailed study of the mechanism of measurement in Quantum Mechanics. On this subject, see references cited in footnote p. 157; see also U. Fano, *loc. cit.*

and the system is represented at time t by the mixture of vectors $|1\rangle_t, |2\rangle_t, \ldots, |m\rangle_t, \ldots$ with the same respective statistical weights $p_1, p_2, \ldots, p_m, \ldots$ as the vectors occurring in the initial mixture. $U(t, t_0)$ is the unitary evolution operator defined in § 8.

From this we derive the causal law of motion of the density operator

$$\begin{aligned}\varrho_t &= \sum_m |m\rangle_t\, p_m\, {}_t\langle m| = \sum_m U(t, t_0)|m\rangle_0\, p_m\, {}_0\langle m|U^\dagger(t, t_0)\\ &= U(t, t_0)\Big(\sum_m |m\rangle_0\, p_m\, {}_0\langle m|\Big)\, U^\dagger(t, t_0)\\ &= U(t, t_0)\,\varrho_0\, U^\dagger(t, t_0).\end{aligned}$$

The density operator at time t arises from the initial density operator by the unitary transformation $U(t, t_0)$.

Taking into account the equation of motion of operator U [eq. (VIII.32)] and its Hermitean conjugate, we have

$$\mathrm{i}\hbar \frac{\mathrm{d}}{\mathrm{d}t}\varrho_t = [H, \varrho_t]. \qquad \text{(VIII.68)}$$

This is the Schrödinger equation of the density operator. Although formally similar to the Heisenberg equation (VIII.40) from which it differs only in the sign of the commutator, it should not be confused with the latter. The quantities which occur in it are operators of the Schrödinger "representation".

One goes over from the Schrödinger "representation" to the Heisenberg "representation" by the unitary transformation $U^\dagger(t, t_0)$. Consequently, the density operator remains stationary in the Heisenberg "representation" ($\varrho_\mathrm{H} = \varrho_0$), whereas the physical quantities are represented by observables moving according to the Heisenberg equation (VIII.40).

23. Characteristic Properties of the Density Operator

The density operator ϱ is a *positive definite, Hermitean* operator (cf. Ch. VII, § 8) of *trace equal to unity.*

Indeed, according to the very definition of ϱ [eq. (VIII.63)], for any $|u\rangle$

$$\langle u|\varrho|u\rangle = \sum_m p_m |\langle u|m\rangle|^2 \geqslant 0 \qquad \text{(VIII.69)}$$

$$\mathrm{Tr}\,\varrho = \sum_m p_m\, \mathrm{Tr}\,(|m\rangle\,\langle m|) = \sum_m p_m = 1. \qquad \text{(VIII.70)}$$

In fact, since all the p_m are positive and since (Schwarz inequality)

$$|\langle u|m\rangle|^2 \leqslant \langle u|u\rangle,$$

one has

$$\langle u|\varrho|u\rangle \leqslant \langle u|u\rangle. \tag{VIII.71}$$

In other words, the operator $(1-\varrho)$ is likewise positive definite.

In the general theory of Hilbert space, one shows that a positive definite Hermitean operator whose trace is finite is an observable and that its eigenvalue spectrum is entirely discrete. The eigenvalues of ϱ in fact all lie between 0 and 1.

Conversely, any positive definite Hermitean operator ϱ of trace 1 may be regarded as a density operator. Indeed, it is an observable which may be written

$$\varrho = \sum_n \varpi_n P_n, \tag{VIII.72}$$

an expression in which $\varpi_1, \varpi_2, \ldots, \varpi_n, \ldots$ designate the non-zero eigenvalues, and $P_1, P_2, \ldots, P_n, \ldots$ the projectors upon their respective subspaces. If none of the eigenvalues is degenerate, each P_n is an elementary projector $P_n = |\bar{n}\rangle\,\langle\bar{n}|$ and

$$\varrho = \sum_n \varpi_n |\bar{n}\rangle\,\langle\bar{n}|. \tag{VIII.73}$$

Since $\sum_n \varpi_n = \mathrm{Tr}\,\varrho$, and $\varpi_n = \langle\bar{n}|\varrho|\bar{n}\rangle \geqslant 0$, the ϖ_n have the properties of statistical weights:

$$\varpi_n \geqslant 0, \qquad \sum_n \varpi_n = 1.$$

ϱ is therefore the density operator of the mixture formed with the vectors $|\bar{n}\rangle$, each having the statistical weight ϖ_n, respectively [1]). This argument can be readily extended to the case where some of the eigenvalues of ϱ are degenerate.

24. Pure States

The density operator formalism allows the treatment of pure states as special cases of statistical mixtures.

If one knows with certainty that the system is in the pure state

---

[1]) In expression (VIII.73), the $|\bar{n}\rangle$ are mutually orthogonal. On the other hand, the $|m\rangle$ occurring in the definition (VIII.63) do not necessarily have this property.

$|\chi\rangle$, one can represent that state by a statistical mixture having $|\chi\rangle$ (assumed to be of norm unity) as its sole element; its density operator is the projector

$$\varrho_\chi = |\chi\rangle \langle\chi| \qquad \text{(VIII.74)}$$

and one has:

$$\varrho_\chi^2 = \varrho_\chi. \qquad \text{(VIII.75)}$$

Conversely, if a density operator is a projector, it represents a pure state; the latter is the state on which the projection is carried out.

One can give two other criteria permitting to recognize whether or not a density operator ϱ represents a pure state:

1) A density operator ϱ can be put in the form of a linear combination of projectors in several different ways: the expression (VIII.63) is not unique. But in order that the operator ϱ defined by eq. (VIII.63) represent a pure state, it is *necessary* (and sufficient) that all the $|m\rangle$ be equal to each other to within a phase. They then represent one and the same dynamical state, which is the pure state desired (Problem VIII.7).

2) Any density operator ϱ – i.e. any positive definite Hermitean operator of trace 1 – possesses the property $\mathrm{Tr}\,\varrho^2 \leqslant 1$. In order that it represent a pure state it is (necessary and) *sufficient* that (Problem VIII.8)

$$\mathrm{Tr}\,\varrho^2 = 1. \qquad \text{(VIII.76)}$$

In conclusion, *it is always possible to represent the dynamical state of a system by its density operator, whether that state be completely or incompletely known.* In fact, the specification of this operator is sufficient to determine all physically measurable quantities which the Quantum Theory is in a position to furnish; eq. (VIII.66) plays the role which is played by eq. (VIII.5) when one uses vectors to represent the states. This procedure has the advantage of providing a uniform treatment for the pure states and the mixtures. Moreover, *the density operator representing the state of a system is defined in a unique manner,* while the vector representing a pure state is at best defined only to within a phase factor. Furthermore, the definition of the vector mixture representing an incompletely known dynamical state is fraught with a still higher degree of arbitrariness.

25. Classical and Quantum Statistics

In Classical Mechanics, a dynamical state is defined by a point in phase space; a statistical mixture of states is represented by a fluid in phase space whose density $\varrho_{\text{cl.}}$ at a point is equal to the probability of finding the system in the state defined by that point.

There is a remarkable parallelism between the classical density in phase $\varrho_{\text{cl.}}$ and the density operator ϱ of the Quantum Theory. $\varrho_{\text{cl.}}$ is a real, positive quantity whose integral over all phase space is equal to unity:

$$\iint \varrho_{\text{cl.}} \, \mathrm{d}q \, \mathrm{d}p = 1, \tag{VIII.77}$$

while ϱ is a Hermitean operator whose eigenvalues are all positive (positive definite operator), and whose trace equals unity.

From a knowledge of $\varrho_{\text{cl.}}$ at a given instant, one deduces the mean value $\langle A \rangle_{\text{cl.}}$ of any function $A_{\text{cl.}}$ of the dynamical variables q and p, by integrating $\varrho_{\text{cl.}} \, A_{\text{cl.}}$ over all phase space

$$\langle A \rangle_{\text{cl.}} = \iint \varrho_{\text{cl.}} \, A_{\text{cl.}} \, \mathrm{d}q \, \mathrm{d}p. \tag{VIII.78}$$

The evolution in time of $\varrho_{\text{cl.}}$ is given by the equation

$$\frac{\partial \varrho_{\text{cl.}}}{\partial t} = \{H_{\text{cl.}}, \varrho_{\text{cl.}}\} \tag{VIII.79}$$

[this equation is not to be confused with eq. (VIII.42)].

The equations (VIII.78) and (VIII.79) are the classical analogues of eqs. (VIII.65) and (VIII.68), respectively. One goes over from the expressions of Classical Theory to those of the Quantum Theory by replacing ordinary quantities by observables, Poisson brackets by commutators (to within the coefficient $i\hbar$) and *the integration over all phase space by the trace.*

This new expression of the Correspondence Principle turns out to be extremely valuable when one seeks to extend the fundamental results of Classical Statistical Thermodynamics to the quantum domain. Most of the classical arguments can be taken over without change. We shall merely point out the main results here.

The state of a quantum system *in thermodynamic equilibrium at temperature T* is represented by the operator

$$\varrho = N \, \mathrm{e}^{-H/kT}; \tag{VIII.80}$$

H is the Hamiltonian of the system, k is the Boltzmann constant. N is a normalization constant adjusted in such a way that $\operatorname{Tr} \varrho = 1$. The various thermodynamic functions of such a system are calculated as in Classical Theory, by means of the *partition function*

$$Z(\mu) = \operatorname{Tr} \mathrm{e}^{-\mu H}. \qquad \text{(VIII.81)}$$

Thus the free energy $\mathscr{F}$, the entropy S, and the energy E are respectively given by the following expressions, written with $\mu = 1/kT$:

$$\mathscr{F} = -kT \ln Z, \qquad \text{(VIII.82)}$$

$$S = k\left(\ln Z - \mu \frac{\partial}{\partial \mu} \ln Z\right), \qquad \text{(VIII.83)}$$

$$E \equiv \langle H \rangle = -\frac{\partial}{\partial \mu} \ln Z. \qquad \text{(VIII.84)}$$

More generally, the entropy of a system is given, in accordance with the Correspondence Principle, by the mean value of the operator $-k \ln \varrho$, that is

$$S = -k \operatorname{Tr} (\varrho \ln \varrho). \qquad \text{(VIII.85)}$$

From this the equilibrium distribution (VIII.80) is easily derived: it is the distribution corresponding to a given mean value of the energy for which the entropy is a maximum.

EXERCISES AND PROBLEMS

1. Starting from the fundamental postulate on mean values, derive expression (VIII.5), which gives the probability law for the results of measurement of a given quantity.

2. Consider a quantum system possessing a classical analogue in one dimension. Derive from the commutation relation $[q, p] = \mathrm{i}\hbar$ that the spectrum of p is entirely non-degenerate and continuous, and extends from $+\infty$ to $-\infty$. The suitably normalized eigenvectors of p form a complete orthonormal system in $\mathscr{E}$. Show that with a suitable choice of the phases of the vectors $|p'\rangle$ of this system, the action of the unitary operator $\exp(\mathrm{i}\varpi q/\hbar)$ (ϖ an arbitrary constant) on these vectors yields

$$\exp(\mathrm{i}\varpi q/\hbar)\, |p'\rangle = |p' + \varpi\rangle,$$

and that the operator q has as its matrix element:

$$\langle p'|q|p''\rangle = \mathrm{i}\hbar\, \delta'(p' - p'').$$

Solve the eigenvalue problem of q in this representation.

3. The derivative of an operator $A(\xi)$ depending explicitly on a continuous parameter ξ is by definition

$$\frac{\mathrm{d}A}{\mathrm{d}\xi} = \lim_{\varepsilon\to 0} \frac{A(\xi+\varepsilon) - A(\xi)}{\varepsilon}.$$

Show that:

1) if $A(\xi)$ is a function of an observable or of several commuting observables, its derivative is obtained by means of the ordinary rules of differentiation. In particular, if O is an observable:

$$\frac{\mathrm{d}}{\mathrm{d}\xi}(\mathrm{e}^{\mathrm{i}O\xi}) = \mathrm{i}\, O\, \mathrm{e}^{\mathrm{i}O\xi};$$

2) if two operators are differentiable, then

$$\frac{\mathrm{d}}{\mathrm{d}\xi}(AB) = \frac{\mathrm{d}A}{\mathrm{d}\xi} B + A \frac{\mathrm{d}B}{\mathrm{d}\xi}.$$

In particular:

$$\frac{\mathrm{d}}{\mathrm{d}\xi} A^2 = \frac{\mathrm{d}A}{\mathrm{d}\xi} A + A \frac{\mathrm{d}A}{\mathrm{d}\xi};$$

3) if A is differentiable and possesses an inverse, one has

$$\frac{\mathrm{d}}{\mathrm{d}\xi} A^{-1} = -A^{-1} \frac{\mathrm{d}A}{\mathrm{d}\xi} A^{-1}.$$

4. Show that the operator $B(t)$ defined by the expression

$$B(t) = \mathrm{e}^{\mathrm{i}At} B_0\, \mathrm{e}^{-\mathrm{i}At},$$

where A and B_0 are operators independent of t, is a solution of the integral equation

$$B(t) = B_0 + \mathrm{i}[A, \int_0^t B(\tau)\, \mathrm{d}\tau].$$

Solving this equation by iteration, one obtains the expansion of $B(t)$ in a power series of t. From this, derive the identity between operators

$$\mathrm{e}^{\mathrm{i}A} B\, \mathrm{e}^{-\mathrm{i}A} = B + \mathrm{i}[A,B] + \frac{\mathrm{i}^2}{2!}[A,[A,B]] + \ldots + \frac{\mathrm{i}^n}{n!} \overbrace{[A,[A,\ldots[A,[A,B]]\ldots]]}^{n \text{ brackets}}.$$

N.B. We agree to regard $[A, B]$ as the operator resulting from the action of A upon B, and to denote this operator by $A\{B\}$; $A^n\{B\}$ then denotes the action of A repeated n times; following this notation

$$A^0\{B\} \equiv B, \qquad A\{B\} \equiv [A, B], \qquad A^2\{B\} \equiv [A, [A, B]], \text{ etc.}$$

and the identity is simply written

$$\mathrm{e}^{\mathrm{i}A} B\, \mathrm{e}^{-\mathrm{i}A} = \sum_{n=0}^{\infty} \frac{\mathrm{i}^n}{n!} A^n\{B\}.$$

5. Let $A(\xi)$ be an operator depending upon the continuous parameter ξ, $\mathrm{d}A/\mathrm{d}\xi$ its derivative with respect to ξ. Show the identity between operators

$$\mathrm{e}^{-\mathrm{i}A} \frac{\mathrm{d}}{\mathrm{d}\xi} (\mathrm{e}^{\mathrm{i}A}) = \mathrm{i} \sum_{n=0}^{\infty} \frac{(-\mathrm{i})^n}{(n+1)!} A^n \left\{ \frac{\mathrm{d}A}{\mathrm{d}\xi} \right\}$$

(notation as in note of Problem 4).

6. If the operator $U(t)$, differentiable with respect to t, is unitary, the operator

$$H(t) \equiv \mathrm{i}\hbar \frac{\mathrm{d}U}{\mathrm{d}t} U^{\dagger}$$

is necessarily Hermitean.

Conversely, if $U(t)$ satisfies the equation

$$\mathrm{i}\hbar \frac{\mathrm{d}}{\mathrm{d}t} U(t) = HU,$$

where H is a Hermitean operator possibly depending upon t, $U^{\dagger} U$ is independent of t, whereas $UU^{\dagger}$ is a solution of the equation

$$\mathrm{i}\hbar \frac{\mathrm{d}}{\mathrm{d}t} UU^{\dagger} = [H, UU^{\dagger}].$$

In particular, if U is unitary when $t = t_0$, it remains so for all values of t.

7. Let $|1\rangle, |2\rangle, \ldots, |M\rangle$ be a sequence of vectors of norm 1 but not necessarily orthogonal. Show that the necessary and sufficient condition that the density operator $\varrho = \frac{1}{M} \sum_{i=1}^{M} |i\rangle \langle i|$ represent a pure state is that these M vectors are equal to within a phase.

8. Show that for the operator ϱ, Hermitean definite of trace 1, to represent a pure state, it is necessary and sufficient that $\mathrm{Tr}\, \varrho^2 = 1$.

Wisdom hath builded her house,
She hath hewn out her seven pillars:
She hath killed her beasts;
She hath mingled her wine;
She hath also furnished her table.
She hath sent forth her maidens:
She crieth upon the highest places of the city,
Whoso is simple, let him turn in hither!

(*Proverbs* ix: 1–4.)

PART TWO

SIMPLE SYSTEMS

CHAPTER IX

SOLUTION OF THE SCHRÖDINGER EQUATION BY SEPARATION OF VARIABLES. CENTRAL POTENTIAL

1. Introduction

The study of a physical system essentially consists in solving its time-independent Schrödinger equation. In particular, this eigenvalue equation enters directly in the two types of problems most frequently encountered in quantum physics, namely:

(*i*) the determination of the energy levels of bound states: they are the eigenvalues of the discrete spectrum of the Hamiltonian;

(*ii*) the determination of collision cross sections: as we shall prove further on (Ch. X) they are derived from the asymptotic form of the eigenfunctions for unbound states.

In Wave Mechanics, the Schrödinger equation is a second-order partial differential equation. For a one-dimensional system, the latter reduces to a differential equation; the study of the eigenvalue problem in this simple case was already made (Ch. III). The problem is generally much more difficult when the system has a larger number of dimensions. However, the symmetry properties the Hamiltonian may possess can facilitate its solution. In particular, it may happen that an appropriate change of variables leads to a partial differential equation whose variables are separable. The eigenvalue problem then splits into several simpler eigenvalue problems involving a smaller number of dimensions.

This occurs for a particle in a central potential, that is to say, in a potential depending only upon the distance r of the particle from a center of force, and not upon the direction of the vector $\boldsymbol{r}$ connecting that center with the particle. Since the Hamiltonian has spherical symmetry, the variables separate completely when one treats the problem in spherical polar coordinates. The solution of the Schrödinger equation, after separation of the angular variables, reduces to the solution of a differential equation involving only the radial variable, an equation which can always be integrated numerically.

The major portion of this chapter is devoted to the solution of the

Schrödinger equation of a particle in a central potential. The general treatment is given in Sec. I. Section II deals with the case of the free particle, and the particle in a "square" central potential.

Another very simple example of the separation of variables forms the subject of Sec. III, namely the center-of-mass motion of a system of particles. As in Classical Mechanics, this motion can be separated from the relative motion whenever the interaction depends only upon the relative positions of the particles with respect to each other.

I. PARTICLE IN A CENTRAL POTENTIAL. GENERAL TREATMENT

2. Expression of the Hamiltonian in Spherical Polar Coordinates

In this section, we intend to study the Schrödinger equation of a particle of mass m in a central potential $V(r)$. If $\boldsymbol{p}$ is the momentum of the particle, and $\boldsymbol{r}$ its position vector, the Hamiltonian is

$$H = \frac{\boldsymbol{p}^2}{2m} + V(r) \tag{IX.1}$$

and the time-independent Schrödinger equation is written

$$H\psi(\boldsymbol{r}) \equiv \left[-\frac{\hbar^2}{2m} \triangle + V(r) \right] \psi(\boldsymbol{r}) = E\psi(\boldsymbol{r}). \tag{IX.2}$$

Since the Hamiltonian has spherical symmetry, we study the problem in spherical polar coordinates.

The polar axis we choose is the z axis according to custom, and the cartesian coordinates (x, y, z) are given as functions of the polar coordinates (r, θ, φ) by the formulae (cf. Fig. IX.1)

$$\begin{aligned} x &= r \sin\theta \cos\varphi \\ y &= r \sin\theta \sin\varphi \\ z &= r \cos\theta. \end{aligned} \tag{I}$$

The expression of the potential energy V as a function of the polar coordinates is already given; we are to find the expression for the kinetic energy $\boldsymbol{p}^2/2m$, in other words to express in polar coordinates the differential operator:

$$-\frac{\hbar^2}{2m} \triangle \equiv -\frac{\hbar^2}{2m} \left(\frac{\partial^2}{\partial x^2} + \frac{\partial^2}{\partial y^2} + \frac{\partial^2}{\partial z^2} \right).$$

This can be done directly by means of the transformation formulae (I) by the usual techniques of the differential calculus. The calculation is lengthy, but not particularly difficult; it will not be given here.

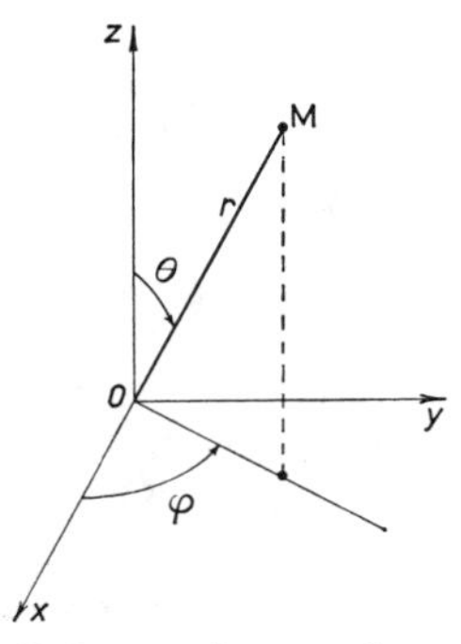

Fig. IX.1. Polar and cartesian coordinates.

However, in order to better grasp the physical significance of the result, we shall seek to express the kinetic energy $\boldsymbol{p}^2/2m$, not as function of the differential operators $\partial/\partial r$, $\partial/\partial\theta$, $\partial/\partial\varphi$ themselves, but as function of Hermitean operators constructed with these operators whose physical interpretation is more apparent.

Hence, rather than using the differential operator $\partial/\partial\varphi$ directly, it is more advisable to use the component along the z axis of the angular momentum which, according to eq. (V.49), has the explicit form:

$$l_z \equiv x\, p_y - y\, p_x = \frac{\hbar}{\mathrm{i}} \frac{\partial}{\partial\varphi}. \tag{IX.3}$$

Since $V(r)$ does not depend upon φ, it is clear that l_z commutes with the potential energy. However, l_z also commutes with the kinetic energy $\boldsymbol{p}^2/2m$ as one can easily verify by using the definition of l_z itself, and the fundamental commutation relations (Problem IX.4) [1])

$$[r_i, p_j] = \mathrm{i}\hbar\delta_{ij}. \tag{IX.4}$$

Thus l_z commutes with the Hamiltonian H. Taking Ox, Oy as polar axes, one arrives at the same conclusions for l_x, l_y. Hence *the three components* l_x, l_y, l_z *of the angular momentum*:

$$\boldsymbol{l} \equiv \boldsymbol{r} \times \boldsymbol{p} = \frac{\hbar}{\mathrm{i}} (\boldsymbol{r} \times \nabla) \tag{IX.5}$$

---

[1]) Henceforth we use the indices $i, j = 1, 2$, or 3 to label the components of the vectors along the axes Ox, Oy, or Oz, respectively; thus $r_1 \equiv x$, $p_1 \equiv p_x$, etc.

commute with the Hamiltonian. For this reason, we shall use these operators rather than the differential operators $\partial/\partial\theta$, $\partial/\partial\varphi$ themselves.

In the same spirit, we use the radial momentum

$$p_r \equiv \frac{\hbar}{\mathrm{i}} \frac{1}{r} \frac{\partial}{\partial r} r = \frac{\hbar}{\mathrm{i}} \left(\frac{\partial}{\partial r} + \frac{1}{r} \right) \tag{IX.6}$$

rather than the operator $(\hbar/\mathrm{i})\,(\partial/\partial r)$ which is not Hermitean (cf. Problem IX.1).

To establish this Hermitean property of p_r, let us examine under what condition the expression $\langle\psi, p_r\psi\rangle$, in which $\psi(\boldsymbol{r})$ is any square-integrable function, is real. One must have

$$\begin{aligned} 0 = \langle\psi, p_r\,\psi\rangle - \langle\psi, p_r\,\psi\rangle^* &\equiv \int [\psi^*(p_r\,\psi) - (p_r\,\psi)^*\psi]\, \mathrm{d}\boldsymbol{r} \\ &= \frac{\hbar}{\mathrm{i}} \int_0^\pi \sin\theta\, \mathrm{d}\theta \int_0^{2\pi} \mathrm{d}\varphi \int_0^\infty \left[\frac{\partial}{\partial r} |r\,\psi|^2 \right] \mathrm{d}r. \end{aligned}$$

Since $r\psi$ vanishes as $r \to \infty$, the integral with respect to r is equal to its value taken at the origin. The operator p_r is therefore Hermitean only if one restricts oneself to square-integrable functions subject to the condition [1]):

$$\lim_{r\to 0} r\,\psi(\boldsymbol{r}) = 0. \tag{IX.7}$$

It follows from the definition that p_r commutes with any function of θ and φ as well as with the three components of $\boldsymbol{l}$, but that, on the other hand,

$$[r, p_r] = \mathrm{i}\hbar. \tag{IX.8}$$

One has the operator identity

$$\boxed{\boldsymbol{p}^2 = p_r{}^2 + \frac{\boldsymbol{l}^2}{r^2} \qquad (r \neq 0)} \tag{IX.9}$$

according to which the action of the operators p^2 and $p_r{}^2 + (\boldsymbol{l}^2/r^2)$ on

---

[1]) p_r is Hermitean but is not an observable. Whatever the constant ϖ, the solution of the differential equation

$$p_r\, f(r) \equiv \frac{\hbar}{\mathrm{i}} \frac{1}{r} \frac{\mathrm{d}}{\mathrm{d}r} [r\, f(r)] = \varpi\, f(r)$$

is, to within a constant, equal to $\exp(\mathrm{i}\varpi r/\hbar)/r$; it never satisfies condition (IX.7); the eigenvalue problem of p_r has therefore no solution.

a function $\psi(\boldsymbol{r})$ yields the same result for all non-zero values of r.

To prove this, we use for the calculation of $\boldsymbol{l}^2$ the identity

$$(\boldsymbol{A}\times\boldsymbol{B})^2 = A^2B^2 - (\boldsymbol{A}\cdot\boldsymbol{B})^2,$$

by substituting for the vectors $\boldsymbol{A}$ and $\boldsymbol{B}$ the operators $\boldsymbol{r}$ and $\boldsymbol{p}$. However, since the components of $\boldsymbol{r}$ and $\boldsymbol{p}$ do not necessarily commute with each other, the above identity holds true only if one preserves the order of the operators, namely

$$\boldsymbol{l}^2 \equiv (\boldsymbol{r}\times\boldsymbol{p})\cdot(\boldsymbol{r}\times\boldsymbol{p}) = \sum_{i,j} (r_i p_j r_i p_j - r_i p_j r_j p_i).$$

Repeated application of the commutation relations (IX.4) allows us to rewrite this identity

$$\boldsymbol{l}^2 = r^2\boldsymbol{p}^2 - (\boldsymbol{r}\cdot\boldsymbol{p})^2 + \mathrm{i}\hbar(\boldsymbol{r}\cdot\boldsymbol{p}). \tag{IX.10}$$

However, since $r = (x^2+y^2+z^2)^{\frac{1}{2}}$,

$$x\frac{\partial}{\partial x} + y\frac{\partial}{\partial y} + z\frac{\partial}{\partial z} = r\frac{\partial}{\partial r},$$

hence

$$\boldsymbol{r}\cdot\boldsymbol{p} = r\frac{\hbar}{\mathrm{i}}\frac{\partial}{\partial r} = r\,p_r + \mathrm{i}\hbar$$

and, taking into account the commutation relation (IX.8),

$$(\boldsymbol{r}\cdot\boldsymbol{p})^2 - \mathrm{i}\hbar(\boldsymbol{r}\cdot\boldsymbol{p}) \equiv [(\boldsymbol{r}\cdot\boldsymbol{p}) - \mathrm{i}\hbar](\boldsymbol{r}\cdot\boldsymbol{p}) = rp_r(rp_r + \mathrm{i}\hbar) = r^2{p_r}^2.$$

The right-hand side of the identity (IX.10) is thus equal to

$$r^2(\boldsymbol{p}^2 - {p_r}^2).$$

Dividing through by r^2 term by term, one obtains the identity (IX.9), valid everywhere except possibly for $r=0$.

From (IX.9) we obtain the expression for the kinetic energy by dividing term by term by $2m$. We thus obtain the expression for the Hamiltonian in spherical polar coordinates:

$$H = \frac{{p_r}^2}{2m} + \frac{\boldsymbol{l}^2}{2mr^2} + V(r). \tag{IX.11}$$

Like the energy of the corresponding classical particle, it is the sum of three terms, the "radial kinetic energy" $p_r^2/2m$, the "kinetic energy of rotation" $\boldsymbol{l}^2/2mr^2$ (mr^2 is the moment of inertia with respect to the origin), and the potential energy $V(r)$.

The direct calculation mentioned at the beginning of the paragraph would evidently have led to the same expression, $\boldsymbol{l}^2$ then appearing in its explicit form

$$\boldsymbol{l}^2 = -\frac{\hbar^2}{\sin^2\theta}\left[\sin\theta\frac{\partial}{\partial\theta}\left(\sin\theta\frac{\partial}{\partial\theta}\right)+\frac{\partial^2}{\partial\varphi^2}\right]. \qquad \text{(IX.12)}$$

In conclusion, the Schrödinger equation in spherical polar coordinates is written

$$\left[\frac{p_r^2}{2m}+\frac{\boldsymbol{l}^2}{2mr^2}+V(r)\right]\psi(r, \theta, \varphi) = E\psi(r, \theta, \varphi). \qquad \text{(IX.13)}$$

However, the solutions of this equation can be retained as solutions of the Schrödinger equation only after examination of their behavior at the origin. In fact, the validity at the origin of expression (IX.11) is not automatically assured for any function upon which operator H may act. We merely state here without proof that equation (IX.13) is equivalent to the Schrödinger equation in all of space, *including the origin*, as long as ψ statisfies condition (IX.7), the condition of hermiticity of p_r.

3. Separation of the Angular Variables. Spherical Harmonics

It is easily seen from their expressions (IX.11) and (IX.12) that H and $\boldsymbol{l}^2$ commute. This was to be anticipated; once H commutes with l_x, l_y, l_z, it commutes with any function of these operators, and with $\boldsymbol{l}^2$ in particular. The observables H and $\boldsymbol{l}^2$ have (at least) one common basis. We are thus led to solve the eigenvalue problem of H in two steps: first, to solve the eigenvalue problem of $\boldsymbol{l}^2$, and second, to search for eigensolutions of $\boldsymbol{l}^2$ satisfying the Schrödinger equation. The particular form of the potential $V(r)$ enters only into the second step of this method of solution.

In the search for a complete system of eigenfunctions of $\boldsymbol{l}^2$, the variable r plays the role of a simple parameter and may be momentarily omitted, since the operator $\boldsymbol{l}^2$ acts exclusively upon the angular variables θ and φ.

$\boldsymbol{l}^2$ commutes with each component of the angular momentum [eq. (V.70)], and in particular with l_z. One shows in function theory that the eigenfunctions common to the differential operators $\boldsymbol{l}^2$ and l_z, defined respectively by expressions (IX.12) and (IX.3) are the *spherical harmonics* $Y_l^m(\theta, \varphi)$. The main properties of these functions are summarized in Appendix B (§ 10). Their construction will be given in detail when we come to the systematic study of angular momentum in Quantum Mechanics (Ch. XIII). They are labelled by the indices l and m; l can take on all integral positive or zero values, and m all integral values in the interval $(-l, +l)$, limits included. One has

$$\boldsymbol{l}^2 Y_l^m(\theta, \varphi) = l(l+1)\hbar^2 \, Y_l^m(\theta, \varphi) \tag{IX.14}$$

$$\left.\begin{array}{c} l_z Y_l^m(\theta, \varphi) = m\hbar \; Y_l^m(\theta, \varphi) \\ (l = 0, 1, 2, \ldots, \infty; \qquad m = -l, -l+1, \ldots, +l). \end{array}\right\} \tag{IX.15}$$

In the function space of square-integrable functions of (θ, φ), i.e. the space of square-integrable functions defined on the unit sphere, the spherical harmonics constitute a complete orthonormal set. One must realize that the scalar product is defined in that case as the integral over the unit sphere, and that the surface element is [1])

$$\mathrm{d}\Omega = \sin\theta \; \mathrm{d}\theta \; \mathrm{d}\varphi;$$

the orthonormality relations thus read

$$\int Y_l^{m*} Y_{l'}^{m'} \mathrm{d}\Omega \equiv \int_0^{2\pi} \mathrm{d}\varphi \int_0^{\pi} \sin\theta \, \mathrm{d}\theta \, Y_l^{m*}(\theta, \varphi) \, Y_{l'}^{m'}(\theta, \varphi) = \delta_{ll'} \, \delta_{mm'}. \tag{IX.16}$$

To each pair of quantum numbers (lm) corresponds a single spherical harmonic. Hence, the angular dependence of a function $\psi(r, \theta, \varphi)$ is completely determined if one requires this function to be a common eigenfunction of $\boldsymbol{l}^2$ and l_z corresponding to the respective eigenvalues $l(l+1)\hbar$ and $m\hbar$; $\psi(r, \theta, \varphi)$ is necessarily of the form $f(r) \, Y_l^m(\theta, \varphi)$.

4. The Radial Equation

We shall now tackle the second step of our method of solution of the Schrödinger equation. It consists in forming the eigenfunctions common to the commuting operators H, $\boldsymbol{l}^2$ and l_z. They are the solutions of the Schrödinger equation of the type

$$\psi_l^m(r, \theta, \varphi) = Y_l^m(\theta, \varphi) \, \chi_l(r). \tag{IX.17}$$

---

[1]) See the discussion at the end of Ch. VII, § 13. Similar precautions must be taken when writing the closure relation in the (θ, φ) representation [eq. (B.88)].

Since ψ_l^m is a solution of eq. (IX.13) and Y_l^m is an eigenfunction of $\boldsymbol{l}^2$ [eq. (IX.14)], $\chi_l(r)$ is a solution of the second-order, linear differential equation

$$\left[\frac{p_r^2}{2m} + \frac{l(l+1)\hbar^2}{2mr^2} + V(r) - E\right] \chi_l(r) = 0 \tag{IX.18}$$

where $p_r^2 \equiv -\hbar^2 \dfrac{1}{r} \dfrac{\mathrm{d}^2}{\mathrm{d}r^2} r$.

It is convenient to set

$$y_l(r) = r\chi_l(r) \tag{IX.19}$$

and to replace eq. (IX.18) by the equivalent radial equation

$$\left[-\frac{\hbar^2}{2m}\frac{\mathrm{d}^2}{\mathrm{d}r^2} + l(l+1)\frac{\hbar^2}{2mr^2} + V(r) - E\right] y_l(r) = 0; \tag{IX.20}$$

note the resemblance to the one-dimensional Schrödinger equation. Note also that the norm of ψ_l^m is given, after integration over angles, by the expression

$$\langle \psi_l^m, \psi_l^m \rangle = \int_0^\infty r^2 \, |\chi_l(r)|^2 \, \mathrm{d}r = \int_0^\infty |y_l(r)|^2 \, \mathrm{d}r \tag{IX.21}$$

and that the Hermiticity condition (IX.7) of p_r is equivalent to the condition

$$y_l(0) = 0. \tag{IX.22}$$

Not all the solutions of the radial equation (IX.20) should be kept. In order that ψ_l^m be acceptable as an eigenfunction, y_l must satisfy certain conditions of regularity. Indeed, it is necessary:

(*a*) to make sure by examining the behavior of y_l at the origin that ψ_l^m is actually a solution of the Schrödinger equation in all of space, including the origin;

(*b*) that this solution be normalizable (in the general sense defined in Ch. V, § 9).

In order to make these conditions of regularity specific, we shall carry out a systematic study of the behavior of the solutions of eq. (IX.20) at the origin.

We suppose that $V(r)$ is bounded in any finite interval, except possibly at the origin, where it may have a singularity in $1/r$.

These conditions are fulfilled in all cases of practical interest. With this assumption, eq. (IX.20) possesses a "regular" solution R_l (defined to within a constant) which vanishes at the origin as r^{l+1}, and all other solutions of this equation behave at the origin as $(1/r)^l$. [1])

To show this, we assume $V(r)$ to be analytic in the vicinity of the origin, and we seek a particular solution of the form

$$r^s(1+a_1r+a_2r^2+\ldots).$$

Substituting this expansion, the second derivative which can be derived from it, and the Taylor expansion of $V(r)$ into eq. (IX.20), and writing down that the coefficients of successive powers of r occurring in the left-hand side vanish, one obtains an infinite set of equations whose first (characteristic equation)

$$s(s-1)-l(l+1)=0$$

yields s, and the following equations yield $a_1, a_2, \ldots$ successively. In the present case, the equation in s has two solutions, $l+1$, and $-l$. If $s=l+1$, the calculation of the coefficients of the series can be carried out to all orders and leads to the "regular" solution R_l. If $s=-l$, one faces an impossibility and the method fails. However, if R_l is a solution of eq. (IX.20), it is easy to see that the function

$$R_l(r)\int^r \frac{\mathrm{d}r'}{R_l^2(r')}$$

which behaves as $(1/r)^l$ at the origin, is also a solution of this equation. The general solution is a linear combination of these two particular solutions.

Now, any solution in $(1/r)^l$ must be rejected since it does not satisfy at least one of the conditions (*a*) and (*b*).

Indeed, if $l\neq 0$, the integral of the square of its modulus diverges at the origin and, according to eq. (IX.21), the function ψ_l^m constructed with this solution does not belong to Hilbert space [condition (*b*)]. This divergence at the origin persists when one forms the eigendiffer-

[1]) This result was to be expected since the general solution of eq. (IX.20) in the vicinity of the origin is an approximate solution of the equation

$$\left[\frac{\mathrm{d}^2}{\mathrm{d}r^2}-\frac{l(l+1)}{r^2}\right]v(r)=0,$$

an equation whose general solution is $v(r)=ar^{l+1}+br^{-l}$ (a and b are arbitrary constants).

ential of $\psi_l{}^m$. This solution must therefore be rejected in any case, whether the eigenvalue E belongs to the discrete or the continuous spectrum.

The foregoing argument does not apply when $l=0$. But in that case, the corresponding wave function ψ_0 does not satisfy the Schrödinger equation [condition (*a*)]. In fact, ψ_0 behaves as $(1/r)$ at the origin, and since [eq. (A.12)] $\triangle(1/r) = -4\pi\delta(\mathbf{r})$,

$$(H-E)\,\psi_0 = \frac{2\pi\hbar^2}{m}\,\delta(\mathbf{r}).$$

One must therefore keep only the so-called "regular" solutions, that is, the solutions satisfying the condition at the origin (IX.22). With such a solution we can be sure that the function $\psi_l{}^m$ is a solution of the Schrödinger equation everywhere, including the origin [condition (*a*)]; moreover, since the normalization integral converges at the origin, the condition that $\psi_l{}^m$ or its eigendifferential belong to Hilbert space [condition (*b*)] depends solely upon the behavior of this solution at infinity.

Supplemented by the condition at the origin

$$y_l(0) = 0 \tag{IX.22}$$

the radial equation (IX.20) is the Schrödinger equation of a particle in one dimension of mass m, subjected to the potential $V(r) + [l(l+1)\hbar^2/2mr^2]$ in the region $(0, \infty)$, and to an infinitely repulsive potential in the region $(-\infty, 0)$. The solution of the three-dimensional Schrödinger equation is thus reduced to the solution of a one-dimensional Schrödinger equation. The properties stated in Ch. III (property of the Wronskian, asymptotic behavior of solutions, orthogonality relations, etc.) remain valid in that case in spite of the singularity in $l(l+1)/r^2$ of the "equivalent potential" at the origin.

5. Eigensolutions of the Radial Equation. Nature of the Spectrum

The nature of the energy spectrum and of the eigensolutions of the radial equation (IX.20) belonging to a given value of l may be deduced from the asymptotic behavior of the solutions of this equation which are regular at the origin. The discussion of Ch. III, § 10 may be taken over unchanged.

Suppose, for instance, that $V(r)$ approaches zero asymptotically more rapidly than $1/r$:

$$\lim_{r\to\infty} rV(r) = 0.$$

The energy spectrum contains two parts:

(*i*) if $E<0$, the solution which is regular at the origin grows indefinitely in absolute value as $\exp(\varkappa r)$ $(\varkappa=\sqrt{-2mE}/\hbar)$, except for certain discrete values $E_l^{(1)}, E_l^{(2)}, \ldots$, for which

$$y_l \underset{r\to\infty}{\sim} \mathrm{e}^{-\varkappa r}.$$

These values are the only possible eigenvalues. To each of them corresponds a radial function of finite norm.

(*ii*) if $E>0$, the solution which is regular at the origin oscillates indefinitely according to the law

$$y_l \underset{r\to\infty}{\sim} \sin(kr - \tfrac{1}{2}l\pi + \delta_l) \qquad \left(k = \frac{\sqrt{2mE}}{\hbar}\right).$$

It is acceptable as an eigensolution for any E, and represents an unbound eigenstate. The constant δ_l is called the *phase shift* (the additional term $-\frac{1}{2}l\pi$ has been added so that $\delta_l=0$ when $V(r)=0$; cf. § 7 below). δ_l is a very important quantity: it characterizes the asymptotic behavior of the regular solution and, as such, often enters into collision problems (Ch. X).

If V approaches zero as $1/r$ or more slowly (but monotonically) when $r\to\infty$, it is not possible to write the asymptotic forms as simply, but the essential result concerning the nature of the spectrum persists. It is an entirely non-degenerate spectrum consisting of a continuous portion, the half-axis of positive energies, and a (denumerably infinite) set of negative, discrete values.

It remains to be shown that for a given value of l, the ensemble of eigenfunctions $y_l(r)$ thus constructed forms a complete set, in the sense that any square-integrable function of r defined on the half-axis $(0, \infty)$ can be expanded in a series of these eigenfunctions. We shall assume that this actually holds true for all potential shapes we shall consider; otherwise, the Hamiltonian H would not be an observable.

6. Conclusions

In conclusion, we note that the observables H, $\boldsymbol{l}^2$, and l_z form a complete set of commuting observables. The problem of constructing the eigenfunctions common to H, $\boldsymbol{l}^2$ and l_z amounts to separating in the Schrödinger equation, the angular variables from the radial variable. If one fixes the eigenvalues $l(l+1)\hbar^2$ and $m\hbar$, of $\boldsymbol{l}^2$ and l_z, respectively, these eigenfunctions are of the form

$$\psi_l^m = r^{-1} y_l(r)\, Y_l^m(\theta, \varphi) \tag{IX.23}$$

where $y_l(r)$ is the solution of the radial equation (IX.20) which vanishes at the origin and remains bounded in all space.

One usually says that such an eigenfunction represents a state of angular momentum l or, more specifically, that the particle has an angular momentum l with a component m along the z axis. We recall that l and m are integers and that

$$l \geqslant 0, \qquad -l \leqslant m \leqslant l.$$

According to the traditional terminology of spectroscopy, l is the *azimuthal quantum number*, and m the *magnetic quantum number*. Tradition also requires that one label the states of lowest angular momenta by particular letters of the alphabet rather than by the value of their azimuthal quantum number: to the values 0, 1, 2, 3, 4, 5, ... of the angular momentum correspond the letters s, p, d, f, g, h, ..., respectively.

The nature of the spectrum of H depends upon the behavior at infinity of $V(r)$. In particular, if $V(r)$ approaches zero (monotonically), the energy spectrum consists of a certain number of discrete, negative values, and of the (continuous) ensemble of positive values.

Each of the values of the *continuous spectrum* is *infinitely degenerate*. Indeed, there exists an eigenfunction of positive energy E for all possible values (lm) of the angular momentum.

The energies of the *discrete spectrum* E_{kl} can be labelled by means of two indices, the azimuthal quantum number l and a radial quantum number k which serves to label the various eigenvalues of the radial equation belonging to a given angular momentum. There exists no *a priori* reason why the radial equations belonging to different values of l should have common eigenvalues: *in general*, the eigenvalues E_{kl} are all distinct, and they are respectively $(2l+1)$*-fold degenerate* since to each of them there correspond as many linearly independent eigenfunctions as there are possible values for the magnetic quantum number:

$$-l,\ -l+1,\ \ldots,\ +l.$$

For some very special shapes of the potential $V(r)$, it may happen that certain ones of these values E_{kl} coincide; in that case, the degeneracy is greater. We shall encounter this type of accidental degeneracy in the treatment of the hydrogen atom (Ch. XI) and of the three-dimensional, isotropic harmonic oscillator (Ch. XII).

II. CENTRAL SQUARE WELL POTENTIAL. FREE PARTICLE

7. Spherical Bessel Functions

If there exist regions of the interval $(0, \infty)$ where the potential $V(r)$ is constant:

$$V(r) = V_0 = \text{const.},$$

the radial equation assumes a particularly simple form, and its general solution is a linear combination of functions well-known in function theory, the spherical Bessel functions.

Let us first of all suppose that $E > V_0$. If one puts

$$k = \frac{\sqrt{2m(E-V_0)}}{\hbar}, \quad \varrho = kr, \tag{IX.24}$$

eq. (IX.20) can be written in the form

$$\left[\frac{\mathrm{d}^2}{\mathrm{d}\varrho^2} + \left(1 - \frac{l(l+1)}{\varrho^2}\right)\right] y_l = 0.$$

Likewise, the radial function $f_l = y_l/r$, considered as a function of φ, is a solution of the "spherical Bessel differential equation":

$$\left[\frac{\mathrm{d}^2}{\mathrm{d}\varrho^2} + \frac{2}{\varrho}\frac{\mathrm{d}}{\mathrm{d}\varrho} + \left(1 - \frac{l(l+1)}{\varrho^2}\right)\right] f_l = 0. \tag{IX.25}$$

The general solution of eq. (IX.25) is a linear combination of two particular solutions. The most commonly used particular solutions are described in Appendix B (§ 6). They are the functions j_l, n_l [1]), $h_l^{(+)}$, $h_l^{(-)}$. j_l is the only solution that is regular at the origin (behaving as ϱ^l); the three others have a pole of order $l+1$ there. j_l and n_l are real functions and behave like standing waves at infinity

$$j_l(\varrho) \underset{\varrho\to\infty}{\sim} \frac{\sin(\varrho - \frac{1}{2}l\pi)}{\varrho}, \qquad n_l(\varrho) \underset{\varrho\to\infty}{\sim} \frac{\cos(\varrho - \frac{1}{2}l\pi)}{\varrho}. \tag{IX.26}$$

---

1) Most authors designate by n_l the same function with the opposite sign.

The functions $h_l^{(+)} \equiv n_l + \mathrm{i} j_l$ and $h_l^{(-)} \equiv n_l - \mathrm{i} j_l$ behave asymptotically as outgoing and incoming waves, respectively:

$$h_l^{(+)} \underset{\varrho\to\infty}{\sim} \frac{\mathrm{e}^{\mathrm{i}(\varrho - \frac{1}{2}l\pi)}}{\varrho}, \qquad h_l^{(-)} \underset{\varrho\to\infty}{\sim} \frac{\mathrm{e}^{-\mathrm{i}(\varrho - \frac{1}{2}l\pi)}}{\varrho}. \tag{IX.27}$$

In the case where $E < V_0$, one puts

$$\varkappa = \frac{\sqrt{2m(V_0 - E)}}{\hbar}, \tag{IX.28}$$

and everything stated thus far remains valid provided we replace k everywhere by $\mathrm{i}\varkappa$. In particular, the asymptotic forms (IX.26) and (IX.27) remain valid. The only radial solution bounded at infinity is the function $h_l^{(+)}(\mathrm{i}\varkappa r)$; it approaches zero exponentially. More precisely, the function $\mathrm{i}^l h_l^{(+)}(\mathrm{i}\varkappa r)$ is a real function, equal to the product of a polynomial of degree l in $(1/\varkappa r)$ and $\exp(-\varkappa r)/\varkappa r$, and its asymptotic form is

$$\mathrm{i}^l h_l^{(+)}(\mathrm{i}\varkappa r) \underset{r\to\infty}{\sim} \frac{\mathrm{e}^{-\varkappa r}}{\varkappa r}. \tag{IX.29}$$

8. Free Particle. Plane Waves and Free Spherical Waves

The foregoing considerations apply in particular to the free particle. In that case $V(r) = 0$ over the entire interval $(0, \infty)$, and the Hamiltonian reduces to its kinetic energy term:

$$H = \frac{\boldsymbol{p}^2}{2m}.$$

Let us therefore look for eigensolutions common to H, $\boldsymbol{l}^2$ and l_z. A solution of angular momentum (lm) and energy E is a function of the form

$$Y_l^m(\theta, \varphi) f_l(r)$$

in which f_l is the solution of eq. (IX.25) bounded over the entire interval $(0, \infty)$.

If $E < 0$, the only solution bounded at infinity, $h_l^{(+)}(\mathrm{i}\varkappa r)$, has a pole of order $l+1$ at the origin. The eigenvalue problem has no solution; as was to be expected, there exists no eigenstate of negative energy.

If $E > 0$, equation (IX.25) has one and only one solution bounded everywhere: the function $j_l(kr)$. There thus exists an eigensolution of angular momentum (lm) for every positive value $E = \hbar^2 k^2/2m$ of the energy, namely the function

$$Y_l^m(\theta, \varphi)\, j_l(kr). \tag{IX.30}$$

Every eigensolution thus formed can be labelled by the two discrete indices l, m and by the continuous index k which can assume any value in the interval $(0, \infty)$; the ensemble of these spherical waves forms a complete orthonormal set (Problem IX.3).

The collection of plane waves $\exp(\mathrm{i}\boldsymbol{k}\cdot\boldsymbol{r})$ forms another complete, orthonormal set of eigenfunctions of the energy of the free particle. They are the eigenfunctions common to the observables p_x, p_y, p_z, that is to say the solutions corresponding to a well-defined value of the momentum $\boldsymbol{p}$. Every plane wave is defined by three continuous parameters, the three components of the vector $\boldsymbol{k}$, which can take on all values between $-\infty$ and $+\infty$.

The wave $\exp(\mathrm{i}\boldsymbol{k}\cdot\boldsymbol{r})$ represents a free particle of momentum $\hbar\boldsymbol{k}$ and of energy $E = \hbar^2 k^2/2m$. On the other hand, it does not represent a well-defined angular momentum state, just as the spherical wave (IX.30) does not represent a well-defined momentum state. This is not surprising since the three components of the momentum p_x, p_y, p_z do not simultaneously commute with $\boldsymbol{l}^2$ and l_z.

9. Expansion of a Plane Wave in Spherical Harmonics

Every energy eigenvalue of the free particle is infinitely degenerate. Since the spherical waves (IX.30) form a complete set, the denumerable set of spherical waves corresponding to a given value of the wave number k, spans the space of the eigenfunctions of energy $E = \hbar^2 k^2/2m$; therefore, the plane wave $\exp(\mathrm{i}\boldsymbol{k}\cdot\boldsymbol{r})$ may be expanded in a series of these functions:

$$\mathrm{e}^{\mathrm{i}\boldsymbol{k}\cdot\boldsymbol{r}} = \sum_{l=0}^{\infty} \sum_{m=-l}^{+l} a_{lm}(\boldsymbol{k})\, Y_l^m(\theta, \varphi)\, j_l(kr). \tag{IX.31}$$

If one chooses the z axis along $\boldsymbol{k}$, this wave can be written in the form $\exp(\mathrm{i}kr\cos\theta)$; it is independent of φ and the expansion (IX.31) contains only the terms $m = 0$ [1]). Let

$$\varrho = kr, \qquad u = \cos\theta.$$

---

[1]) Indeed:

$$l_z \exp(\mathrm{i}kr\cos\theta) = \frac{\hbar}{\mathrm{i}} \frac{\partial}{\partial\varphi} \exp(\mathrm{i}kr\cos\theta) = 0.$$

The expansion of the plane wave reduces to an expansion in a series of Legendre polynomials [cf. eq. (B.94)]

$$\mathrm{e}^{\mathrm{i}\varrho u} = \sum_{l=0}^{\infty} c_l \, j_l(\varrho) \, P_l(u). \tag{IX.32}$$

To determine the coefficients c_l, one may proceed as follows. Differentiating the series (IX.32) term by term with respect to ϱ, we obtain

$$\mathrm{i}u\mathrm{e}^{\mathrm{i}\varrho u} = \sum_l c_l \frac{\mathrm{d}j_l}{\mathrm{d}\varrho} P_l. \tag{IX.33}$$

But this expansion is also written, taking into account the recursion relation (B.78) of the Legendre polynomials,

$$\mathrm{i}u\mathrm{e}^{\mathrm{i}\varrho u} = \mathrm{i} \sum_l c_l j_l u P_l = \mathrm{i} \sum_l \left(\frac{l+1}{2l+3} c_{l+1} j_{l+1} + \frac{l}{2l-1} c_{l-1} j_{l-1} \right) P_l. \tag{IX.34}$$

Equating the coefficients of P_l in the expansions (IX.33) and (IX.34) and using the recursion relations (B.53) and (B.54) of the spherical Bessel functions, one obtains the relations

$$l \left(\frac{1}{2l+1} c_l - \frac{\mathrm{i}}{2l-1} c_{l-1} \right) j_{l-1}(\varrho) = (l+1) \left(\frac{1}{2l+1} c_l + \frac{\mathrm{i}}{2l+3} c_{l+1} \right) j_{l+1}(\varrho).$$

In order that they be satisfied for any ϱ, it is necessary and sufficient that the expressions between brackets all vanish, or that

$$\frac{1}{2l+3} c_{l+1} = \frac{\mathrm{i}}{2l+1} c_l \qquad (l = 0, 1, 2, \ldots, \infty),$$

that is

$$c_l = (2l+1) \mathrm{i}^l c_0.$$

The coefficient c_0 is obtained by writing the expansion $\exp(\mathrm{i}\varrho u)$ for $\varrho = 0$; since $j_l(0) = \delta_{l0}$, $c_0 = 1$.

In conclusion, the expansion of the plane wave may be written

$$\boxed{\mathrm{e}^{\mathrm{i}kz} \equiv \mathrm{e}^{\mathrm{i}kr\cos\theta} = \sum_{l=0}^{\infty} (2l+1) \, \mathrm{i}^l j_l(kr) \, P_l(\cos\theta).} \tag{IX.35}$$

To obtain this expansion in any polar coordinate system, one notes that the angle θ occurring in the expansion (IX.35) is the angle between the vectors **k** and **r**. Let us denote by $\hat{\mathbf{k}}$ and $\hat{\mathbf{r}}$ the respective angular coordinates of these vectors. According to the addition theorem of spherical harmonics [eq. (B.98)]

$$P_l(\cos\theta) = \frac{4\pi}{2l+1} \sum_{m=-l}^{+l} Y_l^{m*}(\hat{\mathbf{k}})\, Y_l^m(\hat{\mathbf{r}}).$$

Substitution of this expression in the expansion (IX.35) yields

$$e^{i\mathbf{k}\cdot\mathbf{r}} = 4\pi \sum_{l=0}^{\infty} \sum_{m=-l}^{+l} i^l j_l(kr)\, Y_l^{m*}(\hat{\mathbf{k}})\, Y_l^m(\hat{\mathbf{r}}). \tag{IX.36}$$

10. Study of the Spherical Square Well

By way of an illustration of the problem of a particle in a central force field, we treat the "square well" (Fig. IX.2)

$$V(r) = \begin{cases} -V_0 & r<a \\ 0 & r>a. \end{cases} \tag{IX.37}$$

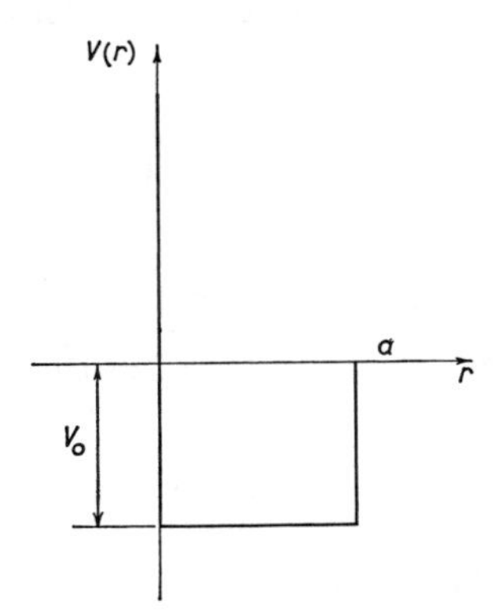

Fig. IX.2. Spherical square well.

The solution of the radial equation is entirely analogous to that of the one-dimensional square well. One knows how to write down the general solution of the Schrödinger equation in each of the two regions $(0, a)$ and (a, ∞): it is a linear combination of spherical Bessel functions. The conditions of regularity at the origin and at infinity, and the continuity condition of the function and of its logarithmic derivative at the point $r=a$ allow us to determine the acceptable solutions.

Let E be the energy of the particle, and put $K = [2m(E+V_0)]^{\frac{1}{2}}/\hbar$. In the inner region $(0<r<a)$ the radial equation reads

$$\left[\frac{\mathrm{d}^2}{\mathrm{d}r^2} + \frac{2}{r}\frac{\mathrm{d}}{\mathrm{d}r} + \left(K^2 - \frac{l(l+1)}{r^2}\right)\right] f_l(r) = 0.$$

Setting $\varrho = Kr$, one again arrives at equation (IX.25). There is only one solution that is regular at the origin:

$$A j_l(Kr) \qquad (A \text{ is a normalization constant}).$$

In the outer region $(r>a)$, the Schrödinger equation is that of a free particle. Two cases must be considered according to whether $E \lessgtr 0$.

A) $E<0$. *Discrete spectrum, bound states.*

Let us put $\varkappa = (-2mE)^{\frac{1}{2}}/\hbar$. The only solution bounded at infinity is the exponentially decreasing solution $Bh_l^{(+)}(\mathrm{i}\varkappa r)$ characteristic of a bound state. The continuity condition of the function for $r=a$ fixes the ratio B/A. The continuity of the logarithmic derivative yields

$$\frac{1}{h_l^{(+)}(\mathrm{i}\varkappa r)}\left(\frac{\mathrm{d}}{\mathrm{d}r}\, h_l^{(+)}(\mathrm{i}\varkappa r)\right)\bigg|_{r=a} = \frac{1}{j_l(Kr)}\frac{\mathrm{d}}{\mathrm{d}r}\, j_l(Kr)\bigg|_{r=a}. \qquad \text{(IX.38)}$$

This condition can be fulfilled only for certain discrete values of E. It determines the energy levels of the bound states of the particle in the well under study. If one deals with s states $(l=0)$, this equation is simply

$$-\varkappa a = Ka \cot Ka. \qquad \text{(IX.38 } a\text{)}$$

This is almost exactly the equation (III.18) for the one-dimensional problem of Ch. III, § 6. The discussion of the number of roots of the equation, and the number of nodes of the solutions can be taken over here without much change. An analogous discussion applies for the levels of higher angular momentum (Problem IX.5).

B) $E>0$. *Continuous spectrum, unbound states.*

Let $k = (2mE)^{\frac{1}{2}}/\hbar$. The general solution of the Schrödinger equation in the outer region is bounded everywhere. It is a linear combination of $j_l(kr)$ and $n_l(kr)$. The continuity conditions at the point $r=a$ fix the coefficients of the linear combination that is acceptable as eigenfunction. To each value of E, therefore, corresponds one and only one wave function (to within a constant).

If we write the outer solution in the form

$$B[\cos \delta_l \, j_l(kr) + \sin \delta_l \, n_l(kr)], \tag{IX.39}$$

the continuity condition of the function at $r=a$ fixes the ratio B/A. δ_l is determined by the continuity condition of the logarithmic derivative

$$\frac{Kj_l'(Ka)}{j_l(Ka)} = k \frac{\cos \delta_l \, j_l'(ka) + \sin \delta_l \, n_l'(ka)}{\cos \delta_l \, j_l(ka) + \sin \delta_l \, n_l(ka)}. \tag{IX.40}$$

It is a real quantity. δ_l is in fact the phase shift of the spherical wave of angular momentum l. Using expressions (IX.26), it is indeed easily verified that the asymptotic form of the solution (IX.39) is

$$B \frac{\sin (kr - \frac{1}{2} l\pi + \delta_l)}{kr}.$$

In the case of an s wave, equation (IX.40) takes on the very simple form

$$K \cot Ka = k \cot (ka + \delta_0). \tag{IX.40 a}$$

III. TWO-BODY PROBLEMS. SEPARATION OF THE CENTER-OF-MASS MOTION

11. Separation of the Center-of-Mass Motion in Classical Mechanics

The study of a system of two particles in Quantum Mechanics is a six-dimensional problem. However, it splits up into two three-dimensional problems, that of a free particle, and that of a particle in a static potential when the two particles feel no interaction other than their mutual interaction, and the latter depends only upon their relative position $\boldsymbol{r} = \boldsymbol{r}_1 - \boldsymbol{r}_2$. Let us denote by m_1, m_2 the masses, by $\boldsymbol{p}_1$, $\boldsymbol{p}_2$ the momenta, and by $\boldsymbol{r}_1$, $\boldsymbol{r}_2$ the respective positions of these two particles. The Hamiltonian of the system under study is of the form

$$H \equiv \frac{\boldsymbol{p}_1^2}{2m_1} + \frac{\boldsymbol{p}_2^2}{2m_2} + V(\boldsymbol{r}_1 - \boldsymbol{r}_2). \tag{IX.41}$$

The treatment consists in separating the motion of the center of mass from the relative motion, in complete analogy with the corresponding classical treatment.

Let us briefly recall how the classical procedure appears in the Hamiltonian formalism. Let us write

$$\begin{aligned} M &= m_1+m_2, & \mathbf{R} &= \frac{m_1\mathbf{r}_1+m_2\mathbf{r}_2}{m_1+m_2}, & \mathbf{P} &= \mathbf{p}_1+\mathbf{p}_2, \\ m &= \frac{m_1 m_2}{m_1+m_2}, & \mathbf{r} &= \mathbf{r}_1-\mathbf{r}_2, & \mathbf{p} &= \frac{m_2\mathbf{p}_1-m_1\mathbf{p}_2}{m_1+m_2}. \end{aligned} \tag{II}$$

According to the change of dynamical variables (II), the motion of the two particles can be pictured as the motion of two fictitious particles. One is the *center of mass*, whose position is $\mathbf{R}$; its momentum $\mathbf{P}$ is the total momentum, and its mass M the total mass of the system. The other is the particle associated with the relative motion; its position $\mathbf{r}$ is the relative position of the first particle with respect to the second, and its velocity $\mathbf{p}/m$ is equal to its relative velocity $(\mathbf{p}_1/m_1)-(\mathbf{p}_2/m_2)$; the mass m of this *relative particle* is called the *reduced mass.*

We indicate in passing some noteworthy properties of the transformation (II):

$$m_1 m_2 = mM \tag{IX.42a}$$

$$\frac{p_1{}^2}{2m_1}+\frac{p_2{}^2}{2m_2}=\frac{p^2}{2m}+\frac{P^2}{2M} \tag{IX.42b}$$

$$m_1 r_1{}^2 + m_2 r_2{}^2 = mr^2 + MR^2 \tag{IX.42c}$$

$$\mathbf{p}_1\cdot\mathbf{r}_1 + \mathbf{p}_2\cdot\mathbf{r}_2 = \mathbf{p}\cdot\mathbf{r} + \mathbf{P}\cdot\mathbf{R} \tag{IX.42d}$$

$$\boldsymbol{l}_1 + \boldsymbol{l}_2 = \boldsymbol{l} + \mathbf{L}. \tag{IX.42e}$$

In equation (IX.42e), we have introduced the angular momenta l_1, l_2 of the two particles, the angular momentum of the relative particle $\boldsymbol{l}=\mathbf{r}\times\mathbf{p}$ and that of the center of mass $\mathbf{L}=\mathbf{R}\times\mathbf{P}$.

One readily verifies that this change of variable conserves the Poisson brackets: it is a canonical transformation. Hence the equations of motion of the new variables are the canonical equations obtained from the Hamiltonian function considered as a function of the new variables, namely:

$$H = \frac{\mathbf{P}^2}{2M}+\frac{\mathbf{p}^2}{2m}+V(\mathbf{r}). \tag{IX.43}$$

One finds

$$\dot{\mathbf{R}} = \frac{\mathbf{P}}{M}, \qquad \dot{\mathbf{P}} = 0,$$

$$\dot{\mathbf{r}} = \frac{\mathbf{p}}{m}, \qquad \dot{\mathbf{p}} = -\operatorname{grad} V.$$

The equations of motion of the center of mass and of the relative particle are completely separated. The center-of-mass motion is a uniform rectilinear motion, namely the motion of a free particle of mass M. The motion of the relative particle is that of a particle of mass m in the potential $V(\mathbf{r})$.

12. Separation of the Center-of-Mass Motion of a Quantized Two-Particle System

In order to treat the same problem in Quantum Mechanics, one likewise introduces the new dynamical variables $\mathbf{r}$, $\mathbf{R}$, $\mathbf{p}$, and $\mathbf{P}$ defined as functions of the old variables by equations (II). The Hamiltonian, given by expression (IX.41) as function of the old variables, takes the form (IX.43) as function of the new variables. In fact, the new variables obey the same commutation relations as if they represented two particles of positions $\mathbf{r}$ and $\mathbf{R}$ and momenta $\mathbf{p}$ and $\mathbf{P}$, respectively, the only non-zero commutators being

$$[r_j, p_j] = i\hbar, \quad [R_j, P_j] = i\hbar. \qquad (j = x, y, z)$$

All these properties are algebraic properties which can be easily checked starting from equations (II).

One may also verify that eqs. (IX.42) remain valid in the Quantum Theory – including eq. (IX.42e) – without the necessity of changing the order of the operators occurring there.

When expressed as a function of the new variables, H is the sum of two terms:

$$H = H_r + H_R,$$

the first of which,

$$H_R = \frac{\mathbf{P}^2}{2M},$$

depends only on the center-of-mass variables, and the second,

$$H_r = \frac{\mathbf{p}^2}{2m} + V(r),$$

only on the variables of the relative particle. The vectors formed by the tensor product of the eigenvectors of H_r and the eigenvectors of H_R constitute a complete set of eigenvectors of H.

Hence the Schrödinger equation in the representation $\{\mathbf{R}, \mathbf{r}\}$ reads

$$\left[\left(-\frac{\hbar^2}{2M}\triangle_R\right)+\left(-\frac{\hbar^2}{2m}\triangle_r+V(\mathbf{r})\right)\right]\Psi(\mathbf{R}, \mathbf{r}) = E\Psi(\mathbf{R}, \mathbf{r}), \quad \text{(IX.44)}$$

where $\triangle_R$ and $\triangle_r$ designate the Laplacians relative to the coordinates $\mathbf{R}$ and $\mathbf{r}$, respectively [1]). This equation possesses a complete set of eigensolutions of the form

$$\Psi(\mathbf{R}, \mathbf{r}) = \Phi(\mathbf{R})\,\varphi(\mathbf{r}),$$

the functions Φ and φ respectively satisfying the separate equations

$$H_R\,\Phi(\mathbf{R}) \equiv \left(-\frac{\hbar^2}{2M}\triangle_R\right)\Phi(\mathbf{R}) = E_R\,\Phi(\mathbf{R})$$

$$H_r\,\varphi(\mathbf{r}) \equiv \left(-\frac{\hbar^2}{2m}\triangle_r+V(\mathbf{r})\right)\varphi(\mathbf{r}) = E_r\,\varphi(\mathbf{r}).$$

The energy eigenvalue of the overall system is the sum of the energy eigenvalues of the partial systems:

$$E = E_R + E_r.$$

More generally, if at the initial time t_0 the wave function is a product of two factors, $F(\mathbf{R})\,f(\mathbf{r})$, this factorization property is conserved in the course of time. The function $F(\mathbf{R})$ moves like a wave packet representing a free particle of mass M, and the function $f(\mathbf{r})$ like the wave function representing a particle of mass m in the potential $V(\mathbf{r})$. [2])

In practice, the solution of our original two-body problem therefore

---

[1]) This equation could just as well have been obtained directly from the Schrödinger equation in the representation $\{\mathbf{r}_1, \mathbf{r}_2\}$, by making in this partial differential equation the change of variables

$$(\mathbf{r}_1, \mathbf{r}_2) \to (\mathbf{r}, \mathbf{R}).$$

[2]) If $m_1 \ll m_2$, $m \approx m_1$ and $M \approx m_2$ (example: hydrogen atom, $m_1 =$ electron mass, $m_2 =$ proton mass); if $m_1 \approx m_2$, $m \approx \frac{1}{2}m_1$ and $M \approx 2m_1$ (example: the deuterium nucleus, $m_1 =$ proton mass, $m_2 =$ neutron mass).

reduces to that of a single particle in the potential $V(\mathbf{r})$, a problem we have learned to solve for the case where this potential is a central potential.

13. Extension to Systems of more than Two Particles

The separation of the center-of-mass motion very generally applies to systems of several particles whenever the interaction potential V depends upon the relative position of the particles and not on their absolute position; in other words, *whenever the interaction is invariant under an overall translation of all particles.*

Indeed, let us consider a quantum system of $(N+1)$ particles whose Hamiltonian H has the stated invariance property. It is always possible to carry out the reduction to the center of mass for two among them, i.e. to replace their dynamical variables by those of their center of mass and their relative particle. This procedure may be continued; having carried out a first reduction with two particles, one can perform this same reduction for the center of mass of these two particles and a third particle. After these two successive changes of variables, the three particles are replaced by: the "relative particle" of particles 1 and 2, the "particle" associated with the relative motion of the center of mass of particles 1 and 2 with respect to particle 3, and the center of mass of all three particles. In general, it is possible, after N such successive changes, to replace the $(N+1)$ particles by N "relative particles" and by the center of mass of the assembly of the $(N+1)$ particles. This reduction to the center of mass of the assembly can be achieved in a variety of ways, either by proceeding step by step, particle by particle – with $\frac{1}{2}(N+1)!$ possible variants – or by splitting the $(N+1)$ particles into two groups of N_1 and N_2 particles, respectively; one then performs the reduction to the center of mass in each of the two groups, and then replaces their centers of mass $\mathbf{R}_1$, $\mathbf{R}_2$ by the "relative particle" $\mathbf{R}_1-\mathbf{R}_2$ and the center of mass of the ensemble $(M_1\mathbf{R}_1+M_2\mathbf{R}_2)/(M_1+M_2)$; or else by splitting up the $(N+1)$ particles into three groups, and so forth.

Denote by $\mathbf{r}_i$, $\mathbf{p}_i$, and m_i the position, momentum, and the mass of the ith particle, respectively; by $\mathbf{R}$, $\mathbf{P}$, and M the corresponding quantities associated with the center of mass of the assembly of $(N+1)$ particles:

$$M=\sum_{i=1}^{N+1} m, \qquad \mathbf{P}=\sum_{i=1}^{N+1} \mathbf{p}_i, \qquad \mathbf{R}=\frac{1}{M}\sum_{i=1}^{N+1} m_i\,\mathbf{r}_i;$$

and by $\boldsymbol{\rho}_j$, $\mathbf{k}_j$, μ_j $(j=1, 2, \ldots, N)$, the corresponding quantities associated with the jth "relative particle" introduced during one of the reductions to the center of mass described above. Since such a reduction is the succession of N reductions to the center of mass of two particles, the equalities (IX.42) may be easily extended to give

$$m_1 m_2 \dots m_{N+1} = M\mu_1 \mu_2 \dots \mu_N \tag{IX.45 a}$$

$$\sum_{i=1}^{N+1} \frac{p_i^2}{2m_i} = \frac{P^2}{2M} + \sum_{j=1}^{N} \frac{\mathbf{k}_j^2}{2\mu_j} \tag{IX.45b}$$

$$\sum_{i=1}^{N+1} m_i r_i^2 = MR^2 + \sum_{j=1}^{N} \mu_j \varrho_j^2 \tag{IX.45c}$$

$$\sum_{i=1}^{N+1} \mathbf{p}_i \cdot \mathbf{r}_i = \mathbf{P} \cdot \mathbf{R} + \sum_{j=1}^{N} \mathbf{k}_j \cdot \boldsymbol{\rho}_j \tag{IX.45d}$$

$$\sum_{i=1}^{N+1} (\mathbf{r}_i \times \mathbf{p}_i) = \mathbf{R} \times \mathbf{P} + \sum_{j=1}^{N} (\boldsymbol{\rho}_j \times \mathbf{k}_j). \tag{IX.45e}$$

On the other hand, and for the same reason, the new dynamical variables obey the commutation relations characteristic of the dynamical variables of a quantum system of $(N+1)$ particles. Finally, it is clear that the potential V depends only on the relative coordinates $\boldsymbol{\rho}_1, \boldsymbol{\rho}_2, \dots, \boldsymbol{\rho}_N$, and that the total kinetic energy is the sum of the kinetic energy of the center of mass $P^2/2M$ and the kinetic energies of the relative particles [eq. (IX.45*b*)]; namely

$$H = \frac{P^2}{2M} + \left[\sum_{j=1}^{N} \frac{\mathbf{k}_j^2}{2\mu_j} + V(\boldsymbol{\rho}_1, \dots, \boldsymbol{\rho}_N)\right].$$

Thus the Hamiltonian separates as in the case of two particles, and the solution of the $(N+1)$-body problem reduces to that of an N-body problem.

All the properties which were just stated are independent of the particular way adopted for the reduction to the center of mass. In particular, whatever the choice of the N "relative particles" we introduce, the product of their reduced masses $\mu_1 \mu_2 \dots \mu_N$, the sum of their kinetic energies $\sum_j(\mathbf{k}_j^2/2\mu_j)$, and the sum of their angular momenta $\sum_j(\boldsymbol{\rho}_j \times \mathbf{k}_j)$ remain unchanged [eqs. (IX.45*a*), (IX.45*b*) and (IX.45*e*)] (Problem IX.7).

EXERCISES AND PROBLEMS

1. Show that the Hermitean radial momentum p_r defined by eq. (IX.6) satisfies the equation

$$p_r = \tfrac{1}{2}\left[\frac{\mathbf{r}}{r} \cdot \mathbf{p} + \mathbf{p} \cdot \frac{\mathbf{r}}{r}\right].$$

2. Given a particle in a central potential $V(r)$ having a certain number of

bound states, show that the ground state is necessarily an s state. More generally, show that if there exists a bound state of angular momentum L, there exists at least one bound state corresponding to each of the values l of the angular momentum such that $l < L$, and that, E_l being the lowest energy level one can obtain with angular momentum l, one necessarily has

$$E_0 < E_1 < \ldots < E_L.$$

3. Write the orthogonality relation and closure relation explicitly, and show that the eigenfunctions of the free particle

$$k\left(\frac{2}{\pi}\right)^{\frac{1}{2}} Y_l^m(\theta, \varphi)\, j_l(kr)$$

depending upon the continuous index $k (0 < k < \infty)$ and the integral indices l and m ($l \geqslant 0, -l \leqslant m \leqslant l$), form an orthonormal and complete set.

For this purpose, derive the relation

$$\int_0^\infty j_l(kr)\, j_l(k'r)\, r^2\, \mathrm{d}r = \tfrac{1}{2}\frac{\pi}{k^2}\,\delta(k-k')$$

being aware that the function $\delta(\mathbf{r} - \mathbf{r}') \equiv \delta(x - x')\,\delta(y - y')\,\delta(z - z')$ is expressed in polar coordinates by:

$$\delta(\mathbf{r}-\mathbf{r}') = [(r^2 \sin\theta)^{-1}\,\delta(r-r')\,\delta(\theta-\theta')\,\delta(\varphi-\varphi')].$$

Show that if (k, θ_k, φ_k) are the polar coordinates of the vector $\mathbf{k}$,

$$\int \exp(-\mathrm{i}\mathbf{k}\cdot\mathbf{r})\, Y_l^m(\theta, \varphi)\, j_l(k'r)\, \mathrm{d}\mathbf{r} = \frac{2\pi^2}{k^2}(-\mathrm{i})^l\, Y_l^m(\theta_k, \varphi_k)\,\delta(k-k').$$

4. Calculate the commutators of each of the components of $\mathbf{r}$ and of $\mathbf{p}$ with the component $(\mathbf{u}\cdot\boldsymbol{l})$ of the angular momentum $\boldsymbol{l} = \mathbf{r}\times\mathbf{p}$ along the unit vector $\mathbf{u}$. Show that they can be written in the abbreviated form

$$[(\mathbf{u}\cdot\boldsymbol{l}), \mathbf{p}] = \frac{\hbar}{\mathrm{i}}(\mathbf{u}\times\mathbf{p}), \qquad [(\mathbf{u}\cdot\boldsymbol{l}), \mathbf{r}] = \frac{\hbar}{\mathrm{i}}(\mathbf{u}\times\mathbf{r}).$$

From this, show that every component of $\boldsymbol{l}$ commutes with the scalar quantities $\mathbf{p}^2$, $\mathbf{r}^2$, and $\mathbf{r}\cdot\mathbf{p}$.

5. What relation must the characteristic parameters V_0 and a of the square well of § 10 satisfy in order that there be no bound s state? That there be a given number of bound s states? Is there a connection between the number of bound s states, and the number of nodes of the radial function belonging to zero energy? Answer the same questions for an arbitrary l.

6. Let us treat the radial equation of a particle in a central potential $V(r)$ by the WKB-method. In order that the method be justified, it is necessary not only that the variation of $V(r)$ over a region of the order of the wavelength be small, but also that $l \gg 1$. One notes empirically — and one can give some theoretical justification for this (cf. Langer, *loc. cit.*, footnote p. 231) -- that the method can also yield good results for small values of l if one replaces $l(l+1)$ by $(l+\frac{1}{2})^2$ in the centrifugal-barrier term $l(l+1)/r^2$ of the radial equation. Show that if this modification of the radial equation is carried out beforehand, the WKB method gives correctly, for any l:

(*i*) the asymptotic form $\sin(kr - \frac{1}{2}l\pi)/r$ of the free spherical wave [$V(r) = 0$];

(*ii*) the spectrum of the hydrogen atom [$V(r) = -e^2/r$];

(*iii*) the spectrum of the isotropic harmonic oscillator [$V(r) = \frac{1}{2}m\omega^2 r^2$].

(The rigorous solutions of the eigenvalue problems (*ii*) and (*iii*) are given in Chapters XI and XII, respectively.)

7. One performs on a quantum system of $(N+1)$ particles the separation of the center-of-mass motion in two different ways. The position vectors $\boldsymbol{\rho}_1', \ldots, \boldsymbol{\rho}_N'$, of the N "relative particles" introduced in the second case are deduced from the position vectors $\boldsymbol{\rho}_1, \ldots, \boldsymbol{\rho}_N$ of the N "relative particles" introduced in the first case by a linear transformation:

$$\boldsymbol{\rho}_j' = \sum_{k=1}^{N} A_{jk}\, \boldsymbol{\rho}_k.$$

Let $\mu_1, \ldots, \mu_N$ and $\mu_1', \ldots, \mu_N'$ be the reduced masses of these relative particles. Show that the $(N \times N)$ matrix having matrix elements

$$U_{jk} = \mu_j'^{\frac{1}{2}}\, A_{jk}\, \mu_k^{-\frac{1}{2}}$$

is an orthogonal matrix.

8. Consider the two-dimensional Schrödinger equation for the case where the potential energy $V(r)$ depends only upon the radial variable ($x = r\cos\theta$, $y = r\sin\theta$). Prove the identity

$$\frac{\partial^2}{\partial x^2} + \frac{\partial^2}{\partial y^2} = \frac{\partial^2}{\partial r^2} + \frac{1}{r}\frac{\partial}{\partial r} + \frac{1}{r^2}\frac{\partial^2}{\partial \theta^2}.$$

Deduce from this that there is a complete set of eigenfunctions of the form

$$\psi(r, \theta) = f(r)\exp(il\theta)$$

whose radial part is that solution of the equation

$$\left[\frac{\mathrm{d}^2}{\mathrm{d}r^2} + \frac{1}{r}\frac{\mathrm{d}}{\mathrm{d}r} - \frac{l^2}{r^2} + \frac{2m}{\hbar^2}[E - V(r)]\right] f(r) = 0$$

which vanishes at $r = 0$.

N.B. — If $V(r) = 0$, this regular solution is the Bessel function $J_{|l|}(kr)$, with

$$k = (2mE)^{\frac{1}{2}}/\hbar.$$

CHAPTER X

SCATTERING PROBLEMS. CENTRAL POTENTIAL AND PHASE SHIFT METHOD

1. Introduction

This chapter is devoted to the elementary concepts concerning collision problems. The results of collision experiments are expressed by means of quantities which are called cross sections and are directly related to the asymptotic behavior of the stationary solutions of the Schrödinger equation. After giving a general definition of cross sections, we devote the major portion of Sec. I to establishing this *connection between cross sections and asymptotic form*, for the simple case of the scattering of a particle by a potential approaching zero sufficiently rapidly at large distances (faster than $1/r$). We then show how the technique of separating the center-of-mass motion permits the extension of this treatment to collisions of two interacting particles.

All the remainder of this chapter is devoted to the scattering of a particle by a *central* potential, and to the method of solution known as the *phase-shift method.* The method is outlined in Sec. II. It is particularly useful when the potential has a finite range; very important properties of the phase shifts, which are particularly easy to demonstrate in this case, are discussed in Sec. III. Resonance phenomena may occur in quantum-mechanical collisions as in any wave-propagation problem; the discussion and interpretation of scattering resonances form the subject of Sec. IV. In the fifth and last section some expressions for phase shifts of current interest are given, in particular two approximate formulae, the Born formula and the Bethe formula, or effective range formula.

I. CROSS SECTIONS AND SCATTERING AMPLITUDES

2. Definition of Cross Sections

Consider a typical scattering experiment. A target is struck by a beam of monoërgic particles. Let J be the magnitude of the incident flux, i.e. the number of incident particles crossing per unit time a unit surface placed perpendicular to the direction of propagation, and at rest with respect to the target. If P is the number of particles per unit volume in the incident beam, and v the velocity of the incident particles relative to the target,

$$J = Pv.$$

Under the conditions of the experiment, P is so small that one can neglect the mutual interaction of the incident particles; they therefore undergo their collisions independently of each other. By means of appropriate counters one measures the number $\mathscr{N}$ of particles scattered per unit time into the solid angle $\mathrm{d}\Omega$ located in the direction $\Omega \equiv (\theta, \varphi)$. $\mathscr{N}$ is directly proportional to the incident current:

$$\mathscr{N} = J\Sigma(\Omega)\, \mathrm{d}\Omega.$$

The quantity $\Sigma(\Omega)$ which has the dimensions of a surface, is a characteristic parameter of the collision of the particle with the target; it is the scattering cross section of the particle by the target in the direction Ω.

In most practical cases, the target is made up of a large number N of atomic or nuclear scattering centers, and the distances between these atoms or atomic nuclei are sufficiently large with respect to the wavelength of the incident particles so that one may neglect all coherence between the waves scattered by each of them [1]). Each scattering center then acts as if it were alone. Moreover, if the target is sufficiently thin so that one may neglect multiple scattering, $\mathscr{N}$ is directly proportional to N and we have

$$\mathscr{N} = JN\, \sigma(\Omega)\, \mathrm{d}\Omega.$$

The area $\sigma(\Omega)$ is called the scattering cross section of the particle by the scattering center in the direction Ω or, for short, *differential scattering cross section.*

The total number of particles scattered in unit time is obtained by integrating over angles. It is equal to $JN\sigma_{\text{tot.}}$, where

$$\sigma_{\text{tot.}} = \int \sigma(\Omega)\, \mathrm{d}\Omega$$

is the *total scattering cross section.*

---

[1]) This circumstance is not always realized. Important exceptions to this rule are the diffraction phenomena in crystals: electron diffraction, thermal neutron diffraction, or X-ray diffraction.

In nuclear physics, where the scattering centers have linear dimensions of the order of 10^{-13} to 10^{-12} cm, the cross sections are usually measured in barns or millibarns:

$$1 \text{ barn} = 10^{-24} \text{ cm}^2, \qquad 1 \text{ mb} = 10^{-27} \text{ cm}^2.$$

In the foregoing, we have implicitly assumed that the only possible collisions are elastic collisions, i.e. collisions during which the quantum state of the scatterer does not change and where, *a fortiori*, there is no energy transfer to the internal degrees of freedom of the scatterer. We confine ourselves to this type of collision for the moment. Moreover, rather than treat the scatterer, atom or nucleus, in all its complexity, we represent it by a static potential $V(\mathbf{r})$ depending on the coordinate $\mathbf{r}$ of the particle.

3. Stationary Scattering Wave

We shall therefore consider the scattering of a particle of mass m by a potential $V(\mathbf{r})$. In this chapter we shall limit ourselves to potentials $V(\mathbf{r})$ which tend to zero more rapidly than $1/r$ as $r \to \infty$. The scattering by a Coulomb potential will be treated in Chapter XI.

Let E be the energy, and $\mathbf{p} = \hbar\mathbf{k}$ the initial momentum of the particle. One can relate the cross section $\sigma(\Omega)$ to the solution of the Schrödinger equation

$$\left[-\frac{\hbar^2}{2m}\triangle + V(\mathbf{r})\right]\psi_{\mathbf{k}}(\mathbf{r}) = E\psi_{\mathbf{k}}(\mathbf{r})$$

whose behavior at infinity is of the form

$$\mathrm{e}^{\mathrm{i}\mathbf{k}\cdot\mathbf{r}} + f(\Omega)\,\frac{\mathrm{e}^{ikr}}{r}. \tag{X.1}$$

We assume [1]) without further discussion that one and only one solution of this type exists for each value of $\mathbf{k}$. We shall call this solution $\psi_{\mathbf{k}}(\mathbf{r})$ the stationary scattering wave, with wave vector $\mathbf{k}$.

The two terms of the asymptotic form are easily interpreted

---

[1]) We shall prove it in Sec. II for the particular case where V is a central potential sufficiently regular at the origin.

if one uses the definition (IV.9) of the current density vector [1])

$$\boldsymbol{J}(r) = \frac{\hbar}{2mi} [\psi^*(\boldsymbol{r})(\nabla \psi(\boldsymbol{r})) - (\nabla \psi(\boldsymbol{r}))^* \psi(\boldsymbol{r})].$$

The plane wave term $\exp(i\boldsymbol{k}\cdot\boldsymbol{r})$ represents a wave of unit density and of current density $\hbar\boldsymbol{k}/m$. Retaining only the lowest order in $1/r$, the term $f(\Omega)\exp(ikr)/r$ represents a wave of density $|f(\Omega)|^2/r^2$ and of current density directed along the direction Ω *toward increasing* r (outgoing wave) and equal to $(\hbar k/m)(|f(\Omega)|^2/r^2)$. In fact, since the effect of the potential $V(\boldsymbol{r})$ can be neglected in the asymptotic region, one can, according to the classical approximation (cf. Ch. VI, § 4) interpret the term $\exp(i\boldsymbol{k}\cdot\boldsymbol{r})$ as a beam of mono-energetic particles of momentum $\hbar\boldsymbol{k}$ and of density 1, representing the incident beam. The term $f(\Omega)$ $\exp(ikr)/r$ is interpreted as a beam of particles emitted radially from the scattering center and represents the beam of scattered particles.

In accordance with this interpretation, one can calculate the number of particles emitted per unit time into the solid angle $\mathrm{d}\Omega$ located in the direction Ω: it is equal to the flux of scattered particles through a spherical surface element of very large radius which subtends the solid angle $(\Omega, \Omega+\mathrm{d}\Omega)$, namely $(\hbar k/m)|f(\Omega)|^2\mathrm{d}\Omega$. Upon dividing by the incident flux $J=\hbar k/m$, one obtains the scattering cross section

$$\boxed{\sigma(\Omega) = |f(\Omega)|^2} \tag{X.2}$$

$f(\Omega)$ is called the *scattering amplitude.*

4. Representation of the Scattering Phenomenon by a Bundle of Wave Packets [2])

The very intuitive argument given above is incorrect for two reasons.

---

[1]) One can define a Hermitean operator $\boldsymbol{J}(\boldsymbol{r}_0)$ representing the current at the point $\boldsymbol{r}_0$:

$$\boldsymbol{J}(\boldsymbol{r}_0) = \frac{1}{2m} [\boldsymbol{p}\delta(\boldsymbol{r}-\boldsymbol{r}_0) + \delta(\boldsymbol{r}-\boldsymbol{r}_0)\boldsymbol{p}];$$

the current defined above is the average value of this operator with respect to a given quantum state.

[2]) The treatment of the scattering problems we give in §§ 4, 5, and 6 is essentially taken from an outline of the scattering theory by Chew and Low. The calculation of § 16 has also been borrowed from this work.

In the first place, the current density vector is not simply the sum of the current of the incident plane wave and that of the scattered wave. One must add to these contributions the interference term between exp $(i\mathbf{k}\cdot\mathbf{r})$ and $f(\Omega)[\exp(ikr)/r]$. In the foregoing argument the interferences between incident and scattered wave were deliberately ignored.

In the second place, the representation of the physical situation by the stationary scattering wave

$$\psi_{\mathbf{k}}(\mathbf{r})\, e^{iEt/\hbar} \tag{X.3}$$

is an idealization. In reality, each particle participating in the scattering must be represented by a wave packet formed by the superposition of stationary waves of the type (X.3) corresponding to wave vectors of *magnitude* and *direction* slightly different from $\mathbf{k}$. This packet is constructed so as to correctly fulfill the initial conditions. Because of the spread in momentum directions, it is bounded transversely such that its transverse dimensions do not exceed those of the diaphragm or diaphragms of the apparatus producing the incident beam. Because of the spread in energy, it is limited longitudinally, and its center moves toward the target along a straight line with a velocity equal to its group velocity $\mathbf{v}=\hbar\mathbf{k}/m$.

We specify each such initial trajectory by the position $\mathbf{b}$ of its point of intersection with the plane (S) passing through the scattering center perpendicularly to the direction of propagation. If we designate by t_0 the instant at which the center of the packet would have passed the plane (S) had its motion not been modified by the presence of the potential, the motion of the center of the packet before collision follows the law

$$\langle \mathbf{r} \rangle = \mathbf{b} + \mathbf{v}(t-t_0) \qquad (t \ll t_0).$$

The incident beam is in reality a beam of wave packets of this type moving parallel to each other at the velocity $\mathbf{v}$, and differing from each other only in the values of the parameters $\mathbf{b}$ and t_0 which fix the motion of their respective centers before collision.

In what follows these characteristic lengths enter into play (Fig. X.1)

$\lambda\!\!\!^{-}=(\hbar/mv)=$ mean wavelength of the incident packet;
$d, l=$ transverse and longitudinal dimensions of the incident packet;
$a=$ extension of the scattering region;
$D=$ distance from the counting instruments to the scattering region.

Through the uncertainty relations, d and l are related, respectively, to the spread in direction and in energy of the incident wave packet.

Since we assume good definition of the incident direction of propagation and energy, we must have

$$\lambda\!\!\!^{-} \ll d \quad \text{and} \quad \lambda\!\!\!^{-} \ll l. \tag{X.4}$$

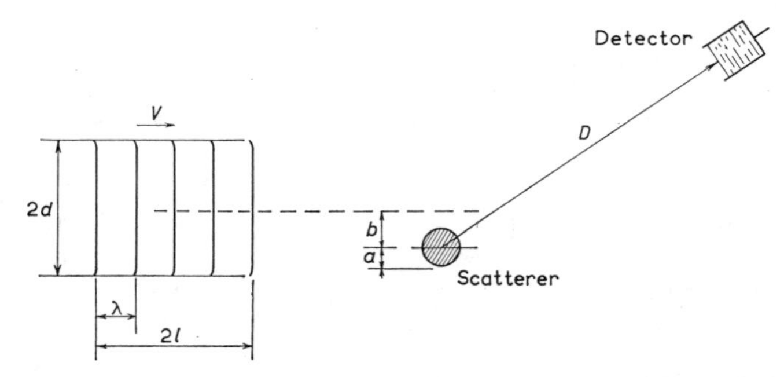

Fig. X.1. Schematic of a scattering experiment and of the characteristic lengths entering into this phenomenon.

In order that the collision phenomenon not depend critically upon the particular form of the wave packet, it is furthermore necessary that its dimensions be very much larger than those of the scattering region, i.e. of the region of space where a non-zero potential prevails, namely

$$a \ll d, l \tag{X.5}$$

($a \approx 10^{-8}$ cm if the scatterer is an atom; $a \approx 10^{-12}$ cm if it is an atomic nucleus).

Now, if the impact parameter is greater than the lateral extension of the packet ($b > d$) the incident wave packet never enters the scattering region and propagates during the entire duration of the phenomenon as a free wave packet. If on the other hand $b < d$, it enters the scattering region at some time $t_1 \approx t_0 - (l/v)$. The collision proper begins at that moment. After a sufficiently long time, the wave packet is again entirely outside the scattering region; it is then generally composed of two additive terms: a transmitted wave packet whose form and law of propagation are essentially the same as those of the incident wave packet, and a wave packet scattered in directions different from the incident direction [1]) (Fig. X.2).

The detection of scattered particles is carried out by placing a suitable counting system (counters, photographic plates, etc.) in a given direction $\Omega = (\theta, \varphi)$ at a distance of order D from the scattering center. This distance must not be too great if one wants

[1]) The phenomenon is analogous to the phenomena of reflection and transmission of one-dimensional waves studied in Ch. III (§§ 3, 6, and 7).

the spreading of the wave packet to remain negligible during the experiment (cf. Ch. VI, § 3):

$$\sqrt{\lambda\!\!\!^{-} D} \ll d, l. \tag{X.6}$$

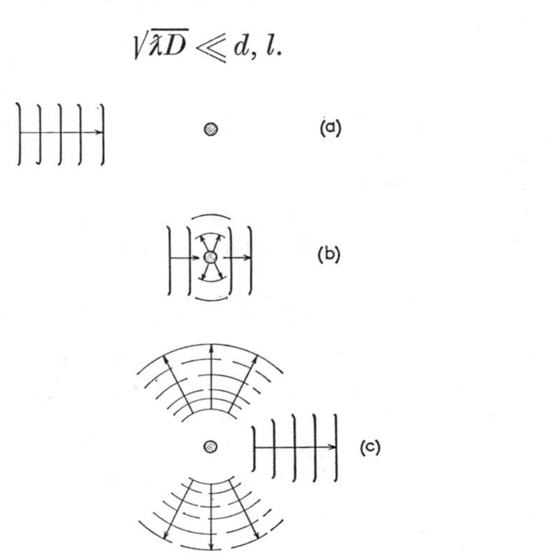

Fig. X.2. Stages of the scattering phenomenon of a wave packet: (*a*) before collision; (*b*) during collision; (*c*) after collision. (Shaded area: region where a non-zero potential prevails.)

With this qualification, the detection must be made sufficiently far away in order that the wave propagation one detects be unaffected by the presence of the scattering center, namely

$$a, \lambda\!\!\!^{-} \ll D \tag{X.7}$$

and that the detector be under no circumstances triggered by the transmitted wave:

$$d \ll D \sin \theta. \tag{X.8}$$

[*N.B.* – The forward-scattered wave ($\theta = 0$) can never be separated from the transmitted wave.]

Combining conditions (X.4)–(X.8), we finally obtain the following inequalities [1]):

$$a, \sqrt{\lambda\!\!\!^{-} D} \ll l \tag{X.9 a}$$

$$a, \sqrt{\lambda\!\!\!^{-} D} \ll d \ll D. \tag{X.9 b}$$

[1]) In atomic or nuclear physics, d is at most equal to the size of the entrance diaphragm of the incident particles, $d \approx 1$ mm; l may be appreciably larger; D is of the order of 1 m. With $a \approx 10^{-8}$ cm and $\lambda\!\!\!^{-} \approx 10^{-8}$ cm, at most, this yields $\sqrt{\lambda\!\!\!^{-} D} \approx 10^{-3}$ cm and $l/a \gtrsim d/a \approx 10^7$, $l/\sqrt{\lambda\!\!\!^{-} D} \gtrsim d/\sqrt{\lambda\!\!\!^{-} D} \approx 10^2$, $d/D \approx 10^{-3}$. The conditions (X.9) are thus amply fulfilled.

In order that the cross section $\sigma(\Omega)$ actually be given by formula (X.2), it is necessary that the measuring device satisfy conditions (X.9).

Moreover, the wave packets must be sufficiently well defined in direction and in energy so that one can assign to them a well-defined scattering amplitude: $f(\Omega)$ must remain practically constant in *modulus* and in *argument* when the energy varies by an amount of order $\delta E \approx \hbar v/l$, and the angle of incidence by an amount of order $\lambda\!\!\!^{-}/d$ about their respective mean values.

The proof of this result forms the subject of the next two sections.

5. Scattering of a Wave Packet by a Potential

Under these circumstances ($l, d \gg a$), the evolution of the wave packets of the incident beam is practically independent of their particular form. We shall assume that they all have the same shape, each being characterized by the parameters $\boldsymbol{b}$, t_0 determining the motion of its center. Without loss of generality, we take $t_0 = 0$.

To define the form of the incident packet, let us introduce the function $\chi(\boldsymbol{\rho})$ normalized to unity:

$$\int |\chi(\boldsymbol{\rho})|^2 \, \mathrm{d}\boldsymbol{\rho} = 1. \tag{X.10}$$

Denote by $A(\boldsymbol{\varkappa})$ its Fourier transform defined with a norm such that:

$$\chi(\boldsymbol{\rho}) = \int A(\boldsymbol{\varkappa}) \, \mathrm{e}^{\mathrm{i}\boldsymbol{\varkappa}\cdot\boldsymbol{\rho}} \, \mathrm{d}\boldsymbol{\varkappa}. \tag{X.11}$$

By hypothesis, $\chi(\boldsymbol{\rho})$ is a real function whose only appreciable values occur when $\boldsymbol{\rho}$ is located in a region of length l and of width d about the point $\boldsymbol{\rho} = 0$. Likewise, $A(\boldsymbol{\varkappa})$ is real and has appreciable values only in a region of length $(1/l)$ and width $(1/d)$ about the point $\boldsymbol{\varkappa} = 0$. To simplify the discussion, we assume $d \approx l$.

A long time before the collision ($t \ll -l/v$) the wave packet under study $\Psi_{\boldsymbol{b}}(\boldsymbol{r}, t)$ must reduce to the free wave packet $\Phi_{\boldsymbol{b}}(\boldsymbol{r}, t)$ whose center moves according to the law

$$\langle \boldsymbol{r} \rangle = \boldsymbol{b} + \boldsymbol{v}t$$

and whose form at the instant $t = 0$ is given by

$$\begin{aligned}\Phi_{\boldsymbol{b}}(\boldsymbol{r}, 0) &= \mathrm{e}^{\mathrm{i}\boldsymbol{k}\cdot(\boldsymbol{r}-\boldsymbol{b})} \, \chi(\boldsymbol{r} - \boldsymbol{b}) \\ &= \int A(\boldsymbol{k}' - \boldsymbol{k}) \, \mathrm{e}^{\mathrm{i}\boldsymbol{k}'\cdot(\boldsymbol{r}-\boldsymbol{b})} \, \mathrm{d}\boldsymbol{k}'.\end{aligned}$$

In fact

$$\Phi_{\boldsymbol{b}}(\boldsymbol{r}, t) = \int A(\boldsymbol{k}' - \boldsymbol{k}) \, \mathrm{e}^{\mathrm{i}[\boldsymbol{k}'\cdot(\boldsymbol{r}-\boldsymbol{b}) - (E't/\hbar)]} \, \mathrm{d}\boldsymbol{k}'. \tag{X.12}$$

If one neglects its spreading, this free wave packet is equally well represented by the expression (cf. Ch. VIII, § 18)

$$\Phi_{\boldsymbol{b}}(\boldsymbol{r}, t) \approx \mathrm{e}^{-\mathrm{i}\boldsymbol{k}\cdot\boldsymbol{b}} \, \mathrm{e}^{\mathrm{i}[\boldsymbol{k}\cdot\boldsymbol{r} - (Et/\hbar)]} \, \chi(\boldsymbol{r} - \boldsymbol{v}t - \boldsymbol{b}) \tag{X.13}$$

obtained by replacing in the integral of the right-hand side of (X.12) the energy $E' = \hbar^2 k^2/2m$ by the first two terms of its expansion in powers of $(\mathbf{k}' - \mathbf{k})$,

$$E' = E + \hbar \mathbf{v} \cdot (\mathbf{k}' - \mathbf{k}).$$

The wave packet Ψ_b is obtained by substituting the stationary scattering wave $\psi_{\mathbf{k}'}(\mathbf{r})$ for the plane wave $\exp(\mathrm{i}\mathbf{k}' \cdot \mathbf{r})$ in the integral (X.12):

$$\Psi_{\mathbf{b}}(\mathbf{r}, t) \approx \int A(\mathbf{k}' - \mathbf{k})\, \mathrm{e}^{-\mathrm{i}\mathbf{k}' \cdot \mathbf{b}}\, \psi_{\mathbf{k}'}(\mathbf{r})\, \mathrm{e}^{-\mathrm{i}E't/\hbar}\, \mathrm{d}\mathbf{k}'. \qquad \text{(X.14)}$$

Indeed, this is actually a solution of the Schrödinger equation since it is a superposition of solutions of that equation. It is therefore sufficient to show that it is identical to the free wave packet $\Phi_{\mathbf{b}}$ before the collision.

Since the function $A(\mathbf{k}' - \mathbf{k})$ possesses a pronounced peak about the point $\mathbf{k}' = \mathbf{k}$, the only important contribution to the integral (X.14) comes from the region surrounding that point. When $t \ll -l/v$, the phase of the integrand varies very rapidly in this region because of the exponential $\exp(\mathrm{i}E't/\hbar)$, and the integral is practically zero except for values of r for which this phase is stationary; this can only happen for r of order $v|t|$, in other words for regions of configuration space where $\psi_{\mathbf{k}'}(\mathbf{r})$ can be replaced by its asymptotic form

$$\psi_{\mathbf{k}'}(\mathbf{r}) \sim \mathrm{e}^{\mathrm{i}\mathbf{k}' \cdot \mathbf{r}} + f_{\mathbf{k}'}(\Omega) \frac{\mathrm{e}^{\mathrm{i}k'r}}{r}. \qquad \text{(X.15)}$$

Substitution of this expression in the integral (X.14) yields

$$\Psi_{\mathbf{b}}(\mathbf{r}, t) \underset{t \to -\infty}{\sim} \Phi_{\mathbf{b}}(\mathbf{r}, t) + \Psi_{\mathbf{b}}^{(d)}(\mathbf{r}, t) \qquad \text{(X.16)}$$

with

$$\Psi_{\mathbf{b}}^{(d)} = \int A(\mathbf{k}' - \mathbf{k})\, \mathrm{e}^{-\mathrm{i}\mathbf{k}' \cdot \mathbf{b}}\, f_{\mathbf{k}'}(\Omega) \frac{\mathrm{e}^{\mathrm{i}[k'r - (E't/\hbar)]}}{r}\, \mathrm{d}\mathbf{k}'. \qquad \text{(X.17)}$$

When $t \to -\infty$, the phase of the integrand of the right-hand side of (X.17) cannot be stationary in the region $\mathbf{k}' = \mathbf{k}$ and the integral $\Psi_{\mathbf{b}}^{(d)}$ is practically zero for any r. The wave packet $\Psi_{\mathbf{b}}$ is thus actually identical to the free wave packet, in this limit.

Consider now the motion of the wave packet $\Psi_{\mathbf{b}}$ in the detection region ($r \gtrsim D$). In this region of space the substitution of its asymptotic form (X.15) for $\psi_{\mathbf{k}'}$ is certainly justified. Expression (X.16) therefore holds again.

We assume that the spreads in angle and energy are so small that $f_{\mathbf{k}'}(\Omega)$ remains practically constant in the region of extension $1/d \approx 1/l$ surrounding the point $\mathbf{k}' = \mathbf{k}$, and that we can replace in the integral (X.17) the modulus of $f_{\mathbf{k}'}(\Omega)$ by its value at the point $\mathbf{k}$, and its argument by the first two terms of its expansion:

$$\begin{aligned} \arg f_{\mathbf{k}'}(\Omega) &\approx \arg f_{\mathbf{k}}(\Omega) + (\mathbf{k}' - \mathbf{k}) \cdot \mathbf{s}(\Omega) \\ \mathbf{s}(\Omega) &= \mathrm{grad}_{\mathbf{k}} \, [\arg f_{\mathbf{k}}(\Omega)] \qquad (s \ll d, \, l). \end{aligned} \tag{X.18}$$

Likewise, we replace the arguments of the other factors by the first terms of their expansion

$$k' \approx k + \mathbf{u} \cdot (\mathbf{k}' - \mathbf{k}), \qquad E' \approx E + \hbar \mathbf{v} \cdot (\mathbf{k}' - \mathbf{k}) \tag{X.19}$$

($\mathbf{u} \equiv \mathbf{v}/v =$ unit vector along the incident direction). Considerations of stationary phase analogous to those for $\Phi_{\mathbf{b}}$ [cf. eq. (X.13)] lead us to the equation

$$\Psi_{\mathbf{b}}^{(d)} \approx \mathrm{e}^{-\mathrm{i}\mathbf{k}\cdot\mathbf{b}} \, f_{\mathbf{k}}(\Omega) \, \frac{\mathrm{e}^{\mathrm{i}[kr - (Et/\hbar)]}}{r} \, \chi[\mathbf{u}(r - vt) + \mathbf{s} - \mathbf{b}]. \tag{X.20}$$

Whether $\Psi_{\mathbf{b}}{}^{(d)}$ can be neglected or not depends upon whether the impact parameter b is larger or smaller than d.

If $b > d$, the argument of χ lies at all times and for all r in the region where this function is negligible; $\Psi_{\mathbf{b}}{}^{(d)}$ always remains practically zero, and the wave packet propagates as a simple free wave packet.

If $b < d$, i.e. if the incident packet actually enters the scattering region during its motion, χ is appreciable in a spherical shell of thickness of order l on either side of the sphere $r = vt$: $\Psi_{\mathbf{b}}{}^{(d)}$, which had remained practically zero before the collision, represents an outgoing wave packet of velocity v after the collision, i.e. a packet of spherical waves traveling away from the scatterer with radial velocity v. At the time $t \approx D/v$, the wave $\Psi_{\mathbf{b}}{}^{(d)}$ enters the detection region; it is then completely separate from the transmitted wave $\Phi_{\mathbf{b}}$ except in the forward direction ($\theta < d/D$) where both of these waves are important and will interfere [1]). We thus find the qualitative results stated in Chapter X, § 4.

6. Calculation of Cross Sections

Before calculating the cross sections, let us clearly state what is involved in the operation of detection. Whatever the experimental arrangement used, it essentially consists in placing a diaphragm at a distance D in the direction $\Omega = (\theta, \varphi)$ with respect to the target. The opening of this diaphragm is arranged in such a way that it passes without distortion any wave emitted into a given solid

---

[1]) The interference term must not be neglected since it is this term which insures the conservation of the norm (Problem X.1).

angle $(\Omega, \Omega+\delta\Omega)$ to the exclusion of the rest. A detection system placed behind the diaphragm is triggered whenever a particle traverses the latter. The probability $P_{\boldsymbol{b}}(\Omega)\delta\Omega$ that the detector is triggered by a particle whose motion before detection is represented by the wave $\Psi_{\boldsymbol{b}}(\boldsymbol{r}, t)$, is equal to the integral of the flux of this wave across the diaphragm, evaluated over the entire duration of the collision [1]), or, what is equivalent, to the probability of presence of the scattered particle in the solid angle $(\Omega, \Omega+\delta\Omega)$ when the collision has occurred $(t=T \gg l/v)$.

Since the detector is located sufficiently far away in the transverse direction so that the transmitted wave can never reach it [condition (X.8)], only the scattered wave $\Psi_{\boldsymbol{b}}^{(d)}$ enters into this calculation of the probability of presence. From its expression (X.20), we obtain

$$\begin{aligned} P_{\boldsymbol{b}}(\Omega) &= \int_0^\infty |\Psi_{\boldsymbol{b}}^{(d)}(\boldsymbol{r}, T)|^2\, r^2\, \mathrm{d}r \\ &= |f_{\boldsymbol{k}}(\Omega)|^2 \int_0^\infty |\chi[\boldsymbol{u}(r-vT)+\boldsymbol{s}-\boldsymbol{b}]|^2\, \mathrm{d}r. \end{aligned} \tag{X.21}$$

Since $vT \gg l$, one can make the change of variable

$$z = r - vT$$

and extend the limit of integration to $-\infty$, which yields

$$P_{\boldsymbol{b}}(\Omega) = |f_{\boldsymbol{k}}(\Omega)|^2 \int_{-\infty}^{+\infty} |\chi(\boldsymbol{u}z+\boldsymbol{s}-\boldsymbol{b})|^2\, \mathrm{d}z.$$

Now consider a beam of particles of unit flux: there are $\mathrm{d}\boldsymbol{b}$ incident particles upon the surface element $(\boldsymbol{b}, \boldsymbol{b}+\mathrm{d}\boldsymbol{b})$ per unit time, and each of these has a probability $P_{\boldsymbol{b}}(\Omega)\delta(\Omega)$ of being scattered in the direction $(\Omega, \Omega+\delta\Omega)$. One obtains the probability of scattering into $(\Omega, \Omega+\delta\Omega)$ per unit time and unit flux by integrating this expression over $\boldsymbol{b}$:

$$\sigma(\Omega) = |f_{\boldsymbol{k}}(\Omega)|^2 \int_{-\infty}^{+\infty} \mathrm{d}z \int \mathrm{d}\boldsymbol{b}\, |\chi(\boldsymbol{u}z+\boldsymbol{s}-\boldsymbol{b})|^2.$$

The region of integration of $\boldsymbol{b}$ is the plane perpendicular to $\boldsymbol{u}$. By carrying out the change of variable $\boldsymbol{\rho}=\boldsymbol{u}z+\boldsymbol{s}-\boldsymbol{b}$, one reduces the

[1]) The instant at which the particle traverses the diaphragm is equal, on the average, to $[D + (\boldsymbol{u}\cdot\boldsymbol{s})]/v$. It can actually not be predicted with a precision greater than l/v, in accordance with the time-energy uncertainty relation. $(\boldsymbol{u}\cdot\boldsymbol{s})/v$ can be interpreted as a delay in the transmission of the scattered wave. However, the experimental arrangement considered here certainly does not permit such a delay to be observed since $s \ll l$ (cf. discussion on the delay upon reflection of Ch. III, § 6). The observation of delays of this kind presupposes a sufficiently poor energy definition (cf. Ch. X, § 16).

triple integral above to the normalization integral of the function χ [eq. (X.10)]. We thus arrive at the expression (X.2):

$$\sigma(\Omega) = |f_{\boldsymbol{k}}(\Omega)|^2.$$

7. Collision of Two Particles. Laboratory System and Center-of-Mass System

The technique of separating the center-of-mass motion permits to extend the treatment of the scattering of a particle by a potential, to the collisions of two particles subjected to a potential $V(\boldsymbol{r})$ which depends only upon their relative position. As was shown in Sec. III of Chapter IX – we take over the notation of that section – the motion of the two particles is decomposed into two separate motions: that of their center of mass which travels like a free particle, and that of the "relative particle" of mass $m = m_1 m_2/(m_1 + m_2)$ subjected to the potential $V(\boldsymbol{r})$.

In a typical scattering experiment, one bombards a target composed of particles of type 2 with a monoërgic beam of particles of type 1, and one counts the number of particles of one type, particles 1 for instance, emitted in a given direction $\Omega_1 = (\theta_1, \varphi_1)$. Before collision, particle 2 is at rest, particle 1 travels with a given velocity $\boldsymbol{v}$, and the center of mass moves with the velocity

$$\boldsymbol{V} = \frac{m_1}{M}\boldsymbol{v} \qquad (M = m_1 + m_2).$$

The total energy of the system is the sum of the energies of the center-of-mass motion and the relative motion

$$E = E_R + E_r$$

$$\left(E = \tfrac{1}{2} m_1 v^2; \qquad E_R = \tfrac{1}{2} M V^2 = \frac{m_1}{m_1 + m_2} E; \qquad E_r = \tfrac{1}{2} m v^2 = \frac{m_2}{m_1 + m_2} E\right).$$

During the collision the center of mass continues its uniform rectilinear motion, the collision phenomenon affecting only the relative coordinate. It is clear that the scattering cross section $\sigma_1(\Omega)$ is related to the asymptotic behavior of the stationary states of energy E_r of the relative coordinate.

In order to establish that relation, it is convenient to change the system of reference and to study the same phenomenon in a frame of reference where the center of mass is at rest. One usually calls the *laboratory system* the frame of reference where the target particle is at rest, and *center-of-mass system* the one where the center of mass is at rest. The first is the one considered above; the second is in uniform translational motion of velocity $\mathbf{V}$ with respect to the former. Passing from one frame of reference to the other affects only the center-of-mass motion; the motion of the relative particle remains unchanged.

The definition of cross sections given in Ch. X, § 2 in no way supposes that the target be initially at rest. We emphasize that the incident flux which enters into that definition is the *relative flux* of the projectile with respect to the scatterer; it is a quantity independent of the frame of reference. Once this is understood, one can define the differential cross section $\sigma(\Omega)$ in the center-of-mass system in the same way one defines the differential cross section $\sigma_1(\Omega)$ of the same process in the laboratory system. $\sigma(\Omega)$ is the number of particles 1 emitted per unit time and per unit solid angle in the direction Ω when one bombards particle 2 with an incident flux of particles 1 of relative flux unity, all observations being carried out in the center-of-mass system, the angles of emission also being measured in that system.

These definitions imply that

$$\sigma(\Omega)\,\mathrm{d}\Omega = \sigma_1(\Omega_1)\,\mathrm{d}\Omega_1, \tag{X.22}$$

an expression in which Ω_1 designates the direction of emission of particle 1 in the laboratory system when this particle travels in the direction Ω in the center-of-mass system. Note the equality of the total cross sections:

$$\sigma_{\text{tot.}} \equiv \int \sigma(\Omega)\,\mathrm{d}\Omega = \sigma_{1\,\text{tot.}}.$$

This in fact was evident *a priori* since the total cross section is the total number of particles scattered per unit incident flux, a quantity that is invariant under a change of reference frame.

$\sigma(\Omega)$ is related more directly than $\sigma_1(\Omega_1)$ to the (three-dimensional) scattering problem of the relative particle by the potential $V(\mathbf{r})$. Indeed, in the center-of-mass system the direction of propagation of particle 1 and that of the relative particle are identical (particle 2 travelling in the opposite direction). Moreover, since the incident flux of the relative particle with respect to the center of force ($\mathbf{r}=0$) is equal to the incident flux of our collision problem, $\sigma(\Omega)$ is the scattering cross section of the relative particle in the direction Ω, that is to say the differential scattering cross section in the direction Ω of a particle of mass m and initial velocity $\mathbf{v}$, by the potential $V(\mathbf{r})$.

In particular, if $V(\mathbf{r})$ tends asymptotically to zero more rapidly than $1/r$, the Schrödinger equation of the relative particle

$$\left[-\frac{\hbar^2}{2m}\triangle + V(\mathbf{r})\right]\psi(\mathbf{r}) = E_r\psi(\mathbf{r})$$

possesses an eigensolution of energy E_r with asymptotic behavior

$$e^{i\mathbf{k}\cdot\mathbf{r}} + f(\Omega)\frac{e^{ikr}}{r}$$

($E_r = \hbar^2k^2/2m$, $\mathbf{k} = m\mathbf{v}/\hbar = m_2\mathbf{k}_1/M$, $\mathbf{k}_1$ being the incident wave vector

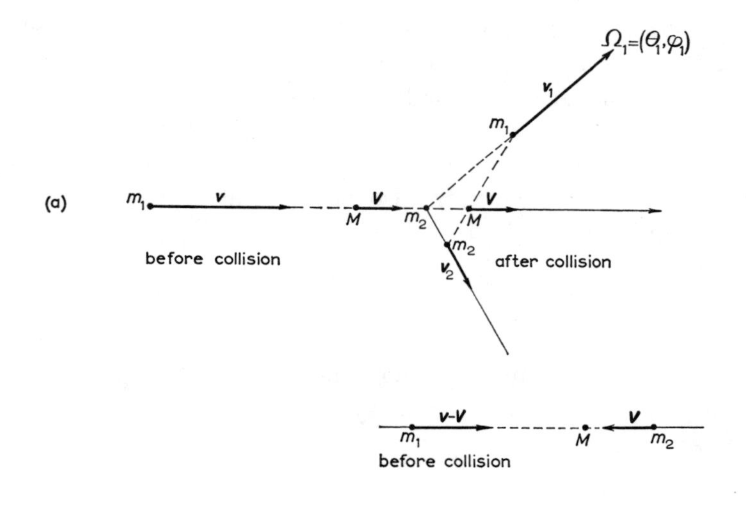

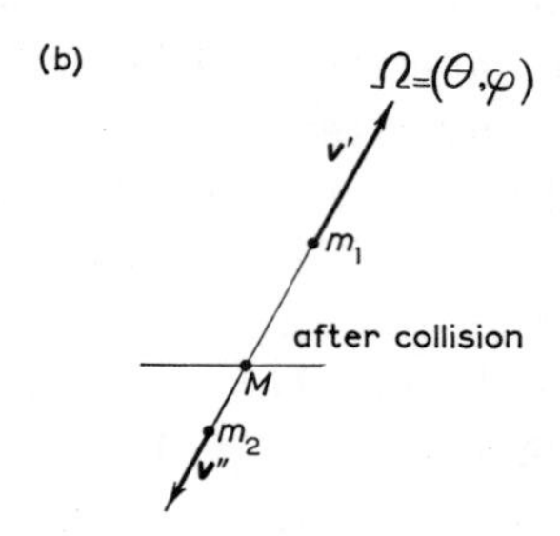

Fig. X.3. (*a*) Collision in the laboratory system ($\mathbf{V} = m_1\mathbf{v}/M$).
(*b*) The same collision in the center-of-mass system.

in the laboratory system) and, applying the results of the foregoing paragraphs

$$\sigma(\Omega) = |f(\Omega)|^2.$$

We have yet to establish the relation between Ω and Ω_1, in order to be able to derive from the above expressions [1]) the cross section in the laboratory system $\sigma_1(\Omega_1)$. To this effect we take as polar axis in the two systems the axis parallel to the direction of propagation. Fig. X.3*a* gives the schematic diagram of a collision in the laboratory system, while Fig. X.3*b* gives the same in the center-of-mass system. The initial and final velocities of the two particles are given in the following table, along with their spherical polar coordinates:

	Initial Velocities		Final Velocities	
	1	2	1	2
Laboratory System	$\mathbf{v}(v, 0, 0)$	0	$\mathbf{v}_1(v_1, \theta_1, \varphi_1)$	$\mathbf{v}_2(v_2, \theta_2, \varphi_2)$
Center-of-Mass System	$\mathbf{v} - \mathbf{V}(v - V, 0, 0)$	$-\mathbf{V}(V, \pi, 0)$	$\mathbf{v}'(v', \theta, \varphi)$	$\mathbf{v}''(v'', \pi - \theta, \varphi + \pi)$

$$\mathbf{V} = \frac{m_1}{m_1+m_2}\mathbf{v}, \qquad v' = v - V = \frac{m_2}{m_1+m_2}v, \qquad v'' = V = \frac{m_1}{m_1+m_2}v$$

(θ_1, φ_1) is defined as function of (θ, φ) by the vector equality [2])

$$\mathbf{v}_1 = \mathbf{v}' + \mathbf{V} \tag{X.23}$$

with

$$\varphi_1 = \varphi, \qquad v_1 \sin\theta_1 = v' \sin\theta, \qquad v_1 \cos\theta_1 = v' \cos\theta + V,$$

---

[1]) All calculations made here are valid only in the non-relativistic approximation. However, the concept of center-of-mass system remains valid in relativistic mechanics: it is the frame of reference where the total momentum (momentum of particle 1 + momentum of particle 2) is zero. One passes from the center-of-mass system to the laboratory system by a Lorentz transformation. In the non-relativistic approximation considered here, this Lorentz transformation reduces to a Galilean transformation, namely

$$\mathbf{r}_i \to \mathbf{r}_i + \mathbf{V}t, \qquad \mathbf{p}_i \to \mathbf{p}_i + m_i\mathbf{V}.$$

[2]) Likewise the relation between (θ_2, φ_2) and (θ, φ) is defined by the equality $\mathbf{v}_2 = \mathbf{v}'' + \mathbf{V}$. Since $v'' = V$, it is easy to show that $\theta_2 = \frac{1}{2}(\pi - \theta)$, $\varphi_2 = \varphi + \pi$.

from which

$$\tan \theta_1 = \frac{\sin \theta}{\cos \theta + \tau},$$

or else

$$\cos \theta_1 = \frac{\cos \theta + \tau}{(1 + 2\tau \cos \theta + \tau^2)^{\frac{1}{2}}}, \tag{X.24}$$

expressions in which we have put

$$\tau = \frac{V}{v'} = \frac{m_1}{m_2}. \tag{X.25}$$

The vector sum (X.23) can actually be made graphically; relation (X.24) can be read directly from the figure and the discussion of the

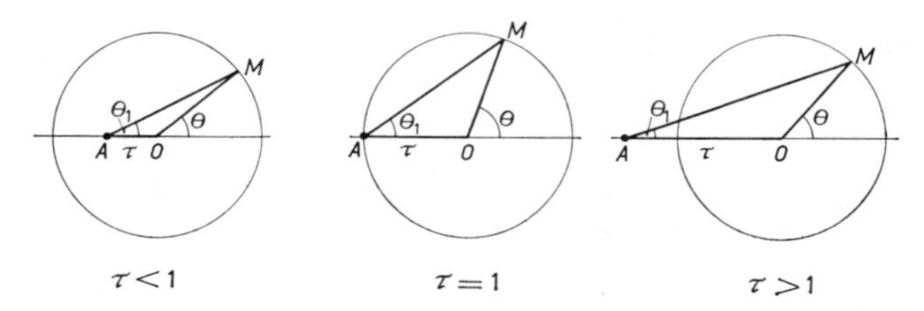

Fig. X.4. Geometrical construction of θ_1 as function of θ:
$OA = \tau \qquad OM = 1 \qquad (\tau = V/v' = m_1/m_2)$.

results is easier in terms of the figure than in terms of formula (X.24) itself. This graphical construction is reproduced in Fig. X.4. Two cases may arise:

(*i*) $\tau < 1$ $(m_1 < m_2)$.—The angle θ_1 grows monotonically from 0 to π as θ grows from 0 to π. Note that $\frac{1}{2}\theta < \theta_1 < \theta$ for any θ. In the limit where $m_1 \ll m_2$, $\theta_1 \approx \theta$ (the center of mass remains practically identical with particle 2 during the collision, hence it is practically at rest in the laboratory system).

(*ii*) $\tau > 1$ $(m_1 > m_2)$.—As θ grows from 0 to π, θ_1 at first increases from 0 to some maximum value smaller than $\frac{1}{2}\pi$, namely $\theta_{1\,\max.} = \sin^{-1}(1/\tau)$, then it decreases from $\theta_{1\,\max.}$ to zero. To each value of θ_1 thus correspond two possible values of θ, $\theta_<$ and $\theta_>$, related by: $\theta_1 = \frac{1}{2}(\theta_< + \theta_> - \pi)$; to each of these correspond two distinct values of v_1, the larger one corresponding to the smaller value of θ.

[When $\tau = 1$ $(m_1 = m_2)$, one simply has $\theta_1 = \frac{1}{2}\theta$.]

From relation (X.24) we obtain

$$\frac{\mathrm{d}\,(\cos\theta_1)}{\mathrm{d}\,(\cos\theta)} = \frac{1+\tau\cos\theta}{(1+2\tau\cos\theta+\tau^2)^{3/2}}$$

and since

$$\frac{\mathrm{d}\Omega_1}{\mathrm{d}\Omega} = \left|\frac{\mathrm{d}\,(\cos\theta_1)}{\mathrm{d}\,(\cos\theta)}\right|,$$

we obtain, applying relation (X.22),

$$\sigma_1(\Omega_1) = \sigma(\Omega)\,\frac{\mathrm{d}\Omega}{\mathrm{d}\Omega_1} = \frac{(1+2\tau\cos\theta+\tau^2)^{3/2}}{|1+\tau\cos\theta|}\,|f(\Omega)|^2. \qquad \text{(X.26)}$$

II. SCATTERING BY A CENTRAL POTENTIAL. PHASE SHIFTS

8. Decomposition into Partial Waves. Phase-Shift Method

Consider the scattering of a particle by a *central potential* $V(r)$. To calculate the cross section, one needs the asymptotic form of the stationary scattering wave ψ. To this effect we solve the Schrödinger equation in spherical coordinates.

The direction of the incident wave vector $\boldsymbol{k}$ is an axis of rotational symmetry of the problem; if we take it as polar axis, the wave ψ and the scattering amplitude f are independent of φ. Expanding in a series of Legendre polynomials, we have

$$\psi(r,\theta) = \sum_l \frac{y_l(r)}{r}\,P_l\,(\cos\theta) \qquad \text{(X.27)}$$

$$f(\theta) = \sum_l f_l P_l\,(\cos\theta). \qquad \text{(X.28)}$$

Let us put

$$\varepsilon = k^2 = \frac{2m}{\hbar^2}\,E, \qquad U(r) = \frac{2m}{\hbar^2}\,V(r).$$

y_l is a regular solution of the radial equation

$$\left[\frac{\mathrm{d}^2}{\mathrm{d}r^2} + \left(\varepsilon - U(r) - \frac{l(l+1)}{r^2}\right)\right] y_l = 0. \qquad \text{(X.29)}$$

Its asymptotic form reads

$$y_l \underset{r\to\infty}{\sim} a_l \sin\,(kr - \tfrac{1}{2}l\pi + \delta_l). \qquad \text{(X.30)}$$

All regular solutions of eq. (X.29) yield the same phase shift δ_l; they differ by the normalization constant a_l. Here, a_l must be adjusted in such a way that $\psi(r, \theta)$ has the desired asymptotic behavior. Using the expansions (IX.35) and (X.28), we can write the asymptotic form of ψ in the form of a series of Legendre polynomials:

$$\mathrm{e}^{\mathrm{i}\boldsymbol{k}\cdot\boldsymbol{r}} + f(\theta)\frac{\mathrm{e}^{\mathrm{i}kr}}{r} = \sum_l \left((2l+1)\,\mathrm{i}^l\, j_l(kr) + f_l\frac{\mathrm{e}^{\mathrm{i}kr}}{r}\right) P_l(\cos\theta).$$

If one takes the asymptotic form of $j_l(kr)$ into account we can rewrite this expression, separating the incoming and outgoing waves

$$r\psi(r, \theta) \sim \sum_l \left[(-)^{l+1}\frac{2l+1}{2\mathrm{i}k}\,\mathrm{e}^{-\mathrm{i}kr} + \left(\frac{2l+1}{2\mathrm{i}k} + f_l\right)\mathrm{e}^{\mathrm{i}kr}\right] P_l(\cos\theta).$$

The asymptotic form of y_l must be equal to the quantity in brackets of the right-hand side. This condition fixes a_l uniquely and enables us to write f_l as function of the phase shift. We have successively

$$a_l = \mathrm{i}^l\,\frac{2l+1}{k}\,\mathrm{e}^{\mathrm{i}\delta_l},$$

$$f_l = \frac{2l+1}{k}\,\mathrm{e}^{\mathrm{i}\delta_l}\sin\delta_l.$$

Substituting this expression in eq. (X.28) we obtain $f(\theta)$ as a function of the phase shifts δ_l:

$$\boxed{f(\theta) = \lambdabar \sum_{l=0}^{\infty} (2l+1)\,\mathrm{e}^{\mathrm{i}\delta_l}\sin\delta_l\, P_l(\cos\theta).} \qquad \text{(X.31)}$$

$\lambdabar$ here refers to the incident wavelength ($\lambdabar = 1/k$).

It is instructive to compare the asymptotic form of y_l/r, the component of angular momentum l of the stationary state of scattering, with the corresponding expression of the function $(2l+1)\,\mathrm{i}^l\, j_l(kr)$, the component of the same angular momentum of the plane wave $\exp(\mathrm{i}\boldsymbol{k}\cdot\boldsymbol{r})$:

$$\frac{y_l}{r} \sim \frac{(2l+1)}{2\mathrm{i}kr}\left[(-)^{l+1}\mathrm{e}^{-\mathrm{i}kr} + \mathrm{e}^{2\mathrm{i}\delta_l}\,\mathrm{e}^{\mathrm{i}kr}\right],$$

$$(2l+1)\,\mathrm{i}^l\, j_l(kr) \sim \frac{2l+1}{2\mathrm{i}kr}\left[(-)^{l+1}\mathrm{e}^{-\mathrm{i}kr} + \mathrm{e}^{\mathrm{i}kr}\right].$$

They are both superpositions of an incoming and an outgoing wave of the same intensity. The incoming wave is obviously the same for the two functions. The outgoing wave of the stationary state of scattering differs from that of the plane wave by the presence of the phase factor $\exp(2i\delta_l)$: *the effect of the scattering potential is to shift the phase of each outgoing partial wave.*

The differential cross section is obtained by forming the square modulus of $f(\theta)$:

$$\sigma(\Omega) = \lambda\!\!\!^{-2} \sum_{ll'} (2l+1)(2l'+1)\, e^{i(\delta_l - \delta_{l'})} \sin\delta_l \sin\delta_{l'}\, P_l(\cos\theta)\, P_{l'}(\cos\theta). \quad (X.32)$$

Integration over the angles (θ, φ) yields the total cross section $\sigma_{\text{tot.}}$. Taking into account the orthogonality relations of the Legendre polynomials, the latter is written in the form of a series:

$$\boxed{\sigma_{\text{tot.}} = 4\pi\lambda\!\!\!^{-2} \sum_{l=0}^{\infty} (2l+1) \sin^2\delta_l} \quad (X.33)$$

each term of which,

$$\sigma_l = 4\pi(2l+1)\,\lambda\!\!\!^{-2} \sin^2\delta_l \quad (X.34)$$

represents the contribution to the scattering of the corresponding angular momentum l. Note the inequality:

$$\sigma_l \leqslant 4\pi(2l+1)\lambda\!\!\!^{-2}. \quad (X.35)$$

The maximum value of σ_l is reached when the phase shift is a "half-integral" multiple of π:

$$\delta_l = (n+\tfrac{1}{2})\pi \qquad (n \text{ integer}).$$

9. Semi-Classical Representation of the Collision. Impact Parameters

Consider the collision of a *classical* particle with a central force field. The energy of the incident particle being fixed at $E = p^2/2m$, every trajectory may be characterized by its impact parameter b, defined as the distance from the center of force C to the straight line containing the initial momentum $\boldsymbol{p}_0$ (Fig. X.5). In such a collision, the angular momentum L is a constant of the motion. b is directly proportional to L:

$$L = bp.$$

If the force field has a finite range r_0,

$$V(r) = 0 \quad \text{for} \quad r > r_0,$$

the incident particle does or does not suffer a deflection according to whether $b < r_0$ or $b > r_0$. The deflection is limited to particles whose angular momentum is sufficiently small.

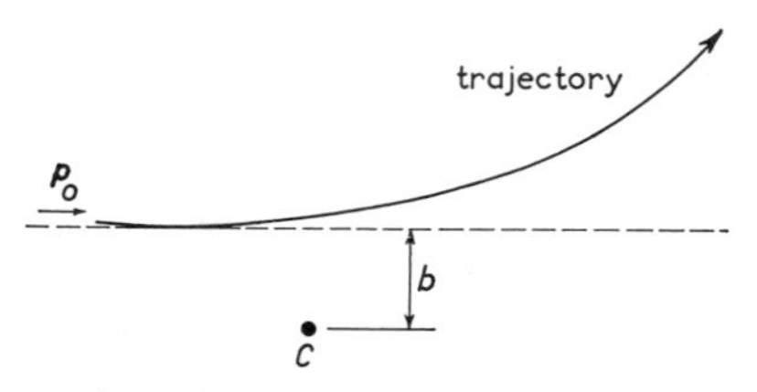

Fig. X.5. Scattering of a classical particle by a center of force C. $\boldsymbol{p}_0$ = initial momentum; b = impact parameter.

A collision in quantum theory is very different from a classical collision: it is essentially a wave phenomenon. However, when the scattering potential $V(r)$ is negligible – without necessarily being zero – beyond a certain distance r_0, the phenomenon is somewhat analogous to the scattering of a beam of classical particles by a potential of finite range r_0. *As a general rule, the contribution σ_l of the* lth *partial wave is negligible* [1]) *if* $l\lambda\!\!\!^{-} \gtrsim r_0$; on the other hand, if $l\lambda\!\!\!^{-} < r_0$, it may take on all values between zero and its maximum value $4\pi(2l+1)\lambda\!\!\!^{-2}$. In the classical scattering problem the part of the beam which corresponds to the lth partial wave is given by the particles whose impact parameters range from $l\lambda\!\!\!^{-}$ to $(l+1)\lambda\!\!\!^{-}$, hence whose angular momenta range from $l\hbar$ to $(l+1)\hbar$. The contribution σ_l obviously vanishes if $l\lambda\!\!\!^{-} > r_0$, takes on the value $\pi(2l+1)\lambda\!\!\!^{-2}$ if $(l+1)\lambda\!\!\!^{-} < r_0$, and takes on some intermediate value if r_0 lies between $l\lambda\!\!\!^{-}$ and $(l+1)\lambda\!\!\!^{-}$.

The foregoing rule for quantum-mechanical scattering is based upon the following semi-quantitative argument. The incident wave is a superposition of spherical waves of given angular momentum. The radial portion of the term corresponding to the partial wave of angular momentum l is proportional to $j_l(r/\lambda\!\!\!^{-})$, hence a relative probability density of presence in the spherical shell $(r, r+\mathrm{d}r)$ equal to $r^2 j_l^2(r/\lambda\!\!\!^{-})$. This density is very low as long as $r < \sqrt{l(l+1)}\lambda\!\!\!^{-}$ and oscillates between 0 and approximately 1 when $r > \sqrt{l(l+1)}\lambda\!\!\!^{-}$ (cf. Fig. X.6). If $r_0 < l\lambda\!\!\!^{-}$, the

[1]) This rule is not absolute; we shall encounter exceptions in § 14 in connection with resonance phenomena.

wave practically does not penetrate into the region of strong potential; there is therefore little chance that the wave be greatly affected by the presence of the potential.

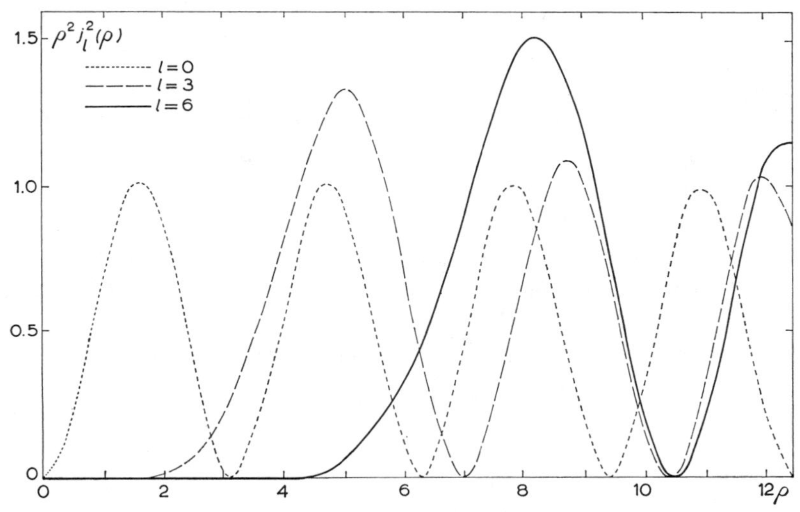

Fig. X.6. Diagram of the function $\varrho^2 j_l^2(\varrho)$ for $l = 0, 3, 6$.

The argument given here is not rigorous. More precise indications concerning the rapidity of convergence of the series (X.31) and (X.32) will be given in § 12. Be that as it may, *the phase-shift method is particularly well suited to the calculation of cross sections whenever the range of the scattering potential does not extend beyond several wavelengths.*

III. POTENTIAL OF FINITE RANGE [1]

10. Relation between Phase Shift and Logarithmic Derivative

Suppose that $V(r)$ is entirely concentrated in a finite region of space, so that one rigorously has

$$V(r) = 0 \quad \text{when} \quad r > r_0.$$

[1] Range is used here in a somewhat restricted sense; it is the value of the radial distance beyond which the potential rigorously vanishes. More generally, the range of a potential is the value of the radial distance beyond which it takes on negligible values. The results obtained below hold essentially true for potentials of finite range in this broader sense; the effect of the tail of the potential may then be estimated with the help of the generalized Born formula. (See § 19 below.)

Denote by q_l the value taken at the point r_0 by the logarithmic derivative of the solution regular at the origin, of the radial equation (X.29)

$$q_l = \left.\frac{r(\mathrm{d}y_l/\mathrm{d}r)}{y_l}\right|_{r=r_0} \tag{X.36}$$

(the definition adopted here differs by a factor r from the usual definition). q_l is known to be a monotonically decreasing function of the energy (Ch. III, § 8); the exact form of this function evidently depends upon the potential $V(r)$ under consideration.

Since the latter vanishes in the region $r > r_0$, there exists between q_l and δ_l a relation which is independent of the particular form of $V(r)$; specifying q_l suffices to determine the asymptotic behavior of the solution.

In the following, we assume y_l to be normalized in such a way that

$$y_l \underset{r\to\infty}{\sim} \sin\left(kr - \tfrac{1}{2}l\pi + \delta_l\right). \tag{X.37}$$

We therefore have in the external region

$$y_l = kr[\cos\delta_l\, j_l(kr) + \sin\delta_l\, n_l(kr)] \qquad (r > r_0).$$

For purposes of later discussions, it is convenient to put

$$\xi = kr$$

and to introduce the outgoing and incoming waves

$$u_l^{(\pm)}(\xi) = \xi[n_l(\xi) \pm \mathrm{i} j_l(\xi)] = \xi h_l^{(\pm)}(\xi)$$

whose Wronskian, a quantity independent of ξ, is

$$u^{(-)}\frac{\mathrm{d}}{\mathrm{d}\xi}u^{(+)} - u^{(+)}\frac{\mathrm{d}}{\mathrm{d}\xi}u^{(-)} = 2\mathrm{i}. \tag{X.38}$$

In the external region,

$$\begin{aligned} y_l &= \frac{1}{2\mathrm{i}}\left(u_l^{(+)}\,\mathrm{e}^{\mathrm{i}\delta_l} - u_l^{(-)}\,\mathrm{e}^{-\mathrm{i}\delta_l}\right) \\ &= \operatorname{Im} u_l^{(+)}\,\mathrm{e}^{\mathrm{i}\delta_l}. \end{aligned} \qquad (r > r_0) \tag{X.39}$$

The continuity condition on the logarithmic derivative at $r = r_0$ yields the relation

$$q_l = \xi \left.\frac{\operatorname{Im} \mathrm{e}^{\mathrm{i}\delta_l}\,(\mathrm{d}u_l^{(+)}/\mathrm{d}\xi)}{\operatorname{Im} \mathrm{e}^{\mathrm{i}\delta_l}\,u_l^{(+)}}\right|_{\xi=kr_0}. \tag{X.40}$$

This is the relation between q_l and δ_l we were seeking.

To put this relation into a more tractable form, it is convenient to put

$$\begin{aligned} |u_l^{(\pm)}(kr_0)| &= v_l^{-\frac{1}{2}}, \\ u_l^{(\pm)}(kr_0) &= v_l^{-\frac{1}{2}}\, \mathrm{e}^{\mp \mathrm{i}\tau_l} \\ \left.\frac{\xi\,(\mathrm{d}u_l^{(\pm)}/\mathrm{d}\xi)}{u_l^{(\pm)}}\right|_{\xi=kr_0} &= q_l^{(\pm)}. \end{aligned} \qquad (\mathrm{X}.41)$$

From equation (X.38) we obtain the relation (Problem X.3)

$$\mathrm{Im}\, q_l^{(+)} = kr_0 v_l;$$

v_l is a positive quantity less than 1, which is smaller, the smaller kr_0 and the larger l; it is called the *penetration factor*.

With these notations the continuity relation (X.40) is written

$$\mathrm{e}^{2\mathrm{i}\delta_l} = \mathrm{e}^{2\mathrm{i}\tau_l}\, \frac{q_l - q_l^{(-)}}{q_l - q_l^{(+)}}$$

or

$$\delta_l = \tau_l + \varrho_l \qquad (\mathrm{X}.42)$$

with

$$\varrho_l = \arg\,(q_l - q_l^{(-)}) = \tan^{-1} \frac{kr_0\, v_l}{q_l - \mathrm{Re}\, q_l^{(+)}}. \qquad (\mathrm{X}.43)$$

In conclusion, δ_l can be put in the form of a sum of two terms the first of which, τ_l, is independent of the particular form of the scattering potential, and the second, ϱ_l, depends on it through q_l, in accordance with eq. (X.43).

Note that

$$y_l(r_0) = \frac{\sin \varrho_l}{\sqrt{v_l}}. \qquad (\mathrm{X}.44)$$

11. Behavior of the Phase Shift at Low Energies ($\lambda \to \infty$)

Knowing the behavior of the spherical Bessel functions for very small values of the argument [eq. (B.52)], one may deduce from eqs. (X.42) and (X.43) the behavior of δ_l when $kr_0 \ll l$, that is at very low energies or for very large values of the angular momentum. Indeed, if $kr_0 \ll l$,

$$\begin{aligned} \tau_l &\approx -\frac{(kr_0)^{2l+1}}{(2l+1)!!\,(2l-1)!!}, \\ v_l &\approx \frac{(kr_0)^{2l}}{[(2l-1)!!]^2} \qquad \mathrm{Re}\, q_l^{(+)} \approx -l + \mathrm{O}(k^2 r_0^2) \end{aligned} \qquad (\mathrm{X}.45)$$

(these expressions are also valid for $l=0$ when $kr_0 \ll 1$).

In this section we examine the behavior of the cross sections as the energy approaches zero.

In that limit, the real quantity q_l increases toward a certain limit $\hat{q}_l$. In general $\hat{q}_l \neq -l$ and the phase shift δ_l tends to zero as k^{2l+1}. One finds

$$\delta_l \underset{k\to 0}{\sim} \frac{l+1-\hat{q}_l}{l+\hat{q}_l} \frac{(kr_0)^{2l+1}}{(2l+1)!!\,(2l-1)!!}. \qquad \text{(X.46)}$$

The amplitude f_l, proportional to δ_l/k, thus tends toward zero as k^{2l}. Therefore, in the limit of very low energies, the cross section becomes isotropic, since all partial cross sections σ_l vanish as k^{4l} [eq. (X.34)], except for the s-wave cross section σ_0, which generally approaches a constant different from zero.

By definition, one calls *scattering length* the length

$$\begin{aligned} a &= -\lim_{k\to 0} f_0 = -\lim \frac{\delta_0}{k} \\ &= \left(1-\frac{1}{\hat{q}_0}\right) r_0 . \end{aligned} \qquad \text{(X.47)}$$

a is obtained by solving the radial equation for zero energy:

$$\left[\frac{\mathrm{d}^2}{\mathrm{d}r^2} - U(r)\right] y_0 = 0.$$

It is the abscissa of the point where the asymptote of y_0 cuts the r axis. In the limit of very low energies

$$\sigma_{\text{tot.}} = \sigma_0 = 4\pi a^2.$$

If $\hat{q}_l = -l$ accidentally, f_l behaves as k^{2l-2} when the energy tends to zero (except for $l=0$ where $\hat{q}_0=0$ implies: $f_0 \sim \mathrm{i}/k$). There is said to be a resonance at zero energy in the state l. The foregoing conclusions must then be modified as follows. If it is an s resonance ($l=0$), a is infinite and $\sigma_{\text{tot.}}$ approaches infinity as $1/E$. If it is a p resonance ($l=1$), $f(\theta)$ approaches the form $-a+b\cos\theta$; the cross section remains finite but never becomes isotropic. If the resonance is of higher order, the zero-energy limit of the cross sections is not affected.

12. Partial Waves of Higher Order. Convergence of the Series ($l \to \infty$)

The asymptotic forms (X.45) likewise enable us to obtain asymptotic expressions for the phase shifts of higher order when one picks a well-defined value of the energy. If l is sufficiently large, the expression

$$k^2 - U(r) - \frac{l(l+1)}{r^2}$$

always remains negative in the interval $(0, r_0)$; hence the solution y_l of eq. (X.29) has an "exponential" behavior in that interval, and q_l is certainly positive. Moreover, if $l \gg kr_0$, the expressions (X.45) can be used and since $q_l + l$ is certainly not zero, we find, for l sufficiently large, an asymptotic expression similar to expression (X.46):

$$\delta_l \underset{l\to\infty}{\sim} \frac{l+1-q_l}{l+q_l} \frac{(kr_0)^{2l+1}}{(2l-1)!!\,(2l+1)!!}. \qquad \text{(X.48)}$$

This expression provides us with a measure of the rapidity of convergence of the expansion in a series of partial waves when the potential has a finite range. Thus the order of magnitude predictions stated in § 9 are confirmed.

13. Scattering by a Hard Sphere

If the potential of finite range is that of a hard sphere, that is to say, if $V(r)$ is infinitely repulsive in the region $r < r_0$, and zero in the region $r > r_0$,

$$V(r) = \begin{cases} +\infty & \text{if} \quad r < r_0 \\ 0 & \text{if} \quad r > r_0 \end{cases}$$

all formulae of § 10 simplify. The wave function must vanish on the surface of the sphere, hence $y_l(r_0) = 0$ for any $l (q_l = -\infty)$, which yields [eq. (X.44)] $\varrho_l = 0$, or

$$\delta_l = \tau_l = \arg u_l^{(-)}. \qquad \text{(X.49)}$$

From this one deduces after some calculation that

$$\sigma_l = \frac{4\pi(2l+1)}{k^2} \frac{j_l^2(kr_0)}{j_l^2(kr_0) + n_l^2(kr_0)} \qquad \text{(X.50)}$$

and in particlar,

$$\sigma_0 = 4\pi r^2{}_0 \left(\frac{\sin kr_0}{kr_0}\right)^2.$$

At very low energies, in accordance with the study of § 11, the differential cross section becomes isotropic and the total cross section has as its limit

$$\lim_{k \to 0} \sigma_{\text{tot.}} = \lim_{k \to 0} \sigma_0 = 4\pi r_0^2, \tag{X.51}$$

corresponding to a scattering length $a = r_0$.

As the energy increases, the contribution of higher-order partial waves becomes more and more important, and the anisotropy of the scattering becomes more and more marked. At very high energies ($\lambda\!\!\!^{-} \ll r_0$), the differential and total cross sections can be evaluated by making use of the asymptotic properties of the Bessel functions for large values of the order l. One finds[1])

$$\sigma(\Omega) \underset{k \to \infty}{\sim} \tfrac{1}{4} r_0^2 \left(1 + \cot^2 \frac{\theta}{2} J_1^2(kr_0 \sin \theta)\right) \tag{X.52}$$

$$\lim_{k \to \infty} \sigma_{\text{tot.}} = 2\pi r_0^2. \tag{X.53}$$

Let us give here a simplified proof of relation (X.53). Knowing the general shape of the functions $j_l(\xi)$ and $n_l(\xi)$ we can deduce the behavior of the function

$$g_l(\xi) = \frac{j_l^2(\xi)}{j_l^2(\xi) + n_l^2(\xi)}.$$

It vanishes as $\xi^{4l+2}/[(2l+1)!!\,(2l-1)!!]^2$ for $\xi = 0$, increase regularly as ξ increases toward the vicinity of $\xi = l$, then oscillates indefinitely according to the law

$$g_l(\xi) \underset{\xi \to \infty}{\sim} \sin^2 (\xi - \tfrac{1}{2} l)\pi. \tag{X.54}$$

Hence, in the summation

$$\sigma_{\text{tot.}} = \sum_{l=0}^{\infty} \sigma_l = \frac{4\pi}{k^2} \sum_{l=0}^{\infty} (2l+1)\, g_l(kr_0)$$

---

[1]) Cf. Morse and Feshbach, *loc. cit.*, (footnote in Ch. VI, p. 231), p. 1551. $J_1(x)$ is the first-order Bessel function. When x increases from 0 to $+\infty$, $J_1(x)$ increases from 0 [$J_1(x) \sim \frac{1}{2}x$] to a first maximum $J_1(1.84) \approx 0.58$, then decreases, and vanishes for the first time at $x \approx 3.83$; it then oscillates indefinitely according to the asymptotic expression

$$J_1(x) \underset{x \to \infty}{\sim} \left(\frac{2}{\pi x}\right)^{\frac{1}{2}} \cos (x - \tfrac{3}{4}\pi).$$

the contribution of the terms $l > kr_0$ can be neglected and that of the terms $l < kr_0$ roughly evaluated by replacing $g_l(kr_0)$ by its asymptotic form (X.54), whence

$$\sigma_{\text{tot.}} \underset{k\to\infty}{\sim} \frac{4\pi}{k^2} \sum_{l=0}^{kr} (2l+1) \sin^2 (kr_0 - \tfrac{1}{2} l\pi).$$

One evaluates the summation of the right-hand side by grouping successive terms two by two, which gives in the limit of very large values of k

$$\int_0^{kr_0} l \, \mathrm{d}l = \tfrac{1}{2} k^2 r_0^2,$$

from which we obtain equation (X.53).

In contrast to what one might expect, in the limit of short wavelengths ($kr_0 \gg 1$) one does not find the scattering cross sections of a classical particle by a hard sphere of radius r_0. The classical total cross section,

$$\sigma_{\text{cl.}} = \pi r_0^2$$

is just half of the quantum-mechanical result in the limit of very short wavelengths. Likewise, the classical differential cross section is isotropic and equal to $\frac{1}{4} r_0^2$: it is equal to the first term of the asymptotic form (X.52) of $\sigma(\Omega)$.

In fact, the wave aspect of the phenomenon can never be neglected, because even at the very short wavelengths, the potential can never be considered as slowly varying because of the sudden discontinuity at the point $r = r_0$. The observed wave phenomenon is entirely analogous to the diffraction phenomena found in optics, as is revealed by an examination of the asymptotic form (X.52) of the differential cross section. The latter contains two terms. The first is an isotropic "*reflection*" term identical to the classical differential cross section. The second:

$$\tfrac{1}{4} r_0^2 \cot^2 (\tfrac{1}{2}\theta) \, J_1^2(kr_0 \sin \theta),$$

strongly anisotropic since its contribution is essentially limited to small angles of the order of $\lambda\!\!\!^{-}/r_0$, is the "*diffraction*" term (shadow scattering) resulting from the shadow cast by the perfectly reflecting sphere upon the path of the incident wave.

IV. SCATTERING RESONANCES

14. Scattering by a Deep Square Well

As another example of a potential of finite range, we once again consider the square well of Chapter IX, § 10. We put

$$E = \frac{\hbar^2 k^2}{2m}, \qquad V_0 = \frac{\hbar^2 K_0^2}{2m}, \qquad K^2 = K_0^2 + k^2$$

and we shall study the behavior of the various partial waves as function of energy when the well is very deep. More precisely, we assume:

$$Kr_0 \gg l \tag{X.55}$$

$$K \gg k. \tag{X.56}$$

In that case, the value of q_l to be substituted in the right-hand side of eq. (X.43) is to a good approximation

$$q_l \approx Kr_0 \cot (Kr_0 - \tfrac{1}{2}l\pi). \tag{X.57}$$

The general behavior of $\delta_l(E)$ at low energies – namely $E \ll V_0$ according to condition (X.56) – is easily obtained by inspection.

In accordance with formulae (X.42) and (X.43), δ_l depends on the energy through the quantities τ_l, $kr_0 v_l$, $\text{Re}\, q_l^{(+)}$, and q_l. The first three are monotonic functions of kr_0 (τ_l a decreasing function, the two others increasing functions), whose behavior at the origin is given by the expressions (X.45) and in the asymptotic region ($kr_0 \gg l$) by the expressions:

$$\begin{gathered} \tau_l \underset{kr_0 \to \infty}{\sim} -(kr_0 - \tfrac{1}{2}l\pi), \\ \lim_{kr_0 \to \infty} v_l = 1 \qquad \lim_{kr_0 \to \infty} \text{Re}\, q_l^{(+)} = 0, \end{gathered} \tag{X.58}$$

In the energy region of interest, these three functions vary rather slowly. On the other hand, the logarithmic derivative q_l, a monotonically decreasing function of the energy, varies very rapidly and possesses a succession of vertical asymptotes located at the values of the energy for which $Kr_0 = \frac{1}{2}l\pi + n\pi$ (n integer). The energy difference between two neighboring asymptotes is approximately

$$D \approx \pi \frac{\hbar^2 K}{mr_0} \approx \pi \frac{V_0}{K_0 r_0}. \tag{X.59}$$

When the energy varies by this amount, $|q_l|$ is almost everywhere of the order or larger than Kr_0, and one has over almost the entire interval

$$|q_l - \mathrm{Re}\, q_l^{(+)}| \gg kr_0 v_l.$$

The second term of the phase shift, ϱ_l, remains very small (to within $n\pi$):

$$\varrho_l \approx \frac{kr_0 v_l}{q_l - \mathrm{Re}\, q_l^{(+)}} \lesssim \frac{k}{K} v_l,$$

while there exists no *a priori* limitation on τ_l. One can therefore write

$$\delta_l \approx \tau_l.$$

The phase shift is practically the same as that which would be produced by a hard sphere of the same radius. Over the major portion of the range of variation of the energy, therefore, the potential scatters each partial wave in much the same way as a hard sphere: this is the so-called "potential scattering"; the incident wave practically does not penetrate into the internal region.

However, there exists a small energy domain surrounding the point E_r, where $q_l = \mathrm{Re}\, q_l^{(+)}$, for which

$$|q_l - \mathrm{Re}\, q_l^{(+)}| \lesssim kr_0 v_l.$$

Let us define the quantities

$$\left.\begin{aligned} \gamma &= -\left.\frac{\mathrm{d}E}{\mathrm{d}q_l}\right|_{E=E_r} \qquad (\gamma > 0) \\ \Gamma &= 2kr_0 v_l \gamma \end{aligned}\right\} \tag{X.60}$$

Γ is the width of the energy domain in question. One notes that [1])

$$\frac{\Gamma}{D} \approx \frac{2}{\pi}\frac{k}{K} v_l \ll 1.$$

---

[1]) Indeed, according to equation (X.57)

$$\frac{\mathrm{d}q_l}{\mathrm{d}(Kr_0)} = -Kr_0\left(1 + \frac{q_l(q_l-1)}{(Kr_0)^2}\right).$$

Since $Kr_0 \gg l$ and $|\mathrm{Re}\, q_l^{(+)}| \lesssim l$, $\mathrm{d}q_l/\mathrm{d}K \approx -Kr_0^2$ when $E = E_r$, hence

$$\gamma \approx \frac{\hbar^2}{mr_0^2} \quad \text{and} \quad \Gamma \approx 2v_l \frac{\hbar^2 k}{mr_0}.$$

For an energy variation of the order of several Γ on either side of E_r, $q_l - \operatorname{Re} q_l^{(+)}$ decreases from values very much larger than kr_0v_l to values very much smaller than $-kr_0v_l$, and ϱ_l suddenly passes from values close to $n\pi$ to values in the neighborhood of $(n+1)\pi$. The partial cross section σ_l undergoes violent variations during which it reaches its maximum value $4\pi(2l+1)\lambda\!\!\!^{-2}$: there is said to be an l *resonance*. By definition, E_r is the resonance energy [1]), Γ is the width of the resonance. Γ is the product of a quantity depending upon the general shape of the potential in the internal region (practically independent of l) and the factor kr_0v_l depending upon the behavior of the wave in the external region (which is smaller, the smaller kr_0 and the larger l).

The resonance region is sufficiently narrow that one may replace the curve $q_l(E)$ by its slope at the point $E = E_r$ in the expression of ϱ_l, hence

$$\varrho_l \approx \tan^{-1} \frac{\Gamma}{2(E_r - E)}. \tag{X.61}$$

Using the continuity condition (X.44) to normalize the radial function in the internal region, one finds to this same approximation (valid over an energy interval surrounding the resonance such that $\Gamma \ll \Delta E \ll D$):

$$y_l = \frac{1}{\sqrt{v_l}} \frac{\Gamma}{\sqrt{4(E - E_r)^2 + \Gamma^2}} K_r j_l(Kr). \qquad (r < r_0) \tag{X.62}$$

The crossing of an l resonance is therefore accompanied by a sudden increase of the intensity of the partial wave of order l in the internal region.

The foregoing analysis simplifies considerably in the case of an s wave. One rigorously has

$$\tau_0 = -kr_0, \; v_0 = 1, \; \operatorname{Re} q_0^{(+)} = 0, \; q_0 = Kr_0 \cot Kr_0$$

$$\delta_0 = -kr_0 + \tan^{-1}\left(\frac{k}{K} \tan Kr_0\right).$$

---

[1]) Because of the presence of the potential scattering term, the maximum value of σ_l is reached for an energy value that is somewhat different from the resonance energy, the latter being defined as the energy for which $\varrho_l = \frac{1}{2}\pi$ (to within $n\pi$). It may even happen that $\tau_l = \frac{1}{2}\pi$ at resonance and therefore that σ_l vanishes when $E = E_r$; the crossing of the resonance region manifests itself by a sudden drop to zero of the function $\sigma_l(E)$.

The solution y_0 reads explicitly

$$y_0 = \begin{cases} \sin(kr+\delta_0) & r > r_0 \\ \dfrac{k}{\sqrt{k^2+K_0^2\cos^2 Kr_0}} \sin Kr & r < r_0. \end{cases}$$

Except for some differences in notation ($L \to r_0$, $K \to K_0$, $\eta K \to k$, $\xi K \to K$, $\varphi_1 \to \varrho_0 - \frac{1}{2}\pi$) the problem of s-wave scattering is identical to the problem of wave reflection by the one-dimensional square well treated in Ch. III, § 6 (case b), and the discussion of the resonance effect can be carried over without modification.

15. Study of a Scattering Resonance. Metastable States

Resonance phenomena are frequently encountered in microscopic physics. Scattering resonances of the type we have discussed in the case of the square well occur more generally with potentials, which are strongly attractive in a limited region of space. Because of the importance of phenomena of this type, we shall make a detailed study of an l resonance of scattering. The particular study carried out here refers to the square well, but it can be generalized almost point for point to more general potentials, the form of the potential entering only through the law of variation of the logarithmic derivative q.

For simplicity, we assume the resonances to be sufficiently narrow and sufficiently well separated so that only a single partial wave exhibits a resonance in the energy region under consideration. Moreover, we assume the resonance energy to be so low that

$$kr_0 \ll 1 \tag{X.63}$$

and that, consequently, the contribution to the scattering of the potential scattering terms can be completely neglected [1]. In other words, all the phase shifts are practically zero with the exception of the phase shift δ_l, which varies as a function of the incident energy E according to the law

$$\delta_l \approx \varrho_l = \tan^{-1} \frac{\Gamma}{2(E_r - E)}.$$

---

[1] The contribution of these terms to the cross section is of the order $4\pi r_0^2$, the contribution of an l resonance at the resonance energy is of the order of $4\pi(2l+1)\lambda\!\!\!^{-2} = 4\pi(2l+1)/k^2$.

Therefore

$$e^{i\delta_l} \sin \delta_l = \frac{\tan \delta_l}{1 - i \tan \delta_l} \approx - \frac{\Gamma}{2(E - E_r) + i\Gamma}$$

and the scattering amplitude may be simply written

$$f(\theta) \approx - \frac{2l+1}{k} P_l (\cos \theta) \frac{\Gamma}{2(E - E_r) + i\Gamma}. \tag{X.64}$$

Upon traversing the resonance, the modulus and the derivative of the argument of $f(\theta)$ exhibit very pronounced peaks. One has

$$\sigma(\Omega) = |f(\theta)|^2 = (2l+1)^2 P_l^2 (\cos \theta) \lambda\!\!\!^{-2} \times \frac{\Gamma^2}{4(E - E_r)^2 + \Gamma^2} \tag{X.65}$$

$$\frac{d}{dE} [\arg f(\theta)] = \frac{d\delta_l}{dE} = \frac{2}{\Gamma} \times \frac{\Gamma^2}{4(E - E_r)^2 + \Gamma^2}. \tag{X.66}$$

Equation (X.65) shows that in the vicinity of the resonance — to the extent that the effect of potential scattering is negligible — the angular distribution of scattering does not depend upon the energy, but only upon l, and that the total cross section varies as a function of the energy according to the "Lorentz law":

$$\sigma_{\text{tot.}} = 4\pi(2l+1) \lambda\!\!\!^{-2} \times \frac{\Gamma^2}{4(E - E_r)^2 + \Gamma^2}. \tag{X.67}$$

To interpret eq. (X.66), one must refer to the study of scattering of a wave packet made in Ch. X, §§ 4 to 6. Indeed, using the notations of these sections, one observes that

$$\frac{\boldsymbol{u} \cdot \boldsymbol{s}}{v} = \hbar \frac{d}{dE} [\arg f(\theta)]$$

and that, consequently, expression (X.66) gives the delay in the transmission of the scattered wave (cf. footnote on p. 379). This delay depends upon the energy according to the same Lorentz law as the total cross section, and reaches its maximum value, $2\hbar/\Gamma$, at the resonance energy.

One can therefore picture the resonance phenomenon as follows. Far from the resonance energy, the incident wave practically does not penetrate into the internal region [cf. eq. (X.62)]; everything proceeds as if the wave met a hard sphere. Only a relatively negligible fraction of that wave is scattered, and this scattering takes place practically without delay (delay of the order $-r_0/v$). In the vicinity of the resonance energy, the incident wave penetrates deeply into the internal region; a large fraction of the incident wave packet thus remains in the internal region during a time of the order of $\hbar/\Gamma$ before being re-emitted in the form of a scattered wave. Thus the existence of a large resonance scattering cross section is explained. During the entire period preceding the re-emission, the probability of presence of the particle inside, or in the vicinity of the internal region is very large, as in a bound state. But whereas a bound state is a stationary state whose lifetime is infinite, the *metastable state* considered here has a lifetime of order $\hbar/\Gamma$. Consequently, instead of being a state of rigorously defined energy, it must be represented by a wave whose spread in energy is of order Γ according to the time-energy uncertainty relation. We are thus led to associate with each resonance a metastable state whose lifetime is $\hbar/\Gamma$ and whose energy is equal on the average to the resonance energy E_{r} with a spread equal to the width of the resonance Γ.

16. Observation of the Lifetime of Metastable States

Strictly speaking, this semi-classical picture of the resonant scattering phenomenon cannot be pushed too far without contradictions. In fact, under the normal conditions of observation of cross sections such as they were described in §§ 4 to 6, it is completely impossible to detect the metastable state which was just discussed. Indeed, to perform a cross-section measurement at a given energy, the energy spread ΔE of the incident wave packet must be so small that the scattering amplitude remains practically constant over the interval ΔE; in the resonance region this implies

$$\Delta E \ll \Gamma.$$

If this condition is realized, it is possible to detect the law of variation of the cross section as a function of the energy in the resonance region. On the other hand, the collision time $\hbar/\Delta E$, that is, the time necessary for the entire wave packet to penetrate into the scattering region, is much longer than the lifetime $\hbar/\Gamma$ of the metastable state. This latter is therefore completely unobservable (cf. footnote p. 379).

In order to detect the metastable state, one has to operate under complementary experimental conditions (in the sense of Bohr), namely

$$\Delta E \gg \Gamma. \tag{X.68}$$

To be more precise (cf. footnote on p. 372), consider a wave packet of the type contemplated in § 5 which simultaneously satisfies conditions (X.9) and condition (X.68). We further suppose that $E_r \gg \Delta E \gg \Gamma$ [1]). We adopt the notations of §§ 4 to 6, and suppose for simplicity that $\mathbf{b}=0$ and $t_0=0$. Substituting expression (X.64) of $f(\theta)$ into (X.17), we obtain the following asymptotic form for the scattered wave packet:

$$\Psi^{(d)} \sim -(2l+1)\, P_l(\cos\theta)\, \frac{\Gamma}{2}\, \frac{e^{i[k_r r-(E_r t/\hbar)]}\, I}{r}$$

$$I = \int \frac{A(\mathbf{k}'-\mathbf{k})}{E'-E_r+\frac{1}{2}i\Gamma} \exp\left[i(k'-k_r)\,r - i(E'-E_r)t/\hbar\right] \frac{d\mathbf{k}'}{k'}.$$

(we use the notation $E_r = \frac{1}{2}\hbar^2 k_r^2/m = \frac{1}{2}mv_r^2$).
In the integral I, the main contribution comes from the region where $A(\mathbf{k}'-\mathbf{k})/(E'-E_r+\frac{1}{2}i\Gamma)$ is large. Going over to polar coordinates, we put

$$d\mathbf{k}' = k'^2\, d\Omega'\, dk' = \frac{mk'}{\hbar^2}\, d\Omega'\, dE'$$

and $\mathbf{k}'=k'\mathbf{u}'$. The angular integration involves only $A(\mathbf{k}'-\mathbf{k})$. As for the integration over dE', according to hypothesis (X.68), the only important region is the region $|E'-E_r| \lesssim \Gamma$, region in which $A(\mathbf{k}'-\mathbf{k})$ can be replaced by $A(k_r\mathbf{u}'-\mathbf{k})$, and k' by the first two terms of its Taylor expansion (we have assumed that $\Gamma \ll E_r$):

$$k' \approx k_r + \frac{1}{\hbar v_r}(E'-E_r).$$

We can therefore write

$$I \approx \frac{m}{\hbar^2} A_r F\left(t - \frac{r}{v_r}\right),$$

with the definitions:

$$A_r = \int A(k_r\mathbf{u}'-\mathbf{k})\, d\Omega'$$

$$F(\tau) = \int_0^\infty \frac{e^{-i(E'-E_r)\tau/\hbar}}{E'-E_r+\frac{1}{2}i\Gamma}\, dE'.$$

---

[1]) This restriction and condition (X.63) are not essential, but they permit to put the final result into a simpler form. In order that they may be realized at the same time as conditions (X.9) and (X.68), one must have

$$v_l \ll kr_0 \ll 1.$$

The reader may easily convince himself that this expression for I is justified only if $|\tau| \gg \hbar/\Delta E$. Note that

$$F(\tau) = \int_{-(2E_\mathrm{r}/\Gamma)}^{\infty} \frac{\exp[-\mathrm{i}\Gamma\tau z/2\hbar]}{z+\mathrm{i}}\,\mathrm{d}z.$$

Since $|\tau| \gg \hbar/E_\mathrm{r}$, one can replace the lower limit of the integral by $-\infty$, and the integral $F(\tau)$ is easily calculated by the method of residues, namely

$$F(\tau) = \begin{cases} 0 & \text{if } \tau \ll -\hbar/\Delta E \quad <0 \\ -2\pi\,\mathrm{i}\,\mathrm{e}^{-\Gamma\tau/2\hbar} & \text{if } \tau \gg \hbar/\Delta E \quad >0. \end{cases} \qquad \text{(X.69)}$$

One finally has

$$\Psi^{(d)} \sim -(2l+1)\,P_l(\cos\theta)\,\frac{mA_\mathrm{r}}{2\hbar^2}\,\Gamma\times F\left(t-\frac{r}{v_\mathrm{r}}\right)\frac{\mathrm{e}^{\mathrm{i}[k_\mathrm{r}r-(E_\mathrm{r}t/\hbar)]}}{r}. \qquad \text{(X.70)}$$

The general behavior of this wave derives from the properties of the function $F(\tau)$ given by expression (X.69). It is an outgoing wave bounded by a wave front moving according to the law $r=v_\mathrm{r}t$; at a given point, the intensity of the wave is zero at first, then passes suddenly from 0 to some positive value. This transition corresponds to the passage of the wave front; it lasts for a time of order $\hbar/\Delta E$, i.e. a time interval very short compared to $\hbar/\Gamma$; afterwards, the intensity decreases according to the law $\exp(-\Gamma t/\hbar)$.

In a typical experiment designed to observe this law of exponential decrease, one turns on, during a very short time, a beam of wave packets satisfying the foregoing conditions, and one counts the particles scattered into the solid angle $(\Omega, \Omega+\delta\Omega)$ by placing a detector in that direction at a certain distance D from the scattering center. Since the energy spread [1] of the incident wave packets is very large, ($\Delta E \gg \Gamma$), the instant $t=0$ at which the collisions take place is very well defined: $\Delta t \ll \hbar/\Gamma$. The counting rate is given by the value assumed by $|\Psi^{(d)}|^2$ at the place where the counter is located. According to eq.(X.70), this is proportional to $F^2(t-D/v_\mathrm{r})$. No particle is detected up to the instant D/v_r at which the wave front reaches the counter. D/v_r is the time needed for a particle emitted by the scattering center at the "resonance velocity" v_r to reach the counter. Afterwards, the counting rate is given by the law $\exp(-\Gamma t/\hbar)$ corresponding to the formation at the instant $t=0$ of a metastable state of mean life $\hbar/\Gamma$.

The foregoing experimental conditions are usually realized in the disintegration of radioactive nuclei (α and β radioactivity, γ radioactivity of isomeric nuclei).

---

[1] We are dealing here with the energy spread of *each wave packet taken individually*.

V. VARIOUS FORMULAE AND PROPERTIES

17. Integral Representations of Phase Shifts

Certain properties or methods of calculation of phase shifts may be obtained by starting from suitable integral representations of the phase shifts. There are many such integral representations. Most of them are obtained simply by applying the Wronskian theorem to suitably defined solutions of the corresponding radial equations. We shall give one of them in this section. Another one will be given in § 20.

The expression we are seeking serves to compare the phase shifts δ_l and $\hat{\delta}_l$ corresponding respectively to the potentials $V(r)$ and $\hat{V}(r)$ at the same energy. We use the notation of § 8 and put $\hat{U}=2m\hat{V}/\hbar^2$. y_l is the regular solution of eq. (X.29) whose asymptotic form is given by expression (X.37). Likewise, we designate by $\hat{y}_l$ the regular solution of the radial equation

$$\left[\frac{\mathrm{d}^2}{\mathrm{d}r^2} + \left(\varepsilon - \hat{U} - \frac{l(l+1)}{r^2}\right)\right]\hat{y}_l = 0, \tag{X.71}$$

having for its asymptotic expression

$$\hat{y}_l \underset{r\to\infty}{\sim} \sin\,(kr - \tfrac{1}{2}l\pi + \hat{\delta}_l).$$

The Wronskian $W(y_l, \hat{y}_l)$ is zero at the origin and asymptotically approaches the limit

$$\lim_{r\to\infty} W(y_l, \hat{y}_l) = k \sin\,(\delta_l - \hat{\delta}_l).$$

According to the Wronskian theorem:

$$W(y_l, \hat{y}_l)\Big|_a^b = -\int_a^b \hat{y}_l(U - \hat{U})y_l\,\mathrm{d}r.$$

Letting a and b approach 0 and ∞, respectively, one has

$$\sin\,(\delta_l - \hat{\delta}_l) = -\frac{2m}{\hbar^2 k}\int_0^\infty \hat{y}_l(V - \hat{V})y_l\,\mathrm{d}r. \tag{X.72}$$

This important relation is valid for any form of the potentials V and $\hat{V}$, provided that they vanish at infinity more rapidly than $1/r$ and that they have no singularity as strong as $1/r^2$ at the origin.

If $\hat{V}=0$, one has $\hat{\delta}_l=0$, $\hat{y}_l=krj_l(kr)$; relation (X.72) in that special case is written:

$$\sin \delta_l = -\frac{2m}{\hbar^2}\int_0^\infty j_l(kr)\, V y_l r\, \mathrm{d}r. \tag{X.73}$$

18. Dependence upon the Potential. Sign of the Phase Shifts

Equation (X.72) allows to draw certain conclusions concerning the effects on the phase shifts when one changes the potential. Indeed, if $\Delta V \equiv V-\hat{V}$ is infinitesimal, $\Delta\delta_l=\delta_l-\hat{\delta}_l$ is also infinitesimal; on the other hand, the difference between the solutions y_l and $\hat{y}_l$ may be neglected in the integral of the right-hand member of eq. (X.72), whence

$$\Delta\delta_l = -\frac{2m}{\hbar^2 k}\int_0^\infty y_l^2 \Delta V\, \mathrm{d}r. \tag{X.74}$$

If the variation of the potential $\Delta V(r)$ has the same sign over the entire interval $(0, \infty)$ the variation of the phase shift $\Delta\delta_l$ has the opposite sign. Hence any increase of the potential (greater repulsion) decreases the phase shift, any decrease of the potential (greater attraction) increases the phase shift.

Heretofore the phase shift δ_l was defined to within $2n\pi$. In order to remove this ambiguity, we imagine a continuous modification of the potential from 0 up to $V(r)$; in this operation, the phase shift varies in a continuous manner from 0 up to a certain value δ_l; one can show that this value is independent of the path followed to go from zero potential to potential $V(r)$. It is this value we adopt henceforth as the definition of the phase shift.

Having established this, if the potential $V(r)$ is repulsive everywhere, one can go over from potential zero to potential $V(r)$ by adding up all the repulsive infinitesimal contributions. According to eq. (X.74), each of these contributions diminishes the phase shift; consequently δ_l is negative. Likewise, if $V(r)$ is attractive everywhere, δ_l is positive.

More generally:

$$\text{if } V(r)>\hat{V}(r) \text{ for any } r,\ \delta_l<\hat{\delta}_l;$$
$$\text{if } V(r)<\hat{V}(r) \text{ for any } r,\ \delta_l>\hat{\delta}_l.$$

19. The Born Approximation

In order to know the phase shift δ_l exactly, one must in principle integrate equation (X.29). However, if $V(r)$ is sufficiently small, the regular solution y_l of this equation differs very little from the free spherical wave $krj_l(kr)$, and the phase shift δ_l is close to zero. One can then without much error replace y_l by the free wave relation in (X.73), which yields

$$\delta_l \approx -\frac{2m}{\hbar^2} k \int_0^\infty j_l^2(kr)\, V(r)\, r^2\, \mathrm{d}r. \tag{X.75}$$

This is the expression for the phase shift in "*Born approximation*".

The error is small if $V(r)$ is sufficiently small compared to $E - [l(l+1)\hbar^2/2mr^2]$ over most of the region of variation of r. One expects therefore that the Born approximation is good at high energies or, provided that $V(r)$ decreases sufficiently rapidly at infinity, for large values of l. In fact, expression (X.75) is the first term of an expansion in powers of V; it is therefore possible to estimate the error by roughly evaluating the term of immediately higher order. This question will be treated in Chapter XIX.

Starting from relation (X.72), one likewise obtains an approximate expression for $\delta_l - \hat{\delta}_l$, namely

$$\delta_l - \hat{\delta}_l \approx -\frac{2m}{\hbar^2 k} \int \hat{y}_l^2 (V - \hat{V})\, \mathrm{d}r. \tag{X.76}$$

This "generalized Born formula" is useful when one knows the regular solution $\hat{y}_l$ of the radial equation for a potential $\hat{V}$ which differs only slightly from the potential V. It then enables us to obtain δ_l to a good approximation, without the necessity of solving exactly the radial equation for the potential V [1]).

20. Effective Range Theory. The Bethe Formula

The formulae of § 17 allow us to study the variation suffered by the phase shift when one modifies the potential while keeping the

---

[1]) One can also make use of this formula to study the effect of the tail of the potential V. It suffices to take for the potential $\hat{V}$ the potential

$$\hat{V} = \begin{cases} V(r) & \text{if } r < r_0 \\ 0 & \text{if } r > r_0 \end{cases}$$

r_0 having a suitably chosen value. $\hat{V}$ is a potential of finite range and possesses all the properties of potentials of this type (Sections III and IV). $V - \hat{V}$ is the tail of potential V; its effect can be evaluated by means of formula (X.76), if it is small.

energy constant. The formula we establish in this section concerns the variation of the phase shift as a function of the energy. It is particularly useful in the limit of low energies for a short range potential.

Denote by u one of the regular solutions of eq. (X.29). We do not specify its normalization for the moment. Let $\hat{u}$ be the (irregular) solution of eq. (X.71) corresponding to the same value of the energy and having the same asymptotic form as u, with the same normalization. Now consider two different energies E_1 and E_2; we label by the index 1 the functions and quantities related to the energy E_1, and by the index 2 those related to energy E_2. According to the Wronskian theorem [eq. (III.27)],

$$W(u_1, u_2)\Big|_a^b = (\varepsilon_1 - \varepsilon_2) \int_a^b u_1 u_2 \, \mathrm{d}r$$

and an analogous relation for the $\hat{u}$, whence

$$W(\hat{u}_1, \hat{u}_2) - W(u_1, u_2)\Big|_a^b = (\varepsilon_1 - \varepsilon_2) \int_a^b (\hat{u}_1\hat{u}_2 - u_1 u_2) \, \mathrm{d}r.$$

As $b \to \infty$, since the u and $\hat{u}$ have the same asymptotic form, the integral of the right-hand side converges, and the difference of the Wronskians evaluated at b tends to zero. Since furthermore $\lim_{a\to 0} W(u_1, u_2) = 0$, the foregoing relation may be written in the limit where $b \to \infty$ and $a \to 0$,

$$\lim_{a\to 0} \left[W(\hat{u}_1, \hat{u}_2) + (\varepsilon_1 - \varepsilon_2) \int_a^\infty (\hat{u}_1\hat{u}_2 - u_1 u_2) \, \mathrm{d}r\right] = 0. \qquad \text{(X.77)}$$

With an appropriate choice of normalization for u, one can put this fundamental relation into a form in which the value of the difference $\delta - \hat{\delta}$ at energies E_1 and E_2 enters explicitly.

We shall restrict this study to the case of s waves ($l = 0$) [1]). Moreover, we suppose that $\hat{V} = 0$ and denote by v_1, v_2 the functions $\hat{u}_1$, $\hat{u}_2$ in this special case. Let us fix the normalization of the u by the condition $v(0) = 1$:

$$v = \cos kr + \cot \delta \sin kr.$$

---

[1]) When $l \neq 0$, u_1 and u_2 have a singularity in $(1/r)^l$ at the origin. In formula (X.77), the terms $W(\hat{u}_1, \hat{u}_2)$ and $(\varepsilon_1 - \varepsilon_2)\int \hat{u}_1\hat{u}_2 \mathrm{d}r$ approach infinity as $(1/a)^{2l-1}$; however, their sum tends toward a finite limit.

Formula (X.77) is then written

$$W(v_2, v_1)\Big|_{a=0} \equiv k_1 \cot \delta_1 - k_2 \cot \delta_2 = (\varepsilon_1 - \varepsilon_2) \int (v_1 v_2 - u_1 u_2)\, dr. \qquad (X.78)$$

Provided that $V(r)$ tends to zero sufficiently rapidly as $r \to \infty$ so that the integral of the right-hand side converges, (X.78) holds in the limit where $\varepsilon_2 \to 0$. Let us designate by u_0, v_0 the functions u, v corresponding to zero energy. Note that

$$v_0 = 1 - \frac{r}{a} \quad \text{and} \quad \lim_{\varepsilon \to 0} k \cot \delta = -\frac{1}{a},$$

where a is the scattering length defined by equation (X.47). Upon setting $\varepsilon_1 = \varepsilon$ and $\varepsilon_2 = 0$ in relation (X.78) (cf. Fig. X.7), we have (*Bethe formula*)

$$k \cot \delta = -\frac{1}{a} + \varepsilon \int (v v_0 - u u_0)\, dr. \qquad (X.79)$$

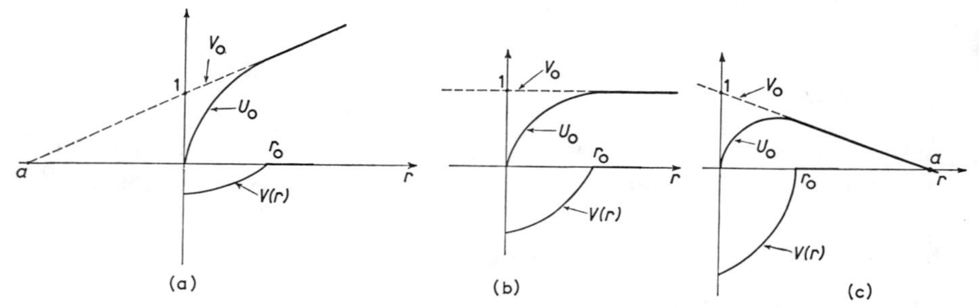

Fig. X.7. s-wave functions of zero energy occurring in effective range theory for increasing depths of the potential of limited range $V(r) = \varpi W(r)$ [ϖ = depth parameter: $\varpi = 1$ yields the depth which is just sufficient for a bound state to occur]

(a) $\varpi < 1$ ($a < 0$) (b) $\varpi = 1$ ($a = \infty$) (c) $\varpi > 1$ ($a > 0$).

N.B. a is a decreasing function of ϖ having a vertical asymptote at each value of ϖ for which there exists a bound state of energy 0.

This relation is exact. It is useful when the integral of the right-hand side varies slowly as a function of energy. This happens when $V(r)$ is a short range potential of the type encountered in nuclear physics, that is to say, when one can divide configuration space into an internal region ($r < r_0$, $kr_0 \ll 1$) for which $|V| \gg E$, and an external region ($r > r_0$) where the potential V is negligible. The contribution to the integral then comes essentially from the internal region, where one can replace, without committing a large error, u by u_0 and v by v_0, since one strictly has $u=u_0=0$ and $v=v_0=1$ at the origin; also, the functions u and u_0 have practically the same relative curvature ($u''/u \approx 2mV/\hbar^2$) in this whole region (Fig. X.7). One thus has to a very good approximation:

$$k \cot \delta \approx -\frac{1}{a} + \varepsilon \int_0^\infty (v_0^2 - u_0^2)\, \mathrm{d}r. \tag{X.80}$$

The quantity $r_{\text{eff.}} \equiv 2\int_0^\infty (v_0^2 - u_0^2)\, \mathrm{d}r$ is a characteristic parameter of the potential $V(r)$, the so-called *effective range*.

The two terms of the right-hand side of (X.80) are the first two terms of the expansion of $k \cot \delta$ in a series of powers of the energy. To write down the terms of higher order, it is necessary to expand u and v in power series of ε and to substitute the expressions thus obtained in the integral of the right-hand side of (X.79) [1]. According to the argument given above, these expansions are rapidly converging in the internal region; consequently, the convergence of the expansion of $k \cot \delta$ as a function of the energy is likewise very rapid.

EXERCISES AND PROBLEMS

1. Consider the scattering of a particle of wavelength $\lambda\!\!\!^{-}$ by a potential $V(r)$ tending to 0 more rapidly than $1/r$ as $r \to \infty$. Let $f(\Omega)$ be the scattering amplitude in the direction $\Omega \equiv (\theta, \varphi)$. Show that

$$\sigma_{\text{tot.}} \equiv \int |f(\Omega)|^2\, \mathrm{d}\Omega = 4\pi\lambda\!\!\!^{-} \operatorname{Im} f(0),$$

where $f(0)$ stands for the forward scattering amplitude ($\theta = 0$) [*Bohr–Peierls–Placzek relation*].

2. By bombarding nucleus A by nucleus a one can form nuclei b and B: $a + A \to b + B$. In the laboratory system the target A is at rest. Let m_a, m_A, m_b, m_B be the masses of the particles involved in this reaction. We make the non-relativistic approximation: $m_a + m_A = m_b + m_B$. Let E_i and E_f be the respective total kinetic energies of the initial state ($a + A$) and of the final state ($b + B$) in the center-of-mass system, and let θ and θ_1 be the emission angles of particle b in the center-of-mass system and the laboratory system, respectively. Show that θ_1 is given as a function of θ by relation (X.24), provided that τ represents the ratio of the velocity V of the center of mass, to the velocity v_b of particle b in the center-of-mass system, namely:

$$\tau = \frac{V}{v_b} = \left[\frac{m_a m_b}{m_A m_B} \frac{E_\mathrm{i}}{E_\mathrm{f}}\right]^{\frac{1}{2}}$$

---

[1] Cf. G. Chew and M. L. Goldberger, Phys. Rev. **75** (1949) 1637; H. A. Bethe, Phys. Rev. **76** (1949) 38.

3. Prove the following relations between the quantities τ_l, v_l, and $q_l^{(+)}$ introduced in § 10 [definition (X.41)]:

$$v_l = -\frac{\mathrm{d}\tau_l}{\mathrm{d}\xi}, \qquad \mathrm{Im}\, q_l^{(+)} = \xi v_l, \qquad \mathrm{Re}\, q_l^{(+)} = -\frac{1}{2}\frac{\xi}{v_l}\frac{\mathrm{d}v_l}{\mathrm{d}\xi}.$$

4. Show that in the WKB approximation the phase shift δ_l is given by the formula

$$\delta_l = \lim_{R\to\infty}\left[\int_a^R \sqrt{k^2 - U(r) - \frac{l(l+1)}{r^2}}\,\mathrm{d}r - \int_{a_0}^R \sqrt{k^2 - \frac{l(l+1)}{r^2}}\,\mathrm{d}r\right].$$

Take the same definition of δ_l, k and $U(r)$ as in § 8. The lower limits of the integrals, a and a_0 are the zeros of the corresponding integrands [if $k^2 - U(r) - l(l+1)/r^2$ has several zeros, a is the largest amongst them]. Discuss the conditions of validity of this approximation [in order that the method apply to small values of l, one must, following the recommendation of Langer (cf. Problem IX.6), replace $l(l+1)$ by $(l+\frac{1}{2})^2$ in the two integrands].

5. Apply the theory of the effective range to p-wave scattering. Show that it gives the expansion of $k^3 \cot \delta_1$ as a function of the energy, and give the expression for the first two terms of this expansion ($\delta_1 = p$-wave phase shift).

CHAPTER XI

THE COULOMB INTERACTION

1. Introduction

Let r be the distance between two particles of electric charge $Z_1 e$ and $Z_2 e$, respectively; the electrostatic interaction potential of these two particles is the Coulomb potential

$$V(r) = \frac{Z_1 Z_2 e^2}{r}.$$

Let $\boldsymbol{p}_1$, $\boldsymbol{p}_2$ be the momenta, m_1, m_2 the masses, and $\boldsymbol{r}_1$, $\boldsymbol{r}_2$ the respective positions of these two particles ($\boldsymbol{r} = \boldsymbol{r}_1 - \boldsymbol{r}_2$). If their interaction is purely of the Coulomb type, their motion is governed by the Hamiltonian

$$\frac{\boldsymbol{p}_1{}^2}{2m_1} + \frac{\boldsymbol{p}_2{}^2}{2m_2} + \frac{Z_1 Z_2 e^2}{r}.$$

The center-of-mass motion can be separated from the motion of the relative particle, using the method outlined in Chapter IX. The motion of the relative particle is governed by the Hamiltonian

$$H \equiv \frac{\boldsymbol{p}^2}{2m} + \frac{Z_1 Z_2 e^2}{r} \tag{XI.1}$$

where m is the reduced mass:

$$m = \frac{m_1 m_2}{m_1 + m_2}. \tag{XI.2}$$

The treatment of a quantum system of two particles with Coulomb interaction therefore reduces to that of a particle in the central potential $Z_1 Z_2 e^2/r$.

Because of its slow decrease for large values of r, a certain number of the properties of central potentials obtained in Chapters IX and X do not apply to the Coulomb potential. In scattering problems especially, the asymptotic behavior of the stationary solutions is less simple than in the case of potentials with shorter range, and the definition of phase shifts must be changed accordingly; actually, the usefulness of the treatment of scattering by separation of angular and radial variables is not as obvious since the expansion of the scattering amplitude in spherical harmonics converges very slowly.

On the other hand, the solution of the Schrödinger equation of a particle in a Coulomb potential can always be reduced to the solution of a Laplace differential equation, an equation that is well known in the theory of functions. Therefore, the most interesting quantities concerning the Coulomb interaction, such as the energy spectrum of bound states and the scattering cross section, can be calculated exactly.

This chapter consists of two sections. The first is devoted to the study of bound states of the hydrogen atom, a study that is easily extended to hydrogen-like atoms, and more generally to systems of two particles with mutual attraction of the form $1/r$. Section II deals with problems of Coulomb scattering.

I. THE HYDROGEN ATOM

2. Schrödinger Equation of the Hydrogen Atom

The simplest system of two bodies with a Coulomb interaction is the hydrogen atom. The two particles, electron and proton, are subject to the attractive potential $-e^2/r$. The reduced mass m of the electron-proton system is slightly smaller than the electron mass: $(m_e - m)/m_e \approx 5 \times 10^{-4}$. Let E be the energy of the electron-proton system in the center-of-mass system. The wave function $\psi(\mathbf{r})$ of the relative particle is a solution of the Schrödinger equation

$$\left[-\frac{\hbar^2}{2m}\triangle - \frac{e^2}{r}\right]\psi(\mathbf{r}) = E\psi(\mathbf{r}). \tag{XI.3}$$

The nature of the regular solutions of this equation is easily revealed if one carries out the separation of angular and radial variables. Thus the eigensolution of energy E and angular momentum (lm) is the function

$$Y_l^m(\theta, \varphi)\,\frac{y_l(r)}{r},$$

where y_l is the solution of the radial equation [cf. eq. (IX.2)] which vanishes at the origin

$$y_l'' + \left[\varepsilon + \frac{2m}{\hbar^2}\frac{e^2}{r} - \frac{l(l+1)}{r^2}\right] y_l = 0. \tag{XI.4}$$

We have put

$$\varepsilon = \frac{2mE}{\hbar^2}. \tag{XI.5}$$

If $E > 0$, this solution has an oscillatory behavior at infinity and can be accepted as eigensolution for any E. It represents an unbound state and enters into the construction of stationary states of electron-proton collisions at energy E.

If $E < 0$, the asymptotic form of the solution which is regular at the origin is a linear combination of exponentials exp $(\varkappa r)$ and exp $(-\varkappa r)$, where

$$\varkappa = (-\varepsilon)^{\frac{1}{2}} = \frac{(-2mE)^{\frac{1}{2}}}{\hbar}. \tag{XI.6}$$

For this solution to be an acceptable eigensolution, the coefficient in front of exp $(\varkappa r)$ must vanish: this happens only for certain discrete values of E. These values are the energies of the discrete spectrum of the hydrogen atom, and the corresponding wave function represents one of the possible bound states of this atom.

The purpose of this section is to study the bound states of the hydrogen atom. The results of this study may easily be extended to hydrogen-like atoms (He^+, Li^{++}, etc.) in which the proton is replaced by a heavier nucleus. Let M_A be the mass of this nucleus, M_p the mass of the proton. The reduced mass of the hydrogen-like atom:

$$m' = \frac{m_e M_A}{m_e + M_A},$$

differs slightly from that of the hydrogen atom

$$m = \frac{m_e M_p}{m_e + M_p}.$$

On the other hand, if Ze is the charge of the nucleus, the Coulomb attraction is Ze^2/r and not e^2/r. All formulae relating to the hydrogen atom apply to the hydrogen-like atom provided that one replaces m by m', and e^2 by Ze^2.

3. Order of Magnitude of the Binding Energy of the Ground State

Denote by r_0 the "radius" of the atom in its ground state. We understand thereby that the wave function is practically entirely concentrated inside a sphere of radius r_0, in other words that the probability of presence of the electron at a distance r from the proton is very small when $r > r_0$, whereas it has a non-negligible value if $r < r_0$. A very crude model consists in taking for the approximate form of the probability density a uniform distribution over a sphere of radius r_0.

Clearly, the average value of the potential energy is smaller (in algebraic value) the smaller r_0: it is of order $(-e^2/r_0)$. The mean value of the kinetic energy, on the other hand, is larger the smaller r_0. Indeed, if the electron is localized inside a sphere of radius r_0, the uncertainty relations impose a lower limit upon the value of its momentum; the root-mean-square deviation of the momentum cannot be smaller than $\hbar/r_0$, hence there is a minimum mean kinetic energy $\hbar^2/2mr_0^2$. The total energy is thus at least equal to the sum of these two quantities:

$$\frac{\hbar^2}{2mr_0^2} - \frac{e^2}{r_0},$$

whose minimum is attained when $r_0 = \hbar^2/me^2$. One expects the value of this minimum

$$E_1 = -\frac{1}{2}\frac{me^4}{\hbar^2} \qquad (= -13.5 \text{ eV}) \qquad \text{(XI.7)}$$

to be of the order of the energy of the ground state, and the value of the corresponding radius:

$$a = \frac{\hbar^2}{me^2} \qquad (= 0.529 \times 10^{-8} \text{ cm}) \qquad \text{(XI.8)}$$

to give the order of magnitude of the extension of the wave function in the ground state.

In fact – and this is a mere coincidence – the energy of the ground state is rigorously given by eq. (XI.7). The length a is called the *Bohr radius* or "radius of the hydrogen atom".

4. Solution of the Schrödinger Equation in Spherical Coordinates

To solve the Schrödinger equation, we go over into spherical polar coordinates. As is the case for any central potential, the angular and radial variables separate in this coordinate system and the problem reduces to a search for regular solutions of the radial equation (XI.4).

It is also possible to solve the Schrödinger equation in parabolic coordinates, because the variables likewise separate in that coordinate system. We merely mention this characteristic property of the Coulomb potential for future reference, and limit ourselves to a treatment of the problem in spherical polar coordinates.

If one makes the change of variable

$$x = 2\varkappa r \tag{XI.9}$$

equation (XI.4) depends only upon the dimensionless parameter

$$\nu = \frac{1}{\varkappa a} = \frac{e^2}{\hbar c}\sqrt{\frac{mc^2}{-2E}}; \tag{XI.10}$$

$\varkappa$ and a were defined by equations (XI.6) and (XI.8), respectively. Equation (XI.4) is equivalent to the equation

$$\left[\frac{\mathrm{d}^2}{\mathrm{d}x^2} - \frac{l(l+1)}{x^2} + \frac{\nu}{x} - \frac{1}{4}\right] y_l = 0 \tag{XI.11}$$

y_l is the solution which goes as x^{l+1} at the origin. For x very large, it increases exponentially except for certain particular values of ν where it behaves as $\exp(-\frac{1}{2}x)$. We intend to determine these special values of ν and their corresponding eigensolutions.

To this effect, we perform the change of function

$$y_l = x^{l+1}\,\mathrm{e}^{-\frac{1}{2}x}\,v_l(x),$$

which yields

$$\left[x\,\frac{\mathrm{d}^2}{\mathrm{d}x^2} + (2l+2-x)\,\frac{\mathrm{d}}{\mathrm{d}x} - (l+1-\nu)\right] v_l = 0. \tag{XI.12}$$

This differential equation is a Laplace equation (cf. Appendix B, § 1). To within a constant, only one solution is finite at the origin; all others have a singularity in $(1/x)^{2l+1}$ there. This regular solution is the confluent hypergeometric series:

$$F(l+1-\nu\,|\,2l+2\,|\,x) \equiv \sum_{p=0}^{\infty} \frac{\Gamma(l+1+p-\nu)}{\Gamma(l+1-\nu)}\,\frac{(2l+1)!}{(2l+1+p)!}\,\frac{x^p}{p!}. \tag{XI.13}$$

To show this, it is actually sufficient to look for the solution of equation (XI.12) represented by the Taylor series expansion at the origin,

$$v_l(x) = 1 + a_1 x + a_2 x^2 + \ldots + a_p x^p + \ldots$$

Substituting this expansion in eq. (XI.12), one can write the left-hand side in the form of an expansion in a power series of x. All the coefficients of this expansion must vanish; hence

$$\begin{aligned}(2l+2)a_1 &= (l+1-\nu)\\ 2(2l+3)a_2 &= (l+2-\nu)a_1\\ &\vdots\\ p(2l+1+p)a_p &= (l+p-\nu)a_{p-1},\end{aligned}$$

whence

$$a_p = \frac{(p+l-\nu)\,(p-1+l-\nu)\,\dots\,(1+l-\nu)}{(p+2l+1)\,(p-1+2l+1)\,\dots\,(1+2l+1)} \times \frac{1}{p!}.$$

a_p is actually the coefficient of x^p in the hypergeometric series (XI.13).

In general, the series (XI.13) is an infinite series and behaves as $x^{-l-1-\nu}\exp(x)$ [eqs. (B.9) to (B.11)], for large x. Then y_l behaves in the asymptotic region as $x^{-\nu}\exp(\frac{1}{2}x)$; it cannot be an eigensolution.

However, for certain privileged values of ν the coefficients all vanish from a certain order on, and the hypergeometric series reduces to a polynomial. For this to happen, $l+1-\nu$ must be a negative integer, or zero, namely

$$\nu = n = l+1+n' \qquad (n' = 0, 1, 2, \dots, \infty). \tag{XI.14}$$

In that case the hypergeometric series reduces to a polynomial of degree n', the radial function behaves as $x^n \exp(-\frac{1}{2}x)$ for $x \to \infty$, and the regular solution of the Schrödinger equation is acceptable as eigensolution.

The quantum condition (XI.14) thus gives the energy levels of the bound states of angular momentum (lm). Each of them is defined by a particular value of the integer n'. The wave function of the corresponding bound state is the (unnormalized) wave function constructed with the radial solution

$$y_l = x^{l+1}\,\mathrm{e}^{-\frac{1}{2}x} \sum_{p=0}^{n'} (-)^p \frac{n'!\,(2l+1)!}{(n'-p)!\,(2l+1+p)!\,p!}\,x^p. \tag{XI.15}$$

The polynomial of degree n' represented by the summation of the right-hand side is, to within a constant, the associated Laguerre polynomial $L_{n'}^{2l+1}(x)$, the definition and main properties of which are given in Appendix B (§ 2).

5. Energy Spectrum. Degeneracy

Replacing in eq. (XI.14) the parameter ν by its expression (XI.10) as a function of energy, one obtains the energy spectrum of the states of angular momentum l,

$$E_{ln'} = -\left(\frac{e^2}{\hbar c}\right)^2 \frac{mc^2}{2(l+1+n')^2}. \qquad \text{(XI.16)}$$

n', the radial quantum number, is equal to the number of nodes of the radial part of the wave function. The spectrum contains a denumerably infinite set of levels since n' may take on all integral values from 0 to $+\infty$. When $n' \to \infty$, these levels become more and more closely spaced and tend to $E=0$ in the limit, at which point the continuous spectrum begins.

This circumstance is characteristic of long-range potentials. Short-range potentials such as square wells, on the other hand, give a finite (and sometimes zero) number of bound states. One can show very generally that the set of energy levels is denumerably infinite (with a point of accumulation at the value 0) when the potential approaches zero asymptotically through negative values, and less rapidly than $1/r^2$:

$$r^2 V \underset{r\to\infty}{\to} -\infty;$$

one can also show that their number is finite — and possibly zero — if the potential asymptotically approaches zero more rapidly than $1/r^2$.

The ensemble of the spectra belonging to different possible values of l ($l=0, 1, 2, \ldots, \infty$) represents the complete spectrum of the hydrogen atom according to the Schrödinger theory. This complete spectrum is thus formed of the set of numbers $E_{ln'}$ defined by eq. (XI.16), l and n' taking all possible integral, non-negative values. Note that these quantities depend only upon the sum $l+n'$ or, what is equivalent, upon the "principal quantum number"

$$n = l+n'+1.$$

One has

$$E_n = -\left(\frac{e^2}{\hbar c}\right)^2 \frac{mc^2}{2n^2} \qquad (n=1, 2, \ldots, \infty). \qquad \text{(XI.17)}$$

For each energy E_n, that is to say for each value of the integer n, the angular momentum may take on all integral values from 0 up to $n-1$. The order of degeneracy of the level E_n is therefore

$$\sum_{l=0}^{n-1} (2l+1) = n(n-1) + n = n^2.$$

The n^2-dimensional subspace of the eigenfunctions is spanned by n^2 functions, each corresponding to a state of given angular momentum (lm), the "azimuthal quantum number" taking on the n values

$$l = 0, 1, 2, \ldots, n-1,$$

and the "magnetic quantum number" taking on the $(2l+1)$ values

$$m = -l, -l+1, \ldots, +l.$$

According to the tradition of spectroscopy, the various eigenstates thus defined are designated by the positive integer n followed by a letter (s, p, d, f, g, ...) indicating the value of l in conformity with the definition stated in the preceding chapter. The quantum number m which indicates the orientation of the system is simply not mentioned. Thus the ground state is a 1s state, the first-excited state is four-fold degenerate and contains one 2s state and three 2p states; the second-excited state is nine-fold degenerate and contains one 3s state, three 3p states and five 3d states; and so forth (Fig. XI.1).

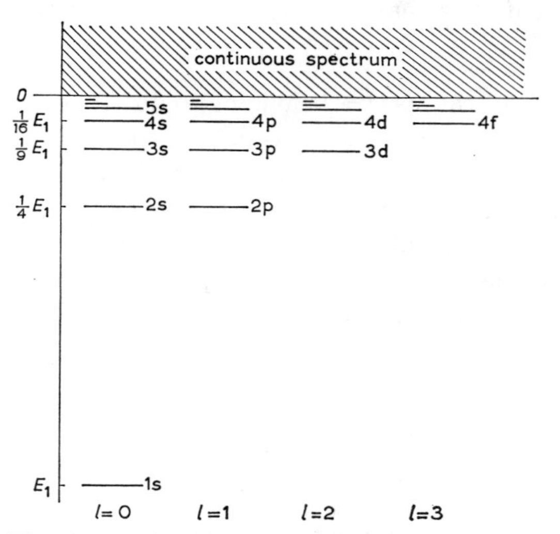

Fig. XI.1. Spectrum of the hydrogen atom.

This spectrum is just the one predicted by the Old Quantum Theory; its excellent agreement with the experimental spectrum was already pointed out. To be more precise, the theory correctly accounts for the position of the spectral lines but not for their fine structure. Its essential shortcoming is to be a non-relativistic theory. The relativistic effects in the positions of the levels are of the order v^2/c^2, or about E_n/mc^2. The relativistic corrections are thus of the order 10^{-4} to 10^{-5}. On the other hand, the Schrödinger theory does not take the electron spin into account: the spin is an internal degree of freedom without classical analogue; we shall have occasion to discuss it in Ch. XIII. The analysis of the fine structure of the hydrogen atom will be taken up again in Ch. XX during the discussion of the Relativistic Quantum Mechanics of the electron.

6. The Eigenfunctions of the Bound States

The eigenfunctions belonging to the energy level E_n are linear combinations of n^2 linearly independent functions. The investigation of § 4 furnishes us n^2 orthogonal eigenfunctions, namely those which correspond to a well-defined value of the angular momentum. Thus the wave function of the quantum state (nlm) is written

$$\psi_{nlm} = a^{-3/2} N_{nl} F_{nl}\left(\frac{2r}{na}\right) Y_l^m(\theta, \varphi) \qquad \text{(XI.18)}$$

with

$$F_{nl}(x) = x^l \, \mathrm{e}^{-\frac{1}{2}x} L_{n-l-1}^{2l+1}(x). \qquad \text{(XI.18}a\text{)}$$

N_{nl} is a normalization constant. One calculates the norm of ψ_{nlm} using the generating function of the Laguerre polynomials [eq. (B.15)]. This norm is equal to unity if one takes

$$N_{nl} = \frac{2}{n^2}\sqrt{\frac{(n-l-1)!}{[(n+l)!]^3}}. \qquad \text{(XI.18}b\text{)}$$

The mean values of successive powers of r in the quantum state (nlm) are of interest. Their calculation will not be undertaken here (Problem XI.1). The results are given in Appendix B (§ 3). One has, in particular

$$\langle r \rangle_{nl} = \tfrac{1}{2}a[3n^2 - l(l+1)]. \qquad \text{(XI.19)}$$

The electron is thus further removed – on the average – from the proton, the larger n. For the ground state, one finds $\langle r \rangle_{1s} = \frac{3}{2}a$ in accord with the rough predictions of § 3.

When l takes its maximum value $(n-1)$, the wave function assumes an especially simple form: it is the product of $Y_l^m(\theta, \varphi)$ and the radial function

$$[(2n)!]^{-\frac{1}{2}} \left(\frac{2}{na}\right)^{\frac{3}{2}} \left(\frac{2r}{na}\right)^{n-1} \mathrm{e}^{-r/na}.$$

The mean value of r in that state is

$$\begin{aligned} \langle r \rangle &= \frac{1}{(2n)!} \left(\frac{2}{na}\right)^3 \int_0^\infty \left(\frac{2r}{na}\right)^{2n-2} \mathrm{e}^{-2r/na}\, r^3\, \mathrm{d}r \\ &= n(n+\tfrac{1}{2})\, a, \end{aligned}$$

in accord with the more general formula indicated above. An analogous calculation yields

$$\langle r^2 \rangle = n^2(n+\tfrac{1}{2})\,(n+1)\, a^2,$$

whence one obtains the expression for the root-mean-square radial deviation

$$\Delta r = \sqrt{\langle r^2 \rangle - \langle r \rangle^2} = \tfrac{1}{2} na\, \sqrt{2n+1} = \frac{\langle r \rangle}{\sqrt{2n+1}}.$$

For very large values of n, $\Delta r/\langle r \rangle$ becomes very small and the electron remains practically localized in the vicinity of a sphere of radius n^2a, whereas the energy of the level, $-\frac{1}{2}e^2/n^2a$, is the same as that of a classical electron describing a circular orbit of radius n^2a.

We verify in this particular example the general correspondence rule according to which one returns to the classical laws of motion in the limit of very large quantum numbers. In order to compare the quantum theory and the classical theory in detail, one would have to study the motion of wave packets. We shall not do so here. We merely note that the states of maximum l $(l = n-1)$ correspond to the classical circular orbits, a fact that is to be compared with the result of the Old Quantum Theory according to which the eccentricity of the quantized orbits is equal to $\sqrt{1-(l^2/n^2)}$ and vanishes when l takes on its maximum value (cf. Ch. I, § 15).

II. COULOMB SCATTERING

7. The Coulomb Scattering Wave

After reduction to the center of mass, the Schrödinger equation of the collision problem of two particles with Coulomb interaction is written, using the notation of § 1:

$$\left[-\frac{\hbar^2}{2m}\triangle+\frac{Z_1 Z_2 e^2}{r}\right]\psi(\boldsymbol{r})=E\psi(\boldsymbol{r}) \tag{XI.20}$$

where E is the energy in the center-of-mass system. The scattering cross section is related to the asymptotic behavior of the eigensolutions of positive energy of eq. (XI.20). Let us put [1])

$$E=\frac{\hbar^2 k^2}{2m}=\tfrac{1}{2}mv^2 \tag{XI.21}$$

$$\gamma=\frac{Z_1 Z_2 e^2}{\hbar v}; \tag{XI.22}$$

equation (XI.20) may also be written

$$\left(\triangle+k^2-\frac{2\gamma k}{r}\right)\psi(\boldsymbol{r})=0. \tag{XI.23}$$

This equation has a regular solution of the form

$$\mathrm{e}^{\mathrm{i}kz} f(r-z). \tag{XI.24}$$

Indeed, if we substitute this expression in eq. (XI.23) and put $u=r-z$, we obtain the differential equation

$$\left[u\frac{\mathrm{d}^2}{\mathrm{d}u^2}+(1-\mathrm{i}ku)\frac{\mathrm{d}}{\mathrm{d}u}-\gamma k\right]f(u)=0,$$

or else, putting

$$v=\mathrm{i}ku=\mathrm{i}k(r-z),$$

$$\left[v\frac{\mathrm{d}^2}{\mathrm{d}v^2}+(1-v)\frac{\mathrm{d}}{\mathrm{d}v}+\mathrm{i}\gamma\right]f(v)=0.$$

---

[1]) γ is the analogue of the parameter ν introduced in the problem of the hydrogen atom. If one sets $a=\hbar^2/Z_1Z_2me^2$, one has $\gamma=1/ka$ [cf. eq. (XI.10)].

This is an equation of the Laplace type whose regular solution at the origin is the confluent hypergeometric series $F(-\mathrm{i}\gamma|1|v)$. The Schrödinger equation thus actually has a regular solution of the form (XI.24), the function

$$\psi_c = A\ \mathrm{e}^{\mathrm{i}kz}\, F[-\mathrm{i}\gamma\,|1|\,\mathrm{i}k(r-z)]. \tag{XI.25}$$

A is an adjustable constant.

In keeping with the study made in Appendix B (§ 1), the hypergeometric series occurring in equation (XI.25) is the sum of two functions whose asymptotic forms for large values of $|v| = 2kr \sin^2 \frac{1}{2}\theta$ are given by equations (B.10) and (B.11), respectively. Let us adopt the notation of Appendix B and put

$$\psi_i = A\ \mathrm{e}^{\mathrm{i}kz}\, W_1(-\mathrm{i}\gamma\,|1|\,\mathrm{i}ku) \tag{XI.26}$$

$$\psi_s = A\ \mathrm{e}^{\mathrm{i}kz}\, W_2(-\mathrm{i}\gamma\,|1|\,\mathrm{i}ku). \tag{XI.27}$$

We have

$$\psi_c = \psi_i + \psi_s. \tag{XI.28}$$

The functions ψ_i and ψ_s are (irregular) solutions of eq. (XI.20).

Upon taking

$$A = \Gamma(1+\mathrm{i}\gamma)\ \mathrm{e}^{-\frac{1}{2}\pi\gamma} \tag{XI.29}$$

one obtains the following asymptotic forms for ψ_i and ψ_s:

$$\psi_i \underset{|r-z|\to\infty}{\sim} \mathrm{e}^{\mathrm{i}[kz+\gamma \ln k(r-z)]} \left[1 + \frac{\gamma^2}{\mathrm{i}k(r-z)} + \dots\right] \tag{XI.30}$$

$$\psi_s \underset{|r-z|\to\infty}{\sim} -\frac{\gamma}{k(r-z)} \frac{\Gamma(1+\mathrm{i}\gamma)}{\Gamma(1-\mathrm{i}\gamma)}\ \mathrm{e}^{\mathrm{i}[kr-\gamma \ln k(r-z)]} \left[1+\frac{(1+\mathrm{i}\gamma)^2}{\mathrm{i}k(r-z)}+\dots\right]. \tag{XI.31}$$

Since $z = r\cos\theta$, the first term of the asymptotic expansion of ψ_s can also be written

$$\psi_s \underset{|r-z|\to\infty}{\sim} \frac{1}{r} \exp\,[\mathrm{i}(kr-\gamma \ln 2kr)]\ f_c(\theta) \tag{XI.32}$$

with

$$f_c(\theta) = -\frac{\gamma}{2k \sin^2 \frac{1}{2}\theta} \exp\,[-\mathrm{i}\gamma \ln(\sin^2 \tfrac{1}{2}\theta) + 2\mathrm{i}\sigma_0] \tag{XI.33}$$

$$\sigma_0 = \arg \Gamma(1+\mathrm{i}\gamma). \tag{XI.34}$$

8. The Rutherford Formula

The wave ψ_c represents the stationary state of collision of a particle of incident momentum $\hbar\mathbf{k}$ directed along the z axis. In the case of a potential tending to zero at least as fast as $1/r^2$ for $r \to \infty$ we know that the same stationary state of collision is represented by the wave function which has the asymptotic form $\exp(\mathrm{i}kz) + f(\theta)\exp(\mathrm{i}kr)/r$; the latter can be interpreted as the sum of the incident plane wave and the outgoing scattered wave. Likewise, the wave ψ_c can be put in the form of a sum of two terms ψ_i, ψ_s whose respective asymptotic forms resemble a plane wave and an outgoing wave.

However, even at infinitely large distances from the origin, ψ_i cannot be likened to a plane wave by reason of the presence of the factor $\exp[\mathrm{i}\gamma \ln k(r-z)]$; in other words, the Coulomb field has such a long range that it affects the incident wave even in the asymptotic region. In spite of this fact, for negative and very large values of z, ψ_i is a wave of unit density whose current density

$$\mathbf{j}_i = \frac{\hbar}{2\mathrm{i}m}[\psi_i^*(\nabla\psi_i) - \psi_i(\nabla\psi_i)^*]$$

is directed along the axis Oz and is equal to $v \equiv \hbar k/m$ (the logarithmic term introduces corrections of the order $1/r$ which one can neglect). This justifies the interpretation of ψ_i as incident wave.

Likewise, the radial dependence of the wave ψ_s never (not even for very large values of r) approaches the form $\exp(\mathrm{i}kr)/r$ characteristic of outgoing waves, but the more complex expression $\exp[\mathrm{i}(kr - \gamma \ln 2kr)]/r$. Nevertheless ψ_s behaves in the asymptotic region (except along the positive z axis where the separation into incident wave and scattered wave is meaningless) like a scattered wave, since the current density $\mathbf{j}_s$ calculated with this wave is actually directed radially and in the direction of increasing r; the effect of the factor $\exp(-\mathrm{i}\gamma \ln 2kr)$ can be neglected to lowest order in $1/r$; in this approximation, the wave ψ_s is a wave of density $|f_c(\theta)|^2/r^2$ and of current density $v|f_c(\theta)|^2/r^2$.

Forming the ratio of the current density scattered into the solid angle $(\Omega, \Omega + \mathrm{d}\Omega)$, to the incident current density, one obtains the differential scattering cross section

$$\sigma_c(\Omega) = |f_c(\theta)|^2. \qquad \text{(XI.35)}$$

This formula is analogous to formula (X.2) for collisions with a potential of shorter range. The proof which was just given is open to the same criticism as the one of Ch. X, § 3. However, a more rigorous proof modeled on the one of Ch. X, §§ 4 to 6 can be worked out easily.

$f_c(\theta)$ is called the Coulomb scattering amplitude. Its explicit form is given by eq. (XI.33). One derives the expression for the Coulomb scattering cross section:

$$\begin{aligned}\sigma_c(\Omega) &= \frac{\gamma^2}{4k^2 \sin^4 \frac{1}{2}\theta} \\ &= \left(\frac{Z_1 Z_2 e^2}{4E}\right)^2 \sin^{-4} \tfrac{1}{2}\theta.\end{aligned} \tag{XI.36}$$

The foregoing expression turns out to be identical to the classical Coulomb scattering cross section calculated in Chapter VI [eq. (VI.29)]. The classical Rutherford formula thus remains true even when the classical approximation ceases to be justified. This fact must be considered as an accident.

In expression (XI.36) of the cross section one notes the following noteworthy properties of Coulomb scattering:

(*i*) it depends solely upon the absolute value of the potential and not upon its sign;

(*ii*) the angular distribution is independent of the energy;

(*iii*) at a given angle, the cross section falls off like $1/E^2$ as the energy increases;

(*iv*) the total cross section is infinite: $\int \sigma_c(\Omega)\, d\Omega$ diverges at small angles.

This divergence is characteristic of the pure Coulomb field. In nature, such a field is never encountered; thus, in the scattering of a charged particle by an atomic nucleus, the Coulomb field due to the nuclei is gradually compensated at increasing distances by the field of the electron cloud, and the potential vanishes at distances sufficiently large compared to the radius of the atom. This screening modifies the scattered wave at small angles in such a way that the differential cross section no longer diverges. One can show that the modification due to screening is negligible at angles large compared to both $2\gamma/ka$ and $1/ka$ (a is the atomic radius). At the energies ordinarily used in nuclear physics, these limiting angles are so small that the screening effect can be completely neglected.

9. Decomposition into Partial Waves

The Schrödinger equation (XI.20) may be solved by separating angular and radial variables. This method is of no great interest for the treatment of pure Coulomb scattering since one has a more direct method at one's disposal. Moreover, since the potential has a long range, the expansion of the scattering amplitude $f_c(\theta)$ in spherical harmonics is known to converge very slowly. However, the decomposition into partial waves is useful in all problems where a short-range interaction is added to the Coulomb interaction proper. The presence of this additional interaction affects only the first few terms of the expansion of the amplitude $f(\theta)$ in spherical harmonics and, consequently, the expansion $f(\theta)-f_c(\theta)$ converges rapidly.

The separation of angular and radial variables has already been performed in connection with the hydrogen atom. With the present notation, equation (XI.4) is written

$$y_l'' + \left[k^2 - \frac{2\gamma k}{r} - \frac{l(l+1)}{r^2}\right] y_l = 0. \tag{XI.37}$$

To construct the solutions of this equation, one proceeds as in the problem of the hydrogen atom, by a change of function and of variable:

$$\begin{aligned} y_l &= \mathrm{e}^{\mathrm{i}kr}\,(kr)^{l+1}\,v_l \\ \xi &= -\,2\mathrm{i}kr; \end{aligned} \tag{XI.38}$$

v_l is a solution of Laplace's equation [cf. eq. (XI.12)]

$$\left[\xi\frac{\mathrm{d}^2}{\mathrm{d}\xi^2} + (2l+2-\xi)\frac{\mathrm{d}}{\mathrm{d}\xi} - (l+1+\mathrm{i}\gamma)\right] v_l = 0. \tag{XI.39}$$

From the asymptotic expansion [eqs. (B.10)–(B.11)] of the two irregular solutions $W_{1,2}(l+1+\mathrm{i}\gamma|2l+2|\xi)$ one deduces the asymptotic form of the general solution of (XI.39) and, after a brief calculation, the asymptotic form of the general solution of (XI.37): it is a linear combination of the two exponentials

$$\mathrm{e}^{\pm\mathrm{i}(kr-\gamma\ln 2kr)}.$$

The solution regular at the origin of (XI.39) is the hypergeometric series $F(l+1+\mathrm{i}\gamma|2l+2|\xi)$; it is the sum of two functions W_1 and W_2

[eq. (B.9)]. The corresponding solution of eq. (XI.37) asymptotically approaches a multiple of $\sin(kr-\gamma \ln 2kr - \frac{1}{2}l\pi + \sigma_l)$, with

$$\sigma_l = \arg \Gamma(l+1+\mathrm{i}\gamma); \tag{XI.40}$$

σ_l is called the *Coulomb phase shift*. By definition, the regular Coulomb wave function $F_l(\gamma; kr)$ is the regular solution of (XI.37) whose asymptotic form is

$$F_l \underset{r\to\infty}{\sim} \sin(kr-\gamma \ln 2kr - \tfrac{1}{2}l\pi + \sigma_l). \tag{XI.41}$$

From the above we obtain

$$F_l(\gamma; kr) = c_l(\gamma)\, \mathrm{e}^{\mathrm{i}kr}\, (kr)^{l+1}\, F(l+1+\mathrm{i}\gamma\,|\,2l+2\,|\,-2\mathrm{i}kr), \tag{XI.42}$$

the constant $c_l(\gamma)$ being adjusted so that F_l satisfies (XI.41), namely

$$c_l = \frac{2^l\, \mathrm{e}^{-\frac{1}{2}\pi\gamma}\, |\Gamma(l+1+\mathrm{i}\gamma)|}{(2l+1)!}; \tag{XI.43}$$

F is a real function. One often denotes it under the name of regular spherical Coulomb function. It is a function of kr depending on the parameter γ.

One also defines "irregular spherical Coulomb functions". They are solutions irregular at the origin of eq. (XI.37). Those most commonly used are defined in Appendix B, § 5. Let us merely mention here the outgoing and incoming waves $u_l^{(+)}$ and $u_l^{(-)}$, of asymptotic form

$$u_l^{(\pm)} \underset{r\to\infty}{\sim} \mathrm{e}^{\pm\mathrm{i}(kr-\gamma\ln 2kr - \frac{1}{2}l\pi)}.$$

These functions are complex conjugates of each other, and one has

$$F_l = \operatorname{Im} \mathrm{e}^{+\mathrm{i}\sigma_l}\, u_l^{(+)}. \tag{XI.44}$$

10. Expansion of the Wave ψ_c in Spherical Harmonics

The Coulomb wave function ψ_c defined in Chapter XI, § 7

$$\psi_c \equiv \mathrm{e}^{-\frac{1}{2}\pi\gamma}\, \Gamma(1+\mathrm{i}\gamma)\, \mathrm{e}^{\mathrm{i}kz}\, F[-\mathrm{i}\gamma\,|\,1\,|\,\mathrm{i}k(r-z)] \tag{XI.45}$$

can be represented by the expansion in a series of Legendre polynomials

$$\psi_c = \frac{1}{kr} \sum_{l=0}^{\infty} (2l+1)\, \mathrm{i}^l\, \mathrm{e}^{\mathrm{i}\sigma_l}\, F_l(\gamma; kr)\, P_l(\cos\theta). \tag{XI.46}$$

The expansion is analogous to the expansion of a plane wave

$$e^{ikz} = \sum_{l=0}^{\infty} (2l+1)\, i^l\, j_l(kr)\, P_l(\cos\theta) \tag{XI.47}$$

to which it reduces in the limit where $\gamma \to 0$.

To prove the relation (XI.46), we use the integral representation (B.6) of the hypergeometric function occurring in the definition of ψ_c, which yields

$$\psi_c = \frac{e^{-\frac{1}{2}\pi\gamma}}{(1-e^{-2\pi\gamma})\,\Gamma(-i\gamma)}\, e^{ikz} \int_{\Gamma_0} e^{ik(r-z)t}\, t^{-i\gamma-1}\,(1-t)^{i\gamma}\, dt$$

$$= \frac{e^{-\frac{1}{2}\pi\gamma}}{(1-e^{-2\pi\gamma})\,\Gamma(-i\gamma)} \int_{\Gamma_0} e^{ikrt}\, e^{ikz(1-t)}\, t^{-i\gamma-1}\,(1-t)^{i\gamma}\, dt.$$

Expanding the exponential $\exp[ikz(1-t)]$ in the integrand in a series of Legendre polynomials, in accordance with formula (XI.47), and interchanging summation and integration, we obtain

$$\psi_c = \sum_{l=0}^{\infty} (2l+1)\, i^l\, \varphi_l(r)\, P_l(\cos\theta) \tag{XI.48}$$

with

$$\varphi_l(r) = \frac{e^{-\frac{1}{2}\pi\gamma}}{(1-e^{-2\pi\gamma})\,\Gamma(-i\gamma)} \int_{\Gamma_0} e^{ikrt}\, j_l[kr(1-t)]\, t^{-i\gamma-1}\,(1-t)^{i\gamma}\, dt. \tag{XI.49}$$

On the other hand, one knows that

$$\xi j_l(\xi) = F_l(0;\xi) = \frac{2^l\, l!}{(2l+1)!}\, \xi^{l+1}\, e^{i\xi}\, F(l+1\,|\,2l+2\,|-2i\xi),$$

or

$$j_l(\xi) = 2^l\, \xi^l\, e^{i\xi} \sum_{p=0}^{\infty} \frac{(l+p)!}{(2l+1+p)!}\, \frac{(-2i\xi)^p}{p!}.$$

Substituting this expression of j_l in the integral of the right-hand side of (XI.49) and once again interchanging summation and integration, one obtains

$$\varphi_l(r) = \frac{2^l\, e^{-\frac{1}{2}\pi\gamma}}{(1-e^{-2\pi\gamma})\,\Gamma(-i\gamma)}\, (kr)^l\, e^{ikr}$$

$$\sum_{p=0}^{\infty} \frac{(l+p)!}{(2l+1+p)!} \left[\int_{\Gamma_0} t^{-i\gamma-1}\,(1-t)^{l+p+i\gamma}\, dt\right] \frac{(-2ikr)^p}{p!};$$

and since, according to relation (B.5)

$$\int_{\Gamma_0} t^{-i\gamma-1}\,(1-t)^{l+p+i\gamma}\,dt = (1-e^{-2\pi\gamma})\,\frac{\Gamma(-i\gamma)\,\Gamma(l+p+1+i\gamma)}{(l+p)!},$$

$$\varphi_l(r) = 2^l\,e^{-\frac{1}{2}\pi\gamma}\,(kr)^l\,e^{ikr}\sum_{p=0}^{\infty}\frac{\Gamma(l+p+1+i\gamma)}{(2l+1+p)!}\,\frac{(-2ikr)^p}{p!}$$

$$= 2^l\,e^{-\frac{1}{2}\pi\gamma}\,\frac{\Gamma(l+1+i\gamma)}{(2l+1)!}\,(kr)^l\,e^{ikr}\,F(l+1+i\gamma\,|\,2l+2\,|-2ikr).$$

Taking into account definitions (XI.40), (XI.42) and (XI.43), this yields

$$kr\varphi_l(r) = e^{i\sigma_l}\,F_l(\gamma;\,kr).$$

One obtains the desired expansion by substituting this expression into equation (XI.48).

Actually, one might have expected a result of this form. Indeed, since ψ_c is a regular solution of the Schrödinger equation (XI.20), $r\varphi_l(r)$ is necessarily a regular solution of the radial equation (XI.37) and therefore proportional to $F_l(\gamma;\,kr)$. The main purpose of the above calculation was to find that constant of proportionality.

It is instructive to express the expansion (XI.46) by means of outgoing and incoming Coulomb waves. Substituting for F_l its expression as function of $u_l^{(+)}$ and $u_l^{(-)}$ [eq. (XI.44)], one finds

$$\psi_c = \frac{1}{2kr}\sum_{l=0}^{\infty}(2l+1)\,i^{l+1}\,[u_l^{(-)} - e^{2i\sigma_l}\,u_l^{(+)}]\,P_l(\cos\theta). \qquad \text{(XI.50)}$$

11. Modifications of the Coulomb Potential by a Short-Range Interaction

When a short-range interaction $V'(r)$ is added to the Coulomb field $V_c(r)$, the stationary state of collision is no longer represented by the pure Coulomb wave, but by a wave ψ whose expansion in a series of Legendre polynomials is of the form

$$\psi = \frac{1}{kr}\sum_{l=0}(2l+1)\,i^l\,\chi_l(r)\,P_l(\cos\theta). \qquad \text{(XI.51)}$$

The phase-shift method which allows us to treat the scattering of a particle by the potential $V'(r)$ may be extended almost step by step to the scattering by the potential $V_c(r)+V'(r)$; it suffices in practice to replace the free waves by the corresponding Coulomb waves at all stages.

The function $\chi_l(r)$ is a solution of the radial equation

$$\left[\frac{\mathrm{d}^2}{\mathrm{d}r^2} + k^2 - \frac{2\gamma k}{r} - \frac{2m}{\hbar^2} V' - \frac{l(l+1)}{r^2}\right] \chi_l(r) = 0. \qquad \text{(XI.52)}$$

One can show (Problem XI.2) that if $V'(r)$ tends to zero asymptotically at least as fast as $1/r^2$, the solutions of this radial equation asymptotically approach linear combinations of the exponentials

$$\mathrm{e}^{\pm \mathrm{i}(kr - \gamma \ln 2kr)}$$

or, what is equivalent, linear combinations of outgoing and incoming Coulomb functions $u_l^{(+)}$ and $u_l^{(-)}$. In particular, the regular solution of this equation tends asymptotically toward a certain linear combination of these two functions. Let

$$A_l[u_l^{(-)} - \mathrm{e}^{2\mathrm{i}\delta_l}\, \mathrm{e}^{2\mathrm{i}\sigma_l}\, u_l^{(+)}]$$

be this linear combination; A_l is an arbitrary constant. The phase shift δ_l is characteristic of the potential $V'(r)$ we have added to the Coulomb potential. It is zero if $V'(r) = 0$ and hereafter plays a role analogous to that of the phase shifts in the theory of scattering by short-range potentials.

The regular solution χ_l of the radial equation must be chosen in such a way that the wave ψ represents a stationary state of collision. Therefore, $\psi - \psi_{\mathrm{c}}$ must behave asymptotically like a purely outgoing wave of the type $\exp\,[\mathrm{i}(kr - \gamma \ln 2kr)]/r$. This condition is fulfilled if $A = \frac{1}{2}\mathrm{i}$ for all values of l, as one can readily see by comparing (XI.50) and (XI.51). In the asymptotic region – that is, for values of r sufficiently large so that $V'(r)$ can be neglected – ψ can thus be expanded as follows:

$$\begin{aligned}\psi &\underset{r\to\infty}{\sim} \frac{1}{2kr} \sum_l (2l+1)\mathrm{i}^{l+1}\, [u_l^{(-)} - \mathrm{e}^{2\mathrm{i}\delta_l}\, \mathrm{e}^{2\mathrm{i}\sigma_l}\, u_l^{(+)}]\, P_l(\cos\theta) \\ &\underset{r\to\infty}{\sim} \psi_{\mathrm{c}} - \frac{1}{2kr} \sum_l (2l+1)\, \mathrm{i}^{l+1}\, \mathrm{e}^{2\mathrm{i}\sigma_l}\, (\mathrm{e}^{2\mathrm{i}\delta_l} - 1)\, u_l^{(+)}\, P_l(\cos\theta).\end{aligned} \qquad \text{(XI.53)}$$

As in § 7, one can put ψ in the form of a sum:

$$\psi = \psi_{\mathrm{i}} + \psi_{\mathrm{s}} \qquad \text{(XI.54)}$$

where ψ_i is the function defined by eq. (XI.26) and represents the incident wave. ψ_s on the other hand differs from the function defined by eq. (XI.27); its asymptotic form, after some calculation, is written

$$\psi_s \underset{r\to\infty}{\sim} \frac{1}{r} \exp\left[i(kr - \gamma \ln 2kr)\right] \cdot f(\theta)$$

with

$$f(\theta) = f_c(\theta) + f'(\theta) \qquad \text{(XI.55)}$$

$$f_c(\theta) = -\frac{\gamma}{2k \sin^2 \frac{1}{2}\theta} \exp\left[-i\gamma \ln(\sin^2 \tfrac{1}{2}\theta) + 2i\sigma_0\right] \qquad \text{(XI.55}a\text{)}$$

$$f'(\theta) = \frac{1}{2ik} \sum (2l+1)\, e^{2i\sigma_l}\, (e^{2i\delta_l} - 1)\, P_l(\cos\theta). \qquad \text{(XI.55}b\text{)}$$

It is easily shown, by an argument analogous to the one given in § 8, that the scattering cross section is

$$\sigma(\Omega) = |f(\theta)|^2. \qquad \text{(XI.56)}$$

One can write it in the form of a sum of three terms if one replaces $f(\theta)$ by its expression (XI.55), namely:

$$\sigma(\Omega) = \sigma_c(\Omega) + 2\text{Re}\, f_c^* f' + |f'(\theta)|^2.$$

Numerous properties of ordinary phase shifts apply without much change to the phase shifts introduced here. In particular, the series (XI.55b) converges more rapidly the shorter the range of the additional potential $V'(r)$. Formulae (X.39) through (X.44) of Ch. X, § 10 remain rigorously valid, it being understood that the functions $u_l^{(\pm)}$ denote Coulomb waves, and not free waves (Problem XI.3). However, the numerical values of the quantities τ_l, v_l, $q_l^{(+)}$ may be very different from the same quantities for free waves, so that the discussions of the behavior at low energies and of the convergence of the series must be rather thoroughly revised. In particular, if the Coulomb potential is repulsive, the penetration factor is smaller the smaller the incident energy, and one has $v_l \ll 1$ for any l as soon as $E \lesssim Z_1Z_2e^2/r_0$ (energy below the Coulomb barrier at the point $r = r_0$). Apart from this, the entire treatment of scattering resonances can be taken over without change. As long as we make a few changes in the definition of the quantities involved (Problem XI.4), the formulae (X.72) and (X.73), starting point of the Born approximation, and formula (X.77), starting point of the "effective range" approximation, remain valid.

EXERCISES AND PROBLEMS

1. Making use of the radial equation of the hydrogen atom, derive the recurrence relation (*Kramers relation*)

$$\frac{s+1}{n^2}\langle r^s\rangle-(2s+1)a\,\langle r^{s-1}\rangle+\frac{s}{4}\,[(2l+1)^2-s^2]\,a^2\langle r^{s-2}\rangle=0$$

in which $\langle r^s\rangle$ stands for the mean value of r^s when the atom is in the quantum state (nlm) $(a > -2l-3)$. Derive the expressions for $\langle r^{-1}\rangle$, $\langle r\rangle$, $\langle r^2\rangle$ given in Appendix B, § 3. [This relation does not enable us to determine $\langle r^{-2}\rangle$].

Show that for any stationary state of a hydrogen-like atom, the mean value of the kinetic energy is equal and opposite to the eigenvalue of the energy:

$$E_n = -\langle p^2/2m\rangle_{nlm}.$$

2. Consider the scattering of a particle of mass m by the central potential $V(r) = Ze^2/r + V'(r)$, where $V'(r)$ tends to zero at least as fast as $1/r^2$ when $r \to \infty$. Show that the solutions of the radial equation tend asymptotically toward linear combinations of the exponentials

$$\exp\,[\pm\mathrm{i}(kr-\gamma\ln 2kr)]$$

$(k = \sqrt{2mE}/\hbar$, $\gamma = Ze^2/\hbar v$; E and v are the incident energy and velocity, respectively).

3. Show that the formulae (XI.39) through (XI.44) of Chapter X (§ 10) remain valid if one adds a Coulomb-interaction term $Z_1Z_2e^2/r$ to the finite-range potential, provided that one uses the appropriate definitions for the functions $u_l^{(+)}$ and $u_l^{(-)}$.

4. How must the integral representations (X.72) and (X.73) be modified when the scattering potential is the sum of a short-range term and a Coulomb-interaction term? Answer the same question for formula (X.77). Examine the effective-range theory for s-wave scattering when the short-range term has the properties stated in Chapter X, § 20 (Cf. H. A. Bethe, *loc. cit.*, footnote p. 409).

CHAPTER XII

THE HARMONIC OSCILLATOR

1. Introduction

In Classical Mechanics, a harmonic oscillator is a particle constrained to move along an axis and subject to a restoring force proportional to a point located on that axis. It is a typical problem whose solution is well known. Let q be the position coordinate of the particle on the axis, taking the center of force as the origin, and let p be its momentum, m its mass, and $-m\omega^2 q$ the restoring force. The equations of motion of the particle are derived from the Hamiltonian $(p^2+m^2\omega^2q^2)/2m$; one easily shows that the motion is an oscillatory, sinusoidal motion of (angular) frequency ω about the origin.

The corresponding quantum-mechanical problem is that of a particle of mass m in one dimension with the Hamiltonian

$$\mathscr{H} = \frac{1}{2m}(p^2+m^2\omega^2 q^2), \qquad \text{(XII.1)}$$

the position variable q and the momentum p being connected by the commutation relation

$$[q, p] = i\hbar. \qquad \text{(XII.2)}$$

One is dealing with a particularly simple quantum system whose Schrödinger equation one knows how to solve rigorously. Moreover, it possesses a certain number of noteworthy properties.

The study of the harmonic oscillator is of great importance in Quantum Theory since the Hamiltonian of the type (XII.1) enters in all problems involving quantized oscillations: one encounters it in Quantum Electrodynamics, and more generally in Quantum Field Theory; one likewise encounters it in the theory of molecular and crystalline vibrations. On the other hand, problems related to the harmonic oscillator furnish an excellent illustration of the general principles and of the formalism of Quantum Theory. All these reasons justify the detailed study which is made of it in this chapter.

The first two sections are devoted to the one-dimensional oscillator. The general solution of the eigenvalue problem of the Hamiltonian is given in Section I. Section II is devoted to various applications: determination of the generating function of the stationary states, solution of Heisenberg's equations of motion, comparison of the classical and quantum-mechanical oscillator; a study of the motion of a wave packet – which constitutes a good illustration both of the correspondence principle as well as of the uncertainty relations – and finally, an examination of some properties of ensembles of harmonic oscillators in thermodynamic equilibrium.

Section III deals with the isotropic harmonic oscillator in several dimensions. The essential characteristic of this problem is the existence of degenerate eigenvalues. The consequences of degeneracy are examined in detail in the two particular cases of the isotropic oscillator in two and three dimensions.

I. EIGENSTATES AND EIGENVECTORS OF THE HAMILTONIAN

2. The Eigenvalue Problem

In order to avoid cluttering the calculations by useless constants, we put

$$\mathscr{H} = H\hbar\omega \tag{XII.3}$$

$$q = \left(\frac{\hbar}{m\omega}\right)^{\frac{1}{2}} Q \tag{XII.4}$$

$$p = (m\hbar\omega)^{\frac{1}{2}} P. \tag{XII.5}$$

Our problem is to find the eigenvalues, and to construct the eigenvectors of the operator

$$H = \tfrac{1}{2}(P^2+Q^2), \tag{XII.6}$$

the Hermitean operators P and Q satisfying the commutation relations

$$[Q, P] = \mathrm{i}. \tag{XII.7}$$

To solve this problem, one can choose a particular representation, the $\{Q\}$ representation for instance, and solve the Schrödinger equation in that representation. Since P is then represented by the differential operator $(-\mathrm{i}\,\mathrm{d}/\mathrm{d}Q)$, we have the one-dimensional Schrödinger equation

$$\frac{1}{2}\left[-\frac{\mathrm{d}^2}{\mathrm{d}Q^2}+Q^2\right]u(Q) = \varepsilon u(Q). \tag{XII.8}$$

The method, due to Dirac, which we follow here, is more direct; it consists in constructing the eigenvectors of H by the application of suitable operators to one of them. One thus manages to solve the eigenvalue problem without referring to a particular representation, basing oneself solely upon the fundamental axioms of Hilbert space and the commutation relation (XII.7). The method described here can in fact be regarded as a method of constructing the vector space $\mathscr{E}$ of the dynamical states of the system, and presents strong analogies with the one described in Chapter VIII, § 6.

3. Introduction of the Operators a, $a^\dagger$ and N

Let us put:

$$a = \tfrac{1}{2}\sqrt{2}(Q+\mathrm{i}P) \qquad \text{(XII.9}a\text{)}$$

$$a^\dagger = \tfrac{1}{2}\sqrt{2}(Q-\mathrm{i}P); \qquad \text{(XII.9}b\text{)}$$

a and $a^\dagger$ are Hermitean conjugates of each other. The commutation relation (XII.7) is equivalent to

$$[a, a^\dagger] = 1. \qquad \text{(XII.10)}$$

If one replaces Q and P by their expressions as functions of a and $a^\dagger$ in equation (XII.6), one finds

$$H = \tfrac{1}{2}(a\, a^\dagger + a^\dagger\, a). \qquad \text{(XII.11)}$$

We put

$$N = a^\dagger\, a. \qquad \text{(XII.12)}$$

From (XII.10) and (XII.11) one deduces

$$H = N + \tfrac{1}{2}. \qquad \text{(XII.13)}$$

From (XII.10) and (XII.12) we extract the important relations

$$Na = a(N-1) \qquad \text{(XII.14}a\text{)}$$

$$Na^\dagger = a^\dagger(N+1). \qquad \text{(XII.14}b\text{)}$$

The eigenvalue problem we seek to solve is equivalent to the problem of constructing the eigenvectors of the operator N defined by expression (XII.12), in which the operators a and $a^\dagger$ are two Hermitean conjugate operators satisfying the relation (XII.10).

For this purpose, we shall prove an important theorem.

THEOREM: *If $|\nu\rangle$ is an eigenvector of N, and ν the corresponding eigenvalue, then:*

(*i*) *necessarily* $\nu \geqslant 0$;

(*ii*) *if* $\nu = 0$, $a|\nu\rangle = 0$; *if not,* $a|\nu\rangle$ *is a non-zero vector of norm*

$$\nu\langle\nu|\nu\rangle,$$

and it is an eigenvector of N *belonging to the eigenvalue* $\nu - 1$;

(*iii*) $a^\dagger|\nu\rangle$ *is certainly not zero; its norm is*

$$(\nu+1)\langle\nu|\nu\rangle,$$

and it is an eigenvector of N *corresponding to the eigenvalue* $\nu + 1$.

By hypothesis,

$$N|\nu\rangle = \nu|\nu\rangle, \quad \langle\nu|\nu\rangle > 0.$$

Making use of the definition (XII.12) and the relation (XII.10) we deduce the respective norms of $a|\nu\rangle$ and of $a^\dagger|\nu\rangle$

$$\langle\nu|a^\dagger\, a|\nu\rangle = \langle\nu|N|\nu\rangle = \nu\langle\nu|\nu\rangle \tag{XII.15a}$$

$$\langle\nu|a\, a^\dagger|\nu\rangle = \langle\nu|(N+1)|\nu\rangle = (\nu+1)\langle\nu|\nu\rangle. \tag{XII.15b}$$

Now, the norm of a vector of Hilbert space is non-negative, and the vanishing of the norm is a necessary and sufficient condition for the vanishing of the vector. In order that this fundamental axiom be fulfilled here, it is necessary and sufficient that $\nu \geqslant 0$ [property (*i*)] [1]. The condition for the vanishing of $a|\nu\rangle$ is a special case of eq. (XII.15*a*). On the other hand, $a|\nu\rangle$ and $a^\dagger|\nu\rangle$ actually satisfy the stated eigenvalue equations since, according to (XII.14*a*) and (XII.14*b*),

$$N\, a|\nu\rangle = a(N-1)|\nu\rangle = (\nu-1)\, a|\nu\rangle$$

$$N\, a^\dagger|\nu\rangle = a^\dagger(N+1)|\nu\rangle = (\nu+1)\, a^\dagger|\nu\rangle.$$

Q.E.D.

4. Spectrum and Basis of N

If $\nu > 0$, the foregoing theorem applies equally well to the vector $a|\nu\rangle$ whose eigenvalue is $\nu - 1$; consequently $\nu \geqslant 1$. If $\nu > 1$, the theorem also applies to the vector $a^2|\nu\rangle$. One thus successively forms the set of eigenvectors

$$a|\nu\rangle,\, a^2|\nu\rangle,\, \ldots,\, a^p|\nu\rangle,\, \ldots$$

belonging respectively to the eigenvalues

$$\nu-1,\, \nu-2,\, \ldots,\, \nu-p,\, \ldots\,.$$

[1]) Cf. Problem VII.9.

This set is certainly limited since the eigenvalues of N have a lower limit of zero. In other words, the vectors of this set all vanish from a certain rank $n+1$ on: the action of a on the non-zero eigenvector $a^n|\nu\rangle$ belonging to the eigenvalue $\nu-n$ yields zero; according to (*ii*), this requires that $\nu=n$.

In the same way we can apply the theorem to the vector $a^\dagger|\nu\rangle$ which is certainly not zero and belongs to the eigenvalue $\nu+1$, then to the vector $a^{\dagger 2}|\nu\rangle$, and so forth. One thus successively forms an unlimited set of non-zero vectors

$$a^\dagger|\nu\rangle,\ a^{\dagger 2}|\nu\rangle,\ \ldots,\ a^{\dagger p}|\nu\rangle,\ \ldots\ ,$$

eigenvectors of N belonging respectively to the eigenvalues

$$\nu+1,\ \nu+2,\ \ldots,\ \nu+p,\ \ldots\ .$$

In conclusion, *the spectrum of eigenvalues of N is formed by the set of non-negative integers.* Moreover, by repeated action of a or $a^\dagger$ on one of them, one obtains a set of eigenvectors, each belonging to one of the eigenvalues of the spectrum. The ratio of the norms of each vector to that of the succeeding one is given by one or the other of the relations (XII.15*a*) or (XII.15*b*). This ensemble of vectors forms a complete set. Indeed, one can show that any function of a and of $a^\dagger$ which commutes with N is a function of N (Problem XII.1). Consequently N by itself forms a complete set of commuting observables, and none of its eigenvalues is degenerate.

The vectors thus constructed are not normalized to unity. To form an orthonormal basis of the observable N, it suffices to multiply each of them by a suitable constant, which is easily deduced from relations (XII.15*a*) and (XII.15*b*). This constant is defined to within an arbitrary phase, which we shall fix in such a way as to obtain formulae that are as simple as possible. We thus form the set of orthonormal vectors

$$|0\rangle,\ |1\rangle,\ \ldots,\ |n\rangle,\ \ldots \tag{XII.16}$$

corresponding, respectively, to the following eigenvalues of N:

$$0,\ 1,\ \ldots,\ n,\ \ldots\ .$$

They can be deduced from each other by the recursion relations

$$a^\dagger|n\rangle = (n+1)^{\frac{1}{2}}|n+1\rangle \tag{XII.17}$$

$$a|n\rangle = n^{\frac{1}{2}}|n-1\rangle \qquad (n\neq 0) \tag{XII.18}$$

$$a|0\rangle = 0. \tag{XII.19}$$

One verifies easily that they can all be deduced from the vector $|0\rangle$ by the relation

$$|n\rangle = (n!)^{-\frac{1}{2}}\, a^{\dagger n}|0\rangle, \tag{XII.20}$$

that they actually obey the eigenvalue equation

$$N|n\rangle = n|n\rangle, \tag{XII.21}$$

and that their norm is actually equal to 1, hence that they satisfy the relations

$$\langle n|n'\rangle = \delta_{nn'}. \tag{XII.22}$$

Since N forms a complete set by itself, the sequence of vectors (XII.16) forms a complete set of vectors orthonormal in the space $\mathscr{E}$, namely the space of the dynamical states of the quantum system under study. The internal consistency of this construction of $\mathscr{E}$ remains to be verified, that is to say one has to make sure that the vectors of $\mathscr{E}$ all satisfy the characteristic axioms of vectors of Hilbert space, and that the physical quantities associated with the system are observables satisfying suitable algebraic relations. We shall not insist here upon these points of mathematical rigor (Problem XII.3).

5. The $\{N\}$-Representation

The vectors of the sequence (XII.16) form the basis of a certain representation which we shall call the $\{N\}$ representation. From equations (XII.17), (XII.18), (XII.19) and (XII.20) one easily derives the representative matrices of the operators N, a and $a^{\dagger}$ in this representation. If one adopts the convention of arranging the rows and columns of these matrices in the order of increasing quantum numbers n (the uppermost row corresponds to $n=0$, the following row to $n=1$, etc.; the extreme left column corresponds to $n=0$, the following column to $n=1$, etc.), one finds for N the diagonal matrix

$$N = \begin{pmatrix} 0 & 0 & \cdots & & \\ 0 & 1 & 0 & & \\ \vdots & 0 & 2 & 0 & \\ & & 0 & 3 & 0 \\ \vdots & & & 0 & \ddots \end{pmatrix},$$

for a the real matrix

$$a = \begin{pmatrix} 0 & \sqrt{1} & 0 & 0 & \cdots \\ 0 & 0 & \sqrt{2} & & \\ \vdots & 0 & 0 & \sqrt{3} & \\ & & 0 & 0 & \sqrt{4} \\ \vdots & & & 0 & 0 & \ddots \end{pmatrix}$$

whose only non-zero elements are those of the diagonal located immediately above the main diagonal, and for $a^\dagger$ the Hermitean conjugate matrix

$$a^\dagger = \begin{pmatrix} 0 & 0 & 0 & \cdots\cdots & & \\ \sqrt{1} & 0 & 0 & \cdots\cdots & & \\ 0 & \sqrt{2} & 0 & & & \\ \vdots & & \sqrt{3} & 0 & & \\ \vdots & & & \sqrt{4} & 0 & \\ \vdots & & & & & \ddots \end{pmatrix}$$

whose only non-zero elements are those of the diagonal located immediately below the main diagonal. Since the observables of the quantum system are all functions of a and of $a^\dagger$, it is easy to form the matrices representing them in the $\{N\}$ representation. In particular, one has

$$\mathscr{H} = (N+\tfrac{1}{2})\hbar\omega \tag{XII.23}$$

$$q = \left(\frac{\hbar}{2m\omega}\right)^{\frac{1}{2}}(a^\dagger + a) \tag{XII.24}$$

$$p = \mathrm{i}\left(\frac{m\hbar\omega}{2}\right)^{\frac{1}{2}}(a^\dagger - a). \tag{XII.25}$$

$\mathscr{H}$ is diagonal in this representation and its eigenvalues are

$$(n+\tfrac{1}{2})\hbar\omega \qquad (n = 0, 1, 2, \ldots, \infty).$$

As q and p are linear functions of a and $a^\dagger$, their only non-zero matrix elements are located on the two diagonals adjacent to the main diagonal. We leave it to the reader to set up these matrices.

6. Creation and Destruction Operators

The operators N, $a^\dagger$ and a were introduced to facilitate the solution of the eigenvalue problem. If $\mathscr{H}$ is the Hamiltonian of a one-dimensional quantized particle, these operators have no immediate physical significance.

But the eigenvalue problem of $\mathscr{H}$ is susceptible of another interpretation. Indeed, since the energy levels are equidistant by $\hbar\omega$, one may consider $\mathscr{H}$ as the Hamiltonian of a system of indistinguishable corpuscles, all being in one and the same dynamical state, whose energy is $\hbar\omega$. Their number N can vary, each eigenstate of $\mathscr{H}$ corresponding to a well-defined value of N and therefore to a well-defined value of the energy of the total system. Thus the vector $|n\rangle$ represents a state constituted of n corpuscles; the vector $|0\rangle$ is the vacuum state, for which the number of corpuscles is zero. When one goes over from the state $|n\rangle$ to the state $|n+1\rangle$, the number of corpuscles increases by one unit and the total energy of the system increases by the amount $\hbar\omega$. One notes that the energy of the vacuum is not zero but equal to $\frac{1}{2}\hbar\omega$; this anomaly may be avoided if one takes $\mathscr{H}-\frac{1}{2}\hbar\omega$, and not $\mathscr{H}$ as the operator defining the energy of the system.

According to this interpretation, the operator N represents the number of corpuscles and can take on all integral values between 0 and $+\infty$. The operator $a^\dagger$ transforms a state constituted of n corpuscles into a state of $(n+1)$ corpuscles: $a^\dagger$ is a *creation operator*. The operator a, on the contrary, diminishes the number of corpuscles present by one unit: a is a *destruction operator*.

This interpretation of the harmonic oscillator is widely used in Quantum Field Theory and in the theory of crystalline and molecular vibrations. The electromagnetic field, for instance, can be put in the form of a superposition of plane waves characterized by their polarization $\boldsymbol{\epsilon}$ and their wave vector $\boldsymbol{k}$; their frequency is $\omega=k/c$. Classically, the intensity of each component can vary in continuous fashion; actually, it varies by light quanta or photons of energy $\hbar\omega$. The Hamiltonian of the quantized electromagnetic field is a superposition of terms, each referring to a particular type of photon characterized by $\boldsymbol{\epsilon}$ and $\boldsymbol{k}$ [we use the subscript s to denote the ensemble $(\boldsymbol{\epsilon}, \boldsymbol{k})$]:

$$\boldsymbol{H} = \sum_s \mathscr{H}_s.$$

Each partial Hamiltonian can be written in the form

$$\mathscr{H}_s = \hbar\omega_s a_s^\dagger a_s.$$

The operators a_s and $a_s^\dagger$ are Hermitean conjugates of each other and satisfy the commutation relations

$$[a_s, a_{s'}^\dagger] = \delta_{ss'},$$

a simple generalization of relation (XII.10). The operators $a_s^\dagger$ and a_s are respectively interpreted as creation and destruction operators of photons of the type s (cf. Ch. XXI).

7. $\{Q\}$ Representation. Hermite Polynomials

In the language of Wave Mechanics the eigenvalue problem of $\mathscr{H}$ consists in determining the values E for which the equation

$$\mathscr{H}\psi(q) \equiv \left(-\frac{\hbar^2}{2m}\frac{\mathrm{d}^2}{\mathrm{d}q^2} + \tfrac{1}{2}m\omega^2 q^2\right)\psi(q) = E\psi(q)$$

possesses a regular solution at the two limits of the interval $(-\infty, +\infty)$. If one applies the discussion of Ch. III (§ 10) to this problem, one finds that the values of E which fulfill this condition form a discrete spectrum, and that to each of them there corresponds one and only one solution (defined to within a constant); moreover, this solution has a finite norm. This is quite in keeping with the foregoing study, according to which the spectrum of $\mathscr{H}$ is entirely discrete and non-degenerate. Upon solving the eigenvalue problem thus stated, we would again find the sequence of eigenvalues of $\mathscr{H}$,

$$\tfrac{1}{2}\hbar\omega,\ \tfrac{3}{2}\hbar\omega,\ \ldots,\ (n+\tfrac{1}{2})\hbar\omega,\ \ldots .$$

The corresponding eigenfunctions $\psi_n(q) \equiv \langle q|n\rangle$ are the functions representing the eigenstates $|n\rangle$ in the $\{q\}$ representation.

In what follows, we adopt the $\{Q\}$ representation deduced from the $\{q\}$ representation by the change of variable (XII.4). The eigenfunctions $u_n(Q)$ and $\psi_n(q)$ which represent the same eigenstate $|n\rangle$ in the $\{Q\}$ and $\{q\}$ representations, respectively, are evidently connected by the relation

$$\langle Q|n\rangle \equiv u_n(Q) = \left(\frac{\hbar}{m\omega}\right)^{\frac{1}{4}}\psi_n(q).$$

Equation (XII.8) is (to within the constant $\hbar\omega$) the Schrödinger equation in the $\{Q\}$ representation.

The eigenfunctions $u_0(Q), u_1(Q), \ldots, u_n(Q), \ldots$ are easily obtained by means of the relations (XII.17) to (XII.19). The eigenfunction of the ground state satisfies eq. (XII.19)

$$\left[\frac{\mathrm{d}}{\mathrm{d}Q} + Q\right] u_0(Q) = 0,$$

whose solution, normalized to unity, is

$$u_0(Q) = \pi^{-\frac{1}{4}}\,\mathrm{e}^{-\frac{1}{2}Q^2}. \qquad \text{(XII.26)}$$

From (XII.17) and (XII.18) we deduce relations between normalized eigenfunctions belonging to neighboring eigenvalues (cf. Appendix B, Sec. III); in particular, the repeated application of (XII.17) allows one to build up all eigenfunctions starting from the function u_0. Rather than use (XII.17), we can equally well make use of relation (XII.20) which is its equivalent; this yields

$$u_n(Q) = [\pi^{\frac{1}{2}} 2^n (n!)]^{-\frac{1}{2}} \left(Q - \frac{\mathrm{d}}{\mathrm{d}Q}\right)^n \mathrm{e}^{-\frac{1}{2}Q^2}. \tag{XII.27}$$

Using the operator identity

$$\left(Q - \frac{\mathrm{d}}{\mathrm{d}Q}\right) \equiv \left(-\,\mathrm{e}^{\frac{1}{2}Q^2} \frac{\mathrm{d}}{\mathrm{d}Q}\, \mathrm{e}^{-\frac{1}{2}Q^2}\right)$$

one can write equation (XII.27) in the form (B.70) in which $H_n(Q)$ is the Hermite polynomial of order n in accordance with the definition (B.59). Thus $u_n(Q)$ is the product of $\exp(-\frac{1}{2}Q^2)$ and an nth order polynomial of parity $(-)^n$. The main properties of these polynomials are listed in Appendix B (§ 7).

II. APPLICATIONS AND VARIOUS PROPERTIES

8. Generating Function for the Eigenfunctions $u_n(Q)$.

As an application, we shall determine a *generating function* of the functions $u_n(Q)$, that is a function $F(t, Q)$ such that

$$F(t, Q) = \sum_{n=0}^{\infty} c_n u_n(Q) t^n,$$

where the c_n are suitable normalization constants. Since $u_n(Q)$ represents the vector $(n!)^{-\frac{1}{2}} a^{\dagger n}|0\rangle$ [eq. (XII.20)], the function $F(t, Q)$ considered as a function of Q represents the vector

$$\sum_n \frac{c_n}{(n!)^{\frac{1}{2}}} (a^\dagger t)^n |0\rangle.$$

With the choice

$$c_n = \frac{1}{(n!)^{\frac{1}{2}}},$$

$F(t, Q)$ represents the vector $\exp(a^\dagger t)|0\rangle$:

$$F(t, Q) = \langle Q| \exp(a^\dagger t)|0\rangle. \tag{XII.28}$$

To calculate this last expression, we shall make use of the following lemma:

LEMMA. – *If the commutator of two operators A, B commutes with each of them:*

$$[A, [A, B]] = [B, [A, B]] = 0,$$

one has the identity

$$e^{A+B} = e^A e^B e^{-\frac{1}{2}[A,B]}. \tag{XII.29}$$

The following proof is due to Glauber.

Let us consider the operator depending on the parameter x:

$$f(x) = e^{Ax} e^{Bx}.$$

One has

$$\begin{aligned}\frac{df}{dx} &= A e^{Ax} e^{Bx} + e^{Ax} B e^{Bx} \\ &= (A + e^{Ax} B e^{-Ax}) f(x).\end{aligned}$$

But since $[B, A]$ commutes with A,

$$\begin{aligned}[B, A^n] &= n A^{n-1}[B, A] \\ [B, e^{-Ax}] &= \sum_n (-)^n \frac{x^n}{n!} [B, A^n] \\ &= \sum_n (-)^n \frac{x^n}{(n-1)!} A^{n-1}[B, A] \\ &= -e^{-Ax}[B, A]x,\end{aligned}$$

hence (cf. Problem VIII.4)

$$e^{Ax} B e^{-Ax} = B - [B, A]x$$

and therefore

$$\frac{df}{dx} = (A + B + [A, B]x) f(x).$$

$f(x)$ is the solution of this differential equation for which $f(0) = 1$. Since the operators $(A+B)$ and $[B, A]$ commute, they can be considered here as quantities of ordinary algebra. The differential equation can be easily integrated and gives

$$f(x) = \exp [(A+B)x] \exp (\tfrac{1}{2}[A, B]x^2).$$

The identity (XII.29) results if one sets $x = 1$.

Q.E.D.

Taking $A = Qt/\sqrt{2}$, $B = -\mathrm{i}Pt/\sqrt{2}$, $[A, B] = \frac{1}{2}t^2$, we apply the identity XII.29) to the operator $\exp(a^\dagger t)$, namely

$$\exp(a^\dagger t) = \exp[Qt/\sqrt{2}] \exp[-\mathrm{i}Pt/\sqrt{2}] \exp(-\tfrac{1}{4}t^2).$$

Inserting this expression in eq. (XII.28),

$$F(t, Q) = \exp(-\tfrac{1}{4}t^2) \exp[Qt/\sqrt{2}] \langle Q| \exp[-\mathrm{i}Pt/\sqrt{2}]|0\rangle.$$

But

$$\begin{aligned} \langle Q| \exp[-\mathrm{i}Pt/\sqrt{2}]|0\rangle &= \exp\left(-\frac{t}{\sqrt{2}}\frac{\mathrm{d}}{\mathrm{d}Q}\right) u_0(Q) \\ &= u_0\left(Q - \frac{t}{\sqrt{2}}\right). \end{aligned}$$

With expression (XII.26) for u_0 we obtain after some calculation

$$\begin{aligned} F(t, Q) &\equiv \sum_{n=0}^{\infty} \frac{u_n(Q)}{(n!)^{\frac{1}{2}}} t^n \\ &= \pi^{-\frac{1}{4}} \exp(-\tfrac{1}{2}Q^2 + tQ\sqrt{2} - \tfrac{1}{2}t^2). \end{aligned} \qquad \text{(XII.30)}$$

9. Integration of the Heisenberg Equations

Let us consider the harmonic oscillator *in the Heisenberg "representation"*. Since all operators occurring in this paragraph are operators in the Heisenberg representation, we shall omit the subscript H which permitted to distinguish them from corresponding operators in the Schrödinger representation in the discussions of Ch. VIII. These operators evolve in time. We attach the subscript 0 to the values they take at the initial instant $t = 0$.

Taking into account eq. (XII.23) and relations (XII.14), the Heisenberg equations of the operators a and $a^\dagger$ are respectively written

$$\mathrm{i}\hbar \frac{\mathrm{d}a}{\mathrm{d}t} = [a, \mathscr{H}] = \hbar\omega a,$$

$$\mathrm{i}\hbar \frac{\mathrm{d}a^\dagger}{\mathrm{d}t} = [a^\dagger, \mathscr{H}] = -\hbar\omega a^\dagger.$$

These equations may be easily integrated and yield

$$a(t) = a_0\, \mathrm{e}^{-\mathrm{i}\omega t} \qquad \text{(XII.31a)}$$

$$a^\dagger(t) = a_0^\dagger\, \mathrm{e}^{+\mathrm{i}\omega t}. \qquad \text{(XII.31b)}$$

Making use of relations (XII.24) and (XII.25) which give q and p as functions of a and $a^{\dagger}$, one has

$$q(t) = \left(\frac{\hbar}{2m\omega}\right)^{\frac{1}{2}}(a_0^{\dagger}\,\mathrm{e}^{\mathrm{i}\omega t} + a_0\,\mathrm{e}^{-\mathrm{i}\omega t}) \tag{XII.32}$$

$$p(t) = \mathrm{i}\left(\frac{m\hbar\omega}{2}\right)^{\frac{1}{2}}(a_0^{\dagger}\,\epsilon^{\mathrm{i}\omega t} - a_0\,\mathrm{e}^{-\mathrm{i}\omega t}). \tag{XII.33}$$

If one replaces in these equations a_0 and $a_0^{\dagger}$ by their expressions as functions of the initial position q_0 and of the initial momentum p_0, one has

$$q(t) = q_0 \cos \omega t + \frac{1}{m\omega}\, p_0 \sin \omega t \tag{XII.34}$$

$$p(t) = p_0 \cos \omega t - m\omega q_0 \sin \omega t. \tag{XII.35}$$

One finds the same sinusoidal functions as for the classical harmonic oscillator. In particular, the mean values $\langle q \rangle_t$, $\langle p \rangle_t$ follow the classical laws of motion

$$\langle q \rangle_t = \langle q \rangle_0 \cos \omega t + \frac{1}{m\omega} \langle p \rangle_0 \sin \omega t \tag{XII.36}$$

$$\langle p \rangle_t = \langle p \rangle_0 \cos \omega t - m\omega \langle q \rangle_0 \sin \omega t. \tag{XII.37}$$

This property of the harmonic oscillator was already pointed out in Chapter VI.

10. Classical and Quantized Oscillator

In order to illustrate the correspondence between Classical Mechanics and Quantum Mechanics, we compare in this and the following section the motion of the classical oscillator with that of the corresponding quantum-mechanical oscillator.

The general solution of the equations of motion of the classical harmonic oscillator can be written

$$q_{\mathrm{cl.}} = A \sin (\omega t + \varphi),$$
$$p_{\mathrm{cl.}} = m\omega A \cos (\omega t + \varphi).$$

It is a sinusoidal, oscillatory motion of (angular) frequency ω. It depends upon two parameters A and φ. The energy of the oscillator is connected with the amplitude of oscillation A by the relation

$$E_{\mathrm{cl.}} = \frac{m\omega^2 A^2}{2}. \tag{XII.38}$$

If one fixes the energy $E_{\text{cl.}}$, the various possible motions differ from each other in the phase constant φ.

Let $F_{\text{cl.}}$ be a dynamical variable of the system. Since it is a function of $q_{\text{cl.}}$ and of $p_{\text{cl.}}$, $F[q_{\text{cl.}}(t), p_{\text{cl.}}(t)]$, it varies periodically (but not necessarily sinusoidally) with time with frequency ω. The time dependence of $F_{\text{cl.}}$ for two motions of equal energy is the same except for the phase. The average $\overline{F}_{\text{cl.}}$ of $F_{\text{cl.}}$ taken over all motions of the same energy (microcanonical ensemble) is obtained by performing the average over the phase constants; $\overline{F}_{\text{cl.}}$ is independent of time and equal to the average over a period $2\pi/\omega$ of the values assumed by $F_{\text{cl.}}$ during any one of these motions. One finds in particular:

$$\bar{q}_{\text{cl.}} = \bar{p}_{\text{cl.}} = 0 \tag{XII.39}$$

$$\overline{q^2_{\text{cl.}}} = \frac{A^2}{2} = \frac{E_{\text{cl.}}}{m\omega^2} \tag{XII.40}$$

$$\overline{p^2_{\text{cl.}}} = m^2\omega^2\,\overline{q^2_{\text{cl.}}} = mE_{\text{cl.}} \tag{XII.41}$$

(the mean kinetic energy and the mean potential energy of an oscillator are equal).

Let us see how this compares with the behavior of the quantized oscillator in a stationary state. In the state $|n\rangle$, the quantized oscillator has a well-defined energy which is constant in time: $(n+\frac{1}{2})\hbar\omega = E_n$. On the other hand, the observables of position q and of momentum p do not have precise values. One may only define the statistical distribution of the results of measurement of one or the other of these quantities in the eventuality that such a measurement is performed. Since the state is stationary, these statistical distributions are constant in time. In particular, the mean values of q and p are respectively equal to the diagonal element of rank n of the observables q and p in the $\{N\}$ representation:

$$\langle n|q|n\rangle = \langle n|p|n\rangle = 0. \tag{XII.42}$$

The mean values of q^2 and p^2 are easily calculated by expressing these operators as functions of the a and $a^\dagger$ [eqs. (XII.24) and (XII.25)] and using relations (XII.17) to (XII.19). We have

$$\langle n|q^2|n\rangle = \frac{1}{2}\frac{\hbar}{m\omega}\langle n|(a^\dagger a + aa^\dagger)|n\rangle = \frac{E_n}{m\omega^2} \tag{XII.43}$$

$$\langle n|p^2|n\rangle = \tfrac{1}{2}m\hbar\omega\langle n|(a^\dagger a + aa^\dagger)|n\rangle = mE_n. \tag{XII.44}$$

The correspondence principle demands (cf. Problem XII.4) that in the limit where $n \to \infty$, the expressions for the average values (XII.42), (XII.43) and (XII.44) become respectively identical to the classical expressions (XII.39), (XII.40) and (XII.41) for the same value of the energy ($E_n = E_{\text{cl.}}$). The fact that this identity is rigorously true for all values of n, even small ones, is a characteristic property of the harmonic oscillator.

We note in passing that in the state $|n\rangle$

$$\Delta p \cdot \Delta q = \frac{E_n}{\omega} = (n + \tfrac{1}{2})\hbar, \qquad \text{(XII.45)}$$

in accordance with the position-momentum uncertainty relations.

11. Motion of the Minimum Wave Packet and Classical Limit

Consider the one-dimensional wave packet

$$f(q) = \left(\frac{m\omega}{\pi\hbar}\right)^{\frac{1}{4}} \exp\left[\frac{\mathrm{i}}{\hbar}\langle p\rangle q - \frac{m\omega}{2\hbar}(q - \langle q\rangle)^2\right]. \qquad \text{(XII.46)}$$

This is a minimum wave packet (Problem IV.4): it represents a particle localized in configuration space about its mean position $\langle q\rangle$ with a root-mean-square deviation $\Delta q = (\hbar/2m\omega)^{\frac{1}{2}}$ and localized in momentum space about its mean position $\langle p\rangle$ with a root-mean-square deviation $\Delta p = (\hbar m\omega/2)^{\frac{1}{2}}$.

If this particle is subject to the Hamiltonian $\mathscr{H}$, one can show (Problem XII.6) that such a packet retains minimum size in the course of time and that it oscillates with the frequency ω. In more precise fashion, the statistical distribution of q, $\varrho(q, t)$, varies according to the law

$$\varrho(q, t) \equiv |f(q, t)|^2 = \left(\frac{m\omega}{\pi\hbar}\right)^{\frac{1}{2}} \exp\left[-\frac{m\omega}{\hbar}(q - \langle q\rangle_t)^2\right].$$

It oscillates without distortion, its center $\langle q\rangle_t$ carrying out the sinusoidal motion predicted by classical theory. The statistical distribution of p exhibits analogous behavior.

The statistical distribution of $\mathscr{H}$, on the other hand, is constant in time. The probability of finding the system in the state of energy $(n + \frac{1}{2})\hbar\omega$ is equal at every instant to [Problem (XII.6)]

$$\mathrm{e}^{-E_{\text{cl.}}/\hbar\omega} \frac{(E_{\text{cl.}}/\hbar\omega)^n}{n!}.$$

We make use of the notation

$$E_{\text{cl.}} = \frac{1}{2m}(\langle p\rangle^2 + m^2\omega^2\langle q\rangle^2).$$

From this probability law, one easily deduces the average value of the energy

$$\begin{aligned}\langle E\rangle &= e^{-E_{\text{cl.}}/\hbar\omega}\sum_{n=0}^{\infty}(n+\tfrac{1}{2})\hbar\omega\,\frac{(E_{\text{cl.}}/\hbar\omega)^n}{n!} \\ &= E_{\text{cl.}} + \tfrac{1}{2}\hbar\omega\end{aligned} \tag{XII.47}$$

and its root-mean-square deviation

$$\Delta E = \sqrt{\langle\mathscr{H}^2\rangle - \langle\mathscr{H}\rangle^2} = \sqrt{\hbar\omega E_{\text{cl.}}} \tag{XII.48}$$

This wave packet provides us with a good illustration of the uncertainty relations.

It was chosen in such a way that the product of the deviations $\Delta p\cdot\Delta q$ remains constant and equal to the minimum value $\frac{1}{2}\hbar$.

As far as the time-energy relation is concerned, we can compare ΔE to the characteristic time τ_q of the rate of evolution of the statistical distribution of q. τ_q is the time required for the center $\langle q\rangle_t$ of the distribution to travel by an amount equal to its extension Δq; since the velocity of the center is $\langle p\rangle_t/m$, one has

$$\tau_q = \frac{m}{\langle p\rangle_t}\,\Delta q = \frac{1}{\langle p\rangle_t}\,(\hbar m/2\omega)^{\frac{1}{2}}.$$

τ_q passes periodically through a minimum when $\langle p\rangle_t$ attains its largest value $(2mE_{\text{cl.}})^{\frac{1}{2}}$. One then has

$$\tau_{q\,\text{min.}} = \tfrac{1}{2}\hbar\left(\frac{1}{\hbar\omega E_{\text{cl.}}}\right)^{\frac{1}{2}}.$$

According to (XII.48),

$$\tau_{q\,\text{min.}}\,\Delta E = \tfrac{1}{2}\hbar \tag{XII.49}$$

in accordance with the time-energy uncertainty relation (VIII.47).

The amplitude of oscillation A of the center of the wave packet is given by the classical relation (XII.38):

$$A = (2E_{\text{cl.}}/m\omega^2)^{\frac{1}{2}}.$$

In the limit where this amplitude is large compared to the extension $(\hbar/2m\omega)^{\frac{1}{2}}$ of the packet, and to the extent one can consider lengths of the order of $(\hbar/2m\omega)^{\frac{1}{2}}$ negligible, the *classical picture* of a point particle oscillating according to the law $\langle q \rangle_t$ provides a statisfactory description of the phenomenon. This limit is just the one of very large quantum numbers as required by the correspondence principle. Indeed, it is realized when $E_{\text{cl.}} \gg \hbar\omega$; now, the number of states of quantized energy contributing appreciably to the composition of the wave packet is of the order of the ratio of the spread ΔE to the level spacing, namely

$$\frac{\Delta E}{\hbar\omega} = \left(\frac{E_{\text{cl.}}}{\hbar\omega}\right)^{\frac{1}{2}} \gg 1.$$

Of course, this classical picture likewise supposes that one treats the dispersion in momentum, $\Delta p = (\hbar m\omega/2)^{\frac{1}{2}}$, and the dispersion in energy, $\Delta E = (\hbar\omega E_{\text{cl.}})^{\frac{1}{2}}$, as negligible quantities. As far as the energy is concerned, one has at this level of precision $\langle E \rangle \approx E_{\text{cl.}}$. Indeed,

$$\langle E \rangle - E_{\text{cl.}} = \tfrac{1}{2}\hbar\omega \ll \Delta E.$$

One is thus well justified in attributing to the system the energy $E_{\text{cl.}}$ of the corresponding classical particle.

12. Harmonic Oscillators in Thermodynamic Equilibrium

Consider a harmonic oscillator in thermodynamic equilibrium with a heat reservoir at temperature T. Its dynamical state is not a pure state but a statistical mixture represented by the density operator

$$\varrho = \frac{e^{-\mathscr{H}/kT}}{\operatorname{Tr} e^{-\mathscr{H}/kT}} \tag{XII.50}$$

in conformity with the Boltzmann law. We shall examine some of the properties of this mixture.

Let us first of all calculate the *partition function*

$$Z(\mu) = \operatorname{Tr} e^{-\mu\mathscr{H}}.$$

This calculation of the trace is easily performed in the representation where $\mathscr{H}$ is diagonal:

$$\begin{aligned} Z(\mu) &= \sum_{n=0}^{\infty} \exp\left[-\mu(n+\tfrac{1}{2})\hbar\omega\right] \\ &= e^{-\frac{1}{2}\mu\hbar\omega} \sum_{n=0}^{\infty} (e^{-\mu\hbar\omega})^n, \end{aligned}$$

from which, after summation of the geometric series of the right-hand member:

$$Z(\mu) = \frac{e^{-\frac{1}{2}\mu\hbar\omega}}{1-e^{-\mu\hbar\omega}}. \tag{XII.51}$$

The mean energy

$$\langle E \rangle \equiv \mathrm{Tr}\, \varrho \mathscr{H}$$

is deduced from the partition function by applying eq. (VIII.84). One has

$$\ln Z = -\tfrac{1}{2}\mu\hbar\omega - \ln(1-e^{-\mu\hbar\omega}),$$

from which

$$\begin{aligned} \langle E \rangle &= -\left.\frac{\partial(\ln Z)}{\partial\mu}\right|_{\mu=1/kT} \\ &= \tfrac{1}{2}\hbar\omega \coth\frac{\hbar\omega}{2kT} \\ &= \tfrac{1}{2}\hbar\omega + \frac{\hbar\omega}{e^{\hbar\omega/kT}-1}. \end{aligned} \tag{XII.52}$$

This is just Planck's formula (to within the constant $\frac{1}{2}\hbar\omega$) for the average energy of a quantized oscillator.

At very low temperatures ($kT \ll \hbar\omega$), the oscillator remains with near certainty in its ground state

$$\langle E \rangle \approx \tfrac{1}{2}\hbar\omega.$$

At very high temperatures ($kT \gg \hbar\omega$), the average energy tends toward the one given by classical Maxwell–Boltzmann statistics:

$$\langle E \rangle \approx kT.$$

As a last property of the quantized oscillator in thermodynamic equilibrium, let us mention the following theorem, due to F. Bloch:

THEOREM. – *The probability law of a given combination $\alpha q + \beta p$ of the momentum and the position is a Gaussian.*

To prove this theorem we calculate the characteristic function $\varphi(\xi)$ of this probability law. $\varphi(\xi)$ is by definition the mean value of $\exp[i\xi(\alpha q+\beta p)]$:

$$\varphi(\xi) = \mathrm{Tr}\, \varrho\, e^{i\xi(\alpha q+\beta p)}. \tag{XII.53}$$

We shall calculate this trace in the $\{N\}$ representation, where ϱ is diagonal.

We first calculate the quantities

$$g_n(\xi) = \langle n|\, \mathrm{e}^{\mathrm{i}\xi(\alpha q+\beta p)}\, |n\rangle. \tag{XII.54}$$

One has [eq. (XII.24) and (XII.25)]

$$\alpha q+\beta p = \gamma a+\gamma^* a^\dagger,$$

with

$$\gamma = \left(\frac{\hbar}{2m\omega}\right)^{\frac{1}{2}}(\alpha-\mathrm{i}m\omega\beta).$$

According to the identity (XII.29),

$$\exp\,[\mathrm{i}\xi(\alpha q+\beta p)] = \exp\,[\mathrm{i}\xi(\gamma a+\gamma^* a^\dagger)] =$$
$$\exp\,(\tfrac{1}{2}\xi^2\gamma\gamma^*)\exp\,(\mathrm{i}\xi\gamma a)\exp\,(\mathrm{i}\xi\gamma^* a^\dagger),$$

whence

$$g_n(\xi) = \exp\,(\tfrac{1}{2}\xi^2\gamma\gamma^*)\,\langle n|\exp\,(\mathrm{i}\xi\gamma a)\exp\,(\mathrm{i}\xi\gamma^* a^\dagger)\,|n\rangle.$$

Expanding the exponentials and taking into account relation (XII.20), one obtains

$$\mathrm{e}^{\mathrm{i}\xi\gamma^* a^\dagger}|n\rangle = \sum_{t=0}^{\infty}\left(\frac{(n+t)!}{n!}\right)^{\frac{1}{2}}\frac{(\mathrm{i}\xi\gamma^*)^t}{t!}\,|n+t\rangle$$

$$\langle n|\,\mathrm{e}^{\mathrm{i}\xi\gamma a} = \sum_{s=0}^{\infty}\left(\frac{(n+s)!}{n!}\right)^{\frac{1}{2}}\frac{(\mathrm{i}\xi\gamma)^s}{s!}\,\langle n+s|.$$

In the scalar product of these two vectors, the cross terms of the double summation are all zero by virtue of the orthogonality relations; hence finally

$$g_n(\xi) = \mathrm{e}^{\frac{1}{2}\xi^2\gamma\gamma^*}\sum_{s=0}^{\infty}\frac{(-\xi^2\gamma\gamma^*)^s}{(s!)^2}\,\frac{(n+s)!}{n!}. \tag{XII.55}$$

Let us put

$$x = -\xi^2\gamma\gamma^*, \qquad y = \mathrm{e}^{-\hbar\omega/kT}. \tag{XII.56}$$

In the $\{N\}$ representation, ϱ is diagonal and its nth diagonal element is equal to

$$\begin{aligned}\varrho_n \equiv \langle n|\varrho|n\rangle &= \frac{\mathrm{e}^{-(n+\frac{1}{2})\hbar\omega/kT}}{Z(1/kT)}\\ &= (1-y)y^n.\end{aligned} \tag{XII.57}$$

From equations (XII.53) to (XII.57) we extract

$$\begin{aligned}\varphi(\xi) &= \sum_{n=0}^{\infty} \varrho_n g_n(\xi) \\ &= (1-y)\, \mathrm{e}^{-\frac{1}{2}x} \sum_{n=0}^{\infty} \sum_{s=0}^{\infty} \frac{(n+s)!}{n!(s!)^2}\, x^s y^n.\end{aligned}$$

This double series can be summed exactly. The summation over n is carried out first by means of the series expansion

$$\begin{aligned}\frac{1}{(1-y)^{s+1}} &= 1+(s+1)\,y+\frac{(s+1)(s+2)}{2!}\,y^2+\ldots \\ &= \sum_{n=0}^{\infty} \frac{(s+n)!}{s!\,n!}\,y^n.\end{aligned}$$

Hence

$$\begin{aligned}\varphi(\xi) &= \mathrm{e}^{-\frac{1}{2}x} \sum_s \frac{1}{s!}\left(\frac{x}{1-y}\right)^s \\ &= \exp\left[x\left(\frac{1}{1-y}-\frac{1}{2}\right)\right].\end{aligned}$$

Taking into account the definitions of x and y, this is written

$$\varphi(\xi) = \mathrm{e}^{-\frac{1}{2}\sigma\xi^2} \qquad \text{(XII.58)}$$

with

$$\sigma = \gamma\gamma^* \coth\left(\frac{\hbar\omega}{2kT}\right). \qquad \text{(XII.59)}$$

Since the characteristic function of the distribution is a Gaussian, the probability law is Gaussian as well: its mean square deviation is σ.

Q.E.D.

III. ISOTROPIC HARMONIC OSCILLATORS IN SEVERAL DIMENSIONS

13. General Treatment of the Isotropic Oscillator in p Dimensions

The isotropic harmonic oscillator in p dimensions is the p-dimensional system with Hamiltonian

$$\mathscr{H} = \sum_{i=1}^{p} \mathscr{H}_i \qquad \text{(XII.60)}$$

$$\mathscr{H}_i = \frac{1}{2m}\,(p_i^2+m^2\omega^2 q_i^2). \qquad \text{(XII.61)}$$

Let $\mathscr{E}_1$ be the space of the dynamical states relating to the pair of variables (p_1, q_1), $\mathscr{E}_2$ the space of the dynamical states relating to the pair of variables (p_2, q_2), etc. The space $\mathscr{E}$ of the dynamical states of the system under consideration is the tensor product of the spaces $\mathscr{E}_1, \mathscr{E}_2, \ldots, \mathscr{E}_p$:

$$\mathscr{E} = \mathscr{E}_1 \otimes \mathscr{E}_2 \otimes \ldots \otimes \mathscr{E}_p. \tag{XII.62}$$

Let us denote by $|n_i\rangle$ (i fixed, $n_i = 0, 1, \ldots, \infty$) the eigenvectors of the Hamiltonian $\mathscr{H}_i$ considered as an operator of the space $\mathscr{E}_i$; they form a complete orthonormal set in $\mathscr{E}_i$. In the following we suppose that their relative phases are chosen in such a way as to satisfy the relations (XII.17) to (XII.20) written in terms of the destruction and creation operators relating to the variables of the type i. The vectors

$$\begin{gathered} |n_1\, n_2 \ldots n_p\rangle \equiv |n_1\rangle\, |n_2\rangle \ldots |n_p\rangle \\ (n_1 = 0, 1, \ldots, \infty;\, n_2 = 0, 1, \ldots, \infty;\, \ldots;\, n_p = 0, 1, \ldots, \infty) \end{gathered} \tag{XII.63}$$

formed by the tensor product of p vectors belonging respectively to the spaces $\mathscr{E}_1, \mathscr{E}_2, \ldots, \mathscr{E}_p$, form a complete orthonormal set in $\mathscr{E}$ [1]). Clearly, these vectors are eigenvectors of $\mathscr{H}$. Moreover, since

$$\mathscr{H}_1|n_1\rangle = (n_1 + \tfrac{1}{2})\,\hbar\omega|n_1\rangle$$

$$\cdot \quad \cdot \quad \cdot \quad \cdot \quad \cdot \quad \cdot \quad \cdot \quad \cdot$$

$$\mathscr{H}_p|n_p\rangle \equiv (n_p + \tfrac{1}{2})\,\hbar\omega|n_p\rangle,$$

one has

$$\begin{aligned} \mathscr{H}|n_1 \ldots n_p\rangle &\equiv (\mathscr{H}_1 + \ldots + \mathscr{H}_p)|n_1 \ldots n_p\rangle \\ &= (n_1 + \ldots + n_p + \tfrac{1}{2}p)\,\hbar\omega|n_1 \ldots n_p\rangle. \end{aligned}$$

The vectors of the basis of $\mathscr{H}$ which we formed are labelled by means of p quantum numbers $n_1, n_2, \ldots, n_p$, which can take on all integral values from 0 to $+\infty$. However, the corresponding eigenvalue of the energy,

$$(n_1 + \ldots + n_p + \tfrac{1}{2}p)\hbar\omega,$$

depends only upon the sum

$$n = n_1 + n_2 + \ldots + n_p$$

---

1) In the representation $\{q\} \equiv \{q_1 q_2 \ldots q_p\}$, the vector $|n_1 n_2 \ldots n_p\rangle$ is represented by the product

$$\langle q_1|n_1\rangle\, \langle q_2|n_2\rangle \ldots \langle q_p|n_p\rangle \equiv \psi_{n_1}(q_1)\, \psi_{n_2}(q_2) \ldots \psi_{n_p}(q_p).$$

of these p numbers. For a given integral value of $n(\geqslant 0)$ there exist

$$C^n_{n+p-1} \equiv \frac{(n+p-1)!}{n!(p-1)!} \tag{XII.64}$$

distinct possible values for the set of numbers $n_1, n_2, \ldots, n_p$. The eigenvalue $(n+\frac{1}{2}p)\hbar\omega$ is thus C^n_{n+p-1}-fold degenerate [1]).

Let us introduce the destruction and creation operators of quanta of the type i:

$$\begin{aligned} a_i &= \left(\frac{m\omega}{2\hbar}\right)^{\frac{1}{2}} q_i + \mathrm{i}(2m\hbar\omega)^{-\frac{1}{2}} p_i \\ a_i^\dagger &= \left(\frac{m\omega}{2\hbar}\right)^{\frac{1}{2}} q_i - \mathrm{i}(2m\hbar\omega)^{-\frac{1}{2}} p_i . \end{aligned} \tag{XII.65}$$

They satisfy the commutation relations [cf. relation (XII.10)]

$$\begin{aligned} [a_i, a_j] &= [a_i^\dagger, a_j^\dagger] = 0 \\ [a_i, a_j^\dagger] &= \delta_{ij} \end{aligned} \qquad (i, j, = 1, \ldots, p). \tag{XII.66}$$

In accordance with the definition of the vectors $|n_i\rangle$ given above, the vectors $|n_1 \ldots n_p\rangle$ satisfy the relations generalizing the relations (XII.17) to (XII.20). In particular, if we designate by $|0\rangle$ the eigenvector of the ground state:

$$|0\rangle \equiv |\overbrace{0 \ldots\ldots\ldots\ldots 0}^{p \text{ times}}\rangle$$

we can write

$$a_1|0\rangle = a_2|0\rangle = \ldots = a_p|0\rangle = 0 \tag{XII.67}$$

$$|n_1 \ldots n_p\rangle = (n_1! \ldots n_p!)^{-\frac{1}{2}} a_1^{\dagger n_1} \ldots a_p^{\dagger n_p}|0\rangle. \tag{XII.68}$$

The observables

$$N_i \equiv a_i^\dagger a_i \qquad (i = 1, 2, \ldots, p) \tag{XII.69}$$

each have a spectrum consisting of the sequence of non-negative integers; they are interpreted respectively as the number of quanta of the type 1, 2, ..., p. Their sum

$$N \equiv \sum_{i=1}^{p} N_i,$$

[1]) In fact, there are as many distinct sets $(n_1, n_2, \ldots, n_p)$ of p integral numbers $\geqslant 0$ with sum equal to n as there are *distinguishable arrangements of putting n indistinguishable objects into p cells.*

is the total number of quanta. One has

$$\mathscr{H} = (N + \tfrac{1}{2}p)\hbar\omega.$$

It is clear that $N_1, N_2, \ldots, N_p$ form a complete set of commuting observables and that their basis is precisely the basis of $\mathscr{H}$ which we have just formed.

The N_i are evidently not the only constants of the motion forming a complete set. Any operator of the form $a_i a_j^\dagger$ commutes with $\mathscr{H}$; by linear combination of operators of this type, and of their adjoints, one can form p^2 independent Hermitean operators in all. Among the functions of these p^2 constants of the motion there exist several complete sets of commuting observables. We shall illustrate this point in the two special cases $p=2$ and $p=3$.

14. Two-Dimensional Isotropic Oscillator

Here we have the two-dimensional system with Hamiltonian

$$\mathscr{H} = \frac{1}{2m}(p_1^2 + m^2\omega^2 q_1^2) + \frac{1}{2m}(p_2^2 + m^2\omega^2 q_2^2).$$

The study of the preceding section applies to this special case. The following table gives the eigenvalues of $\mathscr{H}$ (first column) and the set of eigenvectors common to N_1 and N_2 which span their respective subspaces (third column):

$$\begin{array}{lll} \hbar\omega & 1 & |00\rangle \\ 2\hbar\omega & 2 & |10\rangle,\ |01\rangle \\ 3\hbar\omega & 3 & |20\rangle,\ |11\rangle,\ |02\rangle \\ \ldots & \ldots & \ldots \\ (n+1)\hbar\omega & n+1 & |n\ 0\rangle,\ |n-1\ 1\rangle,\ \ldots,\ |n-s\ s\rangle,\ \ldots\ |0\ n\rangle. \\ \ldots & \ldots & \ldots \end{array} \tag{XII.70}$$

The angular momentum operator L, defined by

$$L \equiv \frac{1}{\hbar}(q_1 p_2 - q_2 p_1) = \mathrm{i}(a_1 a_2^\dagger - a_1^\dagger a_2), \tag{XII.71}$$

is a constant of the motion. We shall show that N and L form another complete set of commuting observables. To this effect we introduce the operators

$$\begin{aligned} A_\pm &= \tfrac{1}{2}\sqrt{2}\,(a_1 \mp \mathrm{i}a_2) \\ A_\pm^\dagger &= \tfrac{1}{2}\sqrt{2}\,(a_1^\dagger \pm \mathrm{i}a_2^\dagger). \end{aligned} \tag{XII.72}$$

These operators satisfy commutation relations identical to the relations (XII.66) between the a and the $a^\dagger$:

$$\begin{aligned} [A_r, A_s] &= [A_r^\dagger, A_s^\dagger] = 0 \\ [A_r, A_s^\dagger] &= \delta_{rs} \end{aligned} \qquad (r = + \text{ or } - ;\ s = + \text{ or } -). \qquad \text{(XII.73)}$$

A_+ and $A_+^\dagger$ can thus be interpreted as destruction and creation operators of quanta of type $+$, A_- and $A_-^\dagger$ as destruction and creation, operators of quanta of type $-$; following this interpretation, the operators

$$N_+ \equiv A_+^\dagger A_+ \quad \text{and} \quad N_- \equiv A_-^\dagger A_- \qquad \text{(XII.74)}$$

represent the numbers of "+ quanta" and of "− quanta", respectively. Since the commutation relations (XII.73) are identical with relations (XII.66), the problem of forming the eigenvectors common to N_+ and N_- is mathematically identical to the problem of forming the eigenvectors common to N_1 and N_2. Therefore, N_+ and N_- each have as their spectrum the sequence of non-negative integers

$$n_+ = 0, 1, 2, \ldots \qquad n_- = 0, 1, 2, \ldots$$

and these two observables form a complete set of commuting observables: to each pair of quantum numbers (n_+, n_-) corresponds a single common eigenvector (to within a constant). In fact, the relations (XII.67) imply

$$A_+|00\rangle = A_-|00\rangle = 0. \qquad \text{(XII.75)}$$

Thus the vector $|00\rangle$ of Table (XII.70) is an eigenvector of the ground state $(n_+ = n_- = 0)$. The vectors

$$|n_+ n_-\rangle \equiv (n_+! n_-!)^{-\frac{1}{2}} A_+^{\dagger n_+} A_-^{\dagger n_-} |00\rangle \qquad \text{(XII.76)}$$

orm a complete orthonormal eigenset common to N_+ and N_-:

$$\begin{aligned} N_+|n_+ n_-\rangle &= n_+|n_+ n_-\rangle, \\ N_-|n_+ n_-\rangle &= n_-|n_+ n_-\rangle. \end{aligned}$$

Now, if one expresses N and L as functions of the A and $A^\dagger$, one has after some calculation

$$\begin{aligned} N &= N_+ + N_-, \\ L &= N_+ - N_-. \end{aligned}$$

Since the observables N_+ and N_- form a complete set of commuting observables, their sum N and their difference L has the same property. This is what we wanted to show.

In conclusion, we have at our disposal another complete orthonormal set of eigenvectors of $\mathscr{H}$, namely the set of vectors $|n_+n_-\rangle$. They satisfy the eigenvalue equations

$$\mathscr{H}|n_+n_-\rangle = (n_+ + n_- + 1)\hbar\omega|n_+n_-\rangle \tag{XII.77}$$

$$L|n_+n_-\rangle = (n_+ - n_-)|n_+n_-\rangle. \tag{XII.78}$$

Let us examine the commutation relations of L with the A and $A^\dagger$. A simple calculation yields

$$[L, A_\pm{}^\dagger] = \pm A_\pm{}^\dagger \tag{XII.79}$$

$$[L, A_\pm] = \mp A_\pm. \tag{XII.80}$$

Consequently, when they act upon an eigenvector of L, $A_+{}^\dagger$ and A_- increase L by one unit, $A_-{}^\dagger$ and A_+ decrease it by one unit. This may be interpreted in various ways. In the quantum theory of charged fields where the field appears as a set of two-dimensional, isotropic oscillators, N_+ is the number of particles with positive charge, N_- that of particles of negative charge, and L the total charge (to within a constant). Following this interpretation $A_+{}^\dagger$ creates a positive charge, A_- destroys a negative charge; both therefore increase the charge by one unit; in similar fashion $A_-{}^\dagger$ and A_+ decrease the charge by one unit.

In the theory of crystalline vibrations, the motions of the lattice are likewise represented by a set of isotropic two-dimensional oscillators; the quanta of oscillation are called *phonons*. The representation in terms of phonons of type 1 and 2 corresponds to standing waves; the representation in terms of phonons of type + and − corresponds to traveling waves "propagating" in one or the other direction. In the problems of scattering (neutrons, X rays, etc.) by a crystal lattice, the representation by traveling waves is the most suitable for purposes of calculation.

15. Three-Dimensional Isotropic Oscillator

The three-dimensional isotropic harmonic oscillator is a particle located in a central potential proportional to the square of the distance from the center. Its Hamiltonian is

$$\mathscr{H} = \frac{\boldsymbol{p}^2}{2m} + \tfrac{1}{2}m\omega^2\boldsymbol{r}^2. \tag{XII.81}$$

$\mathscr{H}$ is the sum of three terms:

$$\mathscr{H} = \mathscr{H}_x + \mathscr{H}_y + \mathscr{H}_z$$
$$\mathscr{H}_i = \frac{1}{2m}(p_i^2 + m^2\omega^2 r_i^2). \qquad (i = x, y \text{ or } z) \qquad \text{(XII.82)}$$

According to our study of Ch. XII, § 13, the eigenvalues of $\mathscr{H}$ are given by the formula

$$(n+\tfrac{3}{2})\hbar\omega \qquad (n = 0, 1, 2, \ldots, \infty) \qquad \text{(XII.83}$$

and are $\frac{1}{2}(n+1)(n+2)$-fold degenerate. The observables N_x, N_y, N_z form a complete set of constants of the motion and the eigenvectors $|n_x n_y n_z\rangle$ of their basis are labelled by the three corresponding eigenvalues n_x, n_y, and n_z. These vectors are deduced by the formula

$$|n_x n_y n_z\rangle = (n_x! n_y! n_z!)^{-\frac{1}{2}} a_x^{\dagger n_x} a_y^{\dagger n_y} a_z^{\dagger n_z} |000\rangle \qquad \text{(XII.84)}$$

from the ground-state vector $|000\rangle$, itself defined (to within a constant) by the three equations

$$a_x|000\rangle = a_y|000\rangle = a_z|000\rangle = 0. \qquad \text{(XII.85)}$$

Let us introduce the angular momentum

$$\boldsymbol{l} \equiv \boldsymbol{r} \times \boldsymbol{p}.$$

According to the well-known property of the Hamiltonian for a central potential (Ch. IX), $\mathscr{H}$, $\boldsymbol{l}^2$ and l_z likewise constitute a complete set of commuting observables. The eigenvectors $|nlm\rangle$ common to these three observables are labelled by the three quantum numbers n, l, m, and the corresponding eigenvalues of $\mathscr{H}$, $\boldsymbol{l}^2$ and l_z are respectively $(n+\frac{3}{2})\hbar\omega$, $l(l+1)\hbar^2$ and $m\hbar$. The vectors $|nlm\rangle$ form a complete orthonormal set of eigenvectors of $\mathscr{H}$. They are derived from the vectors $|n_x n_y n_z\rangle$ by a unitary transformation. The explicit construction of these vectors will not be given here [1]). We shall merely determine

[1]) In the $\{\boldsymbol{r}\}$ representation, $|n_x n_y n_z\rangle$ is represented by the wave function $\psi_{n_x}(x)\, \psi_{n_y}(y)\, \psi_{n_z}(z)$; $|nlm\rangle$ is represented by the function

$$\Psi_{nlm}(\boldsymbol{r}) \equiv \frac{y_{nl}(r)}{r} Y_l^m(\theta, \varphi),$$

$y_{nl}(r)$ is the solution which vanishes at the origin — and is regular at infinity — of the differential equation

$$\left[-\frac{\hbar^2}{2m^2}\frac{\mathrm{d}^2}{\mathrm{d}r^2} + \frac{l(l+1)\hbar^2}{2mr^2} + \tfrac{1}{2}m\omega^2 r^2\right] y_{nl} = (n+\tfrac{1}{2})\hbar\omega y_{nl}.$$

the values which the quantum numbers l and m may take when one fixes n; in other words, we try to find the different possible states of the angular momentum corresponding to each energy level.

The great similarity between the operator l_z considered here and the operator L of the preceding section suggests an analogous change of variable. Let us introduce the operators A_m $(m=1, 0, -1)$ defined by

$$\begin{aligned} A_1 &= \tfrac{1}{2}\sqrt{2}[a_x - \mathrm{i}a_y] \\ A_0 &= a_z \\ A_{-1} &= \tfrac{1}{2}\sqrt{2}[a_x + \mathrm{i}a_y] \end{aligned} \tag{XII.86}$$

and the Hermitean conjugate operators $A_m{}^\dagger$. The A_m and $A_m{}^\dagger$ satisfy commutation relations analogous to relations (XII.73) and can be interpreted respectively as destruction and creation operators of quanta of the type m. The number of quanta of type m is represented by the operator $N_m = A_m{}^\dagger A_m$. Clearly, N_1, N_0, and N_{-1} form a complete set of commuting observables, and

$$\begin{aligned} \mathscr{H} &= (N_1 + N_0 + N_{-1} + \tfrac{3}{2})\hbar\omega, \\ N &= N_1 + N_0 + N_{-1}. \end{aligned}$$

To each triplet of eigenvalues (n_1, n_0, n_{-1}) there corresponds an eigenvector common to these three observables, namely the vector

$$|n_1 n_0 n_{-1}\rangle = (n_1!\, n_0!\, n_{-1}!)^{-\frac{1}{2}} A_1{}^{\dagger n_1} A_0{}^{\dagger n_0} A_{-1}^{\dagger n_{-1}} |000\rangle.$$

The ensemble of these vectors forms a complete set of eigenvectors of $\mathscr{H}$. Indeed,

$$\begin{aligned} \mathscr{H}|n_1 n_0 n_{-1}\rangle &= (n + \tfrac{3}{2})\hbar\omega |n_1 n_0 n_{-1}\rangle, \\ n &= n_1 + n_0 + n_{-1}. \end{aligned}$$

The vectors we have just formed are not, in general, eigenvectors of $\boldsymbol{l}^2$, but they are eigenvectors of l_z, because

$$l_z = (N_1 - N_{-1})\,\hbar, \tag{XII.87}$$

and consequently

$$m = n_1 - n_{-1}. \tag{XII.88}$$

Consider the subspace of the eigenvectors of $\mathscr{H}$ corresponding to the eigenvalue $(n+\frac{3}{2})\hbar\omega$. The $\frac{1}{2}(n+1)(n+2)$ vectors $|n_1 n_0 n_{-1}\rangle$ which span it $(n_1+n_0+n_{-1}=n)$ form a complete orthonormal set of eigenvectors of l_z. According to equation (XII.88), the quantum number m may take on all integral values from $-n$ to $+n$. It is easy to determine the number c_m of linearly independent vectors corresponding to each of these values of m; the result is given by the following table:

$$\begin{array}{lcccccc} |m| = & n & n-1 & n-2 \ldots & n-2s & n-(2s+1) & n-(2s+2) \ldots \\ c_m = & 1 & 1 & 2 \ldots & s+1 & s+1 & s+2 \ldots \end{array} \qquad \text{(XII.89)}$$

Now according to the properties of angular momentum, to each eigenvalue of $\boldsymbol{l}^2$, i.e. to each value of l, corresponds a certain number of sets of $(2l+1)$ vectors of well-defined angular momentum (lm), m in each set taking the $(2l+1)$ integral values contained between $-l$ and $+l$. Let d_l be that number. It is obvious that

$$c_m = \sum_{l \geqslant m} d_l,$$

hence that

$$d_l = c_l - c_{l+1}.$$

Referring to Table XII.89, we see that $d_l = 1$ for $l = n, n-2, \ldots, n-2s, \ldots$, that is to say, for all integral values of l of parity $(-)^n$ contained between 0 and n (limits included), and that $d_l = 0$ for all other values of l.

In conclusion, to each eigenvalue $(n+\frac{3}{2})\hbar\omega$ of the energy correspond $\frac{1}{2}(n+1)(n+2)$ states of well-defined angular momentum (lm). For

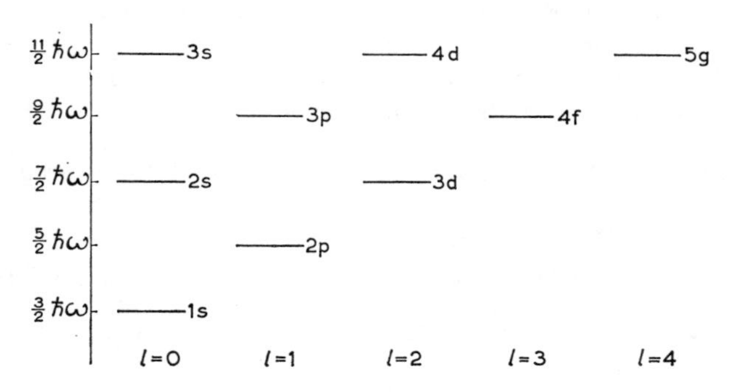

Fig. XII.1. Spectrum of the three-dimensional harmonic oscillator.

each of the possible values of l, there exist $(2l+1)$ eigenstates corresponding, respectively, to the $(2l+1)$ values of m, ranging from $-l$ to $+l$. The values which l may take are

$$n,\ n-2,\ \ldots,\ 0 \quad \text{if} \quad (-)^n = 1 \qquad [\tfrac{1}{2}(n+2) \text{ distinct values}]$$
$$n,\ n-2,\ \ldots,\ 1 \quad \text{if} \quad (-)^n = -1 \qquad [\tfrac{1}{2}(n+1) \text{ distinct values}].$$

The spectroscopic diagram of Fig. XII.1 represents the ground state and first few excited states of the three-dimensional isotropic harmonic oscillator. It is instructive to compare this diagram with the one of the hydrogen atom (Fig. XI.1).

EXERCISES AND PROBLEMS

1. Let a and $a^\dagger$ be two Hermitean conjugate operators such that $[a, a^\dagger] = 1$. We put $N = a^\dagger a$. Show that

(*i*) $[N, a^p] = -pa^p$; $[N, a^{\dagger p}] = +pa^{\dagger p}$ (p integer > 0);

(*ii*) the only algebraic functions of a and $a^\dagger$ which commute with N are the functions of N.

2. Show that the operators a and $a^\dagger$ of Problem XII.1 have no inverse.

3. Form the matrices representing the operators q and p in the $\{N\}$ representation (notation of Ch. XII, § 5). Verify that they are Hermitean and satisfy the commutation relation (XII.2). Set up the eigenvalue problem of q in this representation; show that the spectrum of q is, nondegenerate continuous and extends from $-\infty$ to $+\infty$. Form explicitly the eigenvector corresponding to the eigenvalue 0.

4. One wishes to compare the properties of a quantized oscillator in the state $|n\rangle$ to those of a microcanonical ensemble of classical oscillators of the same energy (Ch. XII, § 10). Show that the statistical distribution of the variable q in that quantum state exhibits oscillations which are tighter the larger n; show also that in the limit where $n \to \infty$ its average value over several oscillations tends toward the corresponding distribution of the ensemble of classical oscillators (use the WKB method).

5. Let χ_0, ϖ_0, η_0 be the respective initial values of the average values

$$\chi = \langle q^2\rangle - \langle q\rangle^2,\ \varpi = \langle p^2\rangle - \langle p\rangle^2,\ \eta = \langle pq+qp\rangle - 2\langle p\rangle\,\langle q\rangle$$

relating to a wave packet of an harmonic oscillator. Establish the law of evolution of these mean values as function of time. Show that they are functions of the form $A + B\cos 2\omega t + C\sin 2\omega t$, and that χ and ϖ remain constant if, and only if

$$\eta_0 = 0, \qquad \varpi_0 = m^2\omega^2\chi_0.$$

6. The state of a harmonic oscillator is represented at time zero by the minimum wave packet

$$f(q) = (2\pi\sigma)^{-\frac{1}{4}} \exp\left[\frac{\mathrm{i}}{\hbar}\langle p\rangle q - \frac{(q-\langle q\rangle)^2}{4\sigma}\right].$$

Show that this packet remains minimum in the course of time if and only if $\sigma = \hbar/2m\omega$ (cf. Problem XII.5). We assume henceforth that this condition is fulfilled. Show that then $f(q)$ is the wave function representing the vector

$$|f\rangle = \exp\left(\frac{\mathrm{i}}{\hbar}\langle p\rangle q\right) \exp\left(-\frac{\mathrm{i}}{\hbar}\langle q\rangle p\right) |0\rangle.$$

Deduce from this [making use of the identity (XII.29)] that the function $f(q, t)$ is equal, to within a phase factor, to the expression obtained by substituting in $f(q)$ the mean values $\langle q\rangle$, $\langle p\rangle$ at time t for their values at time 0. Determine the coefficients c_n of the expansion of $|f\rangle$ in a series of eigenvectors of the Hamiltonian, and show that

$$|c_n|^2 = \mathrm{e}^{-\alpha}\frac{\alpha^n}{n!}$$

$$[\alpha = E_{\mathrm{cl.}}/\hbar\omega;\; E_{\mathrm{cl.}} = \frac{1}{2m}(\langle p\rangle^2 + m^2\omega^2\langle q\rangle^2)].$$

7. Verify that Bloch's theorem (Ch. XII, § 12) also applies to the classical harmonic oscillator and that the statistical distribution of $\alpha q + \beta p$ for a quantized harmonic oscillator in thermodynamic equilibrium approaches the classical distribution in the limit where $kT \gg \hbar\omega$.

8. Show that the Hamiltonian of a particle of mass m and charge e in a constant magnetic field $\mathscr{H}$ directed along Oz can be written in the form

$$H = \frac{p_z{}^2}{2m} + H_\varrho,$$

with

$$H_\varrho = \frac{1}{2m}(p_x{}^2 + p_y{}^2) - \frac{e}{2mc}\mathscr{H} l_z + \frac{e^2}{8mc^2}\mathscr{H}^2(x^2+y^2).$$

Show that the operators p_z, l_z, H_ϱ form a complete set of commuting constants of the motion, and that their common eigenfunctions are written in cylindrical coordinates (z,ϱ,θ) in the form $\exp(\mathrm{i}kz)\exp(\mathrm{i}\lambda\theta)\, v_{\lambda n}(\varrho)$ [k is any real number; $\lambda = 0, \pm 1, \pm 2, \ldots, \pm\infty$; $n = 0, 1, 2, \ldots, \infty$]. The corresponding eigenvalues are, respectively,

$$\hbar k,\; \hbar\lambda,\; (2n+1)\frac{e\hbar}{2mc}\mathscr{H}.$$

Compare these results with those of Problem II.4.

APPENDIX A

DISTRIBUTIONS, δ-"FUNCTIONS" AND FOURIER TRANSFORMATION

PLAN OF THE APPENDIX

I. ELEMENTS OF DISTRIBUTION THEORY [1])

1. Concept of Functional and Rigorous Treatment of the Continuous Spectrum

The δ-"function" of Dirac which enables us to treat the continuous spectrum in a manner analogous to the discrete spectrum is not a well-defined mathematical object. If one wishes to introduce into the theory in a rigorous way observables possessing a continuous spectrum, one must set up the eigenvalue problem in a different manner.

In fact, the eigenfunctions of the observables of Wave Mechanics enter only through their scalar product with the wave functions, that is to say by their scalar product with square-integrable functions. Let F be one of these eigenfunctions, and ψ an arbitrarily chosen wave function; the scalar product $\langle\psi, F\rangle$ (notation of Ch. V) may be regarded as an antilinear functional of ψ, or better, as a linear functional of ψ^*. Let us denote the latter by $\hat{F}$; by definition

$$\hat{F}[\psi] = \langle\psi^*, F\rangle.$$

It is the functionals thus associated with each eigenfunction which enter into the theory, and not the eigenfunctions themselves.

Now these functionals belong to a certain class of functionals, called *distributions*, for which one can define essentially the same operations of algebra and analysis as for the functions. Therefore, it is possible to reformulate Wave Mechanics in rigorous fashion by defining the operators of the theory as *operators acting upon the distributions* and not upon the functions; *the eigensolutions of a Hermitean operator are then particular distributions*: they are *the linear and continuous functionals of bounded, square-integrable functions which satisfy the eigenvalue equation of that operator.*

More generally, let X and Ξ be two observables whose spectra are assumed, for simplicity, to be entirely continuous and non-degenerate; let $\langle\xi|x\rangle$ be the unitary transformation matrix permitting the passage from the $\{X\}$ representation to the $\{\Xi\}$ representation. In a rigorous formulation of Quantum Theory, $\langle\xi|x\rangle$ simultaneously represents:

(i) the ensemble of eigensolutions of X in the $\{\Xi\}$ representation,

---

[1]) Cf. L. Schwartz, *Théorie des distributions* (Paris, Hermann, 1950–1951); see also by the same author, *Les Méthodes Mathématiques de la Physique*, Sorbonne Lectures (Paris, 1955). I. Halperin, *Introduction to the Theory of Distributions*, based on the lectures given by L. Schwartz (Toronto, University of Toronto Press, 1952).

i.e. a certain set of functionals of functions of ξ labelled by the index x;

(*ii*) the ensemble of eigensolutions of Ξ in the $\{X\}$ representation, i.e. a certain set of functionals of functions of x labelled by the index ξ.

In this section, we give the definition of the distributions and state their main properties without proof.

2. Definition of Distributions

Let us denote by $\varphi(x_1, \ldots, x_n)$, or more simply by $\varphi(x)$, a function of n continuous variables $x_1, \ldots, x_n$ whose non-zero values are all contained in a finite domain of these variables, and which has derivatives to all orders with respect to these variables (indefinitely differentiable functions with bounded support).

By definition, *a distribution* $T[\varphi]$ *is a linear and continuous functional of the functions* φ.

Linearity means that for all linear combinations $\lambda_1\varphi_1 + \lambda_2\varphi_2$, one has

$$T[\lambda_1\varphi_1 + \lambda_2\varphi_2] = \lambda_1 T[\varphi_1] + \lambda_2 T[\varphi_2].$$

Continuity means that for any sequence $\varphi_1, \varphi_2, \ldots, \varphi_j, \ldots$ of functions φ such that $\lim_{j\to\infty} \varphi_j = \varphi$, one has

$$\lim_{j\to\infty} T[\varphi_j] = T[\varphi].$$

To any *locally integrable* function f – that is to say any function whose integral [1]) over any finite interval exists – corresponds a distribution $\hat{f}$ defined by the scalar product

$$\hat{f}[\varphi] = \int f(x)\, \varphi(x)\, \mathrm{d}x = \langle \varphi^*, f \rangle. \tag{A.1}$$

Two locally integrable functions define the same distribution if they are equal almost everywhere (i.e. everywhere except on a set of measure zero). In particular, the wave functions of Wave Mechanics (square-integrable functions) define distributions.

[1]) The integrals with which one is dealing throughout this theory are integrals in the sense of Lebesgue. The Lebesgue integral reduces to the integral in the usual sense (Riemann integral) whenever the latter has a meaning; however, the Lebesgue integral exists in cases where the Riemann integral is not defined.

With the function $1/x$ no distribution is associated since this function is not integrable at the point $x=0$. But one can define the distribution

$$\mathrm{PP}\,\frac{1}{x}\,[\varphi] \equiv \mathrm{PP}\int \frac{\varphi(x)}{x}\,\mathrm{d}x \tag{A.2}$$

where PP denotes the Cauchy principal part of the integral

$$\mathrm{PP}\int_{-\infty}^{+\infty} = \lim_{\varepsilon\to 0}\left\{\int_{-\infty}^{-\varepsilon} + \int_{\varepsilon}^{+\infty}\right\}.$$

The "Dirac function" $\delta(x)$ defines the distribution

$$\delta[\varphi] = \varphi(0). \tag{A.3}$$

Likewise, the "function" $\delta(x-x_0)$ defines the distribution

$$\delta_{x_0}[\varphi] = \varphi(x_0). \tag{A.4}$$

Remark. A distribution can eventually be defined over a larger function space than the φ-space. Indeed, if $U[\psi]$ is a linear and continuous functional of the functions ψ of a larger function space than the φ-space, the functional $U[\varphi]$ is well defined and is linear and continuous over φ-space: U is a distribution.

Examples:

δ_{x_0} is defined over the space of the functions $\alpha(x)$ continuous at $x=x_0$:

$$\delta_{x_0}[\alpha] = \alpha(x_0).$$

$\hat{\Psi}$, a distribution corresponding to a square-integrable function, is defined in the space of the square-integrable functions $\psi(x)$:

$$\hat{\Psi}[\psi] = \int \Psi\psi\,\mathrm{d}x = \langle\psi^*, \Psi\rangle.$$

The linear and continuous functionals of the wave functions of Wave Mechanics are particular distributions.

3. Linear Combination of Distributions

$T=\lambda_1 T_1+\lambda_2 T_2$ is a distribution defined by

$$T[\varphi] = \lambda_1 T_1[\varphi]+\lambda_2 T_2[\varphi]$$

(λ_1, λ_2 are given complex constants).

4. Product of Two Distributions

If $\hat{f}$ is the distribution associated with a locally integrable function f, and T an arbitrary distribution, the distribution

$$P = \hat{f} T$$

is well defined if T is a linear, continuous functional of the functions $f\varphi$ and one has, by definition

$$P[\varphi] = T[f\varphi]. \tag{A.5}$$

The product of two distributious *does not always exist.* If f has derivatives of all orders, $\hat{f}T$ exists for any T. If f is continuous at the point x_0,

$$(\hat{f}\delta_{x_0})[\varphi] = f(x_0)\,\varphi(x_0). \tag{A.6}$$

If f and g are square-integrable functions, the product $\hat{f}\,\hat{g}$ is well defined. On the other hand, $[\delta(x)]^2$ has no meaning whatsoever, and neither does $(1/\sqrt{|x|})^2$.

As a special case of eq. (A.6) one has the relation

$$x\delta(x) = 0. \tag{A.7}$$

Conversely, if $xT = 0$, T is a multiple of $\delta: T = c\delta$ (c = const.).

Therefore, if $f(x)$ and $g(x)$ are connected by the relation

$$xf(x) = g(x),$$

one necessarily has

$$f(x) = \mathrm{PP}\,\frac{g(x)}{x} + c\delta(x), \tag{A.8}$$

where c is a constant to be determined.

5. Series and Integrals of Distributions

If a set of distributions $T_1, T_2, \ldots, T_j, \ldots$ is such that when $j \to \infty$, $T_j[\varphi]$ has a limit for any φ, this limit is a distribution (i.e. a linear continuous functional of the functions φ):

$$T = \lim_{j\to\infty} T_j.$$

Equivalent statement: if the infinite series $\sum_i T_i[\varphi]$ is summable for any φ, its sum defines a distribution; one says that the *series* of distributions $\sum_i T_i$ is *summable.*

If $T(\lambda)$ is a distribution depending on a parameter λ which can vary continuously in a domain Λ, and if the integral

$$I[\varphi] = \int_{\Lambda} T(\lambda)[\varphi] \, \mathrm{d}\lambda$$

converges for any φ, it defines a distribution

$$I = \int_{\Lambda} T(\lambda) \, \mathrm{d}\lambda.$$

An analogous definition holds for multiple integrals.

In particular, if $f(x, \lambda)$ is an integrable function of x (locally) and of λ, the distribution $\hat{f}(\lambda)$ is integrable in λ, and its integral is the distribution $\hat{g}$ associated with the function

$$g(x) = \int f(x, \lambda) \, \mathrm{d}\lambda.$$

If the function $a(k)$ remains smaller than a positive power of $|k|$ when $|k| \to \infty$:

$$|a(k)| \leqslant A|k|^{\alpha} \quad (A \text{ and } \alpha, \text{ positive constants})$$

the integral $\int_{-\infty}^{+\infty} \mathrm{e}^{ikx} a(k) \, \mathrm{d}k$ is a distribution.

In particular

$$\int_{-\infty}^{+\infty} \mathrm{e}^{ikx} \, \mathrm{d}k = 2\pi\delta.$$

6. Derivative of Distributions

By definition, the derivative $\partial T/\partial x_i$ of the distribution T is

$$\frac{\partial T}{\partial x_i}[\varphi] = -T\left[\frac{\partial \varphi}{\partial x_i}\right]. \tag{A.9}$$

In particular, if a locally integrable function is differentiable, the derivative of the corresponding distribution is the distribution corresponding to its derivative. Indeed, upon integrating by parts,

$$\hat{f}'[\varphi] = \int f'(x) \, \varphi(x) \, \mathrm{d}x = -\int f(x) \, \varphi'(x) \, \mathrm{d}x = -\hat{f}[\varphi'].$$

All the properties of the derivatives of the functions apply to the distributions. For instance, the derivative of the product $P = \hat{f}T$ is

$$P' = \hat{f}'T + \hat{f}T'. \tag{A.10}$$

Moreover, certain results pertaining to a more or less restricted class of functions apply to *all* distributions without restriction. They are as follows:

1) *The distributions are differentiable to all orders.*

In particular, the locally summable functions

$$\log |x|, \; 1/r \; (r = \sqrt{x^2+y^2+z^2})$$

are differentiable to all orders, since they are distributions:

$$\frac{\mathrm{d}}{\mathrm{d}x} \log |x| = \mathrm{PP} \frac{1}{x}. \tag{A.11}$$

$$\triangle\left(\frac{1}{r}\right) = -4\pi\delta \qquad [\delta \equiv \delta(x)\,\delta(y)\,\delta(z)]. \tag{A.12}$$

2) *Differentiation is a linear, continuous operation in the space of the distributions:*

If $\qquad \lim_{j\to\infty} T_j = T, \qquad \lim_{j\to\infty} T_j' = T'.$

Hence, if a series is summable, it is differentiable term by term under the summation sign $\sum$. Likewise, if $T(\lambda)$ is summable with respect to the parameter λ:

$$I = \int_{\Lambda} T(\lambda)\,\mathrm{d}\lambda,$$

$\partial T(\lambda)/\partial x_i$ is certainly summable in the same domain of λ and its integral is equal to $\partial I/\partial x_i$.

II. PROPERTIES OF THE δ-"FUNCTION"

7. Definition of δ(x)

It is customary in physics to use the notation $\delta(x-x_0)$ rather than the more correct notation $\delta_{x_0}[\varphi]$. This notation proves to be quite convenient in practice. $\delta(x-x_0)$ appears as a function whose manipulation is governed by somewhat peculiar rules. Distribution Theory, which provides the mathematical justification for these rules, need not be mentioned explicitly.

By definition, $f(x)$ being a well-defined function at the point $x=x_0$,

$$\int f(x)\,\delta(x-x_0)\,\mathrm{d}x \equiv \delta_{x_0}[f(x)] = f(x_0). \tag{A.13}$$

One therefore has formally

$$\delta(x-x_0) = \begin{cases} 0 & \text{if } x \neq x_0 \\ +\infty & \text{if } x = x_0 \end{cases} \quad \text{and} \quad \int_{-\infty}^{+\infty} \delta(x-x_0)\,\mathrm{d}x = 1 \tag{A.14}$$

$\delta(x-x_0)$ is a generalization of the Kronecker symbol

$$\delta_{mn} = \begin{cases} 0 & \text{if} \quad m \neq n \\ 1 & \text{if} \quad m = n. \end{cases}$$

8. Representation as the Limit of a Kernel of an Integral Operator

$\delta(x-x_0)$ may be considered as the limit of a function which exhibits a very sharp peak about x_0, and whose integral over all space remains constant and equal to 1. For instance:

$$\delta(x-x_0) = \frac{1}{\pi} \lim_{L\to\infty} \sin \frac{L(x-x_0)}{x-x_0} \tag{A.15a}$$

$$= \frac{1}{\pi} \lim_{\varkappa\to\infty} \frac{1-\cos \varkappa(x-x_0)}{\varkappa(x-x_0)^2} \tag{A.15b}$$

$$= \frac{1}{\pi} \lim_{\varepsilon\to+0} \frac{\varepsilon}{(x-x_0)^2+\varepsilon^2} \tag{A.15c}$$

$$= \lim_{\eta\to 0} \frac{E(x-x_0+\eta)-E(x-x_0)}{\eta}. \tag{A.15d}$$

In the last expression, $E(x)$ is the Heaviside function:

$$E(x) = \begin{cases} 1 & \text{if} \quad x > 0 \\ 0 & \text{if} \quad x < 0. \end{cases}$$

(The distribution δ is the derivative of the Heaviside distribution.)

With the property (A.15c) goes the limiting property:

$$\lim_{\varepsilon\to+0} \frac{1}{x-x_0 \pm i\varepsilon} = \text{PP}\,\frac{1}{x-x_0} \mp i\pi\delta(x-x_0). \tag{A.15e}$$

9. Principal Properties

The main properties of the function $\delta(x)$ are the following:

$$\delta(x) = \delta(-x) \tag{A.16}$$

$$\delta(ax) = \frac{1}{|a|}\,\delta(x) \qquad (a \neq 0) \tag{A.17}$$

$$\delta[g(x)] = \sum_n \frac{1}{|g'(x_n)|}\,\delta(x-x_n) \qquad [g(x_n)=0,\; g'(x_n)\neq 0] \tag{A.18}$$

$$x\delta(x) = 0 \tag{A.19}$$

$$f(x)\,\delta(x-a) = f(a)\,\delta(x-a) \tag{A.20}$$

$$\int \delta(x-y)\,\delta(y-a)\,\mathrm{d}y = \delta(x-a) \tag{A.21}$$

$$\delta(x) = \frac{1}{2\pi}\int_{-\infty}^{+\infty} \mathrm{e}^{ikx}\,\mathrm{d}k. \tag{A.22}$$

All these equalities state that one side can be replaced by the other when it is multiplied by a regular function and integrated over x. One can prove them all rigorously by means of Distribution Theory (cf. Sec. 1). One can also prove them formally (but not rigorously) by showing that the integrals of the product of each term and $f(x)$ are equal for any sufficiently regular function $f(x)$. Thus relations (A.16), (A.17) and (A.18) are proved by performing a suitable change of variable in these integrals. In expression (A.18), the summation must be taken over all zeros of $g(x)$; the expression has a meaning only if $g(x)$ and $g'(x)$ never vanish simultaneously; for example $\delta(x^2)$ is meaningless.

10. Derivatives of $\delta(x)$

The "function" $\delta(x)$ is differentiable to all orders. Its mth derivative $\delta^{(m)}(x)$ is defined by the property

$$\int_{-\infty}^{+\infty} \delta^{(m)}(x)\,f(x)\,\mathrm{d}x = (-)^m\,f^{(m)}(0), \tag{A.23}$$

valid for any function $f(x)$ which is m times differentiable at the point $x=0$. $\delta^{(m)}(x-x_0)$ can be considered as the limit of one or the other of the mth derivatives of the functions occurring on the right-hand side of eqs. (A.15a), (A.15b) and (A.15c). The following properties, which one can deduce formally (but incorrectly) by the usual procedures of the integral calculus, can be proved rigorously by means of Distribution Theory:

$$\delta^{(m)}(x) = (-)^m\,\delta^{(m)}(-x) \tag{A.24}$$

$$\int \delta^{(m)}(x-y)\,\delta^{(n)}(y-a)\,\mathrm{d}y = \delta^{(m+n)}\,(x-a) \tag{A.25}$$

$$x^{m+1}\,\delta^{(m)}(x) = 0. \tag{A.26}$$

In particular, the first derivative $\delta'(x)$ has the properties

$$\int_{-\infty}^{+\infty} \delta'(x)\,f(x)\,\mathrm{d}x = -\,f'(0) \tag{A.27}$$

$$\delta'(x) = -\,\delta'(-x) \tag{A.28}$$

$$\int \delta'(x-y)\,\delta(y-a)\,\mathrm{d}y = \delta'(x-a) \tag{A.29}$$

$$x\delta'(x) = -\,\delta(x) \tag{A.30}$$

$$x^2\,\delta'(x) = 0 \tag{A.31}$$

$$\delta'(x) = \frac{\mathrm{i}}{2\pi}\int_{-\infty}^{+\infty} k\,\mathrm{e}^{\mathrm{i}kx}\,\mathrm{d}k. \tag{A.32}$$

III. THE FOURIER TRANSFORMATION[1]

11. Fourier Transform of a Function. Definition

If $f(x)$ is a (real or complex) function of the variable x, its Fourier transform, if it exists, is the function

$$F(u) \equiv \mathscr{F}[f] = \left(\frac{\alpha}{2\pi}\right)^{\frac{1}{2}}\int_{-\infty}^{+\infty} \mathrm{e}^{-\mathrm{i}\alpha ux}\,f(x)\,\mathrm{d}x \tag{A.33}$$

where α is a constant fixed once and for all (in Wave Mechanics, one takes $\alpha = 1/\hbar$). Provided certain convergence conditions are fulfilled, $f(x)$ is deduced from $F(u)$ by the inverse Fourier transformation

$$f(x) = \mathscr{F}^\dagger[F] = \left(\frac{\alpha}{2\pi}\right)^{\frac{1}{2}}\int_{-\infty}^{+\infty} \mathrm{e}^{\mathrm{i}\alpha ux}\,F(u)\,\mathrm{d}u. \tag{A.33†}$$

More generally, if $f(x_1, x_2, \ldots, x_n)$ is a function of n variables $x_1, \ldots, x_n$, its Fourier transform is

$$F(u_1, \ldots, u_n) \equiv \mathscr{F}[f]$$
$$= \left(\frac{\alpha}{2\pi}\right)^{\frac{1}{2}n}\int_{-\infty}^{+\infty}\cdots\int_{-\infty}^{+\infty} \mathrm{e}^{-\mathrm{i}\alpha(u_1x_1+\ldots+u_nx_n)}\,f(x_1, \ldots, x_n)\,\mathrm{d}x_1\ldots\mathrm{d}x_n \tag{A.34}$$

and the inverse transformation is defined by

$$f(x_1, \ldots, x_n) \equiv \mathscr{F}^\dagger[F]$$
$$= \left(\frac{\alpha}{2\pi}\right)^{\frac{1}{2}n}\int_{-\infty}^{+\infty}\cdots\int_{-\infty}^{+\infty} \mathrm{e}^{\mathrm{i}\alpha(u_1x_1+\ldots+u_nx_n)}\,F(u_1, \ldots, u_n)\,\mathrm{d}u_1\ldots\mathrm{d}u_n. \tag{A.34†}$$

[1] Cf. L. Schwartz, *loc. cit.*, footnote p. 463; also E. C. Titchmarsh, *Introduction to the Theory of Fourier Integrals* (2nd Ed., Oxford University Press, 1948).

Provided the Fourier transform of f exists, one has

$$\mathscr{F}[f(cx_1, \ldots, cx_n)] = \frac{1}{|c|^n} F\left(\frac{u_1}{c}, \ldots, \frac{u_n}{c}\right) \quad (c = \text{arbitrary const.}). \tag{A.35}$$

Likewise, and with the same qualifications, one has

$$\mathscr{F}^\dagger[F(cu_1, \ldots, cu_n)] = \frac{1}{|c|^n} f\left(\frac{x}{c}, \ldots, \frac{x_n}{c}\right) \quad (c = \text{arbitrary const.}). \tag{A.35†}$$

In what follows we give without proof the main properties of the Fourier transforms of functions (or of distributions) of a variable. All these properties of the one-dimensional Fourier transformation may be easily extended to any number of dimensions.

12. Integrable Functions

Any integrable function $f(x)$ $\left(\int_{-\infty}^{+\infty} |f(x)|\, \mathrm{d}x < \infty\right)$ has a Fourier transform:

$$F(u) = \mathscr{F}[f].$$

$F(u)$ is: (*i*) continuous,

(*ii*) bounded: $|F(u)| \leqslant \int_{-\infty}^{+\infty} |f(x)|\, \mathrm{d}x$ for any u,

(*iii*) zero at infinity: $F(u) \xrightarrow[|u|\to\infty]{} 0.$

If $f(x)$ is m times *continuously differentiable*, and if its m derivatives are integrable,

$$\mathscr{F}[f^{(m)}] = (\mathrm{i}\alpha u)^m F(u). \tag{A.36}$$

If $x^m f(x)$ is integrable, $F(u)$ is m times continuously differentiable and

$$F^{(m)}(u) = \mathscr{F}[(-\mathrm{i}\alpha x)^m f(x)]. \tag{A.37}$$

(The properties of the transformation $\mathscr{F}^\dagger$ are deduced from the foregoing by changing i into $-$i in all formulae.)

13. χ Functions

We designate by $\chi(x)$ an indefinitely differentiable function which tends, along with all its derivatives, asymptotically to zero more rapidly than any power of $|x|$:

$$|x|^l \chi^{(m)}(x) \xrightarrow[|x|\to\infty]{} 0 \qquad \text{for any } m \text{ and } l.$$

More generally, $\chi(x_1, \ldots, x_n)$ designates a function of n variables, indefinitely differentiable and such that

$$R^l \frac{\partial^m \chi}{\partial x_1^{\alpha_1} \partial x_2^{\alpha_2} \ldots \partial x_n^{\alpha_n}} \xrightarrow[R\to\infty]{} 0 \qquad [R \equiv (x_1^2 + x_2^2 + \ldots + x_n^2)^{\frac{1}{2}}]$$

for any l, m, and any choice of the indices $\alpha_1, \alpha_2, \ldots, \alpha_n$ $(\alpha_1 + \ldots + \alpha_n = m)$.

The functions φ of Sec. I are particular χ functions; on the other hand, the functions χ are not all φ functions [example: $\exp(-R^2)$].

Since the functions χ are integrable, the properties of the preceding section apply. In addition, we have the following result:

The Fourier transform $\mathscr{F}\chi$ and the inverse Fourier transform $\mathscr{F}^\dagger\chi$ of a function χ are likewise χ functions (of the variables $u_1, u_2, \ldots, u_n$). Moreover, the Fourier transformation has the *reciprocity property*:

$$\mathscr{F}\mathscr{F}^\dagger\chi = \mathscr{F}^\dagger\mathscr{F}\chi = \chi$$

($\mathscr{F}\mathscr{F}^\dagger\chi$ means: Fourier transform of $\mathscr{F}^\dagger\chi$).

14. Fourier Transformation of Distributions. Definition

If T is a distribution, its Fourier transform $\mathscr{F}T$ is the functional defined by

$$\mathscr{F}T[\varphi] = T[\mathscr{F}\varphi],$$

its inverse Fourier transform is the functional

$$\mathscr{F}^\dagger T[\varphi] = T[\mathscr{F}^\dagger\varphi].$$

Since $\mathscr{F}\varphi, \mathscr{F}^\dagger\varphi$ are not necessarily functions of the type φ, it may happen that the functionals $T[\mathscr{F}\varphi], T[\mathscr{F}^\dagger\varphi]$ do not exist; in that case T has neither a Fourier transform nor an inverse Fourier transform.

If $\hat{f}$ is the distribution corresponding to a function $f(x)$, and if $F(u)$ is the Fourier transform of $f(x)$, assuming that it exists,

$$\begin{aligned}\mathscr{F}\hat{f}[\varphi] = \hat{f}[\mathscr{F}\varphi] &= \int_{-\infty}^{+\infty} f(x) \left[\left(\frac{\alpha}{2\pi}\right)^{\frac{1}{2}} \int_{-\infty}^{+\infty} e^{-i\alpha ux}\, \varphi(u)\, du\right] dx \\ &= \int_{-\infty}^{+\infty} \varphi(u) \left[\left(\frac{\alpha}{2\pi}\right)^{\frac{1}{2}} \int_{-\infty}^{+\infty} e^{-i\alpha ux}\, f(x)\, dx\right] du \\ &= \hat{F}[\varphi].\end{aligned}$$

Thus, the Fourier transform of the distribution $\hat{f}$ is the distribution $\hat{F}$ associated with the Fourier transform of the function f.

15. Tempered Distributions

The tempered distributions are by definition the linear and continuous functionals of the functions χ. They are particular distributions.

All the properties of distributions of Sec. I extend to tempered distributions. It suffices to substitute the functions χ for the functions φ in all statements. In particular, the tempered distributions are differentiable to all orders, and *their derivatives are tempered distributions.*

The square-integrable functions, the functions bounded over all space and, more generally, the locally integrable functions $f(x)$ which *increase sufficiently slowly* at infinity (i.e. for which one can find two positive numbers A and α such that $|f(x)| \leqslant A|x|^{\alpha}$ when $|x| \to \infty$) all define tempered distributions. δ, δ_{x_0}, and all their derivatives are tempered distributions.

The solutions of the eigenvalue problems in Wave Mechanics are linear and continuous functionals of the wave functions, that is to say, of the square-integrable functions $\psi(q_1, ..., q_R)$. They are, *a fortiori*, linear and continuous functionals of the functions χ: *they are tempered distributions.*

The interest in tempered distributions stems from their remarkable Fourier transformation properties:

If U_x is a tempered distribution [defined for the functions $\chi(x)$]:

1) *Its Fourier transform V_u and its inverse Fourier transform $V_u{}^\dagger$ always exist and they are tempered distributions* [defined for the functions $\chi(u)$]. They are respectively defined by

$$V_u[\chi] \equiv \mathscr{F} U_x[\chi] = U_x[\mathscr{F}\chi] \tag{A.38}$$

$$V_u{}^\dagger[\chi] \equiv \mathscr{F}^\dagger U_x[\chi] = U_x[\mathscr{F}^\dagger\chi]. \tag{A.38†}$$

2) *The transformations $\mathscr{F}$ and $\mathscr{F}^\dagger$ are reciprocal:*

$$\mathscr{F}^\dagger\mathscr{F} U \equiv \mathscr{F}^\dagger V = U \tag{A.39}$$

$$\mathscr{F}\mathscr{F}^\dagger U \equiv \mathscr{F} V^\dagger = U. \tag{A.40}$$

3) *Differentiation is transformed into multiplication by u, and conversely*, according to the law

$$\mathscr{F}(U_x{}^{(m)}) = (\mathrm{i}\alpha u)^m\, V_u \tag{A.41}$$

$$\mathscr{F}[(-\mathrm{i}\alpha x)^m\, U_x] = V_u{}^{(m)}. \tag{A.42}$$

16. Square-Integrable Functions

The square-integrable functions define tempered distributions.

If one adopts the convention not to consider as distinct, two functions which are equal almost everywhere (i.e. everywhere except on a set of measure zero), the properties of the Fourier transformation of tempered distributions apply to square-integrable functions. To these properties are added specific properties of square-integrable functions. The principal theorems are the following:

Theorem I. If $f(x)$ is *square-integrable,* the integral

$$\left(\frac{\alpha}{2\pi}\right)^{\frac{1}{2}} \int_{-\xi}^{+\xi} \mathrm{e}^{-\mathrm{i}\alpha u x} f(x)\, \mathrm{d}x$$

converges in the quadratic mean [1]) *toward a square-integrable function*

$$F(u) \equiv \mathscr{F} f(x) = \left(\frac{\alpha}{2\pi}\right)^{\frac{1}{2}} \underset{\xi\to\infty}{\mathrm{lqm}} \int_{-\xi}^{+\xi} \mathrm{e}^{-\mathrm{i}\alpha u x} f(x)\, \mathrm{d}x. \tag{A.43}$$

Theorem II. The correspondence between $f(x)$ and $F(u)$ is *reciprocal* in the sense that

$$f(x) = \mathscr{F}^{\dagger} F(u) = \left(\frac{\alpha}{2\pi}\right)^{\frac{1}{2}} \underset{\lambda\to\infty}{\mathrm{lqm}} \int_{-\lambda}^{+\lambda} \mathrm{e}^{\mathrm{i}\alpha u x} F(u)\, \mathrm{d}u. \tag{A.43†}$$

In fact, for any value of x in the vicinity of which $f(x)$ is of bounded variation

$$\mathscr{F}^{\dagger} F(u) = \lim_{\varepsilon\to 0} \frac{f(x+\varepsilon)+f(x-\varepsilon)}{2}. \tag{A.44}$$

Theorem III. Let the square-integrable functions $f(x)$ and $F(u)$ be Fourier transforms of each other. If the derivative $f'(x)$ is square-

---

[1]) Convergence in the quadratic mean (less restrictive than ordinary convergence) of a function $\Phi(u, \xi)$ toward $\Phi(u)$ as $\xi \to X$, denoted here by the symbol

$$\underset{\xi\to X}{\mathrm{lqm}}\, \Phi(u, \xi) = \Phi(u),$$

means that

$$\lim_{\xi\to X} \int_{-\infty}^{+\infty} |\Phi(u, \xi) - \Phi(u)|^2\, \mathrm{d}u = 0,$$

in other words, that $\Phi(u, \xi)$ tends toward $\Phi(u)$ almost everywhere.

integrable, its Fourier transform, $i\alpha u F(u)$, is also square-integrable; conversely, if $i\alpha u F(u)$ is square-integrable, $f(x)$ is differentiable and $f'(x)$ is the inverse Fourier transform of $i\alpha u F(u)$. An analogous property exists for the pair $xf(x)$ and $(i/\alpha)F'(x)$.

N.B. Even if $f(x)$, considered as a function, is not differentiable everywhere, the (tempered) distribution $\hat{f'}_x$ always exists; its Fourier transform is the (tempered) distribution $i\alpha u \hat{F}_u$; the latter, considered as a function, may not be square-integrable.

The Fourier transformation preserves the scalar product of square-integrable functions (Parseval):

Theorem IV. If $F(u)$ and $G(u)$ are the respective Fourier transforms of the square-integrable functions $f(x)$ and $g(x)$, one has

$$\langle g, f \rangle = \langle G, F \rangle, \tag{A.45}$$

or stated differently:

$$\int_{-\infty}^{+\infty} g^*(x)\, f(x)\, \mathrm{d}x = \int_{-\infty}^{+\infty} G^*(u)\, F(u)\, \mathrm{d}u.$$

A particular case ($f=g$) of this theorem is the *conservation of the norm*

$$\int_{-\infty}^{+\infty} |f(x)|^2\, \mathrm{d}x = \int_{-\infty}^{+\infty} |F(u)|^2\, \mathrm{d}u. \tag{A.46}$$

17. Transformation of Folding into Multiplication

By definition, the folding of two functions, if it exists, is the expression

$$f \circ g \equiv \int_{-\infty}^{+\infty} f(x-t)\, g(t)\, \mathrm{d}t. \tag{A.47}$$

One can likewise define (cf. footnote p. 463) — and we shall not state it explicitly here—the folding of two distributions. In particular

$$\delta \circ T = T \quad (T \text{ an arbitrary distribution})$$

$$\delta \circ f = \int_{-\infty}^{+\infty} \delta(x-t)\, f(t)\, \mathrm{d}t = f(x) \quad (f \text{ an arbitrary function}).$$

The folding is a commutative operation: $f \circ g = g \circ f$.

The interest in the folding concept comes from the following theorem:

Theorem V. Provided that it exists, the folding of two functions $f(x)$ and $g(x)$ has as its Fourier transform $\sqrt{2\pi/\alpha}\, F(u)\, G(u)$, an expression in which $F(u)$ and $G(u)$ are the Fourier transforms of $f(x)$ and $g(x)$, respectively; this transformation is reciprocal.

In particular:

Theorem V′. If $f(x)$ is a square-integrable function, $g(x)$ an integrable function, and $F(x)$ and $G(x)$ their respective Fourier transforms, the folding $f \circ g$ and the product $\sqrt{2\pi/\alpha}\, FG$ are square-integrable functions, and the second is the Fourier transform of the first.

TABLE OF FOURIER TRANSFORMS

$f(x) = \left(\frac{\alpha}{2\pi}\right)^{\frac{1}{2}} \int_{-\infty}^{+\infty} e^{i\alpha ux} F(u)\, du$	$F(u) = \left(\frac{\alpha}{2\pi}\right)^{\frac{1}{2}} \int_{-\infty}^{+\infty} e^{-i\alpha ux} f(x)\, dx$
$f\left(\frac{x}{c}\right)$	$\lvert c\rvert\, F(cu)$
$f(-x)$	$F(-u)$
$f^*(x)$	$F^*(-u)$
$F(x)$	$f(-u)$
$xf(x)$	$\frac{i}{\alpha} F'(u)$
$f'(x)$	$i\alpha u\, F(u)$
$f(x-x_0)$	$e^{-i\alpha ux_0} F(u)$
$e^{i\alpha u_0 x} f(x)$	$F(u-u_0)$
$\delta(x)$	$\left(\frac{\alpha}{2\pi}\right)^{\frac{1}{2}}$
$\delta(x-x_0)$	$\left(\frac{\alpha}{2\pi}\right)^{\frac{1}{2}} e^{-i\alpha ux_0}$
$E(x) = \begin{cases} 1 \text{ if } x > 0 \\ 0 \text{ if } x < 0 \end{cases}$	$\frac{1}{\sqrt{2\pi\alpha}}\left[\pi\delta(u) - i\,\mathrm{PP}\,\frac{1}{u}\right]$
$\left(\frac{\varkappa}{\sqrt{\pi}}\right)^{\frac{1}{2}} e^{-\frac{1}{2}\varkappa^2 x^2}$	$\left(\frac{\alpha}{\varkappa\sqrt{\pi}}\right)^{\frac{1}{2}} e^{-\alpha^2 u^2/2\varkappa^2}$ (Re $\varkappa > 0$, Re $\varkappa^2 > 0$)
$\frac{1}{\sqrt{2a}}[E(x+a) - E(x-a)] = \begin{cases} \frac{1}{\sqrt{2a}} \text{ if } \lvert x\rvert < a \\ 0 \text{ if } \lvert x\rvert > a \end{cases}$	$\left(\frac{\alpha a}{\pi}\right)^{\frac{1}{2}} \frac{\sin \alpha a u}{\alpha a u}$ (a real > 0)
$\sqrt{\gamma}\, e^{-\gamma\lvert x\rvert}$	$\left(\frac{2\alpha\gamma^3}{\pi}\right)^{\frac{1}{2}} \frac{1}{\gamma^2 + \alpha^2 u^2}$ (Re $\gamma > 0$)
$i\sqrt{2\gamma}\, e^{-\gamma x} E(x) = \begin{cases} i\sqrt{2\gamma}\, e^{-\gamma x} & \text{if } x > 0 \\ 0 & \text{if } x < 0 \end{cases}$	$\left(\frac{\alpha\gamma}{\pi}\right)^{\frac{1}{2}} \frac{1}{\alpha u - i\gamma}$ (Re $\gamma > 0$)

N.B. The functions of the last four lines are normalized to unity:

$$\int_{-\infty}^{+\infty} \lvert f(x)\rvert^2\, dx = \int_{-\infty}^{+\infty} \lvert F(u)\rvert^2\, du = 1.$$

APPENDIX B

SPECIAL FUNCTIONS AND ASSOCIATED FORMULAE

PLAN OF THE APPENDIX

I. LAPLACE'S EQUATION, LAGUERRE POLYNOMIALS, COULOMB FUNCTIONS

1. Laplace's Equation and Confluent Hypergeometric Series

Laplace's Equation

$$\left[z\frac{\mathrm{d}^2}{\mathrm{d}z^2}+(\beta-z)\frac{\mathrm{d}}{\mathrm{d}z}-\alpha\right]f(z)=0 \tag{B.1}$$

(α, β are any complex constants).

Confluent Hypergeometric Series $F(\alpha|\beta|z)$.

By definition:

$$\begin{aligned}F(\alpha|\beta|z) &= 1+\frac{\alpha}{\beta}\cdot\frac{z}{1!}+\frac{\alpha(\alpha+1)}{\beta(\beta+1)}\cdot\frac{z^2}{2!}+\dots\\ &= \sum_{n=0}^{\infty}\frac{\Gamma(\alpha+n)\,\Gamma(\beta)}{\Gamma(\alpha)\,\Gamma(\beta+n)}\cdot\frac{z^n}{n!}.\end{aligned} \tag{B.2}$$

This series:

(*i*) is well-defined for any α and β provided that $\beta\neq -p$ (p integer $\geqslant 0$);

(*ii*) converges in the entire complex plane z;

(*iii*) is a polynomial of degree p (p integer $\geqslant 0$) if $\alpha=-p$, possesses an essential singularity at infinity if $\alpha\neq -p$;

(*iv*) satisfies the *Kummer relation*:

$$F(\alpha|\beta|z)=\mathrm{e}^z\,F(\beta-\alpha|\beta|-z). \tag{B.3}$$

If they exist [1]), the functions

$$F(\alpha|\beta|z) \quad\text{and}\quad z^{1-\beta}\,F(\alpha-\beta+1|2-\beta|z)$$

are solutions of eq. (B.1).

Solutions in Integral Form (Laplace's Method):

If Γ is a contour of the complex plane t such that $t^{\alpha}(1-t)^{\beta-\alpha}\,\mathrm{e}^{zt}$ takes the same value at both limits, the integral

$$\int_{\Gamma}\mathrm{e}^{zt}\,t^{\alpha-1}\,(1-t)^{\beta-\alpha-1}\,\mathrm{d}t \tag{B.4}$$

---

[1]) If β is not an integer, these two functions exist and are distinct. If $\beta=1$, they are identical. If $\beta=0,-1,-2,\dots$, only the function

$$z^{1-\beta}F(\alpha-\beta+1\,|2-\beta|\,z)$$

exists. If $\beta=2,3,\dots$, only the function $F(\alpha|\beta|z)$ exists.

is a particular solution of eq. (B.1).

We assume from now on:

$$\alpha \text{ non-integer}; \qquad \beta = b \qquad (b \text{ integer} > 0).$$

To the closed contour Γ_0 surrounding the points $t=0$ and $t=1$ (Fig. B.1*a*) [1]) corresponds a solution of the type (B.4). [By convention,

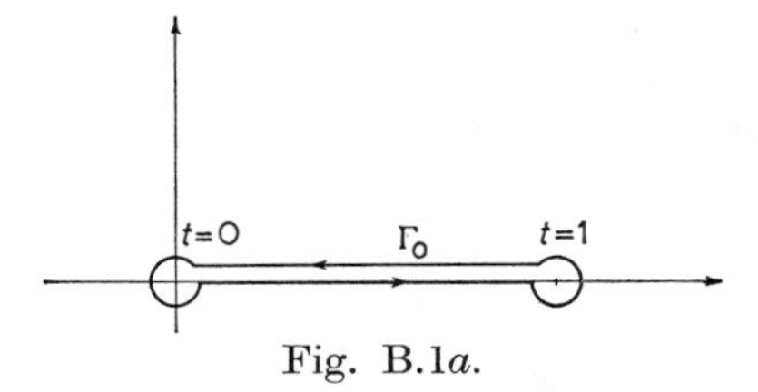

Fig. B.1*a*.

$\arg t - \arg(1-t) = 0$ on that portion of Γ_0 located on the real axis between 0 and 1 and described in the direction of increasing t.] Since this solution is an entire function of z, it is proportional to $F(\alpha|b|z)$. The coefficient of proportionality is obtained by expanding e^{zt} under the integral sign and making use of the formula:

$$B(x, y) \equiv \frac{\Gamma(x)\,\Gamma(y)}{\Gamma(x+y)} = (1-e^{2\pi i y})^{-1} \int_{\Gamma_0} t^{x-1}\,(1-t)^{y-1}\,dt \tag{B.5}$$

$$(x+y = \text{integer}, \quad y \neq \text{integer}).$$

One has:

$$F(\alpha|b|z) = (1-e^{-2\pi i\alpha})^{-1} \frac{\Gamma(b)}{\Gamma(\alpha)\,\Gamma(b-\alpha)} \int_{\Gamma_0} e^{zt}\,t^{\alpha-1}\,(1-t)^{b-\alpha-1}\,dt. \tag{B.6}$$

To the loops Γ_1, Γ_2 surrounding $t=0$ and $t=1$, respectively (Fig. B.1*b*) [2]), correspond two solutions of type (B.4) irregular at

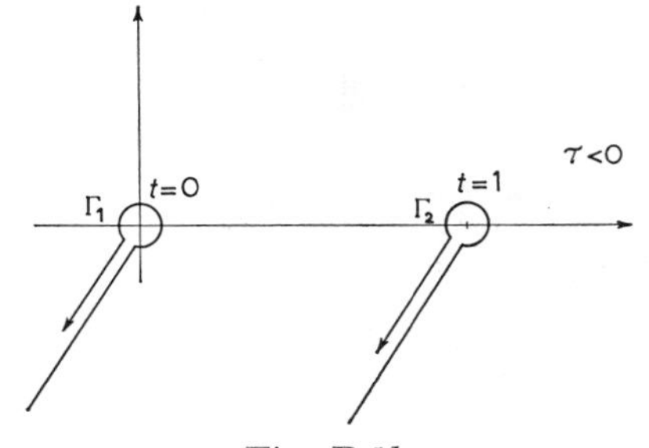

Fig. B.1*b*.

[1]) Γ_0 is described in the positive sense.

[2]) Γ_1 and Γ_2 are described in the positive sense.

the origin, namely:

$$W_r(\alpha|b|z) = (1-e^{-2\pi i\alpha})^{-1} \frac{\Gamma(b)}{\Gamma(\alpha)\,\Gamma(b-\alpha)} \int_{\Gamma_r} e^{zt}\, t^{\alpha-1}\,(1-t)^{b-\alpha-1}\,dt. \tag{B.7}$$

$$(r = 1, 2).$$

The condition of convergence of the integral is:

$$\frac{\pi}{2} < \arg z + \tau + 2n\pi < \frac{3\pi}{2}$$

(τ is the argument of the point at infinity of the loops Γ_1 and Γ_2).

By convention [1]):

$$-\pi < \arg z < +\pi, \qquad -\pi < \tau < +\pi \qquad \text{sign of } \tau = \text{sign of } (\arg z).$$

At the end of loop Γ_1 and at the beginning of loop Γ_2:

$$\arg t = \tau.$$

At the end of loop Γ_1 and at the beginning of loop Γ_2:

$$\arg(1-t) = \begin{cases} \tau-\pi & \text{if} \quad 0<\tau<\pi \\ \tau+\pi & \text{if} \quad -\pi<\tau<0. \end{cases}$$

With these conventions

$$W_1(b-\alpha|b|-z) = e^{-z}\, W_2(\alpha|b|z). \tag{B.8a}$$

$$W_2(b-\alpha|b|-z) = e^{-z}\, W_1(\alpha|b|z) \tag{B.8b}$$

$$[-\pi < \arg z < +\pi; \qquad -\pi < \arg(-z) < +\pi]$$

$$F(\alpha|b|z) = W_1(\alpha|b|z) + W_2(\alpha|b|z). \tag{B.9}$$

Asymptotic expansions of the solutions W_1 and W_2 (determined by the saddle-point method):

$$W_1(\alpha|b|z) \underset{|z|\to\infty}{\sim} \frac{\Gamma(b)}{\Gamma(b-\alpha)} (-z)^{-\alpha} \sum_{n=0}^{\infty} \frac{\Gamma(n+\alpha)}{\Gamma(\alpha)} \frac{\Gamma(n+\alpha-b+1)}{\Gamma(\alpha-b+1)} \frac{(-z)^{-n}}{n!} \tag{B.10}$$

$$[-\pi < \arg(-z) < +\pi].$$

$$W_2(\alpha|b|z) \underset{|z|\to\infty}{\sim} \frac{\Gamma(b)}{\Gamma(\alpha)}\, e^z\, z^{\alpha-b} \sum_{n=0}^{\infty} \frac{\Gamma(n+1-\alpha)}{\Gamma(1-\alpha)} \cdot \frac{\Gamma(n+b-\alpha)}{\Gamma(b-\alpha)} \frac{z^{-n}}{n!} \tag{B.11}$$

$$(-\pi < \arg z < +\pi).$$

---

[1]) These conventions are meaningful provided $\operatorname{Im} z \neq 0$. To define W_1 and W_2 on the real axis, one must perform an analytic continuation; the functions obtained are different according to whether $\operatorname{Im} z \to 0+$ or $\operatorname{Im} z \to 0-$.

2. Laguerre Polynomials

Definition [1]:

$$L_p^0 = \mathrm{e}^z \frac{\mathrm{d}^p}{\mathrm{d}z^p} (\mathrm{e}^{-z} z^p)$$

$$L_p^k = (-)^k \frac{\mathrm{d}^k}{\mathrm{d}z^k} L_{p+k}^0 \tag{B.12}$$

$$(k, p = 0, 1, 2, ..., \infty)$$

L_p^k is a polynomial of degree p having p zeros between 0 and $+\infty$:

$$\begin{aligned} L_p^k(z) &= \frac{[(p+k)!]^2}{p!\,k!} F(-p|k+1|z) \\ &= \sum_{s=0}^{p} (-)^s \frac{[(p+k)!]^2}{(p-s)!\,(k+s)!\,s!} z^s. \end{aligned} \tag{B.13}$$

In particular $L_0^k = k!$

Laplace's Equation

$$\left[z \frac{\mathrm{d}^2}{\mathrm{d}z^2} + (k+1-z) \frac{\mathrm{d}}{\mathrm{d}z} + p\right] L_p^k = 0. \tag{B.14}$$

Generating Function

$$\frac{\mathrm{e}^{-zt/(1-t)}}{(1-t)^{k+1}} = \sum_{p=0}^{\infty} \frac{t^p}{(p+k)!} L_p^k(z) \qquad (|t| < 1). \tag{B.15}$$

"Orthonormality" Relations

$$\int_0^\infty \mathrm{e}^{-z} z^k L_p^k L_q^k \,\mathrm{d}z = \frac{[(p+k)!]^3}{p!} \delta_{pq}. \tag{B.16}$$

3. Eigenfunctions of Hydrogen-like Atoms (Schrödinger Theory)

$$a = \frac{a_0}{Z} = \frac{\hbar^2}{Zm'e^2} \qquad (a_0 \approx \text{radius of the Bohr orbit})$$

Ze = nuclear charge; m' = reduced mass of the electron.

Energy Eigenvalues

$$E_n = -\left(\frac{Ze^2}{\hbar c}\right)^2 \frac{m'c^2}{2n^2} = -\frac{1}{n^2} \frac{Ze^2}{2a}.$$

---

[1] Certain authors designate by L_p^k the polynomial equal to $(-)^k L_{p-k}^k$ in our notation.

Eigenfunctions Normalized to Unity in Spherical Coordinates

$$\psi_{nlm}(r, \theta, \varphi) = a^{-3/2} N_{nl} F_{nl}\left(\frac{2r}{na}\right) Y_l{}^m(\theta, \varphi). \tag{B.17}$$

$$N_{nl} = \frac{2}{n^2}\sqrt{\frac{(n-l-1)!}{[(n+l)!]^3}}. \tag{B.17a}$$

$$F_{nl}(x) = x^l\, \mathrm{e}^{-\frac{1}{2}x}\, L_{n-l-1}^{2l+1}(x) \tag{B.17b}$$

$$(n = 1, 2, \ldots, \infty; \qquad l = 0, 1, \ldots, n-1; \qquad m = -l, -l+1, \ldots, l).$$

Mean Values of Powers of r

Recursion relation between mean values pertaining to the same eigenstate (nlm)

$$\frac{s+1}{n^2}\langle r^s\rangle - (2s+1)\, a\, \langle r^{s-1}\rangle + \frac{s}{4}\left[(2l+1)^2 - s^2\right] a^2 \langle r^{s-2}\rangle = 0$$

$$(s > -2l-1)$$

$$\left\langle\frac{1}{r}\right\rangle = \frac{1}{n^2 a}; \qquad \left\langle\frac{1}{r^2}\right\rangle = \frac{2}{(2l+1)\, n^3\, a^2}$$

$$\langle r\rangle = \tfrac{1}{2}[3n^2 - l(l+1)]\, a; \qquad \langle r^2\rangle = \tfrac{1}{2}[5n^2 + 1 - 3l(l+1)]\, n^2 a^2.$$

Table of the First Few Radial Functions

$$\varrho = \frac{r}{a}, \qquad g_{nl}(\varrho) = N_{nl} F_{nl}\left(\frac{2}{n}\varrho\right)$$

$n = 1 \quad g_{1s} = 2\mathrm{e}^{-\varrho}$

$n = 2 \quad g_{2s} = \frac{\sqrt{2}}{2}\left(1 - \frac{1}{2}\varrho\right)\mathrm{e}^{-\frac{1}{2}\varrho}, \qquad g_{2p} = \frac{\sqrt{6}}{12}\varrho\, \mathrm{e}^{-\frac{1}{2}\varrho}$

$n = 3 \quad g_{3s} = \frac{2\sqrt{3}}{9}\left(1 - \frac{2}{3}\varrho + \frac{2}{27}\varrho^2\right)\mathrm{e}^{-\frac{1}{3}\varrho},$

$$g_{3p} = \frac{8\sqrt{6}}{27}\left(\varrho - \frac{1}{6}\varrho^2\right)\mathrm{e}^{-\frac{1}{3}\varrho}, \qquad g_{3d} = \frac{2\sqrt{30}}{955}\varrho^3\, \mathrm{e}^{-\frac{1}{3}\varrho}.$$

4. Pure Coulomb Wave

$\psi_c(r)$ is the scattering wave of a particle by a pure Coulomb potential $ZZ'e^2/r$.

$$k = \frac{mv}{\hbar} = \text{wave number}$$

v = incident velocity

$$\gamma = \frac{ZZ'e^2}{\hbar v} \tag{B.18}$$

$$\psi_c = e^{-\frac{1}{2}\pi\gamma}\, \Gamma(1+i\gamma)\, e^{ikz}\, F[-i\gamma|1|ik(r-z)] \tag{B.19}$$

$$= \psi_i + \psi_s \tag{B.20}$$

$$\psi_i = e^{-\frac{1}{2}\pi\gamma}\, \Gamma(1+i\gamma)\, e^{ikz}\, W_1\, [-i\gamma|1|ik(r-z)] \tag{B.21}$$

$$\psi_s = e^{-\frac{1}{2}\pi\gamma}\, \Gamma(1+i\gamma)\, e^{ikz}\, W_2\, [-i\gamma|1|ik(r-z)]. \tag{B.22}$$

Asymptotic Form

$$\psi_i \underset{k(r-z)\to\infty}{\sim} \exp\,[ikz + i\gamma \ln k(r-z)] \left\{1 + \frac{\gamma^2}{ik(r-z)} + \ldots\right\} \tag{B.23}$$

$$\psi_s \underset{k(r-z)\to\infty}{\sim} f_c(\theta)\, \frac{e^{i(kr-\gamma \ln 2kr)}}{r} \left\{1 + \frac{(1+i\gamma)^2}{ik(r-z)} + \ldots\right\} \tag{B.24}$$

$$f_c(\theta) = -\frac{\gamma}{2k \sin^2 \frac{1}{2}\theta} \exp\,[-i\gamma \ln (\sin^2 \tfrac{1}{2}\theta) + 2i\sigma_0] \tag{B.25}$$

$$e^{2i\sigma_0} = \frac{\Gamma(1+i\gamma)}{\Gamma(1-i\gamma)}. \tag{B.26}$$

Behavior at the Origin

$$\psi_c(0) = e^{-\frac{1}{2}\pi\gamma}\, \Gamma(1+i\gamma)$$

$$|\psi_c(0)|^2 = \frac{2\pi\gamma}{e^{2\pi\gamma} - 1}. \tag{B.27}$$

5. Spherical Coulomb Functions

Differential Equation

In spherical coordinates, the collision problem of § 4 leads for each value l of the orbital angular momentum, to the *radial equation*:

$$\left[\frac{d^2}{dr^2} + \left(k^2 - \frac{l(l+1)}{r^2} - \frac{2\gamma k}{r}\right)\right] y_l = 0. \tag{B.28}$$

The spherical Coulomb functions are special solutions of this equation. They are functions of the argument

$$\varrho = kr.$$

They depend upon the energy through k and γ. One defines the solution, regular (as r^{l+1}) at the origin, $F_l(\gamma; kr)$, and the irregular solutions, G_l, $u_l^{(+)}$ and $u_l^{(-)}$ [singularity in $(1/r)^l$].

By the change of variable and of function

$$z = -2\mathrm{i}\varrho, \qquad y_l = \mathrm{e}^{\mathrm{i}\varrho}\, \varrho^{l+1}\, v_l,$$

equation (B.28) yields Laplace's differential equation:

$$\left[z\frac{\mathrm{d}^2}{\mathrm{d}z^2} + (2l+2-z)\frac{\mathrm{d}}{\mathrm{d}z} - (l+1+\mathrm{i}\gamma)\right] v_l = 0,$$

of which one knows the solution that is regular at the origin, $F(l+1+\mathrm{i}\gamma|2l+2|z)$, and the two irregular solutions

$$W_{1,2}(l+1+\mathrm{i}\gamma|2l+2|z).$$

Definition and Connecting Relations

$$F_l(\gamma;\, \varrho) = c_l\, \mathrm{e}^{\mathrm{i}\varrho}\, \varrho^{l+1}\, F(l+1+\mathrm{i}\gamma|2l+2|-2\mathrm{i}\varrho) \tag{B.29a}$$

$$= c_l\, \mathrm{e}^{-\mathrm{i}\varrho}\, \varrho^{l+1}\, F(l+1-\mathrm{i}\gamma|2l+2|+2\mathrm{i}\varrho) \tag{B.29b}$$

$$u_l^{(\pm)}(\gamma;\, \varrho) = \pm\, 2\mathrm{i}\, \mathrm{e}^{\mp\mathrm{i}\sigma_l}\, c_l\, \mathrm{e}^{\pm\mathrm{i}\varrho}\, \varrho^{l+1}\, W_1(l+1 \pm \mathrm{i}\gamma|2l+2| \mp 2\mathrm{i}\varrho) \tag{B.30a}$$

$$= \pm\, 2\mathrm{i}\, \mathrm{e}^{\mp\mathrm{i}\sigma_l}\, c_l\, \mathrm{e}^{\mp\mathrm{i}\varrho}\, \varrho^{l+1}\, W_2(l+1 \mp \mathrm{i}\gamma|2l+2| \pm 2\mathrm{i}\varrho) \tag{B.30b}$$

$$G_l(\gamma;\, \varrho) = \tfrac{1}{2}(u_l^{(+)}\, \mathrm{e}^{\mathrm{i}\sigma_l} + u_l^{(-)}\, \mathrm{e}^{-\mathrm{i}\sigma_l}). \tag{B.31}$$

c_l and σ_l (Coulomb phase shift) are the following functions of γ:

$$c_l = 2^l\, \mathrm{e}^{-\frac{1}{2}\pi\gamma}\, \frac{|\Gamma(l+1+\mathrm{i}\gamma)|}{(2l+1)!} \qquad \sigma_l = \arg \Gamma(l+1+\mathrm{i}\gamma), \tag{B.32}$$

or else:

for $l=0$,

$$c_0 = \left(\frac{2\pi\gamma}{\mathrm{e}^{2\pi\gamma}-1}\right)^{\frac{1}{2}} \qquad \sigma_0 = \arg \Gamma(1+\mathrm{i}\gamma) \tag{B.32a}$$

for $l \neq 0$,

$$c_l = c_0\, \frac{1}{(2l+1)!!} \times \prod_{s=1}^{l} \left(1+\frac{\gamma^2}{s^2}\right)^{\frac{1}{2}} \qquad \sigma_l = \sigma_0 + \sum_{s=1}^{l} \tan^{-1}\frac{\gamma}{s} \tag{B.32b}$$

$$F_l \text{ and } G_l \text{ are real}; \quad u_l^{(-)} = u_l^{(+)*}$$

$$F_l = \frac{1}{2\mathrm{i}}\, (u_l^{(+)}\, \mathrm{e}^{\mathrm{i}\sigma_l} - u_l^{(-)}\, \mathrm{e}^{-\mathrm{i}\sigma_l}) \tag{B.33}$$

$$u_l^{(\pm)} = \mathrm{e}^{\mp\mathrm{i}\sigma_l}\, (G_l \pm \mathrm{i}F_l). \tag{B.34}$$

Asymptotic Forms: $r \to \infty \qquad [\varrho \gg l(l+1)+\gamma^2]$

$$F_l \underset{\varrho\to\infty}{\sim} \sin(\varrho - \gamma \ln 2\varrho - \tfrac{1}{2}l\pi + \sigma_l) \tag{B.35}$$

$$G_l \underset{\varrho\to\infty}{\sim} \cos(\varrho - \gamma \ln 2\varrho - \tfrac{1}{2}l\pi + \sigma_l) \tag{B.36}$$

$$u_l^{(+)} \underset{\varrho\to\infty}{\sim} \exp[\mathrm{i}(\varrho - \gamma \ln 2\varrho - \tfrac{1}{2}l\pi)] \text{ (outgoing wave)} \tag{B.37}$$

$$u_l^{(-)} \underset{\varrho\to\infty}{\sim} \exp[-\mathrm{i}(\varrho - \gamma \ln 2\varrho - \tfrac{1}{2}l\pi)] \text{ (incoming wave).} \tag{B.38}$$

Behavior at the origin: $r \to 0$

$$F_l \underset{\varrho\to 0}{\sim} c_l\, \varrho^{l+1}\left[1 + \frac{\gamma}{l+1}\varrho + \dots\right] \tag{B.39}$$

$$G_l \underset{\varrho\to 0}{\sim} \frac{1}{(2l+1)\,c_l}\,\varrho^{-l}\left[1 + \begin{cases} \mathrm{O}(\gamma\varrho \ln \varrho) & \text{if } l=0 \\ \mathrm{O}\left(\frac{\gamma}{l}\varrho\right) & \text{if } l\neq 0 \end{cases}\right]. \tag{B.40}$$

General Behavior of F_l

As ϱ increases from 0 to ∞, the function F_l increases at first as ϱ^{l+1}, then more and more rapidly (exponential behavior) until

$$\varrho = \gamma + \sqrt{\gamma^2 + l(l+1)},$$

where it has a point of inflection; the function then oscillates indefinitely between two extreme values which tend asymptotically toward $+1$ and -1, respectively. The period of these oscillations tends asymptotically toward 2π.

Recursion Formulae

$$(2l+1)\left[\gamma + \frac{l(l+1)}{\varrho}\right] F_l = l\sqrt{\gamma^2+(l+1)^2}\, F_{l+1} + (l+1)\sqrt{\gamma^2+l^2}\, F_{l-1} \quad (l\neq 0) \tag{B.41}$$

$$\left(1+\frac{\gamma^2}{l^2}\right)^{\frac{1}{2}} F_{l-1} = \left(\frac{\mathrm{d}}{\mathrm{d}\varrho} + \frac{l}{\varrho} + \frac{\gamma}{l}\right) F_l. \quad (l\neq 0) \tag{B.42}$$

$$\left(1+\frac{\gamma^2}{l^2}\right)^{\frac{1}{2}} F_l = \left(-\frac{\mathrm{d}}{\mathrm{d}\varrho} + \frac{l}{\varrho} + \frac{\gamma}{l}\right) F_{l-1}. \quad (l\neq 0) \tag{B.43}$$

These relations remain valid if one replaces F_l by $U_l \equiv aF_l + bG_l$ (a, b are arbitrarily fixed coefficients independent of l).

Wronskian Relation

$$G_l \frac{\mathrm{d}F_l}{\mathrm{d}\varrho} - F_l \frac{\mathrm{d}G_l}{\mathrm{d}\varrho} = 1 \tag{B.44}$$

from which ($l \neq 0$):

$$G_l F_{l-1} - F_l G_{l-1} = \frac{l}{\sqrt{l^2+\gamma^2}}. \tag{B.45}$$

If $\gamma = 0$, one obtains the spherical Bessel functions, to within the factor ϱ:

$$\begin{aligned} F_l(0;\varrho) &= \varrho\, j_l(\varrho), & G_l(0;\varrho) &= \varrho\, n_l(\varrho) \\ u_l^{(+)}(0;\varrho) &= \varrho\, h_l^{(+)}(\varrho), & u_l^{(-)}(0;\varrho) &= \varrho\, h_l^{(-)}(\varrho) \end{aligned} \tag{B.46}$$

(for definitions of j_l, n_l, $h_l^{(\pm)}$, cf. following section).

II. SPHERICAL BESSEL FUNCTIONS

6. Spherical Bessel Functions

Differential Equation

In polar coordinates, the Schrödinger equation for the free particle leads, for each value l of the orbital angular momentum, to the *radial equation* [1])

$$\left[\frac{1}{\varrho}\frac{\mathrm{d}^2}{\mathrm{d}\varrho^2}\varrho + 1 - \frac{l(l+1)}{\varrho^2}\right] f_l \equiv \left[\frac{\mathrm{d}^2}{\mathrm{d}\varrho^2} + \frac{2}{\varrho}\frac{\mathrm{d}}{\mathrm{d}\varrho} + 1 - \frac{l(l+1)}{\varrho^2}\right] f_l = 0. \tag{B.47}$$

In the complex plane, f_l exhibits an essential singularity at infinity and, in general, a pole of order $l+1$ at $\varrho = 0$.

The spherical Bessel functions are special solutions of this equation. One defines the solution j_l, which is regular (as r^l) at the origin (proper spherical Bessel function) and the irregular solutions n_l (Neumann function), $h_l^{(+)}$ (Hankel function of the first kind) and $h_l^{(-)}$ (Hankel function of the second kind).

---

[1]) One again finds the radial equation (B.28) in the special case $\gamma = 0$ by putting:

$$\varrho = kr, \qquad y_l = kr\, f_l(kr).$$

Definition [1])

$$j_l(\varrho) = \left(\frac{\pi}{2\varrho}\right)^{\frac{1}{2}} J_{l+\frac{1}{2}}(\varrho), \qquad n_l(\varrho) = (-)^l \left(\frac{\pi}{2\varrho}\right)^{\frac{1}{2}} J_{-l-\frac{1}{2}}(\varrho)$$

$$h_l^{(\pm)}(\varrho) = n_l(\varrho) \pm \mathrm{i} j_l(\varrho)$$

(J_ν designates the ordinary Bessel function of order ν)

$$j_l \text{ and } n_l \text{ are real; } h_l^{(-)} = h_l^{(+)*}. \tag{B.48}$$

Explicit Form

$$j_l = R_l \frac{\sin \varrho}{\varrho} + S_l \frac{\cos \varrho}{\varrho}, \qquad n_l = R_l \frac{\cos \varrho}{\varrho} - S_l \frac{\sin \varrho}{\varrho}$$

$$h_l^{(\pm)} = (R_l \pm \mathrm{i} S_l) \frac{\mathrm{e}^{\pm \mathrm{i}\varrho}}{\varrho}. \tag{B.49}$$

R_l is a polynomial in $1/\varrho$ with real coefficients of degree l and parity $(-)^l$;
S_l is a polynomial in $1/\varrho$ with real coefficients of degree $l-1$ and parity $(-)^{l-1}$.

$$R_l + \mathrm{i} S_l = \sum_{s=0}^{l} \frac{\mathrm{i}^{s-l}}{2^s s!} \frac{(l+s)!}{(l-s)!} \varrho^{-s} \tag{B.50}$$

$$j_0 = \frac{\sin \varrho}{\varrho}, \qquad n_0 = \frac{\cos \varrho}{\varrho}, \qquad h_0^{(\pm)} = \frac{\mathrm{e}^{\pm \mathrm{i}\varrho}}{\varrho}$$

$$j_1 = \frac{\sin \varrho}{\varrho^2} - \frac{\cos \varrho}{\varrho}, \qquad n_1 = \frac{\cos \varrho}{\varrho^2} + \frac{\sin \varrho}{\varrho}, \qquad h_1^{(\pm)} = \left(\frac{1}{\varrho^2} \mp \frac{\mathrm{i}}{\varrho}\right) \mathrm{e}^{\pm \mathrm{i}\varrho}.$$

Asymptotic Forms: $\varrho \to \infty \qquad [\varrho \gg l(l+1)]$

$$j_l \underset{\varrho\to\infty}{\sim} \frac{1}{\varrho} \sin(\varrho - \tfrac{1}{2}l\pi), \qquad n_l \underset{\varrho\to\infty}{\sim} \frac{1}{\varrho} \cos(\varrho - \tfrac{1}{2}l\pi)$$

$$h_l^{(\pm)} \underset{\varrho\to\infty}{\sim} \frac{1}{\varrho} \exp[\pm \mathrm{i}(\varrho - \tfrac{1}{2}l\pi)] \left[1 \pm \mathrm{i} \frac{l(l+1)}{2\varrho} - \cdots\right]. \tag{B.51}$$

---

[1]) Most authors denote by n_l the same function *with a change in sign*, and introduce under the name of spherical Hankel function of the first and second kind, respectively, the functions:

$$h_l^{(1)} = -\mathrm{i} h_l^{(+)}, \qquad h_l^{(2)} = \mathrm{i} h_l^{(-)}.$$

Behavior at the Origin: $\varrho \to 0$

$$j_l \underset{\varrho\to 0}{\sim} \frac{\varrho^l}{(2l+1)!!}\left[1-\frac{\varrho^2}{2\,(2l+3)}+\ldots\right]$$
$$n_l \underset{\varrho\to 0}{\sim} \frac{(2l+1)!!}{(2l+1)}\left(\frac{1}{\varrho}\right)^{l+1}\left[1+\frac{\varrho^2}{2\,(2l-1)}+\ldots\right]. \tag{B.52}$$

General Behavior of j_l

As ϱ increases from 0 to $+\infty$, the function ϱj_l increases first as ϱ^{l+1}, then more and more rapidly (exponential behavior) up to the point $\varrho=\sqrt{l(l+1)}$, where it has a point of inflection. The function then oscillates indefinitely between two extreme values which tend asymptotically toward $+1$ and -1, respectively. The asymptotic form (B.51) is a good approximation when $\varrho \gg \frac{1}{2}l(l+1)$, but the amplitude of the oscillations practically attains its asymptotic value (to within 10 %) as soon as $\varrho \gtrsim 2l$.

Recursion Formulae

Below, $f_l \equiv aj_l+bn_l$, a and b being arbitrarily fixed coefficients independent of l. One has $(l\neq 0)$:

$$(2l+1)f_l = \varrho[f_{l+1}+f_{l-1}] \tag{B.53}$$

$$f_{l-1} = \left[\frac{\mathrm{d}}{\mathrm{d}\varrho}+\frac{l+1}{\varrho}\right]f_l = \frac{1}{\varrho^{l+1}}\frac{\mathrm{d}}{\mathrm{d}\varrho}(\varrho^{l+1}f_l) \tag{B.54}$$

$$f_l = \left[-\frac{\mathrm{d}}{\mathrm{d}\varrho}+\frac{l-1}{\varrho}\right]f_{l-1} = -\varrho^{l-1}\frac{\mathrm{d}}{\mathrm{d}\varrho}\left(\frac{f_{l-1}}{\varrho^{l-1}}\right) \tag{B.55}$$

from which:

$$f_l = \left[\varrho^l\left(-\frac{1}{\varrho}\frac{\mathrm{d}}{\mathrm{d}\varrho}\right)^l\right]f_0. \tag{B.56}$$

Wronskian Relation

$$\varrho^2\left[n_l\left(\frac{\mathrm{d}}{\mathrm{d}\varrho}j_l\right)-j_l\left(\frac{\mathrm{d}}{\mathrm{d}\varrho}n_l\right)\right] = 1 \tag{B.57}$$

from which $(l\neq 0)$

$$\varrho^2[n_l j_{l-1}-j_l n_{l-1}] = 1. \tag{B.58}$$

III. HARMONIC OSCILLATOR AND HERMITE POLYNOMIALS

7. Hermite Polynomials

Definition

$$H_n(z) = (-)^n \, e^{z^2} \left(\frac{d^n}{dz^n} e^{-z^2} \right) \qquad (n = 0, 1, 2, \ldots, \infty) \tag{B.59}$$

H_n is polynomial of degree n, parity $(-)^n$, having n zeros.

$$H_n(z) = \begin{cases} (-)^p \dfrac{(2p)!}{p!} F(-p|\tfrac{1}{2}|z^2) & \text{if } n = 2p \\ (-)^p \, 2 \dfrac{(2p+1)!}{p!} z \, F(-p|\tfrac{3}{2}|z^2) & \text{if } n = 2p+1. \end{cases} \tag{B.60}$$

Differential Equation

$$\left[\frac{d^2}{dz^2} - 2z \frac{d}{dz} + 2n \right] H_n(z) = 0. \tag{B.61}$$

Generating Function

$$\exp(-s^2 + 2sz) = \sum_{n=0}^{\infty} \frac{s^n}{n!} H_n(z). \tag{B.62}$$

Recursion Relations

$$\frac{d}{dz} H_n = 2n \, H_{n-1} \tag{B.63}$$

$$\left(2z - \frac{d}{dz} \right) H_n = H_{n-1} \tag{B.64}$$

$$2z \, H_n = H_{n+1} + 2n \, H_{n-1}.$$

Table of the First Six Hermite Polynomials

$H_0 = 1,$ $\qquad H_1 = 2z$

$H_2 = 4z^2 - 2,$ $\qquad H_3 = 8z^3 - 12z,$

$H_4 = 16z^4 - 48z^2 + 12,$ $\qquad H_5 = 32z^5 - 160z^3 + 120z.$

8. Eigenfunctions of the Harmonic Oscillator

$u_n(Q)$ is the eigenfunction normalized to unity, with the eigenvalue

$$E_n = (n + \tfrac{1}{2}) \, \hbar\omega \qquad (n = 0, 1, \ldots, \infty)$$

(phase chosen in such a way that relation (B. 68) is satisfied, and that $u_0(0)$ is real and positive).

Eigenvalue Equation $[Q=(m\omega/\hbar)^{\frac{1}{2}}q]$.

$$\tfrac{1}{2}\left(Q^2 - \frac{d^2}{dQ^2}\right) u_n(Q) = (n+\tfrac{1}{2})\, u_n. \tag{B.65}$$

Generating Function

$$\pi^{-\frac{1}{4}}\, e^{-\frac{1}{2}Q^2}\, e^{-\frac{1}{2}t^2+\sqrt{2}Qt} = \sum_{n=0}^{\infty} \frac{t^n}{\sqrt{n!}}\, u_n(Q). \tag{B.66}$$

Orthonormality and Closure Relations

$$\int_{-\infty}^{+\infty} u_n u_p \, dQ = \delta_{np}$$

$$\sum_{n=0}^{\infty} u_n^*(Q) u_n(Q') = \delta(Q-Q').$$

Recursion Relations

$$\frac{1}{\sqrt{2}}\left(Q + \frac{d}{dQ}\right) u_n = \sqrt{n}\, u_{n-1} \tag{B.67}$$

$$\frac{1}{\sqrt{2}}\left(Q - \frac{d}{dQ}\right) u_n = \sqrt{n+1}\, u_{n+1} \tag{B.68}$$

$$Q u_n = \sqrt{\frac{n+1}{2}}\, u_{n+1} + \sqrt{\frac{n}{2}}\, u_{n-1}. \tag{B.69}$$

Parity $(-)^n$: $u_n(-Q) = (-)^n\, u_n(Q)$.

Expression as a Function of Hermite Polynomials

$$u_n = (\sqrt{\pi}2^n n!)^{-\frac{1}{2}}\, e^{-\frac{1}{2}Q^2}\, H_n(Q). \tag{B.70}$$

IV. ASSOCIATED LEGENDRE FUNCTIONS AND LEGENDRE POLYNOMIALS; SPHERICAL HARMONICS

9. Legendre Polynomials and Associated Legendre Functions

Definitions

Legendre polynomial P_l $(l=0, 1, 2, \ldots, \infty)$:

$$P_l(u) = \frac{1}{2^l l!} \frac{d^l}{du^l} (u^2-1)^l \tag{B.71}$$

P_l is a polynomial of degree l, parity $(-)^l$, having l zeros in the interval $(-1, +1)$.

Associated Legendre function

$$P_l^m \ (l = 0, 1, 2, \dots, +\infty; \ m = 0, 1, 2, \dots, l):$$

$$\begin{aligned} P_l^m(u) &= (1-u^2)^{\frac{1}{2}m} \frac{\mathrm{d}^m}{\mathrm{d}u^m} P_l(u) \\ &= \frac{(1-u^2)^{\frac{1}{2}m}}{2^l l!} \frac{d^{l+m}}{\mathrm{d}u^{l+m}} (u^2-1)^l \end{aligned} \qquad (-1 \leqslant u \leqslant 1) \qquad \text{(B.72)}$$

$P_l^m(u)$ is the product of $(1-u^2)^{\frac{1}{2}m}$ and a polynomial of degree $(l-m)$ and parity $(-)^{l-m}$, having $(l-m)$ zeros in the interval $(-1, +1)$. In particular:

$$\begin{aligned} m &= l \qquad P_l^l = (2l-1)!!\,(1-u^2)^{\frac{1}{2}l} \\ m &= 0 \qquad P_l^0 = P_l(u) \end{aligned} \qquad \text{(B.73)}$$

$P_l(u)$ is a particular associated Legendre function.

Differential Equations

$$\left[(1-u^2)\frac{\mathrm{d}^2}{\mathrm{d}u^2} - 2u\frac{\mathrm{d}}{\mathrm{d}u} + l(l+1) - \frac{m^2}{1-u^2}\right] P_l^m = 0. \qquad \text{(B.74)}$$

Generating Functions

$$\frac{1}{\sqrt{1-2tu+t^2}} = \sum_{l=0}^{\infty} t^l P_l(u) \qquad \text{(B.75)}$$

$$(|t| < 1)$$

$$(2m-1)!!\,(1-u^2)^{\frac{1}{2}m} \frac{t^m}{[1-2tu+t^2]^{m+\frac{1}{2}}} = \sum_{l=m}^{\infty} t^l P_l^m(u). \qquad \text{(B.76)}$$

Orthonormality Relations

$$\int_{-1}^{+1} P_k^m P_l^m \,\mathrm{d}u = \frac{2}{2l+1} \frac{(l+m)!}{(l-m)!} \delta_{kl}. \qquad \text{(B.77)}$$

Recursion Relations

$$(2l+1)u\, P_l^m = (l+1-m)\, P_{l+1}^m + (l+m)\, P_{l-1}^m \qquad \text{(B.78)}$$

$$(1-u^2)\frac{\mathrm{d}}{\mathrm{d}u} P_l^m = -\,lu\, P_l^m + (l+m)\, P_{l-1}^m \qquad \text{(B.79)}$$

$$= (l+1)u\, P_l^m - (l+1-m)\, P_{l+1}^m \qquad \text{(B.80)}$$

(these relations are also valid when $l=0$, using the convention $P_{-1}=0$).

Particular Values

$$\begin{array}{ll} P_l(1) = 1, & P_l(-1) = (-)^l \\ \text{if } m \neq 0, & P_l^m(1) = P_l^m(-1) = 0 \end{array} \tag{B. 81}$$

$$P_l^m(0) = \begin{cases} (-)^p \dfrac{(2p+2m)!}{2^l\, p!\, (p+m)!} & \text{if } l-m=2p \\ 0 & \text{if } l-m=2p+1. \end{cases}$$

Table of the First Five Legendre Polynomials

$$\begin{array}{lll} P_0 = 1, & P_1 = u & P_2 = \tfrac{1}{2}(3u^2-1), \\ P_3 = \tfrac{1}{2}(5u^3-3u) & & P_4 = \tfrac{1}{8}(35u^4-30u^2+3). \end{array}$$

10. Spherical Harmonics

Operators L_x, L_y, L_z in Polar Coordinates

L_x, L_y, L_z are Hermitean differential operators, defined (in a system of units where $\hbar = 1$) by

$$\mathbf{L} \equiv \frac{1}{\mathrm{i}} (\mathbf{r} \times \nabla).$$

One takes Oz as polar axis; (r, θ, φ) are the polar coordinates of $\mathbf{r}$; $\Omega \equiv (\theta, \varphi)$ designates the set of two angular coordinates ($\varphi = 0$: plane zOx; $\varphi = \pi/2$: plane zOy). The element of solid angle is:

$$\mathrm{d}\Omega = \sin\theta\, \mathrm{d}\theta\, \mathrm{d}\varphi.$$

In polar coordinates:

$$L_z = \frac{1}{\mathrm{i}} \frac{\partial}{\partial \varphi} \tag{B.82}$$

$$L_\pm \equiv L_x \pm \mathrm{i} L_y = \mathrm{e}^{\pm \mathrm{i}\varphi} \left[\pm \frac{\partial}{\partial \theta} + \mathrm{i} \cot\theta \frac{\partial}{\partial \varphi} \right] \tag{B.83}$$

$$\mathbf{L}^2 \equiv L_x{}^2 + L_y{}^2 + L_z{}^2 = -\left[\frac{1}{\sin\theta} \frac{\partial}{\partial \theta} \left(\sin\theta \frac{\partial}{\partial \theta} \right) + \frac{1}{\sin^2\theta} \frac{\partial^2}{\partial \varphi^2} \right]. \tag{B.84}$$

Definition of the Spherical Harmonics $Y_l^m(\theta, \varphi)$:

Eigenfunctions common to the operators $\mathbf{L}^2$ and L_z:

$$\begin{aligned} \mathbf{L}^2\, Y_l^m &= l(l+1)\, Y_l^m \\ L_z\, Y_l^m &= m\, Y_l^m \end{aligned} \tag{B.85}$$

$$(l = 0, 1, 2, \ldots, \infty; \quad m = -l, -l+1, \ldots, l). \tag{B.86}$$

One completes their definition by adopting the conventions:

a) the Y_l^m are normalized to unity on the unit sphere;

b) their phases are such that the recursion relations (B.89) are satisfied, and that $Y_l^0(0, 0)$ is real and positive.

Orthonormality and Closure Relations

$$\int Y_l^{m*} Y_{l'}^{m'} \mathrm{d}\Omega \equiv \int_0^{2\pi} \mathrm{d}\varphi \int_0^{\pi} \sin\theta \, \mathrm{d}\theta \, Y_l^{m*}(\theta, \varphi) Y_{l'}^{m'}(\theta, \varphi) = \delta_{mm'} \delta_{ll'} \qquad \text{(B.87)}$$

$$\sum_{l=0}^{\infty} \sum_{m=-l}^{+l} Y_l^{m*}(\theta, \varphi) Y_l^m(\theta', \varphi') = \frac{\delta(\theta-\theta')\,\delta(\varphi-\varphi')}{\sin\theta} \equiv \delta(\Omega-\Omega'). \qquad \text{(B.88)}$$

The Y_l^m form a complete orthonormal set of square-integrable functions on the unit sphere.

Recursion Relations

$$\begin{aligned} L_\pm Y_l^m &= [l(l+1) - m(m \pm 1)]^{\frac{1}{2}} Y_l^{m\pm 1} \\ &= [(l \mp m)(l+1 \pm m)]^{\frac{1}{2}} Y_l^{m\pm 1} \end{aligned} \qquad \text{(B.89)}$$

$$\cos\theta \, Y_l^m = \left[\frac{(l+1+m)(l+1-m)}{(2l+1)(2l+3)}\right]^{\frac{1}{2}} Y_{l+1}^m + \left[\frac{(l+m)(l-m)}{(2l+1)(2l-1)}\right]^{\frac{1}{2}} Y_{l-1}^m. \qquad \text{(B.90)}$$

Parity $(-)^l$

Under a space reflection $(\theta, \varphi) \to (\pi-\theta, \varphi+\pi)$:

$$Y_l^m(\pi-\theta, \varphi+\pi) = (-)^l \, Y_l^m(\theta, \varphi). \qquad \text{(B.91)}$$

Complex Conjugation

$$Y_l^{m*}(\theta, \varphi) = (-)^m \, Y_l^{-m}(\theta, \varphi). \qquad \text{(B.92)}$$

Connection with the Associated Legendre Functions $(m \geqslant 0)$

$$Y_l^m(\theta, \varphi) = (-)^m \left[\frac{(2l+1)}{4\pi} \frac{(l-m)!}{(l+m)!}\right]^{\frac{1}{2}} P_l^m(\cos\theta) \, \mathrm{e}^{\mathrm{i}m\varphi}. \qquad \text{(B.93)}$$

Y_l^m is the product of $\mathrm{e}^{\mathrm{i}m\varphi} \sin^{|m|}\theta$ and a polynomial of degree $(l-|m|)$ in $\cos\theta$ and parity $(-)^{l-m}$. In particular:

$$m = 0 \qquad Y_l^0 = \sqrt{\frac{2l+1}{4\pi}} P_l(\cos\theta) \qquad \text{(B.94)}$$

$$m = l \qquad Y_l^l = (-)^l \left[\frac{(2l+1)}{4\pi} \frac{(2l)!}{2^{2l}(l!)^2}\right]^{\frac{1}{2}} \sin^l\theta \, \mathrm{e}^{\mathrm{i}l\varphi}. \qquad \text{(B.95)}$$

Harmonic Polynomials and *Spherical Harmonics*

The $(2l+1)$ homogeneous polynomials of degree l in x, y, z

$$\mathcal{Y}_l^m(\mathbf{r}) \equiv r^l\, Y_l^m(\theta, \varphi) \qquad (m = -l, -l+1, \ldots, +l) \tag{B.96}$$

form a set of $(2l+1)$ linearly independent harmonic polynomials of degree l [1]):

$$\triangle\, \mathcal{Y}_l^m(\mathbf{r}) = 0. \tag{B.97}$$

Table of the First Few Spherical Harmonics

$$Y_0^0 = \frac{1}{\sqrt{4\pi}}, \qquad Y_1^0 = \sqrt{\frac{3}{4\pi}} \cos\theta, \qquad Y_2^0 = \sqrt{\frac{5}{16\pi}}\,(3\cos^2\theta - 1),$$

$$Y_3^0 = \sqrt{\frac{7}{16\pi}}\,(5\cos^3\theta - 3\cos\theta),$$

$$Y_1^1 = -\sqrt{\frac{3}{8\pi}} \sin\theta\, \mathrm{e}^{\mathrm{i}\varphi}, \qquad Y_2^1 = -\sqrt{\frac{15}{8\pi}} \sin\theta \cos\theta\, \mathrm{e}^{\mathrm{i}\varphi},$$

$$Y_3^1 = -\sqrt{\frac{21}{64\pi}} \sin\theta\,(5\cos^2\theta - 1)\, \mathrm{e}^{\mathrm{i}\varphi},$$

$$Y_2^2 = \sqrt{\frac{15}{32\pi}} \sin^2\theta\, \mathrm{e}^{2\mathrm{i}\varphi}, \qquad Y_3^2 = \sqrt{\frac{105}{32\pi}} \sin^2\theta \cos\theta\, \mathrm{e}$$

$$Y_3^3 = -\sqrt{\frac{35}{64\pi}} \sin^3\theta\, \mathrm{e}^{3\mathrm{i}\varphi}.$$

11. Various Expansions and Formulae

Addition Theorem

$$\frac{2l+1}{4\pi} P_l(\cos\alpha) = \sum_{m=-l}^{+l} Y_l^{m*}(\theta_1, \varphi_1)\, Y_l^m(\theta_2, \varphi_2) \tag{B.98}$$

[α = angle between directions (θ_1, φ_1) and (θ_2, φ_2)].

---

[1]) (B.97) arises from the operator identity which holds for any function finite at the point $r = 0$:

$$\triangle \equiv \frac{1}{r} \frac{\mathrm{d}^2}{\mathrm{d}r^2} r - \frac{\boldsymbol{L}^2}{r^2}.$$

By definition, $h(x, y, z)$ is a harmonic polynomial if it is homogeneous in x, y, z and satisfies the equation: $\triangle h = 0$. There exist $(2l + 1)$ linearly independent harmonic polynomials of degree l.

Green's Functions of the Operators $\triangle$ *and* $\triangle + k^2$ [1])

$$\frac{1}{|\mathbf{r}_1 - \mathbf{r}_2|} = \sum_{l=0}^{\infty} \frac{r_<^l}{r_>^{l+1}} P_l(\cos\alpha) \tag{B.99}$$

$$\frac{e^{ik|\mathbf{r}_1-\mathbf{r}_2|}}{|\mathbf{r}_1 - \mathbf{r}_2|} = k \sum_{l=0}^{\infty} (2l+1)\, j_l(kr_<)\, h_l^{(+)}(kr_>)\, P_l(\cos\alpha) \tag{B.100}$$

$$\frac{\cos(k|\mathbf{r}_1 - \mathbf{r}_2|)}{|\mathbf{r}_1 - \mathbf{r}_2|} = k \sum_{l=0}^{\infty} (2l+1)\, j_l(kr_<)\, n_l(kr_>)\, P_l(\cos\alpha) \tag{B.101}$$

(α = angle between the directions of $\mathbf{r}_1$ and $\mathbf{r}_2$; $r_<$ = the smaller of the lengths r_1 and r_2; $r_>$ = the larger of the lengths r_1 and r_2).

Formulae (B.100) and (B.101) are valid for any k, even if k is complex.

Expansion of the Plane Wave and the Pure Coulomb Scattering Wave

Polar axis $\equiv z$ axis = direction of the incident wave vector $\mathbf{k}$

$$e^{ikz} = \sum_{l=0}^{\infty} (2l+1)\, i^l\, j_l(kr)\, P_l(\cos\theta) \tag{B.102}$$

$$\psi_c = \frac{1}{kr} \sum_{l=0}^{\infty} (2l+1)\, i^l\, e^{i\sigma_l}\, F_l(\gamma; kr)\, P_l(\cos\theta) \tag{B.103}$$

$$f_c(\theta) = \frac{1}{k} \sum_{l=0}^{\infty} (2l+1)\, e^{i\sigma_l} \sin\sigma_l\, P_l(\cos\theta). \tag{B.104}$$

The definitions of γ, ψ_c, $f_c(\theta)$, F_l, σ_l are those of §§ 4 and 5 [equations (B.18), (B.19), (B.25), (B.29) and (B.32)].

With a different choice of polar axis, the expansions (B. 102), (B.103), and (B.104) remain valid since θ represents the angle between the directions of $\mathbf{k}$ and $\mathbf{r}$.

By applying the addition theorem, one derives the expansions of the same expressions in series of spherical harmonics having the arguments (θ_k, φ_k) and (θ_r, φ_r). For instance:

$$e^{i\mathbf{k}\cdot\mathbf{r}} = 4\pi \sum_{l=0}^{\infty} \sum_{m=-l}^{+l} i^l\, j_l(kr)\, Y_l^{m*}(\theta_k, \varphi_k)\, Y_l^m(\theta_r, \varphi_r). \tag{B.105}$$

---

[1]) $\triangle\left(\frac{1}{r}\right) = -4\pi\delta(\mathbf{r})$, $(\triangle + k^2)\frac{e^{ikr}}{r} = -4\pi\delta(\mathbf{r})$, $(\triangle + k^2)\frac{\cos kr}{r} = -4\pi\delta(\mathbf{r})$.

INDEX

(Page numbers followed by * refer to footnotes)

K

L

M

N

O

P

Q

R